The secret of success in science
is the combination of
a bold imagination and a willingness
to grub for the facts.

Alfred North Whitehead

how to find the meanings of terms

Because we believe the terminology is best
learned by use, all new terms are defined, either
explicitly or by context, at the place where they are
first introduced. If chapters are used out of numerical
sequence, students may have to consult a dictionary
or the index to find definitions. Page numbers in **bold face**
in the index indicate the location of definitions or
major discussions of terms and concepts. An *f* after
a page number indicates a figure or a table.

MAINSTREAMS OF
BIOLOGY

MAINSTREAMS OF BIOLOGY

Gairdner B. Moment, Ph.D.

Professor of Biological Sciences
Goucher College
Baltimore, Maryland
Guest Scientist
National Institutes of Health

Helen M. Habermann, Ph.D.

Professor of Biological Sciences
Goucher College
Baltimore, Maryland

The Williams & Wilkins Company
BALTIMORE

Copyright ©, 1977
The Williams & Wilkins Company
428 E. Preston Street
Baltimore, Md. 21202, U.S.A.

Made in the United States of America

Library of Congress Cataloging in Publication Data

Moment, Gairdner Bostwick, 1905–
  Mainstreams of biology.

  Bibliography:
  Includes index.
  1. Biology. I. Habermann, Helen Margaret, joint author. II. Title. [DNLM: 1.
Biology. QH308.2 M732m]
QH308.2.M66   574   76-25872
ISBN 0-683-06122-4

The Publishers have made every effort to trace the copyright holders for
borrowed material. If they have inadvertently overlooked any, they will be
pleased to make the necessary arrangements at the first opportunity.

Composed and printed at the
Waverly Press, Inc.
Mt. Royal and Guilford Aves.
Baltimore, Md. 21202, U.S.A.

Preface

In the text which follows, the authors hope that every reader will feel the excitement of discovery and gain a lasting sense of biology as an ongoing human enterprise, a part of our cultural heritage. To help students attain a portion of the scientific wisdom which comes from historical perspective, we have included something of the way biological discoveries have been made.

Biology is a science of specialties, some clearly in the mainstreams of human concerns, others somewhat removed. We will begin with cells, the structural units of life; proceed to the metabolic, energy-utilizing processes that are virtually universal in living systems; and then discuss ways in which genetic information needed to make new organisms is passed on from one generation to the next and translated in development. A second theme concerns the strategies of life, the ways in which living organisms (both plant and animal) cope with the problems of capture and release of matter and energy, of nutrition and transport, of internal controls of metabolism and growth, and of biologically appropriate behavior. Finally we discuss the specialties of evolution and ecology which attempt to understand the great web of life of which we are all a part.

There is no single chapter devoted to the human importance of the biological sciences because we believe that to do so is to accept a false separation of biology into two parts, *i.e.*

what is important to the human race and what is not. Our conviction is that the whole of biology is relevant to mankind. Different aspects of it are of greater or lesser importance to different people for different reasons. Consequently, we point out the human relevance with each topic, whether it is an understanding of the role of plants in the world ecosystem or the astonishing possibilities presented by the newer knowledge of the centers of emotion in the brain.

Citizens of the future will be faced with the responsibility of deciding how a whole constellation of discoveries from genetic engineering to the farming of the seven seas can best be used for the welfare of all mankind. It is a staggering responsibility. All of these problems have important social, economic, political and moral dimensions. It is our conviction that without an understanding of the biological bases of these problems, sound value judgments are impossible. As biologists, we will concentrate on the biological dimensions.

We wish to express our gratitude to our colleagues and students, who have provided loyal support and valuable criticism, to our artist, Ann M. Symkowicz, and to the staff of Williams & Wilkins, who have given us the benefit of their special expertise.

Gairdner B. Moment
Helen M. Habermann
Baltimore, Maryland

Contents

Preface v

SECTION ONE
SOME BASIC FACTS OF LIFE

Chapter 1. MAINSTREAMS OF BIOLOGY 3
    Towards the Improvement of the Human Condition 3
    Mainstream of Exploration — Science: Its Nature and Limitations ... 5
    The Stream of Genetic Information 7
    Space-Age Perspectives on Life 7
    Levels of Organization 11

Chapter 2. THE UNITS OF LIFE 13
    The Cell Theory 13
    The Basic Anatomy of Life — The Cell 14
    Modern Cell Research: The Methods 14
    Modern Cell Research — The Results 16
        Prokaryotes 17
        Eukaryotes 18
    Membranes in and around Cells 23
    pH: Acidity-Alkalinity 26
    Cell Size 27
    Cell Division 28
    Viruses (Akaryotes) 32
    Viroids 35

Chapter 3. MOLECULES AND ENERGY: THE ESSENTIAL
INGREDIENTS OF LIFE 36
    The Subatomic Level 36
    The Atomic Level 37
    The Molecular Level — Organic Compounds 37
    Carbohydrates — The Building Blocks and Fuel of Life 39
    Lipids 41
    Proteins and Amino Acids 42
    Nucleic Acids 45

Metabolism 46
Metabolic Pathways 47

SECTION TWO
THE STREAM OF LIFE

Chapter 4. GENETICS I: THE FACTS OF HEREDITY 55
In the Beginning 55
Enter Gregor Mendel 56
Mendel's Fundamental Discoveries 56
Chromosomes and Heredity 60
Mutation 69
Heredity versus Environment 71
Concerning the Human Gene Pool 74
A Biological Look at Races 78

Chapter 5. GENETICS II: READING THE CODE 81
Genes: Their Chemical Basis and the Genetic Code 81
Genes in Action: Genes, Enzymes and Their Regulation 88
Technology of Genetic Engineering 92
Human Genetic Diseases 95
What of Eugenics? 98

Chapter 6. GENES IN ACTION I. DEVELOPMENT IN ANIMALS 100
Removing the Roadblock to Modern Knowledge 100
Gametogenesis and Gametes 101
Ovulation 104
Fertilization 104
Cleavage of the Zygote 107
Nucleus or Cytoplasm? The Paradox of Differentiation 108
Building the Embryo 110
Embryonic Induction—Turning on Genes 114
Sex Differentiation 118
Membranes 119
Development of the Nervous System 121
Aging 123

Chapter 7. GENES IN ACTION II: DEVELOPMENT IN PLANTS 124
Life Cycles: Readouts of the Genetic Code 124
Life Patterns in the Seed Plants 126
The Structure of Seed Plants—the Products of Development 127
Apical Meristems: the Regions of Growth 131
Growth vs. Differentiation 131
Environmental Control of Plant Growth and Development 133
Internal Control of Development 139
Aging, Senescence, and Death 147
A Look to the Future 149

Chapter 8. REPRODUCTION IN ANIMALS 151
Biological Meaning of Reproduction: Sexual and Asexual 151
Adaptation to Ensure the Meeting of Gametes 151
Parental Care 154
The Anatomy of Reproduction 155
Reproductive Hormones 157

Hormonal Control of Mammalian Reproduction 159
Gestation and Birth 160
Lactation 161

Chapter 9. REPRODUCTION IN PLANTS 164
Asexual Reproduction in Plants 164
Sexual Reproduction in Plants 168
Annuals, Biennials, and Perennials 174
Flower Development 174
Plant Dispersal 177

SECTION THREE
THE STRATEGIES OF LIFE: HOW TO COPE
Part A. Capture and Release of Energy

Chapter 10. PHOTOSYNTHESIS: HARVEST OF THE SUN 181
Chloroplasts: The Photosynthetic Machines 182
Effects of Environmental Factors on Rates of Photosynthesis:
    What They Reveal about the Process 185
Primary Reactions of Photosynthesis 188
Van Niel's Unifying Concept: Light-dependent Splitting of Water ... 191
Present Views on The Photochemistry of Photosynthesis 193
Photophosphorylation 196
Carbohydrate Synthesis: The Calvin Cycle 197
Some Conclusions about Photosynthesis 202

**Chapter 11. ENERGY RELEASE: MOLECULES AND ENERGY RE-
VISITED** 203
Enzymes 203
Details of the Energy Pathway 206
    The Anaerobic Phase 206
    The Aerobic Phase 207
Respiration in Animals: Adaptations for Breathing 214

Part B. Nutrition

**Chapter 12. SOIL, WATER, AND AIR: THE MODEST REQUIREMENTS
OF PLANTS** 218
Elements Required by Plants 218
Symptoms of Mineral Deficiency 221
Nitrogen Fixation 224
Recycling—The Way of Nature 225

Chapter 13. HUMAN NUTRITION AND WORLD FOOD PROBLEMS ... 227
Metabolic Food Processing 227
Human Nutritional Requirements 230
World Food Problems and the Green Revolution 235

Part C. Transport

Chapter 14. CIRCULATION IN ANIMALS 237
Harvey's Discovery of Circulation 237
Vascular Systems 238

Vertebrate Heart 240
Lymphatics 242
Blood 242
Oxygen-Carbon Dioxide Transport 243
Coagulation 243

Chapter 15. TRANSPORT OF WATER AND SOLUTES IN PLANTS 245
The Cellular Basis for Water Movement 245
Transpiration: The Mass Flow of Water through Plants 247
Water Movement from the Roots to the Shoots 250
Translocation: Transport of the Products of Plant Metabolism 253
Differences between Transport in Plants and Animals 257

Part D. Internal Controls

Chapter 16. MAINTAINING INTERNAL STABILITY IN ANIMALS: TEMPERATURE, IMMUNITY, AND EXCRETION 258
Excretion and a Constant Internal Environment 259
Body Temperature 262
Immunity 263
Theories of the Origin of Immunocompetent Cells 265

Chapter 17. HORMONAL CONTROL IN ANIMALS (ENDOCRINES) . . . 267
Foundations of Endocrinology: Methods of Study and Hormonal Action 268
Endocrine Glands of Mammals 270

Chapter 18. HORMONAL REGULATION IN PLANTS 279
Naturally Occurring Plant Hormones 279
Auxins: The Molecular Basis for Tropisms 280
The Gibberelins: Promoters of Elongation 285
The Cytokinins: Cell Division Factors that Act Synergistically with Auxin 287
Interactions of Auxins, Gibberelins, and Cytokinins 289
Ethylene: The Gaseous Hormone Involved in Ripening Fruits and Leaf Abscission 290
Abscisic Acid: A Naturally Occurring Inhibitor of Development 291
Morphactins: Recently Synthesized Chemicals Affecting Plant Growth and Development 291
The Future: Made-to-Order Plants 292

Part E. Components of Behavior

Chapter 19. SKELETAL AND MUSCULAR SYSTEMS 293
Major Types and Functions of Skeletons 293
Principles of Muscle Action 297
Muscular Dystrophy 300

Chapter 20. THE NERVOUS SYSTEM 302
Neurons, the Basic Units 302
The Major Divisions 304
The Nervous Impulse 307
Synapses 309
Synaptic Transmission of Nervous Impulses 309
Reflexes 311

Brain and Behavior 312
The Sense Organs 317

Chapter 21. ANIMAL BEHAVIOR 321
The Modern Revolt 321
The Modern Synthesis 323
The Way of Instinct 326
The Way of Learning 329
Hormonal Basis of Behavior 332
Biological Clocks 332
Social Behavior 333
Animal Societies: Sociology 338

SECTION FOUR
THE WEB OF LIFE: INTERDEPENDENCIES

Chapter 22. EVOLUTION 343
The Classic Darwinian Theory 343
Lamarckianism 345
The Modern Synthesis 346
The Evidence of Evolution 348
Origin of Life 350
Three Primary Life Styles Emerge 353
The History of Life on Earth 355
Human Origins 358
Persistent Questions about Evolution 362
The Diversity of Life and the Problem of Relationships 364

Chapter 23. SYMBIOSIS AND PARASITISM 368
Definitions 368
Positive Symbiosis: Mutualism 369
Negative Symbiosis: Parasitism 370

Chapter 24. ECOLOGY 383
Ecology and Current World Problems 383
Environments 385
Atmosphere 388
Soils 388
Ecological Succession 390
Lakes 391
Streams, Rivers, and Estuaries 392
Oceans 392
Matter and Energy Pathways 394
Energy Flow and Productivity 399
The Dynamics of Populations 401
Our Planetary Crisis—Is There a Way Out? 406
Index 411

SOME BASIC
FACTS OF LIFE

[Photograph courtesy of U.S.
Department of Agriculture.]

1

Mainstreams of Biology

The DNA double helix, the Pill, and the environmental crisis have forced themselves on the attention of people everywhere regardless of race, religion, or economic condition. Yet these things are only three of the more conspicuous points where biological knowledge impinges on our lives. So deeply interconnected are the diverse areas of biology that no satisfactory understanding of any one is possible without some understanding of the others. The poet-naturalist sees downy plumed milkweed seeds drifting on the sunlit autumn air as objects of great beauty and symbols of life's perpetuation; the biochemist sees little travelling packets of DNA. The ecologist realizes that both views are correct and appropriate for a complete understanding of life on this planet.

Towards the Improvement of the Human Condition

THE STRUGGLE FOR FOOD

From those remote ages when our ancestors lived in caves or perhaps in skin-covered huts, the need for food has been the driving force of one of the four great streams of ever-increasing knowledge about plants and animals and therefore of power over them. Today, every modern nation supports agricultural research and development stations to discover, originate, and develop new or improved varieties of food plants and animals. It has been well said that there is nothing as effective as hunger to bring a man to his senses.

Many methods have been used to achieve these ends so desperately needed on a planet of limited size with an exploding population. Selection, crossbreeding, followed by more selection has produced the "beefalo" (Fig. 1-1), a happy cross that combines the ability of the North American buffalo (bison) to flourish on dry prairie grass with the "beefiness" of cattle. Similar methods gave us hybrid corn (maize) and the new kinds of rice that made the "Green Revolution" possible.

In the foreseeable future it will very probably be possible to introduce new genes into food plants that will enable them to obtain their nitrogen from the air, as do some bacteria, instead of from expensive fertilizer. More plants will be developed that can carry on C_4 photosynthesis and are therefore able to maintain a high rate of carbohydrate production under conditions when most plants slow down. These and other achievements will assist mankind not only to survive but to lead a better life, while world population is brought under some kind of control.

THE DRIVE FOR HEALTH

One of the most ancient and powerful motives for pursuit of new biological knowledge is the search for cures for illness and pain, which are no respecters of age or sex, wealth or poverty, race or ideology. From remote antiquity plants and plant products of many kinds have been used for their therapeutic properties, both real and imagined. New plants have been sought which might provide cures or even some slight relief. It was no accident that Hippocrates, the father of medicine, was a botanist.

It should not be forgotten (for the penalties

3

Fig. 1-1. Beefalo. (Courtesy of Stephen Frisch.)

of forgetting are potentially very great) that many of the most destructive diseases of man and of his domestic plants and animals on which his life depends are due to fungi, bacteria, viruses, and animal parasites. Bubonic plague threw the Roman world of Justinian (565 A.D.) into confusion. Whole towns and countrysides were depopulated, and crops rotted in the fields. It is estimated that half the population of the western world died, many after suffering so intense that people jumped from housetops to escape their agonies. In the pandemic bubonic plague (Black Death) of the 14th century two out of every three students at Oxford died. Malaria, caused by a blood parasite and transmitted by a mosquito, still brings debilitating illness and death to more millions of people than any other disease. Bilharziasis (schistosomiasis), caused by a blood fluke harbored by a snail, is still a scourge in Southeast Asia and in Egypt as it was in the days of the pharaohs. Southern corn leaf blight, caused by a highly virulent fungus, *Helminthosporium maydis*, to name but one of thousands of plant diseases, is a real and present danger that resulted in over 100 million dollars of crop damage in 1970 due to the susceptibility of widely used strains of hybrid corn.

In modern times it has become clear that so

deep-set is the unity of life that discoveries made on one organism can often, although not always, be applied to others far removed in the plant or animal realms. For example, the results from studying the effectiveness of drugs in experiments with warm-blooded mammals, such as dogs or monkeys, can usually be relied on to furnish valid indications of their probable action on human beings. Sometimes the basic investigations can best be carried out on cold-blooded animals like frogs or mollusks. Many of the recently developed antibiotics used in the treatment of bacterial infections are obtained from the fungi, a group of relatively simple plants.

Most of the modern understanding about how the human body and brain function has come from studies of other animals. Modern theories of the way heartbeat is controlled were greatly aided by work on the heart of *Limulus,* the horseshoe crab (Fig. 1-2). The ideal place to study nerve conduction is not in the nerve of a man or a frog but in certain giant nerve fibers of *Loligo,* the squid. At first sight what animals could be more different from man than one-celled protozoans such as *Amoeba, Paramecium,* and *Tetrahymena*? Yet the requirements for vitamins and other dietary constituents of these microorganisms are very similar, and in some cases virtually identical, to those of man. This is not to say that one-celled animals living in a test tube will soon replace white rats as the standard test animals for nutritional studies; yet studies with these microscopic animals are uncovering important new facts of wide and perhaps universal importance among living things. Thus by searching for controls of plant and animal disease, biological knowledge provides a basis for healthy and abundant living.

MENTAL HEALTH AND THE SOCRATIC IMPERATIVE: KNOW THYSELF

Every decade it becomes more evident that the enormous unsolved problems of human behavior, both individual and social, cannot be understood, much less alleviated, without a knowledge of the vertebrate nervous and hormonal systems, including the decisive underlying biochemical events, and how these systems have been molded during the course of evolution. Indeed without such knowledge, insight into the nature of the human mind will remain superficial. Thinking about the intimate interactions between body and soul has been but a small tributary of the stream of biological knowledge until recent times,

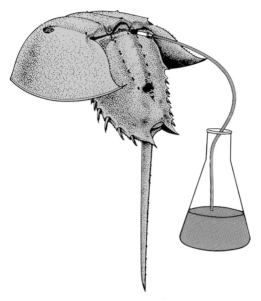

Fig. 1-2. Drawing blood from the heart of Limulus, the horseshoe "crab." The blood is used in the preparation of certain pharmaceutical products, and study of its heart furnished important clues to understanding the physiology of the human heart. (After a photograph from the Worthington Biochemical Corporation.)

when it has become a turbulent river. The depth of modern research into these problems can be neatly illustrated by the discovery that the hallucinogen of the peyote cactus (Fig. 1-3), mescaline, has a molecular structure closely similar to a form of adrenaline, a normal hormone of emotion, about which much is known.

The study of animal models can help clarify human behavior but such studies present dangerous temptations. A number of popular writers do not hesitate to jump easily from the behavior of rats, dogs, or apes to conclusions about human behavior. Even more dangerous, because it can be unconscious, is to see in the behavior of animals justifications for one's own prejudices. As a sharp warning consider, for example, the recent studies on three species of North American bears: the black bear, the grizzly bear, and the polar bear. The Canadian investigator, Charles Jonkel, has shown that when these three species are cornered or caught in a trap or snare each behaves very differently. The black bear will try to hide when approached and, if there is nothing to hide behind, may even cover its face with its paws. The grizzly bear goes into a vicious tantrum. In fact, before the trapper approaches, everything within the grizzly's reach will have been chewed, clawed, or dug up. Polar bears behave in neither of these

ways. They calmly sit and watch as you approach. They may even sneak up on you, but more probably they will turn their heads far to one side, exposing their long necks. It seems likely that this is one of the "I give up" signs that certain animals make when defeated. Which species of bear exhibits the behavior that reveals the most about the behavior of a cornered man? Obviously, conclusions drawn from any one of these behavior patterns can be extrapolated to human behavior only with great caution.

Mainstream of Exploration— Science: Its Nature and Limitations

QUESTIONS: SCIENTIFIC AND UNSCIENTIFIC

Anyone who wants to understand what science can and cannot do for humanity should keep clearly in mind the kinds of questions science can answer and the kinds it cannot. The clue is given by a rule adopted at the founding of the Royal Society of London, one of the oldest scientific societies in the world. That body from the beginning prohibited discussions of religion or politics. In other words, science is not concerned directly with value judgments. A clear expression of this

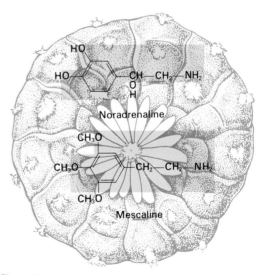

Fig. 1-3. Peyote cactus, *Lophophora williamsii*, also known as mescal button cactus and devil's root, has a biochemistry which produces eight identified alkaloids, the most important of which is mescaline, a mind-bending hallucinogen. Note its similarity to noradrenaline, a biochemical component of normal emotion.

point of view can be found in the recent writings of Garrett Hardin, a prominent American biologist. He maintains that there is a class of problems, such as the population explosion, for which there is no purely technical solution. By this he does not mean that methods of birth control are not technologically possible but rather that the extent to which such techniques should be used, if they are to be used at all, is a question lying outside the field of the natural sciences since it depends on value judgments.

Science is value free in the sense that the multiplication table is value free. Moral and political judgments play no part in the nature of those tables. However, value judgments do affect science in two ways. The problems a scientist chooses to investigate are determined by subjective judgments as to what is most worthwhile to learn, and whether any scientific discovery is used for good or evil purposes is again a value judgment just as is true of the multiplication table. Most biologists, including the authors of this book, take this orthodox position. Put succinctly, scientific study can tell you whether or not, statistically speaking, Swedes have big ears (they do); however, the natural sciences cannot tell you whether big ears are beautiful. Science is a matter of B follows A, not that B is better or worse than A. Such a position certainly does not mean that scientists are freed from the obligation of making value judgments. On the contrary, their responsibility to exercise their role as citizens is, if anything, greater rather than less because they may be better informed. For example, a biologist may foresee environmental effects of pollutants from an internal combustion engine that the automobile manufacturer may not foresee.

It is important to remember, although it is hard for some people to admit, that the truth or falsity of any discovery in science is independent of the beliefs and personality of the discoverer. The validity of Mendel's laws of heredity depends in no way on his religion, nationality, race, economic beliefs, fame, sex, or table manners. Even his motives are irrelevant. We do not know whether Mendel spent those long hours in his garden crossing peas primarily as refreshment from the routines of monastery life, because he was fascinated with the problem of inheritance, as some sublimation of an interest in sex, or through a determination to show that an Augustinian could make scientific discoveries of the first rank. All, some, or none of these factors may have been involved. The only relevant questions are whether his experiments can be repeated and whether his conclusions hold true for organisms other than peas.

PURE AND APPLIED SCIENCE

The stream of pure science, seeking to know for the sake of knowing, arises in an inherent trait we share with our primate relatives—curiosity. Although we can only infer what the inquiring ape thinks about his investigations, mankind is beginning to realize that pure science is extremely difficult to separate from the practical technologies which grow out of it. Who would have predicted that Mendel's work on heredity in garden peas would have led to practical advances of great importance in agriculture and medicine? The few biologists of the time who knew about Mendel's work thought it of no importance whatever. Who would have thought that advances in glassmaking would have led to the conquest of a host of diseases? Yet the invention of the microscope made possible the discovery of bacteria.

SCIENCE: CUMULATIVE AND OPEN-ENDED

Science is indeed like a river. It accumulates from many tributaries and it keeps moving, sometimes through slow level stretches and occasionally over turbulent water falls. Mendel used the techniques developed by earlier hybridizers of peas. Harvey based his discovery of the circulation of the blood on the work of three predecessors. The point to remember is that new discoveries are almost always built upon older ones and only rarely negate them. It simply is not true that science is advancing so fast that by the time a student graduates a large part of what he or she has learned will have been shown to be false. When T. H. Morgan and his students demonstrated that the genes, the mendelian factors of heredity, are located on the chromosomes, they did not prove that Mendel was wrong, even though Mendel knew nothing about chromosomes. In the same way, the recent discoveries about DNA in no way indicate that Morgan was in error in thinking that the chromosomes are the physical basis of inheritance and that the genes are arranged in a linear order along the chromosomes.

EVIDENCE AND THE SOCIAL NATURE OF SCIENCE

All human activities have a social dimension of some kind but in none is it more important than in the sciences, and for two interrelated reasons. Concern for evidence is not only central in the sciences, it extends throughout the entire fabric of this human activity. The evidence must be universally acceptable and independent of the culture of the observer, whether European, Chinese, Islamic, African, Indian, or other. There is no greater concern among scientists than that their work can be confirmed.

The deep-set social nature of science also results from its cumulative nature and as a consequence every investigator, whether a loner or a member of a research team, is in fact engaged in a social enterprise. This aspect of science brings important rewards not only by making scientific progress possible, but, as Bertrand Russell once said: "If knowledge be true and deep it brings with it a sense almost of comradeship with other seekers after truth in other lands and distant times."

The Stream of Genetic Information

Through the entire world of life there flows a continuous stream of genetic information passed on from one generation to the next. Coded in the DNA of the chromosomes within the reproductive cells lie the instructions for producing a new organism, be it a man or a mushroom. Spelled out in the purines and pyrimidines of DNA, this information is translated into enzymatic proteins and they in turn produce all the myriad traits of a new plant or animal. Thus, all the wonders of life from the slow-moving, blue-green alga to the human mind unfold from this informational stream whose origins are largely hidden in the mists of time.

THE STREAM OF MATTER AND ENERGY

The current of biological information does not exist in a vacuum but feeds on matter and energy. Without the energy from sunlight, photosynthesis would fail and life as we know it would end, because photosynthesis is the source of all but a minute portion of the energy on which living things depend. Matter could not be organized into the macromolecular blocks of which living things are built, the organization itself could not be maintained, cellular membranes and cell organelles like chromosomes would disintegrate, and the activities of life would fail.

The study of all the interactions of living things with each other and with the interweaving streams of matter and energy constitutes the young science of ecology. It has to become an essential science for keeping our planet habitable, because as Slobodkin, an American ecologist puts it, saving our environment requires "more than a good heart."

Space-Age Perspectives on Life

The exploration of outer space has dramatized both the limits imposed on life and the possibilities open to it that are governed by the laws of physics, chemistry, and mathematics. Because these laws are valid throughout the known universe, the same basic constraints and possibilities will exist for living organisms on any planet in any solar system, even those among the most distant stars. These factors play a fundamental role in determining the properties of the molecules in living things. They have an equally fundamental role in natural selection, independent of the special conditions which may happen to be present on any particular planet. To examine living things in the light of this enlarged space-age perspective is to gain a clearer understanding of why living organisms are what they are and behave as they do. Some of these inherent cosmic factors that govern life are well understood; others are but dimly perceived or completely unknown.

SIZE AND GRAVITY

Consider a very simple case where prediction has been possible based on considerations of gravity and the possibilities of molecular structure. Female moths of various species such as the gypsy moth and the large North American silk moths like *cecropia* or *luna* give off a sex attractant into the air which attracts males from a distance of well over a mile away. Scientists who investigated these remarkable molecules predicted that they would consist of a core of carbon atoms. They further predicted that the number of carbon atoms in these moth sex attractants would have no fewer than five or six carbon atoms and not more than about 20. Why? Because from the biological point of view each moth must produce its own unique kind of messenger molecule. It would be biologically useless for a male gypsy moth to fly to a female *cecropia*. The differences between one molecule and another depend on the carbon skeleton and the kinds and arrangement of atoms attached to it. With fewer than five or six carbons, there are too few possibilities to provide each species with its own identifying sex attractant. What sets the upper limit of molecular size? Gravity. If the molecules become too large they become so heavy that they soon sink to the ground and are hence ineffective as long distance attractants.

The sex **pheromones** of about two dozen insects are now known. All fall within the

Fig. 1-4. Earth photographed from the Apollo 11 spacecraft 98,000 nautical miles from earth. Most of Africa and portions of Europe and Asia can be seen. (Courtesy of National Aeronautics and Space Administration, Washington, D.C.)

predicted size limits. It seems evident that this prediction would hold for such chemical messengers on any planet possessing an atmosphere. The upper size limit would be controlled by the size of the planet. On very large planets the upper limit of molecular size would be lower, while on small planets it would be higher, for the very simple reason that large planets have a stronger gravitational pull than small ones. A molecule with a given number and kinds of atoms would be "heavier" on a large planet than it would on a small one.

In contrast, inherent limitations on the size of living things have been extensively analyzed and much is known. Paradoxical as it may seem, much more is understood about size limitations due to geometrical and physical factors than about how the code in the genes produces, for example, domestic cats of one size and Siberian tigers of a very different size.

Erwin Schrödinger, a Nobel Prize-winning modern physicist, introduced his discussion of the place of life in the universe by asking why atoms are so very small, so small that even a bacterium is inconceivably gigantic by comparison. He answered his own question by turning the question right side up and asking why living organisms must be so large compared with the atom: "We cannot see or feel or hear single atoms because every one of our sense organs . . . being itself composed of innumerable atoms, is much too

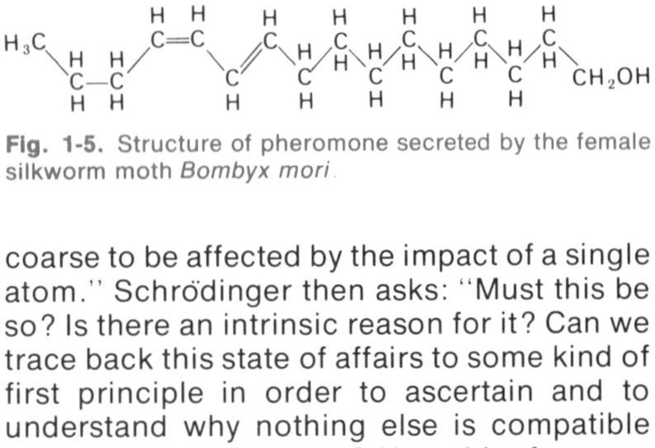

Fig. 1-5. Structure of pheromone secreted by the female silkworm moth *Bombyx mori.*

coarse to be affected by the impact of a single atom." Schrödinger then asks: "Must this be so? Is there an intrinsic reason for it? Can we trace back this state of affairs to some kind of first principle in order to ascertain and to understand why nothing else is compatible with the laws of Nature? Now this, for once [he continues], is a problem which the physicist is able to clear up completely. The answer to all the queries is in the affirmative."

The explanation is that if sense organs were so small as to respond to single atoms, perception would be utterly chaotic due to the randomness of atomic motions. An ultramicro-micro nervous system would not be able to function because its components would be too unstable.

Most biologists are more familiar with the factors which govern the lower limit of cell size. Cells are composed of proteins, lipids, water, and nucleic acids. To construct a cell with at least one chromosome, a nuclear membrane, some surrounding cytoplasm, and a cell membrane obviously requires some minimum number of protein, lipid, nucleic acid, and other molecules. You cannot make a cell any smaller, once you have reduced the size to this minimum number of molecules. You cannot get smaller molecules of any given kind unless you can find smaller atoms from which to make them. But atoms of hydrogen, oxygen, nitrogen, and the other needed elements come only in the standard sizes. To find toy-sized atoms you would have to go not just to another world but to another universe.

So, the old verse, usually attributed to Jonathan Swift, the author of *Gulliver's Travels*,

Great fleas have little fleas
Upon their backs to bite 'em
And little fleas have lesser fleas,
And so *ad infinitum*.

is false when it gets to the *ad infinitum*. There is a lower limit set by the structure of matter itself.

Of more immediate interest are the factors which set the lower limit of size for any **homoiothermic,** i.e. warm-blooded, organism. This limit would exist on any planet because size is related to the metabolic rate and surface area. According to the laws of solid geometry, as any body decreases in size without radically changing shape, the volume decreases much faster than the surface area. Therefore, the smaller the animal, the greater is its surface in proportion to its volume. Since any solid object loses heat to its environment through its surface, it follows that the smaller any warm-blooded animal is, the higher its metabolic rate will have to be to provide the heat that is lost. This means also that the smaller any animal that maintains a constant body temperature above its environment is, the more it will have to eat in proportion to its size. Shrews, which are one of the very smallest of mammals, weighing only 3–4 grams, have to eat almost continuously. Every 24 hours they eat approximately their own body weight (as if a 150-lb man were to eat three 50-lb meals a day).

What about an upper limit to size? When one thinks of the giant sequoias and the whales it might seem that the upper limits are very large indeed. Yet any tree has many of the same geometrical problems that beset tall office buildings. The higher the building, the more elevator shafts and service conduits are required to transport people and utilities to the upper stories. In the tallest skyscrapers, much of the space in the lower floors must be devoted to such nonproductive structures and supportive foundations, and any further increase in height is self-defeating. For trees there are also limits imposed by the feasible length of the water transport system of the stem and the size of the roots which provide the feeder system and support for the trunk.

The upper limit of size in terrestrial animals is restricted by geometrical factors similar to those operative in setting the lower limit of size in warm-blooded animals. As linear dimensions increase, body volume (which means body weight) increases much faster. Thus any animal that is twice as long as another requires legs which are more than twice as thick to support a more than twice as heavy body. So, because the legs have to become not just absolutely larger but larger in proportion to body size, very large animals would be practically all legs, an obvious impossibility. It is interesting to note here a principle already seen with airborne sex attractants. The larger the planet, the greater the pull of gravity. Hence, if a planet the size of Jupiter supported organisms which could walk, they would have to be very small.

Thus, living organisms are restricted within maximum and minimum size limits. These may seem very widely separated limits, but that is only because we ourselves live within them. Compared with the sizes of an atom and of intergalactic space, life is restricted to a very narrow size band indeed.

THREE BIOLOGICAL LIFE STYLES

One of the most conspicuous features of life on our planet is the presence of two major forms of life, plants and animals. A closer look reveals actually three basic forms: green plants, animals, and a group of plants called fungi. What is the explanation for this three-part division? Is it due to something so basic that it would be very likely to appear on other planets? Or is it due to some idiosyncrasy of Earth? On the biochemical and cellular levels all plants and animals are basically alike. Furthermore, they are subject to the same laws of inheritance and the same processes of natural selection. However, it is impossible to overlook the differences between the higher green plants like trees or the grasses and the higher animals like the octopus and man. One group is stationary, while the other is motile and has a neuromuscular system from which emerges intelligence. How did such profound differences arise?

According to current theory, for which there is very convincing evidence, this distinction between green plants on one side and animals and fungi on the other arose very early in the evolution of life. It was based on nutrition, that is to say, on differences in the way energy and the macromolecules involved in cell structures are obtained. Present evidence indicates that living organisms arose in the ocean after a long period when organic molecules of greater and greater complexity were formed under the influence of various energy sources, including ultraviolet radiation from the sun. If such molecules should be formed today they would quickly be taken up by living organisms. However, before life appeared there were no living organisms to devour such organic molecules and they therefore accumulated eon after eon.

The first living things fed on this rich "soup." However, as living things increased and these food molecules became more and more scarce, a crisis in the history of life must have occurred. There were two possible responses to this challenge. One depended on mutations that conferred greater and greater synthetic abilities on the organism. For example, if the original organisms required substances A, B, C, and D to make their own cytoplasm, and one underwent a mutation that enabled it to make D out of A, B, and C, it would clearly be favored by natural selection over the rest of the population. If one of its descendants had a mutation which enabled it to make B or C out of A, it would have a still greater advantage, since it would require only two instead of four kinds of molecules to survive. This line of evolution, suggested here in the simplest terms, obviously led to green plants which have developed the ability to utilize the sun's energy to manufacture their own food from carbon dioxide (CO_2) and water (H_2O). The other viable series of mutations did not confer the ability to synthesize complex molecules from simpler and fewer compounds but instead enabled their possessors to move around in their environment to reach places where useful prefabricated molecules existed. Ultimately they became able to engulf not only other molecules but other organisms. This line of mutations obviously led to animals.

On our planet there is a kind of "third world" of organisms, the fungi, consisting of the bacteria, some strange forms called slime molds, and the more familiar molds and mushrooms. They are probably diverse in evolutionary origin but they are characterized by dependence on other organisms, ultimately green plants, in the same way animals are for nutrition. They have met the nutritional crisis by evolving in the same nutritional direction as the animals but without evolving a neuromuscular system to move about.

What is the probability that a similar divergence in the basic life styles of living things would occur on other planets? There is every reason to believe that if life arises anywhere in the universe, it must obey the same laws of physics and chemistry and utilize the same kinds of atoms that exist throughout the known cosmos. Therefore, the same nutritional crisis would probably arise and the same two basic ways of meeting it might evolve.

The probability that a group of organisms similar to our fungi would exist is very high. Without them the essential cycling of carbon compounds and nitrogen and phosphorus compounds would cease. The three basic ways of life on our planet may not be the result of some odd quirk. Rather, they appear to be due to a fundamental aspect of evolution which would reappear wherever life arose.

SEXUAL REPRODUCTION

In looking at life on this planet, there are several additional features which appear to be due to what Schrödinger called "intrinsic reasons" rather than to chance circumstances peculiar to our world. One of these features is sexual reproduction, a virtually universal phenomenon among all the hundreds of thousands of kinds of plants and animals. Even in bacteria there is a mechanism for the passage of genetic information between individuals. There are of course many organisms which regularly reproduce by budding or some other asexual process—the coelenterates, the group to which the jellyfish belong, for example, and most groups of plants. However, in the life cycles of these organisms there is also a sexual phase. In sexual reproduction there is a great evolutionary advantage. Mutations, or changes in the genes, of course form the raw material of evolution. Sexual exchange makes it possible to bring together advantageous mutations which have occurred in different individuals and thus produce superior offspring. On top of this, in the random assorting of chromosomes (genes) in the formation of eggs and sperms and in the randomness of fertilization, a vast amount of variation, the raw material of evolution, is produced. Given the kind of thing life is, sexual reproduction is to be expected wherever life appears.

INTELLIGENCE

Everyone has at some time wondered if there are intelligent creatures native to planets in outer space. Is intelligence due to some special peculiarities of earth or is it one of those things which would have a very high probability of emerging wherever there is life? We have already seen that the probability of both plants and animals existing is very great. It is characteristic of animals to move around and "seek" food. This means that to have a head with sense organs at the front end is a selective advantage. You only have to look at animals to see that this is so. Only in animals which have become **sessile** (sedentary or fixed) is there no head. But a head with sense organs lacking a nervous system to control and coordinate the input of the sense organs and regulate the responses would be useless. From the biological point of view, intelligence, the ability to solve problems, is a tool for living just as much as is the antenna cleaner of a honeybee or the hand of a man.

Levels of Organization

In recent times a broad new concept has come into prominence as the most inclusive framework for biological thought. This new frame of reference does not negate the theory of organic evolution but goes beyond it to include the nonliving world from which life arose. The new concept is simple on the surface; it postulates that the known universe is composed of a series of levels of organization of such a nature that the units of one level of complexity form the building blocks for the units of the next higher level. Simple as it appears, the concept of a series of levels of organization, each linked to successively higher levels, involves profound philosophical problems. It is by no means a new concept to the philosopher, but has recently come into greatly increased use by biologists as a framework that enables us to know where we are and to find our way among vast mountains of data. It suggests new areas of research and illuminates the whole landscape of biology.

The first, but not necessarily the simplest, level of organization is the realm of the hundred or so subatomic particles or **wave particles** such as electrons, protons, and neutrons. The next level comprises the **atoms** built up from the subatomic particles. The units of this level are the well-known chemical elements hydrogen, oxygen, nitrogen, carbon, phosphorus, iron, etc.

The third or **molecular** level can be roughly equated with the biochemical level. The molecules of this level are composed of the atoms of the previous level organized in hundreds of thousands—in fact, in millions—of different patterns. Some are as simple as water, H_2O, and some as intricate as a molecule of hemoglobin, or as large as DNA, the deoxyribonucleic acid that carries the chemical basis of the genetic code of heredity in the chromosomes.

Above the molecules is the level of living **cells.** Except for the problematic viruses, cells are the smallest and primary independent living units. Each cell is bounded by a complex and physiologically active membrane constructed of large molecules. Within the cell is an assortment of molecules ranging from simple water and salts to complexities like deoxyribonucleic acid and conjugated proteins.

Above the level of single cells lies the level of **multicellular organisms,** the familiar plants and animals which may be composed of millions of cells. It will be noted that organs, e.g. liver, brain, or oak leaf, do not constitute a primary level of organization. A primary level is composed of units that can have an independent existence. Electrons or protons, for example, can exist independently. So also can atoms of nitrogen or iron, cells, and whole multicellular animals. A liver or oak leaf can exist for any length of time only as part of an organism. Such structures plainly belong to a secondary level of organization not comparable to the primary levels discussed here.

Above the level of many-celled organisms is the level of **populations.** A school of fish, a swarm of bees, or a stand of redwoods is regarded as a new sixth level for two important reasons. Clearly each can do things which no single individual can do. It takes more than one goose to migrate in a V formation. The "language" of the bees presupposes a community of bees. Single individuals do not evolve into new species. Only populations existing through long periods of time can evolve. In a larger, and perhaps more important, sense, all plants and animals are members of that great population of living things which inhabit the earth. The lonely albatross flying far out over the trackless ocean and the very local field mouse are each not only members of a population of albatrosses or field mice but active participants in the great web of life which is the subject of ecology.

Each level has its own characteristic properties, laws, and independent validity. A fact established on one level remains true regardless of what is discovered about it on another level, even though the meaning we give to the discovery may change. Consider again the case of the phenomena of heredity. Gregor Mendel worked on the level of whole multicellular organisms. His discoveries of the mathematical laws of heredity remain entirely valid regardless of any discovery about cell structures such as chromosomes. And, of course, biochemical discoveries have their own validity in terms of DNA and will remain so no matter what may be discovered in the future.

In the shift from one level of organization to another, new properties emerge. The properties of water molecules are neither the sum nor the average of the properties of the gaseous hydrogen and oxygen atoms of which water is composed. Freezing point, boiling point, chemical attributes — all these are new. This is what biologists mean when they say that the whole is more than the mere sum of its parts.

New properties also emerge when you break up complex patterns and descend to "lower" levels. Any freshman student of chemistry knows that the pure metallic element sodium has many exciting properties not found in sodium chloride (table salt). An oak tree which has grown all its life as a single individual in the center of a field spreads out and becomes a far different thing from the tall, straight-trunked tree with only a small crown of branches at the top found in a forest.

USEFUL REFERENCES

Dubos, R. 1968. *So Human an Animal.* Charles Scribner's Sons, New York. (Good reading, and the animal is man.)

Ehrlich, P. R., and A. H. Ehrlich. 1970. *Population Resources Environment.* W. H. Freeman and Co., San Francisco.

Ghiselin, B., ed. 1963. *The Creative Process.* Mentor Books, New York.

Glass, B. 1965. *Science and Ethical Values.* University of North Carolina Press, Chapel Hill. (A skilled treatment of an unsolved problem.)

Hardin, G. 1968. The tragedy of the commons. *Science* 162:1243–1248.

Taylor, G. R. 1963. *The Science of Life.* McGraw-Hill Book Co., New York. (A book deservedly praised despite some errors.)

2

The Units of Life

The realization that all living things are built up of semi-independent units called cells has come gradually over a period of more than 400 years. It is the work of many investigators in many countries. The idea itself is simple, but it is difficult to assimilate into everyday thinking. After all, what one sees is an oak tree or a rose, not a beautifully organized collection of millions of cells. No one feels subjectively that he or she is composed of vast numbers of minute, more or less independent, units. Nor is it obvious to common sense that the actual bridge between parents and children is a single cell, the fertilized egg.

The Cell Theory

The realization that life is not a property of the whole intact organism but rather is based on independent, or potentially independent, units, began in the 17th century with a discovery by the versatile English mathematician and microscopist, Robert Hooke. What Hooke saw was the honeycomblike structure in the substance of cork, which is the bark of a species of oak. Not surprisingly, Hooke named these units "cells." The boxlike structure of cells is relatively easy to see in plant material (the skin of a red Italian onion, for instance); but the outlines of the cells which Hooke saw are the nonliving cellulose walls.

No one at the time foresaw the central importance of cells for understanding living things, but almost everyone who had one of the newly invented microscopes observed cells. In Holland, Anton van Leeuwenhoek, the gifted Dutch lensmaker, described many kinds of cells, from blood corpuscles to human sperms. In the south of Europe Marcello Malpighi, a professor on the Faculty of Medicine in Bologna, turned his microscope on parts of large and small animals, and thus became the founder of microscopic anatomy. The innermost cellular layer of the skin is named, after its discoverer, the **Malpighian layer.**

The cell theory as we know it today is mainly the achievement of the 19th century. Microscopists all over Europe began to notice some kind of "mucus" within the boxlike "cells" of Hooke. Then in 1830 Robert Brown, a young Scottish army surgeon and botanist, discovered a nucleus in the cells of orchids and subsequently in many other kinds of cells. By 1839 a Czechoslovakian, Johannes Purkinje, usually known for his studies on vision, showed in a classic paper, "On the Structural Elements of Plants and Animals," that both plants and animals are constructed of cells. It was Purkinje who introduced the term **protoplasm** to denote the living material of the cell.

At about the same time M. J. Schleiden and M. Schwann in Germany popularized similar ideas about the cellular structures of all living organisms, although their belief that cells originate from a noncellular matrix turned out to be very wrong. Finally, in the 1860's, Max Schultze drew together the diverse facts that adult plants and animals are made of cells, that eggs and sperms are cells, and that microscopic animals like *Amoeba* and *Paramecium* are essentially single cells. At the same time he formulated the famous definition of protoplasm as "the physical basis of life."

It remained for a group of investigators in the latter part of the 19th century, mostly in Germany and Belgium, to show that all cells come from previous cells and, in the higher organisms, by a process usually called **mitosis.** Strictly speaking, mitosis is a synonym for the term **karyokinesis** which refers only to the duplication of the nucleus, including, of course, the **chromosomes. Cytokinesis** is a

13

word introduced over a decade later to refer to the division of the cytoplasm. These two terms are coming into use again and are thus worth at least a nodding acquaintance. A survey of recent publications of outstanding cytologists (investigators of cells) shows that mitosis is commonly used by them in the broad sense of the sum total of events connected with cell division. All of this preoccupation with terminology may be very unfortunate but it is a commonplace situation in the biological sciences.

The Basic Anatomy of Life — The Cell

Modern cell research uses powerful new methods which are yielding important new insights over a very wide range of biological problems from the diagnosis of hereditary disease before the birth of a child to the way hormones produce their results or a muscle contracts. Before considering modern cell research, it is essential to have clearly in mind the 19th century or "classical" knowledge of cells because it is the indispensable foundation of present-day biology.

It is not too difficult to see that most cells are much alike. All consist of a **nucleus** and its surrounding living material, the **cytoplasm**. Both nucleus and cytoplasm are surrounded by a membrane so thin that its structure cannot be discerned with the highest powers of a light microscope. Cells fitting this description can easily be scraped off the inside of your cheek.

A conspicuous difference between the cells of most adult plants and adult animals is that plant cells typically are surrounded by a more or less rigid envelope of **cellulose,** called the **cell wall,** and contain within the cytoplasm a large **vacuole.** In many adult plant cells the cytoplasm is only a thin layer pressed against the cellulose wall by the large central vacuole. Parts of a plant cell within the wall are frequently referred to as the **protoplast.** An examination of most animal cells reveals only nuclei and cytoplasm. The cell membranes separating the cytoplasm of the different cells of the liver, thyroid gland, or a muscle cannot be readily seen without special staining techniques, and the large central vacuole is missing. Some animal cells, such as cartilage and bone cells, secrete extracellular material which surrounds them in the manner of the cellulose wall of plant cells. The chromosomes within the nucleus are essentially the same in both plant and animal cells in their structure and in their behavior during cell

division and sexual reproduction. This is why the laws of inheritance are the same in plants and animals. Modern work has revealed that within the cytoplasm is not only a nucleus but also a variety of other organelles and membranes.

Modern Cell Research: The Methods

One of the big differences between modern science and the science of the ancient Greeks, perhaps the crucial difference, is the modern realization of the central importance of experimentation, of laboratory work, of technique. Not only is no theory or supposed fact any better than the evidence on which it is based, but how it can be known is a nondisposable part of that fact.

The indispensable workhorse for the study of cells remains the familiar light microscope. Progress in both optics and methods of preparing material for observation has continued ever since Robert Hooke cut that piece of cork with a penknife "sharpened as keen as a razor" and looked at it under a 1665 model microscope. The latest improvements on the light microscope are the phase contrast and the Nomarski lens assemblies. Both show objects within cells with greater clarity and brilliance than ordinary light microscopes. So far their chief contribution has been to confirm in the living cell what had been discovered by the so-called "classical" or standard methods.

The great problem in investigating cellular structure is that cells and their components are almost transparent. Because cells are fragile and disintegrate very easily, they have to be preserved, i.e. fixed, and stained. Thick tissue must be sliced thin for the same reason Hooke sliced his piece of cork: microscopes can only "see" by light transmitted through paper-thin slices.

The **fixatives** used are mixtures of chemicals such as formaldehyde, alcohol, and acids in such proportions that the tendencies to shrink protein and other protoplasmic constituents are counterbalanced by the tendencies to cause swelling. The result is that the protoplasm is "fixed" approximately as it was in life. The next step is to infiltrate the tissue with paraffin to make it firm enough to cut. It is then sliced into thin sections with a **microtome,** a device that works much like a bacon-slicing machine. The slices are then pasted on a glass slide and stained with aniline or other dyes.

If there is a pressing need for speed, as

when a piece of suspected cancer is to be examined while the patient lies on the operating table, or if the object is to be tested for substances like enzymes that may be removed or changed by the procedures of fixation and infiltration with paraffin, another method is used. The tissue is quickly frozen by evaporating liquid carbon dioxide. It can then be immediately sliced and examined. This method of frozen sections is important in the rapidly developing field of cell chemistry (cytochemistry).

NEW METHODS FOR STUDYING CELLS

Four of the most significant new developments in cell study are, as is commonly the case, not completely new but radical improvements or new combinations of familiar techniques. New ways of growing human and other cells in vitro, i.e. in glass dishes, permit accurate study of the chromosomes. Cells and their components can be labeled with radioactive atoms. The electron microscope is opening up a whole new world within the cell to exploration, and special methods of centrifugation enable us to separate and collect for investigation the ultramicroscopic constituents of cells.

Radioactive Labels

When they are built into the molecules of amino acids, hormones, or whatever substance is to be followed, the labeled molecules undergo reactions that are the same as those of "normal" molecules, except that from time to time a labeled atom will disintegrate, producing radiation which can be detected on a photographic plate or by an instrument, such as the Geiger counter, that can measure the amount of radioactivity. Tritium, i.e. radioactive hydrogen (^3H), is frequently used. It has a half-life of 12 years. This means that of a group of tritium atoms, one-half will have "decayed" to ordinary hydrogen within 12 years. Phosphorus-32 and iodine-131, also frequently used as tracers, have half-lives of 14.3 and 8.1 days, respectively.

After exposure to the radioactively labeled material, a tissue is fixed and sectioned, and the sections are placed in contact with a photographic emulsion. Radioactive disintegrations will occur where the labeled amino acid, hormone, or other material is located. Each disintegration can be detected as a black spot on the developed photographic film. The way chromosomes duplicate themselves has been determined by this method (a chromosome organizes a duplicate of itself along one side

for its entire length; the new and the old then separate; see Fig. 2-1). Much of the behavior and life history of cells can be investigated by such labeling.

Electron Microscopy

In addition to new methods of cell culture to study chromosomes and new ways of labeling cells, the electron microscope has added greatly to modern knowledge in many fields. The use of an electron microscope requires basically the same kind of preparative methods that the light microscope does. The material to be studied is usually fixed in a vacuum at freezing temperatures. The material is then embedded in a special plastic and sectioned ultrathin with a special microtome using a sharp glass or diamond edge. The section is placed on a supporting membrane and then put into the electron microscope.

The source of the electrons, which take the place of the light waves of an ordinary microscope, is a cathode filament, much as in a television tube. The electron beam is focused by magnets rather than by glass or quartz lenses. Since electrons travel only very short distances in air, the entire path of the beam is enclosed in a vacuum. The magnified image is projected either onto a fluorescent viewing screen or a photographic plate. Magnifications of over 10,000 diameters are common and up to 100,000 diameters are possible. The ordinary light microscope can magnify only up to about 1,500 times. Table 2-1 indi-

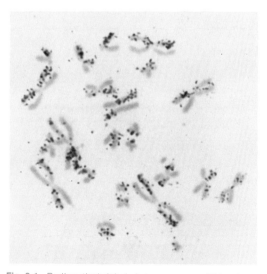

Fig. 2-1. Radioactively labeled chromosomes. Sister chromosomes (chromatids) of the Chinese hamster at the second division after feeding the cells with radioactive thymidine which is incorporated into DNA. Note that only one chromatid of each pair is labeled. (Courtesy of T. C. Hsu. From Moment, G. B. 1967. *General Zoology*, 2nd ed., Houghton Mifflin, Boston).

TABLE 2-1
Sizes of some biological objects (From Moment, G. B., and H. M. Habermann. 1973. *Biology: A Full Spectrum.* The Williams & Wilkins Co., Baltimore.)

Object	Diameter	
	μm	Å
Human egg	100	1,000,000
Red blood cell	10	100,000
Bacterium	1	10,000
Virus	0.1	1,000
Protein molecule	0.01	100
Amino acid	0.001	10

One micrometer, 1 μm, is equal to 1/1000 of a millimeter, i.e. 1 μm = 0.001 mm. One Ångstrom is equal to 1/10,000 of a micrometer, i.e. 1 Å = 0.0001 μm.

cates the size of some biologically important objects.

The light microscope can **resolve,** i.e. distinguish as separate, two lines separated by 0.2 μm; an electron microscope can resolve lines separated by 3 to 5 Å, i.e. 0.0003 to 0.0005 μm. Resolving power depends both on the type of lens and the wavelength of the source of illumination. In general, an object cannot be resolved in the light microscope if its diameter is less than half the wavelength of green light—about 5,500 Å—or in the electron microscope if its diameter is less than half the wave length of the electrons that are used—about 0.05 Å.

A recent advance in electron microscopy has been the development of the scanning electron microscope (SEM), in which an image is produced by secondary electrons from the surface of the specimen. The resolving power of the scanning electron microscope is considerably less than that of the ordinary electron microscope (only 100 to 200 Å) but it has other characteristics which are distinct advantages. Specimens of almost any size can be examined and SEM photomicrographs are remarkably three-dimensional in appearance because of a great depth of field. In the light microscope and in transmission electron microscopy the image is two-dimensional because these instruments must be focused on a single plane.

Analysis of Cell Constituents

The fortunate coincidence which has made so many of the discoveries with the electron microscope meaningful has been the simultaneous development of ultrasensitive methods to separate cell constituents. If cells are homogenized and centrifuged at low tempera-

tures enzyme activity is not destroyed. With relatively light centrifugation the nuclei, being the largest and heaviest constituents, move to the bottom of the tube first and can be separated from the cytoplasmic components. Somewhat higher speeds throw down mitochondria and plastids. Still higher speeds precipitate minute cytoplasmic particles called **ribosomes.** By filling the centrifuge tube with fluids of different densities, a **density gradient** can be produced and very accurate separations achieved. Particles come to rest when they reach a layer with a density closely approximating their own (Fig. 2-3). Each cell component can then be investigated by biochemical methods to study its function and with the electron microscope to study its structure.

Modern Cell Research— The Results

One generalization that has emerged from modern studies of cells is that there are two fundamentally different types. Bacteria and blue-green algae consist of cells which are **prokaryotic** *(pro,* before or early, + *karyon,*

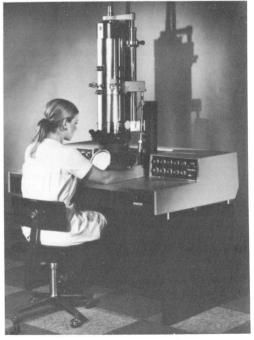

Fig. 2-2. Exterior view of an electron microscope. (Courtesy of Philips Electronic Instruments, Inc.)

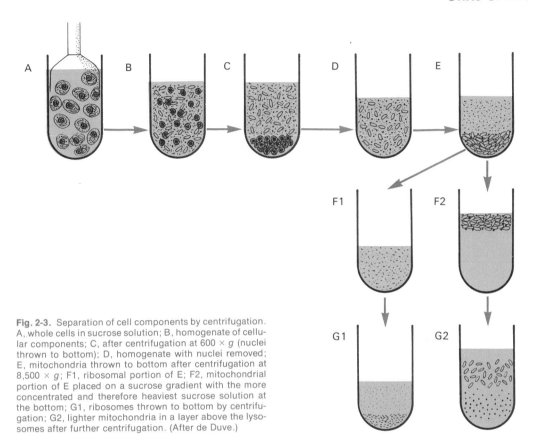

Fig. 2-3. Separation of cell components by centrifugation. A, whole cells in sucrose solution; B, homogenate of cellular components; C, after centrifugation at 600 × g (nuclei thrown to bottom); D, homogenate with nuclei removed; E, mitochondria thrown to bottom after centrifugation at 8,500 × g; F1, ribosomal portion of E; F2, mitochondrial portion of E placed on a sucrose gradient with the more concentrated and therefore heaviest sucrose solution at the bottom; G1, ribosomes thrown to bottom by centrifugation; G2, lighter mitochondria in a layer above the lysosomes after further centrifugation. (After de Duve.)

kernel). All animals and all green plants including the algae other than the blue-greens, and all the plants commonly called fungi are **eukaryotic** *(eu,* true, + *karyon,* kernel). The problematical viruses, which some biologists do not regard as living things, are called **akaryotic** since they are not cells but consist merely of a bit of genetic material, either DNA or RNA, covered with a coating of protein.

PROKARYOTES

Both prokaryotes and eukaryotes contain those cell structures which are essential for life: an outer limiting membrane, internal membrane systems, a nucleus, and ribosomes. Prokaryotic cells differ from eukaryotic cells in a number of ways. Although both depend on DNA as the source of information, the nuclear DNA in prokaryotes is not packaged in a separate membrane-bound organelle. Instead a double circular strand of DNA (the "nucleus") is attached to the inside of the cell membrane. The cell membrane, like all other membranes, is made up of two layers of lipid plus protein. It controls the movement

of materials into and out of cells, i.e. it is **semipermeable.** Specialized enzymes called **permeases** are associated with the cell membrane and are involved in the active transport of materials into the cell. Also associated with the cell membrane in prokaryotic cells are the final enzymes of respiration. These are separately packaged in the mitochondria in eukaryotes. At some points the prokaryotic cell membrane invaginates (folds inward) to form membrane plates. In the photosynthetic blue-green algae chlorophyll and accessory light-absorbing pigments are associated with the membrane plates. Rounded invaginations of the cell membrane are called **mesosomes** and these structures appear to be involved with the formation of cross walls and are closely associated with the chromosome (Fig. 2-4).

The internal cell sap of prokaryotic cells contains sugars, amino acids, fatty acids, vitamins, minerals, and enzymes in solution — constituents very similar to those found in the cytoplasm of eukaryotic cells. Also found in the cell sap are **ribosomes,** structures involved in protein synthesis. Prokaryotic ribosomes have been isolated by centrifugation. Although they are chemically similar to ribosomes isolated from eukaryotic cells, containing protein plus ribonucleic acid, they are somewhat smaller (having a sedimentation constant in the ultracentrifuge of 70 S vs. 80 S for eukaryotic ribosomes).

Also found within the cell sap are structures called **plasmids.** They contain DNA (although they do not function as a nucleus). Genes found in the plasmids are known to code for such varied characteristics as resistance to antibiotics, the ability to fix atmospheric nitrogen, and the formation of sex pilli (surface structures involved in the exchange of DNA between two prokaryotic cells).

External features that can be present in prokaryotic cells include a **slime layer,** a **flagellum** (or flagella) and short tubes called **pilli** (singular, pillus). Both the long flagella and the short pilli are anchored in the cytoplasm. They are composed of protein fibers and do not have the structural pattern characteristic of the flagella and cilia of eukaryotic cells. Flagella of both prokaryotes and eukaryotes are much longer than the cell and function in cell locomotion. Pilli usually are numerous and are involved in the adhesiveness of cells. Surface layers of slime can function in attaching one end of a prokaryotic cell to a surface or can surround the cell, forming a protective capsule. Chemically, the slime layer is composed of long chains of sugars (polysaccharide) or of amino acids (polypeptide).

There is great variety in shape among the prokaryotes. Single cells can be spherical, cylindrical, or spiral. They can be associated in chains, filaments, or aggregates. Reproduction is mainly asexual, and cell division usually results in two daughter cells of equal size (**binary fission**) (Fig. 2-5). **Budding,** in which the two resulting cells are very different in size, produces one daughter cell which receives very little cytoplasm. Sexual reproduction is rudimentary and consists of occasional transfer of genetic material. Cytoplasmic bridges between bacteria of different morphological types (so they could be identified) have been revealed by the electron microscope. These bridges are believed to be the physical basis of the genetic recombinations that are now well established in bacteria.

An extremely wide degree of metabolic traits characterize the prokaryotes. Some cannot live in the presence of oxygen. Others are facultative anaerobes and can live either using oxygen or without it; others are aerobes. A number of species metabolize sulfur, iron, and even atmospheric nitrogen. Some are photosynthetic. Little wonder that C. B. van Niel, whose work with bacteria provided one of the keys to our modern understanding of photosynthesis, decided that the primary division of life was not into plants and animals, but rather between what he terms prokaryotes and eukaryotes.

EUKARYOTES

The most distinguishing feature of eukaryotic cells is that their discrete chromosomes are enclosed in a nuclear membrane. Within the nucleus there is usually a nucleolus, which has an abundance of RNA and certain proteins. Cell division occurs by that complex of chromosomal maneuvers called mitosis. Probably of equal importance, the cytoplasm is endowed with organelles, notably mito-

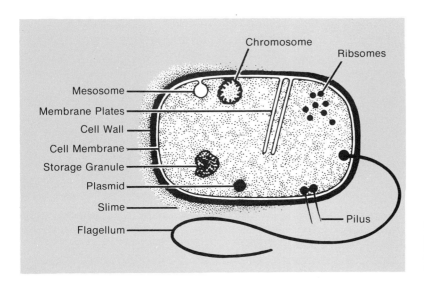

Chromosome
Ribsomes
Mesosome
Membrane Plates
Cell Wall
Cell Membrane
Storage Granule
Plasmid
Slime
Flagellum
Pilus

Fig. 2-4. Diagramatic summary of prokaryotic cell structure. Generally recognized structures are shown and any given prokaryote need not have all of these.

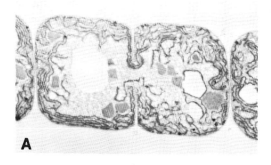

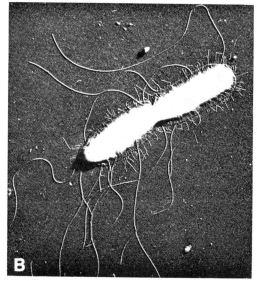

Fig. 2-5. Prokaryotic cells dividing as seen in the electron microscope. A, the blue-green alga, *Anabaena*, × 8,000. (Courtesy of N. J. Lang.) B, *Salmonella typhi*, × 7,600. Note that both flagella and fimbriae can be seen. (Courtesy of P. D. Duguid. From Gilles, R. R., and T. C. Dodds. 1968. *Bacteriology Illustrated*, 2nd ed. E. & S. Livingstone Ltd., Edinburgh and London.)

chondria, which carry on oxidative respiration and provide the cell with a supply of available energy packaged in molecules of adenosine triphosphate (ATP). In green plants the cells are further provided with photosynthetic organelles, the chloroplasts.

How eukaryotic cells arose in the history of life, no one knows. It has been suggested for nearly a century that mitochondria and chloroplasts may be symbiotic bacteria and algae living within the cytoplasm. Although this theory does not explain the origin of mitosis, it explains a lot, and strong evidence has been accumulating for it in recent years. Both mitochondria and chloroplasts are self-duplicating and contain their own DNA. They do not undergo mitosis any more than bacteria do. When the "host" eukaryotic cell divides, they are merely distributed to the two daugh-

ter cells in a more or less random fashion. At present, the outstanding proponent of the theory that eukaryotic cells arose by multiple symbiosis is Lynn Margulis (Fig. 2-6), an investigator of cilia in Protozoa. Intracellular symbiotic algae with nuclei are well known in some of the lower animals like the green species of *Paramecium* and the green hydras and corals.

There was a time when a eukaryotic cell was regarded by some as merely a bag containing a nucleus and various loosely mixed chemicals, but this theory is no longer held. To the discoveries made with the light microscope, the electron microscope has added an almost incredible amount of detail and additional unsuspected structures.

Mitochondria

The mitochondrion is one of the easiest cellular organelles to see because of its size and activity. Under a light microscope mitochondria appear as rods or threads which (in living cells) stain greenish-blue with Janus Green B. They can be observed unstained in living cells by the use of special techniques such as dark field and phase contrast microscopy. Long mitochondria wiggle about as though they were independent creatures—in fact, they resemble rod-shaped bacteria. Under an electron microscope each mitochondrion is revealed as composed of a double membrane, the inner one forming folds called **cristae** that extend into the interior,

Fig. 2-6. Lynn Margulis, present-day champion of the theory that eukaryote cells arose by symbiotic association. The implication of this concept is that mitochondria and chloroplasts may have originated from free-living bacteria and prokaryotic algae. (Courtesy of L. Margulis.)

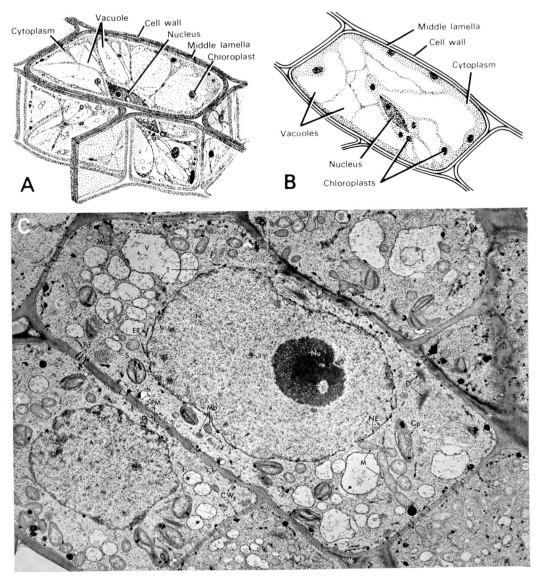

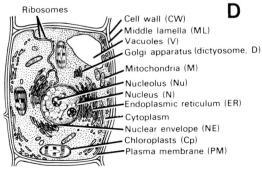

Fig. 2-7. Eukaryotic plant cells. A, three-dimensional view; B, as seen in the light microscope; C, electron photomicrograph; D, artist's diagram of electron microscope image. Letters in parentheses after labels in part D refer to letters in part C. (A and B from Janick, J., R. W. Schery, F. W. Woods, and V. W. Ruttan. 1974. *Plant Science,* 2nd ed. W. H. Freeman, San Francisco, redrawn from Esau, K. 1965. *Plant Anatomy.* John Wiley & Sons, Inc., New York. C from Ledbetter, M. C., and K. R. Porter. 1970. *Introduction to the Fine Structure of Plant Cells.* Springer-Verlag, New York, Heidelberg, Berlin. D, after Brachet. *The Living Cell.* Copyright © 1961 by Scientific American, Inc. All rights reserved.)

Biochemical analysis shows that mitochondria are the "powerhouses" of cells. The enzymes of aerobic respiration are located in the walls of the mitochondria. The final energy-giving steps in respiration take place here, yielding so-called "packaged energy" in ATP molecules. The ATP is then available for muscular contraction, secretion, or the synthesis of new complex molecules, such as proteins.

Endoplasmic Reticulum

A second conspicuous feature of the cytoplasm is the **endoplasmic reticulum.** This system of membranes is found in all eukaryotic cells, but is especially abundant in cells that synthesize protein. For example, lymphocytes, the white blood cells that make protein antibodies, are rich in endoplasmic reticulum. On the outer side of the endoplasmic reticulum in such cells are minute granules of very uniform size. These are the **ribosomes,** the sites of protein synthesis. It is here that the genetic code is translated into the linear order of amino acids in protein dictated by the sequence of code words in a strand of RNA. It is at the ribosome that the amino acids are linked together to form new protein. Not all ribosomes are located on membranes in the endoplasmic reticulum. Some are free in the cytoplasmic matrix, at times in groups termed **polyribosomes.**

Golgi Complex (Dictyosome)

A third universal organelle is the **Golgi complex,** also called a **dictyosome.** This is a series of three, four, or more curved membranes stacked like a pile of saucers. The membranes are much like those of the endoplasmic reticulum but lack ribosomes and hence are smooth. Usually the Golgi complex lies near the nucleus. In Fig. 2-9 it can be seen that the ends of membranes of the Golgi complex appear to be giving off droplets. The Golgi "apparatus," as it is often called, apparently secretes material, and is in fact well developed in secretory cells. In dividing plant

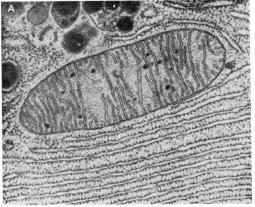

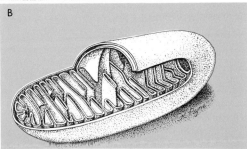

Fig. 2-8. Mitochondrion. A, electron micrograph of mitochondrion from bat pancreas, × 64,000. (Photograph by K. R. Porter. From Fawcett, D. W. 1966. *The Cell*. W. B. Saunders Co., Philadelphia.) B, diagrammatic reconstruction of a mitochondrion.

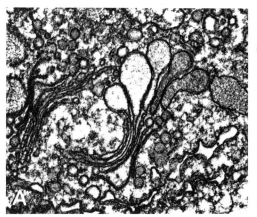

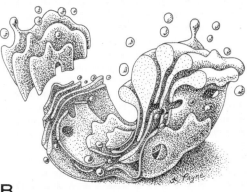

Fig. 2-9. Golgi complex. A, as seen under an electron microscope. × 29,000. (From Mollenhauer, H. H., and W. G. Waley. 1963. An observation on the functioning of the Golgi apparatus. *J. Cell Biol.* 17:223.) B, diagrammatic reconstruction.

cells, droplets of calcium pectate which coalesce to form the middle lamella separating the two daughter cells may be secreted by the Golgi apparatus. It also seems to secrete components of the cell wall.

Lysosomes

A fourth cytoplasmic structure is the **lysosome.** Lysosomes are small, rounded vesicles varying in diameter from about the width of a mitochondrion to much smaller. Lysosomes apparently contain lytic (digestive) and other enzymes which become active when the lysosome membrane is ruptured. Since these enzymes digest the cell, lysosomes are popularly called "suicide bags." Apparently they serve to digest unhealthy or injured cells and return their raw materials to the rest of the organism. Some animal cells engulf food from the environment. The resulting membrane-bound **food vacuole** can fuse with a lysosome to form a digestive vacuole. The products of digestion diffuse into the cytoplasm where they are utilized to build cell constituents or as sources of energy. In some cases at least, lysosomes seem to be produced by the Golgi apparatus.

Plastids

Photosynthetic organelles, or **chloroplasts,** are found in all eukaryotic cells that are capable of converting light into chemical energy. These green, most often elliptical structures, like the mitochondria, are self-replicating. Under the highest magnification available with the light microscope the green chlorophyll pigment appears to be concentrated in discrete regions called **grana** which are embedded in an optically clear matrix called the **stroma.** The electron microscope has revealed details of the ultrastructure of chloroplasts and their highly organized membranes, or lamellae. Each granum is made up of a stack of lamellar (platelike) subunits

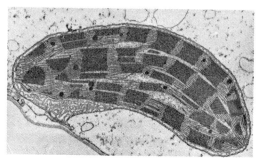

Fig. 2-10. Chloroplast from maize (Indian corn) under an electron microscope. Original magnification, × 10,000. (From Shumway, L. K., and T. E. Weiner. 1967. The chloroplast structure of iojap maize. *Am. J. Bot.* 54:744, 1967.)

thought by some experts to be like flattened sacs, and referred to as **grana lamellae.** The stroma, which appears granular in the electron microscope, also contains layered membranes called **stroma lamellae.**

An array of organelles under the collective term **plastids** can contain chlorophyll, pigments other than chlorophyll, or no pigments at all. A variety of chromoplasts, yellow, orange, or red in color, found in the cells of flower petals, certain fruits, and carrot roots, contain carotenoid pigments (xanthophylls, carotenes, and related compounds). The colorless **leukoplasts,** common in the cells of plant tissues that store food reserves, often contain starch grains. Other leukoplasts function as storage organelles for fats or oils. All of the plastids (chloroplasts, chromoplasts, and leukoplasts) are thought to arise from small and structurally simple precursors called **proplastids.** Both proplastids and mature plastids can divide and are randomly distributed between daughter cells during mitosis.

Microtubules

Many cells in both plants and animals possess structures involved in motion and transport called **microtubules.** Such structures have been known for a decade or so but are now being discovered in many hitherto unsuspected places. An electron microscope is required to see them.

Cilia and Flagella. Most motile and many fixed eukaryotic cells possess threadlike cytoplasmic extensions called **cilia** (if short) or **flagella** (if long) which are capable of whiplike motion. Both are the same in structure, consisting of two central microtubules (often called fibrils) surrounded by a circle of nine similar pairs of microtubules and the whole enclosed by an extension of the cell membrane. At its base in the cell, each cilium or flagellum is connected with a **basal granule** or body. The electron microscope shows that this is really a very short cylinder made up of nine pairs of microtubules.

Centrioles. Closely similar and perhaps identical structures called **centrioles** are found close to the nuclear membrane in most animal and in a few plant cells. They play an essential role in mitosis. During cell division microtubules extend from the centrioles to a special attachment point on each chromosome. All these microtubules together form the mitotic spindle which is necesary for the separation of the chromosomes into the two daughter cells. Once again the actual mechanics of all this is unknown as are the factors which trigger mitosis. The centrioles du-

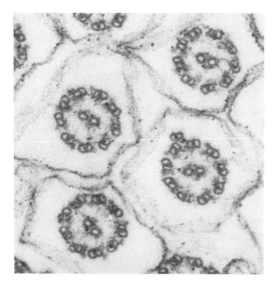

Fig. 2-11. Cross section of cilia in an annelid showing the circle of nine pairs of microtubules (fibrils) plus two central ones. As seen under an electron microscope. (Courtesy of Bjorn Afzelius. From Moment, G. B. 1967. *General Zoology*, 2nd ed., Houghton Mifflin Co., Boston.)

plicate themselves during cell division. The electron microscope reveals that they consist of nine triplet units of microtubules.

Microtubule Function. Microtubules have been investigated in developing plant cells by several workers. There is no certainty about their role, but it seems that the tubules are somehow concerned with transporting polysaccharides from the site of synthesis in the Golgi apparatus to the thickening cell wall. So far no such delivery function has been suggested for developing animal cells.

It now appears that microtubules are found wherever cytoplasmic motion takes place. The long radiating cytoplasmic extensions of freshwater and oceanic protozoans, which make them look like sunbursts, consist of more or less rigid extensions of as many as 200 microtubules. These are arranged in two beautifully regular parallel spirals as seen in cross-section. In various amoebae the pseudopods (as the locomotory extensions are called) contain microtubules. The evidence to date indicates that motion is somehow produced by the microtubules sliding past each other.

Membranes in and around Cells

Most intracellular membranes appear to be double when viewed through the electron microscope and seem to be extensions of invag-

inations of the exterior cell membrane (Fig. 2-12). Why all these membranes? It sometimes seems as though life were mostly a membrane phenomenon. No topic in the whole field of cell structure has caused as much recent interest and controversy as the nature of cell membranes. There is no controversy, however, about their importance. Virtually everything that enters or leaves a cell must pass through the cytoplasmic membrane which surrounds it. It is this membrane which separates the living from the nonliving. If the cell membrane of a frog's egg, for example, is ruptured in calcium-free water, the membrane cannot repair itself and the granular cytoplasm of the cell will flow out and drift away. Many hormones produce their effects by doing something to the cell membrane. There is good evidence that at least some hormones act directly on the genes, but even these hormones have to pass through the cell membrane and through the nuclear membrane as well. The function of that nuclear envelope is still much of a mystery. It is provided with pores large enough to see under an electron microscope and therefore large enough for good size molecules such as proteins and nucleic acids to pass through. The fact that the nuclear membrane is an abso-

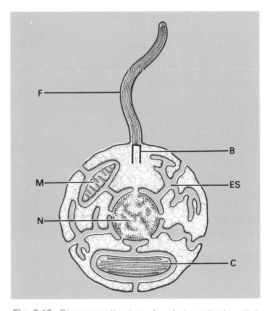

Fig. 2-12. Diagrammatic view of a photosynthetic cell (in this case a flagellate) showing how the mitochondria and chloroplasts may be regarded as "external" structures, their outer membranes being continuous with an interconnected endomembrane system (ES) formed by invaginations of the limiting cell membrane. Mitochondrion, M; chloroplast, C; flagellum, F; basal body, B; nucleus, N. (After Margulis, L., and F. J. R. Taylor. 1974. Origin and evolution of the eukaryotic cell. *Taxon* 23:235.)

lutely constant feature of all eukaryotic cells indicates that its functional significance is great.

Photoreceptors, whether in the eyes of a man or the chloroplasts of a green plant, are composed of layers of membranes. Within cells, the mitochondria and the Golgi apparatus are built of membranes, and much of the cytoplasm is crowded with the membranes of the endoplasmic reticulum adjacent to which lie the ribosomes where proteins are synthesized.

In addition to all this, a nervous impulse passes along a nerve in the form of a self-propagating wave of change in the surface membrane of the nerve fiber. Although we are admittedly only at the beginning of knowledge in this area, all the psychopharmacologically active molecules, whether dangerous artificial mind benders or the normal hormone of emotion, adrenaline, seem to produce their effects by acting on the surface membranes of nerve cells at the synapse where an impulse from one nerve cell is transmitted to another.

MEMBRANE STRUCTURE

Interest in membranes has grown steadily. A breakthrough came when J. F. Danielli and H. Davson proposed a model of what a cell membrane might be like. As is usually the case, their idea is based on earlier work, in this instance extending back to Benjamin Franklin. By the time of Danielli, the Nobel Prize-winning physicist Langmuir had shown that a drop of oil on the surface of water spreads out into a monomolecular layer. This indicated that cell membranes do not have to be many layers of molecules thick and might even be only one layer in thickness. The model proposed by Danielli and Davson is that cellular membranes are like a sandwich four molecules thick. The two outer layers are protein; the two inner layers are phospholipid molecules. The phospholipid, which is the most common lipid found in cells, helps make an effective barrier between the cell's environment which is aqueous and the cell's interior which is also aqueous. The protein covers provide some tensile strength. After all, a cell membrane needs to be stronger than a soap bubble.

Under an electron microscope all cellular membranes, whether on the surface of the cell or in part of an intracellular organelle, appear much the same: there are two dark lines very precisely separated by a clearer layer. The whole membrane is about 90 Å thick. Although measurements vary a bit de-

Fig. 2-13. J. F. Danielli, who with H. Davson proposed an early model of membrane structure (the unit membrane) made up of two inner layers of lipid and outer layers of protein. (Courtesy of J. F. Danielli.)

pending on methods of fixation and other factors, this is about right for a membrane composed of four layers of molecules. In the early 1960's J. D. Robertson drew together all the facts and coined the term **unit membrane** to designate this basic pattern of membrane structure.

Recently a new concept of membrane structure has emerged. According to the current **globular protein mosaic hypothesis,** the plasma membrane and other cell membranes with similar structure are thought to be composed of lipid with proteins distributed through or on either surface of the continuous double lipid layer. As in the earlier unit membrane hypothesis, the lipid bilayer provides a waterproof barrier between the cell interior and its environment (Fig. 2-14). Globular proteins can provide pathways for passage of water-soluble molecules or can act as permeases (see section on active transport) or as enzymes with fixed geometry on the membrane.

INTRACELLULAR ROLES OF MEMBRANES

One function of intracellular membranes may be to separate different and possibly incompatible reactions within the cell. Mitochondria, for example, are permeable to some substances but not to others. It should be noted that both the outer membrane and the inner one which forms the folds or cristae of a mitochondrion have a double appearance in the electron microscope. The inner one can now be seen under the newer electron microscopes to be thinner than the outer one, so all membranes need not be precisely the same.

Another important role of cellular membranes appears to be the ordering of enzymes in sequential functional assemblies. Many metabolic reactions require a series of enzymes to act in a particular sequence. The interior of a cell is an enormous space compared with the size of a molecule. If sequential reactions are to take place with reasonable speed, the necessary enzymes need at least to be near each other, if not lined up. Hence, it is not surprising that many enzymes now appear to be associated with membranes. The best understood case is the mitochondrion, where the enzymes of aerobic respiration are located.

Transport of Molecules and Ions through Membranes

The function of a cell membrane, in regulating everything which enters or leaves a

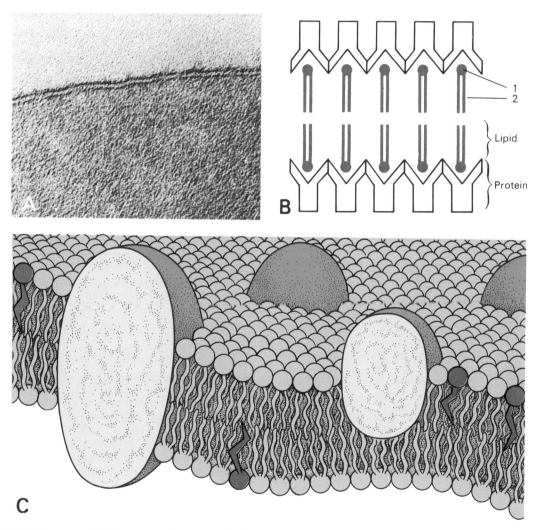

Fig. 2-14. Concepts of cell membrane structure, old and new. A, viewed under the electron microscope all membranes appear as two dark lines separated by a clear layer, × 275,000. (From Robertson, J. D. 1964. Unit membranes. In M. Locke, ed. *Cellular Membranes in Development*. Academic Press, New York.) B, unit membrane model proposed by Danielli and Davson is four molecules thick. It consists of two outer layers of protein and two inner layers of phospholipid (1, hydrophilic end; 2, hydrophobic end of phospholipid molecule). C, current view of cell membrane structure (globular protein mosaic hypothesis) proposes two layers of lipid molecules (light color), their hydrophilic heads aligned at the outermost and innermost surfaces, their hydrophobic tails meeting at the center. Odd shaped molecules associated with the lipid layers (dark color) are steroids. Proteins (gray) may be at either surface, be inserted part way into the membrane or may pass completely through the membrane. (After Capaldi, R. E. 1974. A dynamic model of cell membranes. *Sci. Am.* 230 (3):26. Copyright © 1974 by Scientific American, Inc. All rights reserved.)

cell, has been known for decades. However, the passage of materials through cellular membranes has turned out to be a highly complex phenomenon. Despite an enormous amount of work, the whole process is still far from being fully understood.

At least it is clear that some of the properties of cell membranes are primarily physical and dependent on the structure of the membrane. Such physical properties do not require the continuous expenditure of metabolic energy; thus poisons and anesthetics do not interfere with them. The osmotic properties of cell membranes fit into this category. For example, if a solution of protein or sugar in water is separated from pure water by a nonliving membrane which is permeable to water molecules but not to sugar or protein, i.e. a **semipermeable membrane, osmosis** will occur. Water will pass into the sugar or protein solution and make it more dilute. This is a fact. The common explanation is that the thermodynamic activity of the water molecules is less in the solution than it is in pure water. This is a theory, although there is good evidence for it.

A simple model to show what happens to cells under different osmotic conditions can be made by filling a cellophane bag with the concentrated sugar solution, tying it tightly closed and placing it in a beaker of water. The result? The bag will swell and ultimately burst. On the contrary, if a bag of water is placed in a beaker of concentrated sugar solution, it will shrink.

Comparable phenomena can easily be seen in living cells. If the surrounding medium has a greater osmotic concentration than the cell interior, it is said to be **hypertonic** and will cause the cell to lose water and shrink, a phenomenon called **plasmolysis**. If the medium has a lower osmotic concentration, it is said to be **hypotonic** and will cause the cell to take up water, swell, and burst. Such cell breakage is called **lysis**. If the osmotic concentration of the medium is the same as that of the cell, it is said to be **isotonic** or **isosmotic**.

So far we have considered circumstances where the materials passing through membranes have been passing from regions of greater to regions of lesser concentration, following the laws of diffusion. Cells, or more specifically their membranes, have the ability to move molecules in the reverse direction against a concentration gradient. That this activity requires the continuous expenditure of metabolic energy is indicated by the fact that such **active transport**, as it is often

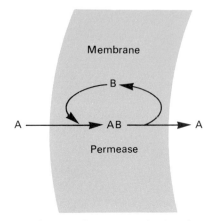

Fig. 2-15. Diagram of permease action in active transport across a membrane. (From Moment, G. B., and H. M. Habermann. 1973. *Biology: A Full Spectrum.* The Williams & Wilkins Co., Baltimore.)

called, is inhibited by metabolic depressants and poisons such as cyanide.

The mechanisms by which cells concentrate materials against a concentration gradient are believed to involve special enzymes called **permeases**, although the evidence is far from complete. A very simple model can be visualized as follows. Molecule A, outside the cell, unites with a carrier molecule B, within the membrane. Complex AB moves through the membrane and on its inner side breaks into A, which is released into the cell, while B remains in the membrane and eventually picks up another A (see Fig. 2-15).

pH: Acidity-Alkalinity

Living cells are extremely sensitive to pH, i.e. the acidity or alkalinity of their environment. pH is measured on a scale which extends from 0 to 14, and represents the negative logarithm of the hydrogen ion concentration of a solution. A pH of 1.0 represents an acidity equivalent to that of $1/10$ normal hydrochloric acid. A pH of 7.0 is neutral, while a pH of 13 represents an alkalinity like that of $1/10$ normal sodium hydroxide. Both internal and external pH are important to cells because most enzymes function only within very restricted ranges of pH.

The pH of a solution is best measured by a pH meter. However, many dyes, including natural plant pigments, assume different colors at different pH's. Consequently they can be conveniently used as indicators. Litmus, a pigment obtained from certain lichens, and anthocyanin, the purple pigment of red cabbage, are such indicators.

Cell Size

In addition to the basic nuclear-cytoplasmic organization of all eukaryotic cells, they are also essentially alike in size. The enormous differences in bulk between a mouse and an elephant, or between a moss and a redwood tree, are due to differences in numbers of cells, not in cell size. The sperms and eggs, as well as the cells of the liver, skin, brain, and other organs, are the same size within very small limits.

Some very potent factors must set both the upper and the lower limits of cell size. Apparently the upper limit is set by the volume of cytoplasm that can be serviced by a single nucleus. Another factor is the geometrical fact that as size increases, volume increases roughly as the cube of the radius of the cell, while surface increases as the square. This means that the relative amount of surface, through which all food and oxygen must enter the cell and through which all wastes must leave, becomes less and less in proportion to the amount of cytoplasm as volume increases. The lower limits of cell size are set because a certain minimum number of molecules are required to construct a eukaryotic cell with membranes and at least one chromosome, and that the atoms used to construct these essential molecules come only in standard sizes.

Much remains to be discovered about what controls the sizes of cells within the basic limits just described. For example, in animal cells that contain only one set of chromosomes (**monoploid** or **haploid**) instead of the two sets (**diploid**) they normally have, both the nuclei of the cells and the cells themselves are smaller than usual. If cells possess three sets of chromosomes (**triploid**), the nuclei and the cells are larger than normal. Exactly how this regulation is accomplished is not known. Even more puzzling, the triploid animals, although having much larger cells,

A

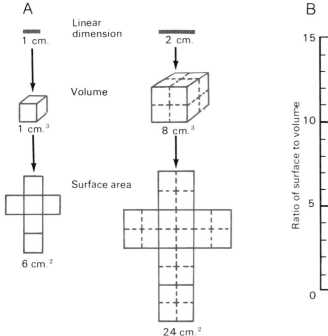

B

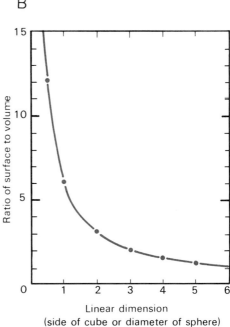

Fig. 2-16. Diagram of surface-volume relationships in cells. Cubes and spheres share the same mathematical relationship of changing surface to volume ratio as linear dimensions vary. For a cube, volume = s^3 (where s = the length of each side) and surface area = $6s^2$. The ratio of surface to volume for a cube = $6s^2/s^3$ or $6/s$. These relationships are shown graphically for cubes of $s = 1$ and $s = 2$ in A. For a sphere, volume = $4/3\ \pi r^3$ or $\pi d^3/6$ (where r = radius and d = diameter) and surface area = $4\ \pi r^2$ or πd^2. The ratio of surface to volume for a sphere = $\pi d^2 \div \pi d^3/6$ or $6/d$. The curve in B is a plot of this relationship. While spherical shape is more characteristic of animal cells and a cube of plant cells, both geometrical shapes have comparable surface to volume relationships which affect the physical limits of cell size. (From Moment, G. B., and H. M. Habermann. 1973. *Biology: A Full Spectrum.* The Williams & Wilkins Co., Baltimore.)

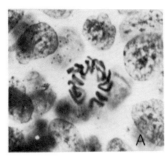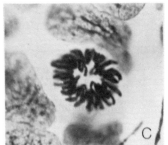

Fig. 2-17. Relationship of the number of chromosomes to cell size in tail fin of a salamander. A, monoploid (haploid) cells; B, diploid cells; C, triploid cells. Note a metaphase group of chromosomes in each case. (From Moment, G. B. 1967. *General Zoology,* 2nd ed., Houghton Mifflin Co., Boston, 1967. Courtesy G. Fankhauser.)

have fewer cells so that their body size is about normal. The photographs in Fig. 2-17 show a portion of the tail fin of each of three larval salamanders with one, two, and three sets of chromosomes. Salamander cells, which are extremely large, are the classic material in which Walter Flemming in the 1870's first described cell division clearly and in detail and coined the word mitosis (*mitos*, thread).

In plants, the relationship between the number of sets of chromosomes in the nucleus and cell size is not so clear-cut as in animals. In some plants, for example, certain of the algae, morphologically identical monoploid plants and diploid plants alternate in the life cycle. In higher plants, where the cells of the plant body are normally diploid, **tetraploid** (four sets of chromosomes) forms are known and frequently are horticulturally desirable because they are larger and more vigorous, with larger flowers and fruits. Some that are triploid, including varieties of pears, hyacinths, and tulips, are sterile and must be propagated vegetatively.

Cell Division

It has been just about a century since Rudolph Virchow (1821–1902) announced his famous aphorism, *Omnis cellula e cellula*, "Every cell from a cell," thus correcting the misconception of Schleiden and Schwann. Since his time, a small army of investigators has sought to describe the facts and explain the mechanism of cell division. Such knowledge is important, because cell division is the method by which the continuity of life is maintained and therefore by which heredity is passed from cell to cell. Furthermore, cell division has an important relationship to cancer. Malignant tumors are essentially populations of cells which have escaped from the normal controls of cell division.

In both animals and plants, cells divide in two quite different although related ways. In embryos and in growing tissues new cells are formed by **mitosis**. This is an asexual process in which the chromosomes of the parent cell are duplicated and passed in identical sets to the two daughter cells. The other method is called **meiosis**, from a Greek word meaning to reduce. The result of this kind of division of the nucleus is that the daughter cells have only one instead of the usual two sets of chromosomes. In animals, meiosis occurs in gonads (sex glands) and results in the formation of sperms and eggs. In plants meiosis usually occurs during the formation of spores.

Despite much work, often ingenious and penetrating, very little is known about the physiology and causation of cell division, and even less about what factors induce cells to divide by mitosis on one occasion and by meiosis on another.

THE EVENTS OF MITOSIS

Interphase is the period between divisions, and is the phase in which cells spend most of their lives. The nucleus is surrounded by a membrane within which only a finely granular material plus a nucleolus are visible. In adult animal cells the cytoplasm may be full of some special product of the cell's synthetic activity, perhaps muscle fibers or animal starch, in which case cell division is difficult or impossible and interphase is permanent. Differentiated plant cells often have elaborately thickened cell walls and very special circumstances are required for such cells to divide.

It has now been firmly demonstrated that the duplication of chromosomes takes place during interphase and occurs by the building of a duplicate chromosome (chromatid) alongside the parent chromosome. This conclusion has been reached by feeding cells radioactive thymidine, which is taken up and incorporated into the DNA of chromosomes. All the radioactivity appears on one and only one of each pair of daughter chromatids. If chromosomes reproduced by enlarging and then splitting down the middle, the radioactivity would be found randomly placed in both members of the resulting chromatid pairs (Fig. 2-1).

Prophase is the preliminary stage of division, during which the nuclear membrane disappears and the chromosomes become visible, first as long, thin threads and then finally as short, thick threads or rods. It is because of the threadlike appearance of the chromosomes that cell division was named mitosis. Chromosomes in late prophase and subsequent stages of mitosis can be clearly seen in living cells with a phase contrast microscope. In preserved cells they are easily stained with dyes like hematoxylin or various aniline stains. The fact that they can be stained gave their name (*chroma,* color, + *soma,* body).

It was learned early in the study of cells that the number of chromosomes in every cell of a given species is the same, and it was soon realized that each cell normally contains two similar sets of chromosomes. This means that each chromosome has its own permanent individuality. If the chromosomes in a set from one cell are lined up from the longest chromosome to the shortest and compared with the set from any other cell, provided that it is from the same species, this one-to-one correspondence becomes very evident. In *Drosophila,* the little fruit fly widely used in genetic research, there is always one long chromosome in a set, two middle-sized chromosomes, and one short one. Such an inventory of the chromosomes of an organism is known as its **karyotype**.

A conspicuous feature of a chromosome is its **kinetochore** (sometimes called **centromere**). This is a small rounded region that stains differently from the rest of the chromosome and appears to be the kinetic or controlling point in chromosomal movements within the cell. It may be in the middle of a chromosome, near one end, or at some intermediate point. During cell division the kinetochores become attached to the **spindle fibers,** i.e. the microtubles which extend from the poles of the mitotic spindle. It has been well established that the hereditary factors are lined up on the chromosomes in single file and in fixed order.

During early prophase the chromosomes appear as very thin threads which gradually shorten and thicken. In well-preserved material it is easy to see that each chromosome is really double along its entire length. Each member of such a pair is called a **sister chromatid** (although it would be more accurate to call them mother and daughter), and each pair is held together by a single kinetochore.

At the same time that the chromosomes are condensing, the nuclear membrane disappears. In animal cells the centrioles, described earlier along with cilia, duplicate and move toward opposite sides of the cell to form the astral rays of microtubules and a spindle-shaped structure on which the chromosomes become attached at their kinetochores. The star-shaped astral rays are well developed only in large cells such as those of cleaving eggs. In most plant cells, centrioles and astral rays are absent but microtubules forming the spindle are present.

Following prophase is **metaphase**, the period in mitosis when the chromosomes have moved to the equator of the spindle. Actually it is only the kinetochore of each chromosome that is on the equator. The rest of the chromosome may dangle at any angle. Each chromosome is still double along its entire length.

The doubling of the kinetochores signals the end of metaphase and the beginning of **anaphase**. Each of the double chromosomes separates, one chromatid going to each pole of the spindle. The action requires but a few minutes. As the chromosomes move toward the poles, their kinetochores lead the way. It looks as though the kinetochore were being pulled by a spindle fiber. In animal cells, the division of the cytoplasm is initiated during anaphase when the surface membrane begins to constrict between the two poles of the spindle at the level of the equator.

In the final phase of mitosis, known as **telophase,** the cell is divided into two daughter cells. The important point is that as a result of mitosis, each of the two daughter cells has two sets of chromosomes, just as the original cell had. During telophase the nuclear membrane reappears, and the chromosomes lose their stainability as they return to the interphase condition.

In plant cells, **cytokinesis**, or division of the cytoplasm, usually begins at the end of nuclear division. In telophase droplets appear at the center and later toward the edges of the equatorial plate. Sometimes these pectic substances appear as thickenings of the spindle fibers. The droplets coalesce and form a **cell plate** which divides the protoplasm into two daughter cells. The cell plate later be-

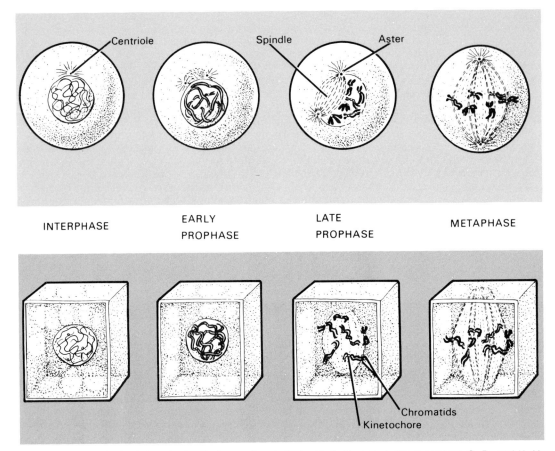

Fig. 2-18. Stages of mitosis in animal cells (top row) and plant cells (bottom row). (From Moment, G. B., and H. M. Habermann. 1973. *Biology: A Full Spectrum.* The Williams & Wilkins Co., Baltimore.)

comes an intercellular structure, the **middle lamella**. Plasma membranes are formed at the edge of the daughter protoplasts and become continuous with the remaining segments of plasma membrane of the original cell. The cytoplasm secretes cellulose cell wall material and later, after cell enlargement, additional cellulose is secreted to form secondary cell wall layers.

ANOTHER WAY OF VIEWING THE MITOTIC CYCLE

Many investigators of cell division now divide the mitotic cycle into four divisions: **M** (mitosis), with prophase, metaphase, anaphase, and telophase; G_1, the first gap or growth period; **S**, the period during which DNA is being synthesized; and G_2, a second gap or growth period. The events occurring and the relative amount of time spent in each period can be learned by feeding a radioactively labeled precursor of DNA such as tritiated thymine and then recording the stage of

the mitotic cycle when it has been incorporated (Table 2-2).

It has been found that the susceptibility of cells to radiation damage varies markedly with the stage of this cycle. Mitosis is delayed when cells receive relatively small doses of radiation during the G_2 stage. Also, radiation during the S phase can partially or completely block DNA synthesis, although the G_1 and S phases are relatively less sensitive to radiation damage than G_2.

MEIOSIS

When the facts of mitosis and fertilization became known, particularly that each egg and each sperm carries chromosomes, and that fertilization is the fusion of two such cells, August Weismann (1834–1914) made an important prediction. He foresaw that a stage would be found in the life history of every animal when the number of chromosomes is reduced from the double set or diploid condition to the single set or monoploid (often

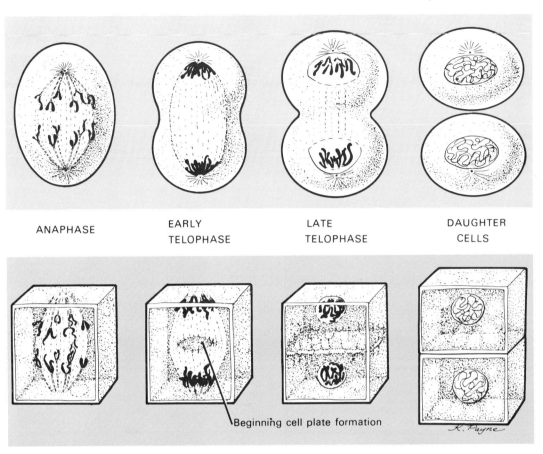

ANAPHASE   EARLY   LATE   DAUGHTER
                TELOPHASE   TELOPHASE   CELLS

Beginning cell plate formation

Fig. 2-18 (cont.)

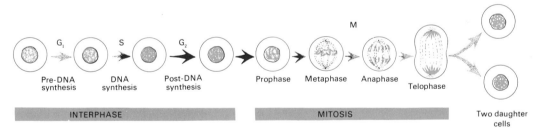

Fig. 2-19. General cell division cycle. (From Moment, G. B., and H. M. Habermann. 1973. *Biology: A Full Spectrum*. The Williams & Wilkins Co., Baltimore.)

called haploid) condition. Were this not true, he pointed out, the number of chromosomes would double with each fertilization, i.e. with every generation. This generalization applies to plants as well. There are various times and places where this process, known as meiosis, could conceivably take place. Weismann was nearly blind when he made this prediction and could not himself discover where meiosis does in fact occur. Others investigated this problem and found the following facts.

In animals, meiosis takes place in testes or ovaries during **gametogenesis**, that is, during the formation of the gametes, whether sperms or eggs. This is true of all animals. In plants it is difficult to generalize about where and when meiosis takes place. In the seed plants, the monoploid nuclei are formed in special cells within the male and female parts of the cone or flower. The end products are pollen grains and a cell within the ovary which develops into the embryo sac. Nuclear

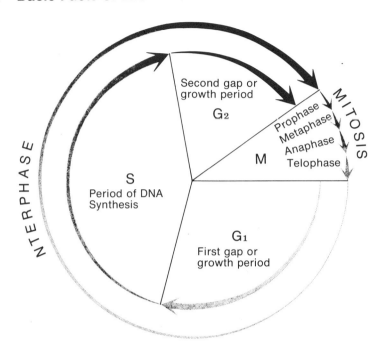

Second gap or growth period

G_2

Prophase
Metaphase
Anaphase
Telophase

MITOSIS

M

S

Period of DNA Synthesis

INTERPHASE

G_1

First gap or growth period

Fig. 2-20. An alternate way of viewing the cell cycle. The approximate duration of each stage is proportional to the length of the arrow. Note that the time required for DNA synthesis is much longer than for mitosis.

TABLE 2-2
Approximate duration of stages in cell division for several organisms (From Moment, G. B., and H. M. Habermann. 1973. *Biology: A Full Spectrum.* The Williams & Wilkins Co., Baltimore.)

Stage of Cell Cycle	Duration in hours			
	Mouse fibroblasts	Chinese hamster fibroblasts	Pea root tip	*Vicia faba* bean
G_1	2.7	9.0	2.66	5.0
S	5.8	10.0	4.0	7.5
G_2	2.1	2.3	2.5	5.0
M	0.4	0.7	3.0	2.0

divisions occur both in pollen grains and the developing embryo sac and both become multinucleate. The monoploid nuclei of the pollen grains and the embryo sac are involved in fertilization.

In all animals and plants meiosis requires two cell divisions. This is just as true between the gills of a mushroom or in a one-celled protozoan as it is in the gonads of a mouse or a man. Neither of these two cell divisions is a normal mitosis, although the cell goes through prophase, metaphase, anaphase, and telophase.

Discussion of the details of meiosis will be postponed until the chapter on genetics, where a knowledge of meiosis is essential for an understanding of the laws of inheritance.

Viruses

These smallest of living things, the viruses, differ from the prokaryotes and eukaryotes in that they are all intracellular parasites. They cannot multiply outside living cells, and therefore cannot be cultured in chemically defined media as can the blue-green algae and the bacteria. The viruses are unique among living things. They were originally described as invisible, filterable through membranes that retain bacteria, and noncultivable. However, since the development of the electron microscope in the 1930's, viruses have been observable. With specially prepared filters they can be removed from liquids and, provided with appropriate living cells as culture media, they may even be propagated.

DISCOVERY OF THE VIRUSES

By the latter part of the 19th century, a great many microorganisms had been discovered and studied. It was generally agreed at that time that the bacteria represented the lower limits of life in terms of size and simplicity of cellular structure. As happens so often in science, an apparently sound generalization was proved erroneous by later discoveries.

The first virus to be discovered is the causative agent of a disease of tobacco plants. In 1892, Iwanowski, a Russian botanist, showed that sap from plants with tobacco mosaic disease could be used to infect healthy plants. Sap from diseased plants remained infective even after filtration through materials known to trap all bacteria. However, it was not until 1935 that the first virus particles were purified. In that year Stanley obtained tobacco mosaic virus (TMV) in crystalline form (Fig. 2-21). This preparation, which remained infective after crystallization, was mostly protein in chemical composition, and for a time viruses appeared to be specialized protein molecules. Several years later it was shown that a constant proportion of nucleic acid is always present, and it is now clear that viruses are complexes of two kinds of molecules: protein and nucleic acid.

The first recognized virus-caused disease of vertebrates, **rabies**, was already under investigation by Pasteur at the time of Iwanowski's discovery of tobacco mosaic virus. By 1900, additional virus-caused diseases were being studied, including foot-and-mouth disease of farm animals, yellow fever in man, and a viral disease of the silkworm.

VIRUS STRUCTURE

Viruses are not cellular in structure. Chemically, they are composed of only two kinds of

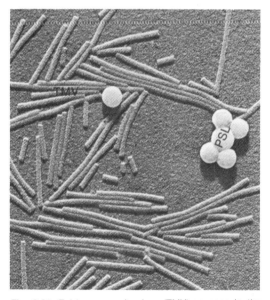

Fig. 2-21. Tobacco mosaic virus (TMV) as seen in the electron microscope, × 52,000. Polystyrene calibration spheres (PSL) 880 Å in diameter make it possible to calculate size of TMV rods. (From Jensen, W. A., and R. B. Park. 1967. *Cell Ultrastructure*. Wadsworth Publishing Co., Inc., Belmont, Calif.)

molecules: protein and nucleic acid. DNA *or* RNA but not both makes up the core of the **virion** (virus particle). The presence of only one kind of nucleic acid is a unique feature of the viruses; in all other living things both kinds of nucleic acid are present. The nucleic acid center is surrounded by a protein shell called the **capsid**. The capsid is made up of protein units called **capsomeres**. The nucleic acid core plus its protein capsid is called the **nucleocapsid**. In some viruses an outer limiting membrane (the envelope or mantle) surrounds the nucleocapsid. The envelope or mantle appears to be derived from membranes of the host cell and can contain lipids.

In their overall morphology, the viruses exhibit a variety of shapes. Some are rod-shaped, others are spherical, cuboidal, or one of many possible polyhedral shapes. Virions vary in size from the large smallpox or vaccinia virus, which is about 0.25 μm long and just resolvable by light microscopy, to 0.020 μm or less. The Japanese type B encephalitis virus has a diameter less than twice that of a molecule of egg albumin protein.

REPLICATION OF VIRUSES

Multiplication of viruses has been studied most extensively in bacterial viruses (the **bacteriophages**) because of the relative ease and rapidity with which their host cells can be cultured. *Escherichia coli,* the colon bacterium most widely used in these studies, has a generation time of only 20 to 30 min. A number of bacteriophages parasitic on *E. coli* are known, and they are usually referred to as T1, T2, T3, etc. (T stands for type). The T phages are spermlike in shape with a polyhedral head (the nucleocapsid) and a long tail several head diameters in length. DNA is found at the center of the head. The phage tail is an extension of the protein coating of the head. This hollow tube, enclosed in a retractable sheath, functions in attaching the virus particle to the host cell and in its penetration into the cell.

The life cycle of a virus involves penetration of the host cell, replication of virus nucleic acid and protein using the enzymatic machinery of the host, reassembly of new virus particles, and their release by lysis of the host cell.

VIRUS-CAUSED DISEASES

Many viral diseases are known, and it is estimated that at least half the infectious diseases of man are caused by viruses. Included in this long list of afflictions are colds, measles, polio, rabies, smallpox, infectious hepatitis, influenza, and probably cancer.

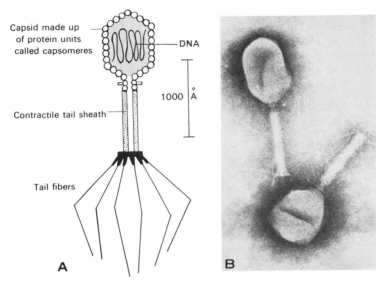

Capsid made up of protein units called capsomeres

DNA

1000 Å

Contractile tail sheath

Tail fibers

A

B

Fig. 2-22. T-even phage structure. A, diagrammatic representation; B, electron photomicrograph of *Escherichia coli* T4 bacteriophage stained with phosphotungstic acid, × 54,000. (From Rhodes, A. J., and C. E. van Rooyan. 1968. *Textbook of Virology*, 5th ed. The Williams & Wilkins Co., Baltimore, 1968.)

Fig. 2-23. A, viroid-caused disease of tomato. Left, healthy young tomato plant; right, tomato plant with spindle tuber disease thought to be caused by a viroid (Courtesy of T. O. Diener, U.S. Department of Agriculture.) B, Dr. T. O. Diener (foreground). Ultraviolet absorption profiles and electrophoresis are being used in viroid studies. (Courtesy of U.S. Department of Agriculture.)

The kind of viral cycle exhibited by the bacteriophages occurs also when a disease-causing virus invades a receptive nonbacterial cell. This is a key event in the onset of all infectious diseases caused by viruses. We do not always catch all the diseases to which we are exposed, so it would appear that cells can be resistant to invading viruses.

Since the last century, vaccines have been developed that utilize our natural defense mechanisms to fight virus-caused diseases. Vaccines are still prepared in much the same way that Pasteur made the first effective vaccine against rabies. Injecting an avirulent or inactivated virus can stimulate the production of antibodies which protect against infection by the virulent form of the virus. The most

recently developed vaccines include those against measles and polio.

Although vaccines have been widely used and are an effective means of controlling certain kinds of virus-caused diseases, there are some viral infections that cannot be controlled by vaccination. Examples of the latter include those where the causative agent has not been isolated (e.g. infectious hepatitis and most cancers), others caused by many different kinds of viruses (such as the common cold, which can be caused by any of 100 or more different viruses), or still others that are caused by viruses so unstable that vaccination provides limited protection (e.g. influenza). Where vaccination is impossible, one means of treatment is to use chemicals that alter or block the virus reproductive cycle. Effective chemical therapy has been developed for only a few virus-caused diseases.

Viroids

Very recently a Swiss-born investigator in the U.S. Department of Agriculture, Theodor Diener, has discovered particles much smaller than viruses which produce diseases in tomatoes, potatoes, and perhaps in animals and man. He calls these pathogens **viroids**. They are composed of only RNA. By the familiar techniques of electrophoresis through acrylamide gel and density gradient centrifugation, Diener has shown that they have a molecular weight of about 50,000.

Viroid research is still in its beginning phases and it is too early to conclude that yet another life form has been discovered. We can only wonder whether once again an apparently sound generalization about the ultimate size of living things is about to be challenged.

USEFUL REFERENCES

Buvat, R. 1969. *Plant Cells: An Introduction to Plant Protoplasm*. McGraw-Hill Book Co., New York. (Paperback.)

Cohen, S. C. 1970. Are/were mitochondria and chloroplasts miroorganisms? *Am. Sci.* 58(3): 281–289.

Compaldi, R. A. 1974. A dynamic model of cell membranes. *Sci. Am.* 230(3):27.

De Robertis, E. D. P., W. W. Nowinski, and F. A. Saez. 1970. *General Cytology*, 5th ed., W. B. Saunders Co., Philadelphia.

Dubos, R. 1962. *The Unseen World*. The Rockefeller Institute Press, New York.

Jensen, W. A., and R. B. Park. 1967. *Cell Ultrastructure*. Wadsworth Publishing Co., Inc., Belmont, California. (Paperback.)

Ledbetter, M. C., and K. R. Porter. 1970. *Introduction to the Fine Structure of Plant Cells*. Springer-Verlag, New York.

McElroy, W. D., and C. P. Swanson. 1976. *Modern Cell Biology*, 2nd ed. Prentice-Hall, Inc., Englewood Cliffs, N. J. (Paperback.)

Swanson, C. P. 1969. *The Cell*, 3rd ed. Prentice-Hall, Inc., Englewood Cliffs, N. J. (Paperback.)

Thomas, L. 1974. *The Lives of a Cell, Notes of a Biology Watcher*. Viking Press, New York.

Wilson, G. B., and J. H. Morrison. 1966. *Cytology*, 2nd ed. Reinhold Publishing Corp., New York.

3

Molecules and Energy: The Essential Ingredients of Life

A surprisingly large number of well intentioned but unimaginative people feel that to study anything chemical is to study death. They suppose that the only way to gain an understanding of living things is by wandering deep into the woods or perhaps out over the endless prairies under the big sky. Such excursions can confer valid insights, but what such individuals forget is that Friedrich Miescher, for example, was as fully immersed in the mysteries of life when he was in his laboratory extracting a strange substance, now called DNA, from the heads of fish sperm as any nature lover on the high trails of Glacier Park or beside the tide pools of the sea coast.

The truth is that the ancient comparison of life to a flame is not wide of the mark. The old analogy points to the central fact that life is a steady state which maintains an essential constancy of form amid a continually changing flow of matter and energy. Obviously the flame of life is a very unusual flame, highly structured and complex. Biochemistry has made it possible to penetrate many of the secrets of that flame which, in burning, maintains its structure and constructs the molecules of life.

Modern workers have broken the genetic code in the DNA, the blueprint for making a cornstalk or a man. They have discovered the molecules responsible for day-length control of flowering, those which control the cycle of sexual activity in a female mammal, those involved in capturing the sun's energy in green plants, and those passing a nervous impulse from one nerve cell to another. Even the way that mind-bending drugs act is beginning to be understood. Mescaline, for example, a psychoactive substance extracted from a kind of cactus, has a chemical structure closely related to that of adrenaline, a hormone associated with strong emotions.

Whether you are concerned with heredity, nutrition, the neurological basis of learning and memory, or the world's food and energy cycles, a solid knowledge of the basic facts and principles of biochemistry is essential. Without that, your thinking is stalled in the late 1920's, for there are no biological problems that lack a biochemical dimension.

The Subatomic Level

An understanding of life on the biochemical level requires some knowledge of the relationship of living things to the subatomic and the atomic levels of organization. The outstanding fact about events on the level of electrons, protons, photons, and other wave particles is that they do not seem to follow the rules which hold for the physics and chemistry of our familiar macroscopic world. The pertinent question for the biologist is what roles, if any, do events on the subatomic level play in the lives of plants and animals?

Electrons are perhaps the best examples of subatomic particles that play a vital role in biological processes, specifically in oxidation-reduction reactions. The loss of electrons is termed **oxidation** and the gain of electrons, **reduction**. Thus, if a substance is an electron donor, it is called a **reducing agent**. If a substance is an electron acceptor, it is an **oxidizing agent**.

The emissions of radioactive materials which are so very useful in biological tracer studies in living organisms are all subatomic. Alpha particles are the nuclei of helium atoms (two protons and two neutrons), beta particles are fast moving electrons, and gamma rays are quanta of higher frequency and shorter wavelength than X rays.

The Atomic Level

On the atomic level, a number of questions — some old, some new — arise. Are living organisms composed of the same chemical elements that are found in the nonliving world of rocks and rivers, oceans and clouds, and even the sun and other planets? If we are so made, which elements are involved? Most of the 100-odd or but a few?

The answer to the first question has been clear for many decades. If a plant or animal is killed and dried, and the water that evaporates then measured, it will be found that a very large portion of living matter is water and that the hydrogen and oxygen in this water are the same as any other hydrogen and oxygen. The dried corpse can be burned. Carbon dioxide and water vapor will be given off, revealing a third element, carbon. The remaining ash, on chemical analysis, yields the following elements in decreasing order of amount: nitrogen, calcium, phosphorus, potassium, sodium, sulfur, chlorine, magnesium, and iron, plus several more such as copper and manganese. It is worth noting that the list represents only a handful of the known elements. Why? No one really knows. The big four as far as quantity goes are oxygen, carbon, hydrogen, and nitrogen, but phosphorus is now known to be so important in both energy-carrying compounds and in nucleic acids that one should really speak of the big five.

The most interesting as well as most important problem in this field today concerns what are called trace elements, or better, **micronu-**trients. These are elements essential for survival but required only in minute amounts. For animals, some essential trace elements have been known for many years. Iodine, the lack of which produces goiter, is one. That traces of cobalt are necessary was discovered as a result of an attack of "bush disease" in sheep on Australian ranges in 1895. Since then other micronutrients have been discovered: copper, zinc, and manganese. As seems reasonable for material essential only in minute amounts, these elements are either part of enzyme molecules or behave as catalysts for enzyme action. Copper, for example, is part of the enzyme cytochrome oxidase which is involved in oxidative respiration in the mitochondria. This entire subject is one of great practical importance about which much remains to be learned.

The structure of an individual atom bears an interesting resemblance to a minute solar system. There is a central nucleus consisting of a tightly packed group of protons and neutrons having a net positive charge. Revolving around the nucleus are one or more electrons carrying a negative charge and of negligible weight compared with the nucleus. The simplest atom, hydrogen, consists of one proton and one electron in orbit around it. A carbon atom consists of six protons and six neutrons with two electrons in an inner orbit and four in an outer. Oxygen atoms have eight protons and eight neutrons and two electrons in the inner orbit and six in the outer (see Fig 3-1) The chemical properties of an atom are determined very largely by the number of electrons in its outermost orbit. These are the ones involved in valence and shared in covalent bonds.

The Molecular Level — Organic Compounds

On the molecular level living organisms are characterized by **organic compounds**, or molecules containing the element carbon. Until about a century ago it was believed that all compounds containing carbon could be synthesized only by some unanalyzable vital force in living organisms; hence, the names organic compounds and organic chemistry. This misconception was corrected by a 28-year-old German chemist, Friedrich Wohler. He had no intention of producing an organic compound in his laboratory but he did so in the course of trying to synthesize something else, so his discovery is one of the most important instances of serendipity on record. The substance was urea, $NH_2-CO-NH_2$, a

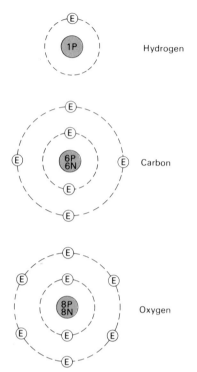

Fig. 3-1. Structure of hydrogen, carbon, and oxygen atoms.

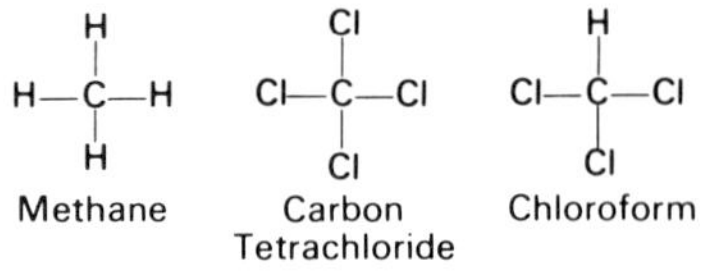

H—C—H

H
Methane

Cl—C—Cl

Cl
Carbon
Tetrachloride

Cl—C—Cl

Cl
Chloroform

Fig. 3-2. Structure of methane, carbon tetrachloride, and chloroform molecules.

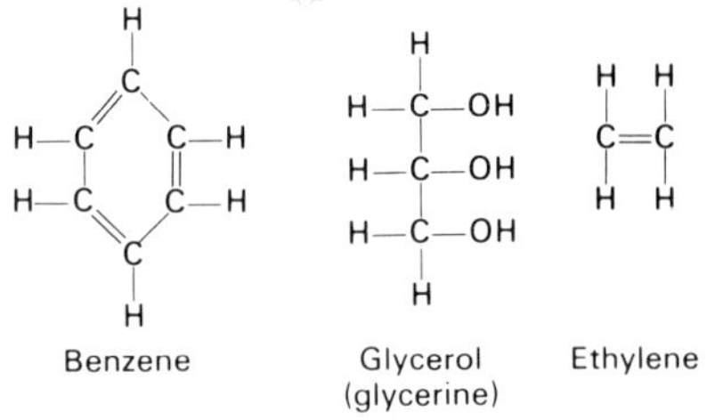

Benzene

Glycerol
(glycerine)

Ethylene

Fig. 3-3. Structure of benzene, glycerol, and ethylene molecules.

colorless, tasteless, odorless compound which is a waste product of protein breakdown. Of course Wohler did not produce urea the way animals do, but that is not the point. He had killed the old myth and convinced the most skeptical chemists that organic chemicals can be analyzed, synthesized, and understood. Since then enormous numbers of carbon compounds have been synthesized: dyes, drugs, explosives, nylon, hormones, and nucleic acids. In 1976 Nobel Prize winner Har Khorana using familiar chemicals in his laboratory at M.I.T. synthesized a gene which functions when introduced into a bacterium.

The carbon atom has a number of properties that make it biologically important. Each carbon atom has four covalent bonds, chemical "links" that enable it to unite with four other atoms by sharing electrons with them. However, it may unite with fewer than four atoms by using two of its bonds to link to another. This is a double bond. The **covalent bonds** are determined by the number of electrons in the outermost orbit around the nucleus of the carbon atom. They are represented in written formulas by short bars (Fig. 3-2). Each atom in a molecule is represented by the letter symbol for its name: H is hydrogen, C is carbon, Cl is chlorine, and so on. An example of a familiar carbon compound is

methane (CH_4) or marsh gas, a waste product of bacteria living in decaying debris on the bottom of quiet ponds and often seen bubbling to the surface. Carbon tetrachloride has a similar structure except that in place of four hydrogen atoms, four chlorine atoms are attached to the carbon. Carbon tetrachloride has been widely used in fire extinguishers and as a cleaning fluid, although it is highly poisonous. Substituting a hydrogen atom for one of the chlorine atoms produces chloroform.

Carbon atoms also have the ability to unite with each other and in this way make extremely large molecules. The carbons may join in short or long single chains, in rings as in benzene, or in many other configurations. Familiar examples include glycerol (also called glycerine), a sweetish lubricant that can be obtained from fats and oils, and ethylene, a sweetish, colorless, highly explosive component of coal gas and a great menace to miners. This compound is also produced by plants and its role in fruit ripening and leaf fall is discussed in Chapter 18. Benzene is obtained from coal tar and is used as a solvent and cleaning fluid and also as the chemical base for the synthesis of thousands of other substances, from dyes to explosives (see Fig. 3-3).

The existence of functional groups is of considerable biochemical importance. **Functional groups** or **radicals** behave as units in chemical reactions. One of the most familiar is the carboxyl group, COOH, which is also written

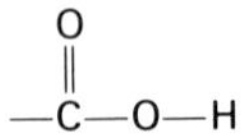

$$\begin{array}{c} O \\ \parallel \\ -C-O-H \end{array}$$

This group may look like a base, but in water it is the hydrogen which is released so that it is in fact acidic. The carboxyl group is present in amino acids and fatty acids. The **amino group**—NH₂, is basic and is found in amino acids and other important compounds. It behaves like ammonia, which forms ammonium hydroxide with water: $NH_3 + H_2O \rightarrow NH_4OH$. Thus, the amino group is alkaline. Other important groups are the **methyl group**, the highly reactive **keto**, and **aldehyde groups**:

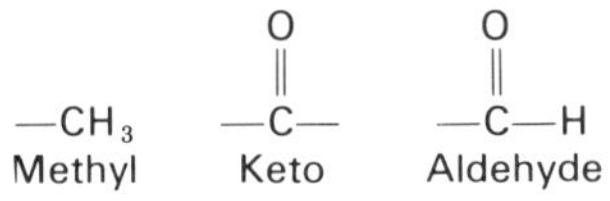

$$-CH_3 \qquad -\overset{\overset{\textstyle O}{\|}}{C}- \qquad -\overset{\overset{\textstyle O}{\|}}{C}-H$$
Methyl   Keto   Aldehyde

Carbohydrates—The Building Blocks and Fuel of Life

The familiar sugars and starches are **carbohydrates**. A simple sugar is composed solely of carbon, hydrogen, and oxygen, e.g. $C_6H_{12}O_6$, glucose. Chemically, a carbohydrate often is defined as a simple sugar or a substance which yields simple sugars by hydrolysis, that is, by splitting the compound with the addition of water. Carbohydrates may also be defined as those substances with the general formula $C_x(H_2O)_y$. The values of x and y may range from 3 to several thousand. Yet another way of looking at carbohydrates is to think of them as polyalcohols made up of chains of $H-\overset{\displaystyle |}{\underset{\displaystyle |}{C}}-OH$ units.

If the values of x and y in the formula of a carbohydrate are low, between 3 and 7, the sugar is called a **monosaccharide**. Two monosaccharides linked together comprise a **disaccharide**, e.g. sucrose, our common table sugar. More than two monosaccharides joined together form a **polysaccharide**. The link that holds one sugar molecule to another is known as a **glycoside linkage** (Fig. 3-4). Such joining of similar units is called **polymerization**, an extremely important process that we will meet frequently, since it is the way proteins and DNA, as well as the whole array of carbohydrate-related materials, are made.

Monosaccharides, i.e. simple sugars with three carbons ($C_3H_6O_3$), are called triose sugars or merely **trioses**. Those with four carbons are **tetroses**; with five, **pentoses**; and with six, **hexoses**. Glucose, a simple hexose, $C_6H_{12}O_6$, is a highly important sugar because it is the structural unit from which many more

complex sugars as well as the starches and cellulose are built.

Another extremely important sugar is a pentose called ribose and the related sugar deoxyribose; deoxy- because it has less oxygen than ribose. While ribose fits into the general formula, $C_x(H_2O)_y$ (i.e. $C_5H_{10}O_5$), deoxyribose ($C_5H_{10}O_4$) does not. Both of these five-carbon sugars are important constituents of nucleic acids: deoxyribose in the DNA of the chromosomes, and ribose in the RNA of the ribosomes and other components of a cell's machinery for making proteins.

Sugars, especially glucose, are commonly and correctly called the "fuels of life," but it would be incorrect to think that they are only fuels. As just mentioned, the pentoses ribose and deoxyribose form essential parts of nucleic acid molecules. Sugars in general have a marked ability to combine with other molecules and other sugars. The combination of a sugar and a nonsugar is called a **glycoside**, no matter what kind of sugar is involved. If the sugar is glucose, the term glucoside can be used. Many glycosides have a strong action on the heart. The non-sugar part of these molecules is usually a steroid similar to the sex and adrenal hormones. An exception to this generalization is digitalis, the best known glycoside, where the non-sugar part is digitonin.

The non-sugar part of a glycoside may be an amino acid or simply an amino group, —NH₂. It may be recalled that proteins are built up of amino acid chains. An amino-containing glycoside is called a glucosamine (if the sugar is glucose) or simply an **amino sugar**. Amino sugars are polymerized, i.e.

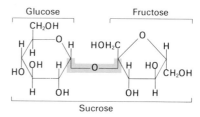

Fig. 3-4. The two monosaccharides, glucose and fructose, joined by a glycoside linkage (color) form in the disaccharide, sucrose.

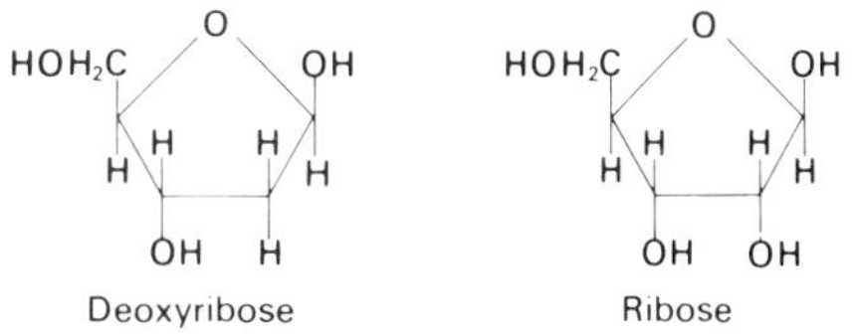

Deoxyribose   Ribose

Fig. 3-5. Structural formulas of deoxyribose and ribose, important constituents of nucleic acids.

united into long chains, to form **chitin**. This is the highly protective skeletal material of insects, lobsters, and related animals. Interestingly enough, chitin is present in the cell walls of certain molds. Glucosamines are also important in the connective tissues of vertebrates.

Sugar molecules are asymmetrical, i.e. they can exist in two forms that are mirror images like right and left gloves or shoes. This remarkable property of carbohydrates is also characteristic of amino acids. The chemical behavior of the molecule depends not only on the kind and number of atoms in it but on their arrangement (configuration). One of Pasteur's first discoveries, made when he was a very young man, was that simple organic compounds like sugar (he was actually working with a simple organic acid, tartaric) are asymmetrical, and that two kinds of crystals can be sorted out under a microscope. As working with model atoms will show convincingly, such asymmetry is inevitable whenever a carbon atom is attached to four atoms that are each different. Such pairs of molecules are called **isomers** (Fig. 3-6). Since they will rotate the plane of polarized light in a clockwise or counterclockwise direction according to their asymmetry, they are commonly called **dextro-** or **levorotary optical isomers**. Such isomers are extremely difficult for a chemist to separate but cells do so readily. Virtually all naturally occurring sugars are dextro-sugars, but why this should be is unknown. It may be a mere accident of evolution or it may have some great significance. In any case, digestive enzymes can digest dextro- but not levo-sugars.

By more polymerization, that is, by sugars being linked to other sugars and still other compounds, a vast array of important substances are produced by plants and animals. What is the significance of all this polymerization? First, the process of forming glycoside linkages removes water (dehydration synthesis) which means less oxygen, and that means that the molecule is left richer in energy. Second, these big polymers have biologically useful properties that simple sugars lack, e.g. great tensile strength in fibers, protective qualities in gums and waxes, and lubricating ability in various sorts of mucus. Third, making one large molecule from many small ones (as in making one molecule of starch from many molecules of glucose) reduces the osmotic concentration of the cell in which this polymerization occurs. Ordinarily, green plant cells carrying out photosynthesis rapidly convert their surplus sugars into starch.

Starch and **glycogen**, which is often called animal starch, are two of the more familiar simple polysaccharides. Glycogen is found not only in the vertebrate liver and the cytoplasm of protozoa, but also in blue-green algae and bacteria, so it is produced by both prokaryotes and eukaryotes and is presumably of very ancient origin. Plant starches such as those from potatoes, rice, etc., differ from glycogen. Glycogen molecules are composed of many-branched chains of glucose while plant starch molecules consist of both linear and branched portions. **Cellulose**, a principal component of wood and plant fibers, is a highly indigestible and apparently unbranched glucose polymer. Many of the gums which are produced by plants after injury are **heteropolysaccharides,** i.e. polymers of two or more different sugars. So also is the ground substance of cartilage, chondroitin, and hyaluornic acid, the "glue" which holds animal cells together. Heparin,

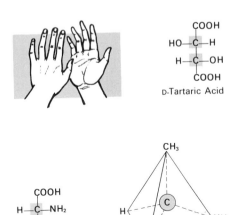

D-Tartaric Acid   L-Tartaric Acid

Alanine

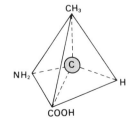

Fig. 3-6. Stereoisomers have the same structural relationship as right and left hands or as an object and its mirror image. Asymmetric carbons in structural formulas are indicated in color. (From Moment, G. B., and H. M. Habermann. 1973. *Biology: A Full Spectrum.* The Williams & Wilkins Co., Baltimore.)

a powerful anticoagulant found in the liver and other vertebrate organs, is a mucopolysaccharide.

Lipids

Lipids are a heterogeneous group including the fats, oils, steroids, waxes, and related compounds. They are so varied that a satisfactory definition is difficult to formulate. In general, lipids are composed of the same three elements as carbohydrates—carbon, hydrogen, and oxygen, although some also contain phosphorus and other elements. Unlike most carbohydrates, lipids are insoluble in water, but are soluble in various organic solvents such as chloroform, ether, and alcohol. Many lipids, including the common animal fats, can be split into two components, glycerol and fatty acids.

The most obvious function of lipids is energy storage. Since lipids generally have less oxygen per molecule, they carry more energy than starches, but it is less easily available. Migratory birds develop large deposits of fat before they begin their long flights. On arrival hundreds or even thousands of miles from their start, the fat is used up. Hibernating animals also begin the winter with large deposits of fat. This is used partly during the winter and partly when they first emerge and food is scarce. Lipids stored in seeds are utilized during the process of germination.

Lipids are located in both plant and animal cell membranes. In many animals, lipids form heat-insulating layers; the blubber of a whale is nearly a foot thick. Certain vitamins are lipoidal in character and are commonly found associated with other lipids. The sex hormones are lipids, as are cortisone and other hormones. The epidermal cells of plants usually have a coating of wax, also a lipid.

Historically, the chemical nature of fats was understood before that of either carbohydrates or proteins. In fact, a rule-of-thumb knowledge of fat chemistry has been used for centuries in soap making. Fats can be broken down into glycerol (glycerine) and a weak organic or fatty acid. No matter what the fat or oil, the glycerol is always the same, although the fatty acids are different for each fat or oil. At the acid end of a fatty acid is a carboxyl group, COOH. From one to over a dozen carbon atoms can be attached to this group. **Formic acid,** secreted by ants and other insects as a chemical weapon, is the simplest possible organic acid, $H-COOH$. **Acetic acid** or vinegar is CH_3-COOH. Stearic acid of beef fat is a veritable train, usually

written $C_{17}H_{35}COOH$. Glycerol is a three-carbon compound (see Fig. 3-7). If the three OH groups are removed and NO_3^- radicals substituted, the result is trinitroglycerine, the explosive component of dynamite.

The basic reaction by which glycerol and fatty acid are formed from a fat involves splitting the fat with a molecule of water. This process is called **hydrolysis** *(hydro*, water, + *lysis,* to loosen) or hydrolytic splitting. It is one example of digestion, for hydrolysis is what happens to carbohydrates and proteins as well as fats in the digestive tract. The reverse process, by which two chemical substances are combined, is much more complicated and involves a series of steps. Living things can convert carbohydrates into fats, some of us all too easily.

When every carbon in the fatty acid chain is holding two hydrogens, the fat or oil is said to be **saturated**. When two adjacent carbons are linked by a double bond, the lipid is said to be **unsaturated**. Butyric and crotonic acids illustrate this contrast. The structural formulas of these two acids are shown in Fig. 3-8.

The melting point, which determines whether a lipid is an oil or a solid fat at any given temperature, depends on two factors. The shorter the chain of fatty acids, and the greater their unsaturation, the lower the melting point. Unsaturated fats are thought to be much less easily converted into cholesterol than saturated fats. Although cholesterol is heavily implicated in the cause of heart and other vascular diseases, it should be remembered that all the normally occurring steroid hormones, including the stress hormone of the adrenal gland, adrenaline, as well as cortisone and the like, are derived from cholesterol.

Fig. 3-7. Structure of glycerol, trinitroglycerol, and the generalized structures of a lipid. When R_1, R_2, and R_3 are all fatty acids, the molecule is a triglyceride, a fat, or an oil.

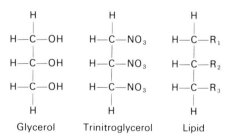

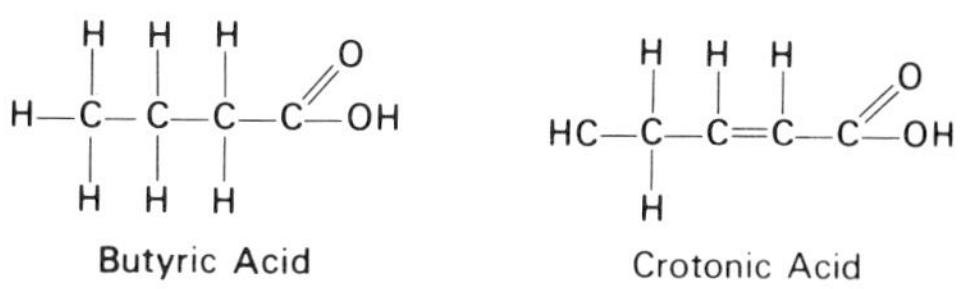

Fig. 3-8. Structure of butyric and crotonic acids.

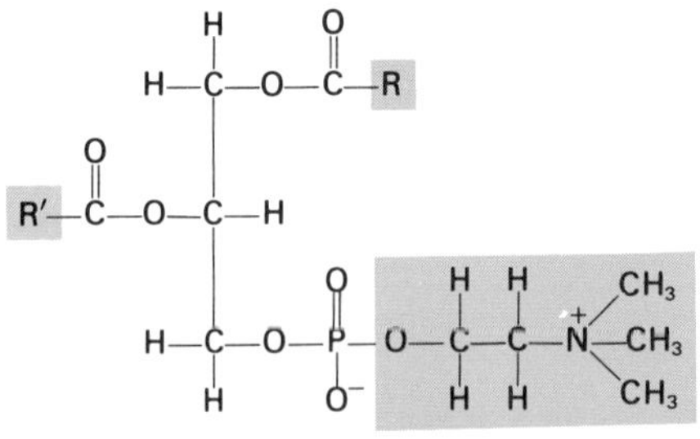

Choline

Fig. 3-9. Structure of lecithin, a phosphodiglyceride with two fatty acids (R and R'). A phosphate group links choline to the third carbon of glycerol.

When a lipid has three fatty acid chains, as does stearin or beef fat, it is known as a **triglyceride**. Most of the familiar oils and fats are triglycerides and in most of them not all of the three fatty acids are the same. In fact, each of the three may be different. When a lipid has only two fatty acid chains, it is called a **diglyceride**.

We can now take a second look at the lipids in cell membranes. These extremely important lipids are diglycerides that have a phosphate group attached to the glycerol where the third fatty acid might have been. Attached to this phosphate group is usually an additional molecule. In the case of **lecithin**, the most abundant **phosphodiglyceride** in cell membranes, this last group is choline (Fig. 3-9). Choline is a small molecule consisting of a nitrogen atom to which are attached three methyl groups, $-CH_3$, plus two carbons with hydrogens. Choline will be discussed again in connection with the conduction of a nerve impulse, which is a membrane phenomenon. The phosphate group of lecithin is an energy-rich group similar to those found in the energy-carrying compounds of respiration. This suggests a role of lecithin in pumping molecules across cell membranes. Thus, like the carbohydrates, the lipids play many roles in the economy of life.

Proteins and Amino Acids

Proteins, together with the nucleic acids, DNA and RNA, form the triumvirate of molecular biology. The key information that the genetic code in the nucleic acids spells out concerns the millions of kinds of catalytic proteins, i.e. **enzymes**. Every organism is what its proteins make it because all cellular activities, synthetic, digestive, locomotor, reproducitve, and the rest, are either directly controlled by enzymes or by the products of enzymes. The proteins confer species and individual uniqueness. The distinction between self and non-self has its biochemical basis in the proteins characteristic of each organism. The rejection of organ transplants and the identification of blood both depend on proteins. In small amounts they are essential components of the human diet; without them physical and mental deterioration and finally death result.

Many kinds of tissues have high concentrations of protein. Certain seeds such as beans and peanuts are rich sources. A number of animal tissues are constructed largely of protein—especially horns, fingernails, hair, tendons, and feathers. Egg white is almost pure protein and water. Proteins are components of all living cells. They are part of the cellular membranes as well as constituting important parts of the cellular machinery. Grass contains the proteins which steers build into beef steaks.

Proteins have been recognized for over a century as a distinct group of compounds containing nitrogen in addition to the carbon, oxygen, and hydrogen of the carbohydrates and lipids. Many proteins also contain small amounts of sulfur and other elements. Hundreds of kinds of proteins from a variety of plants and animals have been isolated and purified. Their molecular weights are enormous, ranging from several thousand up to several million. It will be recalled that the molecular weight of water is only 18. All the millions of proteins can be broken down by boiling in acid (or gently by enzymatic digestion) into about two dozen amino acids. All the millions of plant and animal proteins are made from this same little group of amino acids. They have been compared to the 26 letters of the alphabet which can spell all the words of all the languages of Europe.

All **amino acids** have a so-called alpha carbon atom to which are attached four other atoms or groups of atoms. There is always a hydrogen, an acidic carboxyl group, $-COOH$, a basic amino group, $-NH_2$, plus an additional group designated as R. Both the carboxyl and amino groups present in all amino acids are very reactive. In neutral solution they exist as charged $-COO^-$ and $-NH_3^+$ radicals. Such an amino acid with its two charged groups is called a zwitterion (Fig. 3-11).

In the simplest case possible, glycine, the fourth group or R is merely a hydrogen atom. If R is a methyl group, $-CH_3$, the amino acid is alanine. If a phenol ring is substituted for one of the H's in the methyl group, phenylalanine is the result. Phenylalanine is an extremely important substance because it can

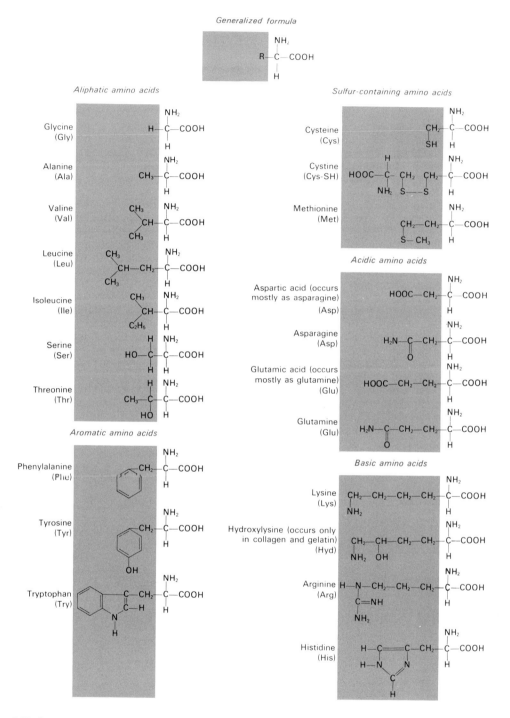

Generalized formula

Aliphatic amino acids

Sulfur-containing amino acids

Acidic amino acids

Aromatic amino acids

Basic amino acids

Fig. 3-10. Generalized formula for an amino acid and structures of all naturally occurring amino acids. Color indicates structure equivalent to R in the generalized formula.

be made into the amino acid tyrosine merely by adding an OH group to the phenol ring. Tyrosine is used by cells to synthesize thyroxin (the hormone of the thyroid gland), the hormone adrenaline, and the brown, reddish and black melanin pigments of potatoes, skin, hair, and feathers. Tyrosine also forms part of many proteins.

The nutritional value of a protein, whether from meat, fish, peanuts, corn, or any other source, depends on the amino acids it contains. If an animal cannot make a particular amino acid by modifying other amino acids, its diet must provide it. For men and rats, for example, tryptophan (or indolealanine) is essential in the diet or growth cannot take place. Most plants differ from animals in being able to synthesize all the different amino acids that they need.

About 50 years ago a German chemist, Emil Fischer, who devoted much of his life to the study of proteins, discovered how amino acids are joined together into proteins. One can visualize the process by imagining that an −OH is lost from the carboxyl group of one amino acid and an −H from the amino group of another. A molecule of water is released and the two amino acids are left, linked by a bond between the nitrogen of one amino group and the carbon of the carboxyl group of the other. Note that when two amino acids are thus joined into what is called a dipeptide, the resulting molecule still possesses a carboxyl group at one end and an amino group at the other. Consequently, a dipeptide can unite with another amino acid and form a tripeptide, and this in turn with other amino acids until a long **polypeptide** or, more descriptively, a polymer of amino acids, and finally a protein is formed. The linkage is always the same, passing through the alpha carbon to the carboxyl carbon to the nitrogen to the next alpha carbon. Such a bond is called a **peptide linkage** (Fig. 3-12).

The sequence of amino acids in the chain determines the character of a protein. This sequence, termed the **primary structure** of a protein, is spelled out by the DNA code in the chromosomes, as will be further clarified in the chapter on genetics. The asymmetrical shapes of the amino acids cause a twisting of the long polypeptide chain, usually into a corkscrew-like helix. This helix is called the **secondary structure** of a protein. The helical shape is maintained by **hydrogen bonds,** weak links formed when a hydrogen atom is shared by two other atoms, one of which is usually oxygen. However, few proteins remain as long helical chains. They become folded and bent on themselves in characteristic ways. This **tertiary structure** of a protein is determined by several factors. One of these is the disulfide bonds (−S−S−) that form between the sulfhydryl (−SH) groups of two sulfur-containing amino acids. Since the position of the sulfur-containing amino acids is determined by the genetic code, the folding of any particular kind of protein is determined by heredity, i.e. the structure of DNA.

The primary structure of a protein, i.e. the actual sequence of amino acids, was determined for the first time by Frederick Sanger in Cambridge, England. He used insulin partly because its molecules are relatively small for a protein that can be obtained commercially, and partly because of its great medical importance. The method is both laborious and tricky. It consists of breaking up the molecule by several different methods into pieces of different overlapping lengths, determining

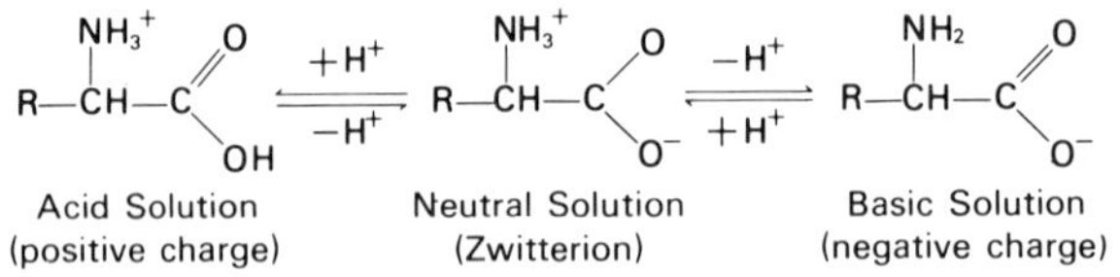

Fig. 3-11. Amino acids are charged zwitterions in neutral solution. In basic solution the positively charged ammonium ion loses a proton to become the uncharged amino group; in acid solution the negatively charged carboxyl ion gains a proton to become the uncharged carboxyl group.

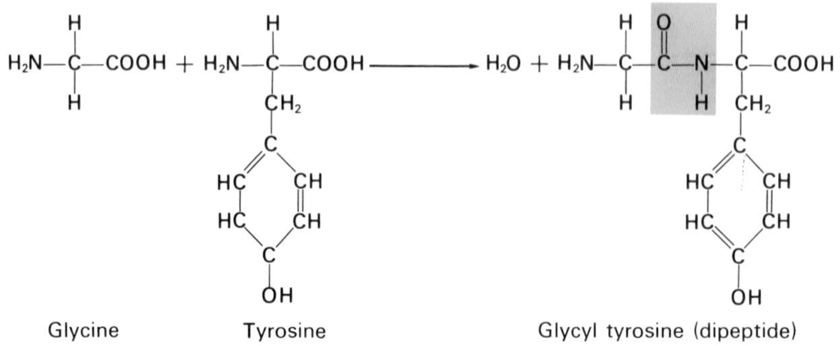

Glycine   Tyrosine   Glycyl tyrosine (dipeptide)

Fig. 3-12. The carboxyl group of glycine and the amino group of tyrosine are involved in the peptide bond (color) joining the two parts of the dipeptide.

Fig. 3-13. Frederick Sanger determined the sequence of amino acids in insulin, an important molecule in animal physiology and the first protein so characterized. (Courtesy of the Burndy Library of Science and Technology and Frederick Sanger.)

the amino acids in each piece, and then finally fitting all the data together into a consistent picture.

The importance of the correct amino acids in the correct sequence has been dramatically shown in sickle cell anemia (see Chapter 5). In this disease the hemoglobin is abnormal; there is a single error in two of the four chains which together make up a hemoglobin molecule. In place of one of the valines is glutamic acid. This is one error out of about 300 amino acids. A consequence of this small change in primary structure is a change in the shape of the red blood cells and in their oxygen-carrying properties.

Nucleic Acids

The most famous case of an important biological discovery that was neglected for decades is probably Mendel's discovery of the laws of heredity. The discovery of nucleic acids, which constitute the physical basis of those laws, is another. This emphasizes how difficult it is to recognize which scientific discoveries are truly important.

An understanding of the molecular structure of nucleic acids, and proof that their structure carries the genetic code from cell to cell and from generation to generation, was achieved within the past two decades. However, a century has passed since Friedrich Miescher (Fig. 3-14) began to explore the biochemistry of nuclei. The experimental materials for these early studies were the white blood cells of pus (obtained from wounded soldiers), which have large nuclei (lymphocytes almost lack cytoplasm altogether) and also the heads of salmon sperms which consist of little more than condensed nuclei. Miescher discovered that nuclei were composed almost entirely of an acid (later called nucleic acid). It was unusual in containing not only nitrogen but also phosphorus, a fact

Fig. 3-14. Friedrich Miescher, discoverer of DNA, which he extracted from sperm of salmon caught in the Rhine River. (From *Die histochemischen und physiologischen Arbeiten von Friedrich Miescher, Gesammelt und Herausgegeben von seine Freunden*. F. C. W. Vogel, Leipzig, 1897. Courtesy of The Rockefeller University Archives and Springer-Verlag, Heidelberg.)

so peculiar that his major professor refused to allow him to publish his discovery until after it had been very carefully confirmed more than two years later.

From the 1870's until the 1920's little work was done on nucleic acids. In 1924, Feulgen and others found that the chromosomes could be stained a brilliant red after proper treatment of the deoxyribose located there. This reaction not only afforded a very elegant method for staining but it demonstrated the presence of deoxyribose along the entire length of every chromosome.

Analysis by a variety of chemical and physical methods has now revealed the actual structure of nucleic acids and shown clearly that there are two kinds, deoxyribonucleic acid (DNA) and ribonucleic acid (RNA). DNA is found associated with proteins in the chromosomes. RNA is found in both the nucleus and the cytoplasm. Nucleic acids resemble proteins in that both types of compounds are polymers, i.e. long chains of small molecules-nucleotides and amino acids, respectively. Each **nucleotide** consists of a sugar (ribose or deoxyribose), a phosphoric group (PO_4H), and a nitrogen-containing base. Chains of these units are called **polynucleotides.**

There are only four kinds of nitrogen-con-taining bases in DNA and also only four common ones in RNA; three of these occur in both. In DNA, two of the bases are **purines** (adenine and guanine) and two are **pyrimidines** (cytosine and thymine). In RNA, uracil substitutes for thymine. This fact offers a useful research tool. Radioactively labeled thymine will appear in the DNA in the chromosomes. Labeled uracil will appear in RNA.

When the nucleotides of DNA are linked together, they form two chains with links between them, and resemble a ladder twisted like the stripes on a barber pole. The sides of the ladder are built of alternating phosphoric acid and ribose sugar units. The rungs of the ladder are made of the nitrogen bases, two bases in each rung, one purine and one pyrimidine. In DNA, adenine is always paired with thymine, and guanine with cytosine. The rungs are attached to the uprights of the helical ladder at the sugars. How these four bases constitute a code for making proteins will be discussed in the chapter on genetics.

In living cells the DNA in the chromosome makes more DNA like itself. It also serves as the template for making **messenger RNA.** The messenger RNA moves out into the cytoplasm where it furnishes the instructions for making proteins, including the most important proteins, enzymes. Thus life is regulated by three key compounds: DNA, RNA, and proteins.

Metabolism

The proteins, carbohydrates, lipids, and even nucleic acids in a living organism are continually undergoing change and renewal. The old comparisons of life to a flame or to a fountain turn out to be even truer than the ancients suspected. Much of our knowledge of metabolism has come from experiments with radioactive tracers. Tracers have been used not only to unravel complex biochemical processes but also to follow the fate of individual compounds taken up by cells. **Tritium,** a radioactive isotope of hydrogen, has been used to label the nitrogenous bases incorporated into nucleic acids. The use of tritiated thymine or uracil has provided information about the replication of chromosomes and the site of RNA synthesis. ^{14}C-labeled amino acids have demonstrated that there is a constant exchange of amino acids between the proteins of which animal tissues are built and the amino acids in the bloodstream. Tracer studies show that the very structure of living things is in flux, with a definite turnover rate. However diverse it may appear, the stream of matter and energy which flows through a living organism is utilized in very few ways. The principle uses of energy are to make big molecules from small ones, to transport material across membranes, or to produce motion.

The sum of all the chemical activities within a living organism is called **metabolism**. Synthetic metabolism (biosynthesis), which constructs more complex molecules, is known as **anabolism**. Destructive metabolism, such as hydrolysis, breaks up large molecules and is called **catabolism**. The most obvious single aspect of metabolism in animals and ourselves is the mechanical inspiration and expiration of air. Although respiration in this gross mechanical sense has always been the criterion of life, it is only within the last four centuries that anyone has known what breathing does for an animal.

BEGINNINGS OF OUR
UNDERSTANDING OF RESPIRATION

The growth of knowledge about respiration marks one of the most long continued and important series of achievements in the history of science. In the 17th century a group of young investigators, including Robert Boyle in London, discovered what is still an amazing thing: a breathing mouse and a burning candle both do the same thing to air. By enclosing candles and mice together and separately under glass bell jars, these men were able to show that a candle cannot burn in air in which a mouse has suffocated, nor can a mouse live long in a closed place in which a candle flame has burned out. In 1796 the Dutch physician, Jan Ingen-Housz, published his findings demonstrating that in the dark, plants, like mice and burning candles, remove a component from air that is needed for combustion. In the light, however, adding a plant to the mouse in the bell jar may prevent his suffocation indefinitely.

With the growth of chemistry in the 18th century, Antoine Lavoisier (1743–1794) and Joseph Priestley (1733–1804) proved that in respiration oxygen is taken from the air and carbon dioxide returned to it. By keeping small mammals in confined vessels and measuring accurately both the CO_2 and the heat (i.e. calories) given off, they further proved that animals obey what are now known as the **laws of conservation of matter and of energy**. A breathing guinea pig and burning charcoal give off the same number of calories of heat energy when the same amount of oxygen is consumed or carbon dioxide given off.

The 19th century researchers refined the discoveries of their predecessors and added some of their own. The simple sugar, glucose, is the material usually burned in a living organism. The over-all equation

$$C_6H_{12}O_6 + 6O_2 \rightarrow 6CO_2 + 6H_2O + energy$$

conceals a vast complexity but needs to be assimilated first. One molecule of glucose combines with six molecules of oxygen to yield six molecules of carbon dioxide and six of water. The energy released is measured in calories (cal). A **small calorie**, the unit usually used in cellular metabolism, is defined as the amount of heat necessary to raise 1 g of water 1°C (Celsius). (Note that the **large calorie** or kilocalorie is the usual dietetic unit.) At the same time studies were being made on the caloric value of different foodstuffs, extensive knowledge was gained of how oxygen and carbon dioxide are absorbed and transported in the blood, and also about the way respiration is controlled. Many of these facts and principles established in the 19th century are of great importance today in aviation medicine, deep sea diving, dietary programs, etc.

The achievement of the present century has exceeded the wildest dreams of the 19th. It is nothing less than the successful exploration of the inner workings of cells. The living cell is a complex chemical factory with many diverse and interconnected production lines. The details of respiration in cells are basically the same in plants and animals. This realization has contributed a unity to biology that had not existed before. Knowledge of how oxygen and sugar power life's activities, and how they are controlled by enzymes, has led directly to important practical results. This is especially true in understanding how to choose antibiotics and antimetabolites that will block living processes at specific points. It is a large topic to which we will return.

Metabolic Pathways

There have been two massive achievements in this century in the biological sciences. One was the discovery and then the breaking of the genetic code. The other has been the gradual unraveling of the intricacies of cell metabolism. Not only have the general contours of energy flow been sharply delineated, but even the fine details have been traced in an amazingly intricate and extensive system of change and interchange of matter and energy. This new knowledge confers a profound understanding of living processes and their relationship to the nonliving world. It also means a vastly increased possibility of control over living processes in many areas — anesthetics, therapeutic medicine, fermentation industries, agriculture, the nervous system, and of animal and human behavior.

At the same time this new knowledge has resulted in a very real simplification. What once seemed a puzzling jumble of unrelated processes now is revealed as a single system. Aerobic and anaerobic respiration, fermentation, and photosynthesis all fit together smoothly. The chief energy-yielding processes are everywhere the same, in bacteria, plants and animals, whether in fermenting yeast or respiring potatoes, in beef hearts, in the flight muscles of honeybees, or in the cilia of a clam.

ANAEROBIC RESPIRATION — AN OVERVIEW

The main pathway by which energy is made available for living organisms has two distinct stages. The first stage does not require free oxygen and is therefore called **anaerobic respiration**. Since sugar is broken down it is also called **glycolysis**, and because alcohol and related substances are end products it is also referred to as **fermentation**, although this term is sometimes restricted to the very final steps of glycolysis. This anaerobic process is also named the **Embden-Meyerhof glycolytic pathway** after the men who established its existence.

Glycolysis begins with glucose. In the first step a phosphate group is added. This phosphorylated sugar is then further modified and torn apart by a series of enzymes until it is converted into pyruvic acid. During glycolysis, two molecules (three, in the case of muscle cells) of ATP are produced. It will be recalled that ATP, the nucleotide adenosine triphosphate is the molecule in which energy is "packaged" and then distributed within cells.

Pyruvic acid, the end product of the Embden-Meyerhof glycolytic pathway, is highly reactive. It consists of three active groups:

a methyl, $-CH_3$, keto, $-\overset{\displaystyle O}{\overset{\|}{C}}-$, and carboxyl,

$-COOH$, connected thus: $CH_3-\overset{\displaystyle O}{\overset{\|}{C}}-COOH$. Pyruvic acid stands at a metabolic crossroads. If no oxygen is present but certain enzymes are, the pyruvic acid is converted into lactic acid (as happens in muscular exercise), or a fatty acid, or some kind of alcohol (as happens in certain yeasts and bacteria),

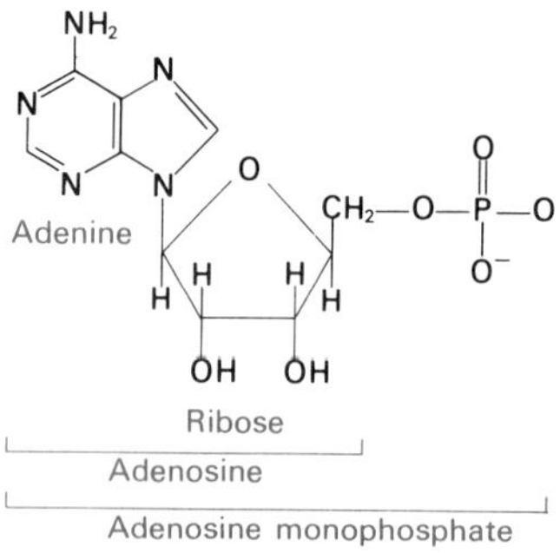

Fig. 3-15. Structure of adenosine 5′-monophosphate, a nucleotide.

or one of the various other products of the process of fermentation.

AEROBIC RESPIRATION — AN OVERVIEW

A second stage in the release of energy is the **citric acid** or **Krebs cycle**, which receives the products of glycolysis (and often input from other processes) and breaks down these compounds to CO_2. At the very end of the process, the electrons removed by the action of enzymes on several intermediate substrates are transported along a chain of carrier molecules in the mitochondria and finally united with oxygen and protons to form water.

If oxygen is present and also the proper enzymes, then pyruvic acid loses a carbon and CO_2 is released as a waste product. The acetic acid thus formed unites with coenzyme A and enters the citric acid or Krebs cycle. It is at the level of acetic acid and coenzyme A that fat metabolites enter the main energy flow. Residues from amino acids also enter the citric acid cycle at this point, although this is not the only place where they come into it, and are used as energy sources. The cycle itself consists of nine different and relatively small organic acids. The two-carbon fragment from pyruvic acid is carried by coenzyme A and enters the cycle by uniting with the last of the nine acids in the cycle, thereby forming citric acid which is the first intermediate in the cycle. As each acid is formed from its immediate predecessor by a series of splittings, coalescences, and modifications, CO_2 is released as waste, more ATP's are produced, and finally the last acid in the cycle is reached. It then unites with more acetyl coenzyme A to form more citric acid and the cycle begins again. The hydrogens from the original glucose are passed along a chain of carriers to molecular oxygen with which they form water. The reactions of the Krebs cycle are carried out only by the **mitochondria**. That all this second or aerobic part of the respiratory pathway occurs in the mitochondria can be demonstrated by breaking cells, separating mitochondria from the other cytoplasmic materials by centrifugation, and then testing the metabolic capabilities of the cytoplasm and of the mitochondria.

As a result of terminal respiration, the citric acid cycle is kept functioning and three ATP's are produced per pair of electrons entering the electron transport system.

ATP is the key product of both anaerobic and aerobic respiration and is the way energy is packaged and transported within cells, regardless of whether the energy will be used in the synthesis of more complex molecules, the contraction of a muscle, the performance of osmotic work, or the lighting of a firefly.

All of the enzymes of the aerobic portion of carbohydrate metabolism are located within the mitochondria. The enzymes of **electron transport** are located on the inner membrane, where they are lined up in a way that facilitates these sequential reactions.

The release of energy through the utilization of oxygen and the "storage" of the energy in ATP is called **oxidative phosphorylation**. The relative amounts of energy packaged into ATP molecules by glycolysis and by the aerobic portion of the metabolic pathway are very different. For each molecule of glucose, glycolysis yields only two (or three) molecules of ATP. The aerobic portion yields a total of 36. Actually this varies a bit with the substrate and the organism—yeast cells get 38 ATP's from the aerobic process. The efficiency of the total process is roughly 45 per cent. If 1 mole (1 g molecular weight) of glucose, which is 180 g of glucose, is burned to $6CO_2 + 12H_2O$, about 680,000 small calories of heat energy are released. Making 1 mole of ATP from adenosine diphosphate + inorganic phosphate stores about 8,000 cal in chemical bond energy. Thus, the 2 moles of ATP formed in the glycolytic breakdown of 1 mole of glucose equal about 16,000 cal. The 36 moles of ATP formed in the aerobic part equal about 288,000 cal. Hence, the entire process from glucose to CO_2 and water stores approximately 304,000 cal packaged as ATP, out of a total of 680,000 cal released in the complete combustion of one mole of glucose. The energy that is not conserved in ATP is used in breaking down the glucose and

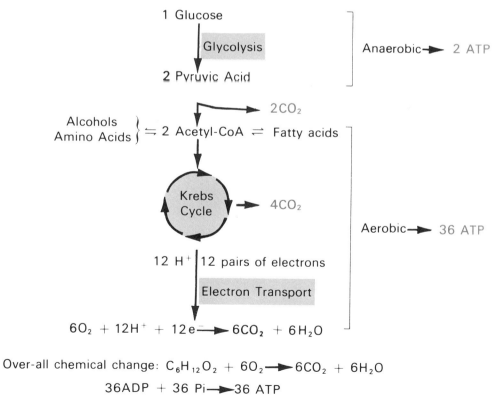

Over-all chemical change: $C_6H_{12}O_2 + 6O_2 \longrightarrow 6CO_2 + 6H_2O$

$36ADP + 36\ Pi \longrightarrow 36\ ATP$

Fig. 3-16. Energy-yielding pathways of carbohydrate breakdown. (From Moment, G. B., and H. M. Habermann. 1973. *Biology: A Full Spectrum.* The Williams & Wilkins Co., Baltimore.)

50   Basic Facts of Life

producing the ATP's or lost as heat. Evidently there is a packaging charge of over 50 per cent.

METHODS AND EVIDENCE

Anaerobic Respiration

When you look beneath the dogmatic assertions about glucose uniting with phosphate as the first step in the utilization of sugar, or about a cycle of small organic acids going round and round grinding up acetic acid and producing energy in little ATP bundles, or beneath any of the dozens of other assertions about what is supposed to go on inside the cell, what do you find? How convincing is the whole complicated story, or even any of its parts? How could anyone ever get so much as a finger in a crack to gain access to such deeply hidden events?

A natural starting point is the achievement of Eduard Buchner. It was he who, by grinding yeast cells in sharp sand and then filtering and testing the cell-free extract on sugar, found that such "killed" material could convert sugar to alcohol. This was the beginning. The active agent was not itself living but was produced in living cells.

Buchner died in World War I, but his discovery excited Arthur Harden, then teaching and writing textbooks in Manchester, England. Harden studied the cell-free extracts of yeast and found that they converted sugar into alcohol and CO_2 very rapidly at first and

then more and more slowly until action finally stopped. He sought to discover why. After a long search, Harden found that the addition of blood serum or of boiled fresh yeast extract (which would not in itself ferment since it had been boiled) would restore activity. He finally found that the active ingredient in these restorative agents was inorganic phosphate. Fermentation could be virtually at a standstill but then be fully restored by the addition of phosphate. Clearly the enzyme was not wearing out. The surprising thing about this discovery was that sugar, the enzyme zymase, and alcohol do not contain any phosphate. Harden at last discovered that the phosphate combines with the sugar. This is now recognized as the first step in glycolysis.

The discoveries about fermentations in yeasts came to be seen, after some years, to be related to the metabolism of muscle contraction. It became known that muscles contain glycogen, the "animal starch" discovered long ago in livers by Claude Bernard at the Sorbonne. Further, F. G. Hopkins, one of the discoverers of vitamins, showed that a working muscle accumulates lactic acid, an organic acid also produced by various bacteria, notably those which cause milk to sour. It also became known that as a muscle works, it uses up its glycogen.

At this point Otto Meyerhof, at the University of Kiel where Buchner had worked a generation earlier, succeeded in showing first that there is an exact quantitative relationship between the amount of glycogen that disap-

Fig. 3-17. Carl and Gerti Cori, a husband and wife team working at Washington University in St. Louis, identified the phosphorylated intermediates of glycolysis. (Courtesy of Washington University Photographic Service.)

pears and the amount of lactic acid produced just as there is between the sugar used and the alcohol produced in fermentation. Second, Meyerhof showed that in the absence of oxygen this relationship between glycogen and accumulated lactic acid remains the same. In other words, the energy metabolism of muscle in the first stage is anaerobic. When oxygen is again admitted to muscle, the lactic acid disappears and oxygen is utilized. Muscle metabolism thus appeared to consist of two parts: a first anaerobic part, glycolysis, and a second aerobic portion. The working out of the actual steps and enzymes in the anaerobic pathway, of which there are many, between glucose at one end and pyruvic and lactic acids at the other, was largely the work of a husband and wife team, Carl and Gerti Cori. Among other accomplishments they identified the actual structure of the phosphate-sugar compounds which Harden had described only in vague terms.

No series of discoveries in the whole field of biology has resulted in a more brilliant or more important achievement than the unraveling of the mysteries of cellular respiration. Nor has the solution of any problem of comparable significance ever been to such a marked extent the result of the combined contributions of so many investigators in so many parts of the world.

USEFUL REFERENCES

Barker, G. R. 1968/70. *Understanding the Chemistry of the Cell*. Institute of Biology, Studies in Biology No. 13. Edward Arnold (Publishers) Ltd., London. Distributed in the U.S. by Crane, Russak and Co., New York. (Paperback.)

Barry, J. M., and E. M. Barry. 1969. *An Introduction to the Structure of Biological Molecules*. Prentice-Hall Biological Science Series. Prentice-Hall, Inc., Englewood Cliffs, N.J., (Paperback.)

Haynes, R. H., and P. C. Hanawalt. 1968. *The Molecular Basis of Life: Readings from the Scientific American*. W. H. Freeman and Co., San Francisco and London. (Available in paperback.)

Lehninger, A. L., ed. 1975. *Biochemistry*, 2nd ed. Worth Publishers, New York. (An advanced treatise.)

Loewy, A. G., and P. Siekevitz. 1969. *Cell Structure and Function*, 2nd ed. Holt, Rinehart and Winston, Inc., New York. (Treats both cell structure and cell chemistry.)

McElroy, W. D. 1971. *Cell Physiology and Biochemistry*, 3rd ed. Prentice-Hall, Inc., Engelwood Cliffs, N.J. (Paperback.)

THE STREAM OF LIFE

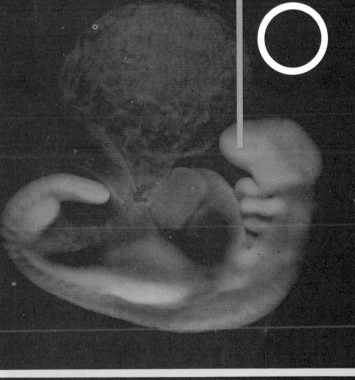

[San Francisco, Embryo photographs courtesy of Carnegie Institute of Washington, Department of Embryology, Davis Division]

4

Genetics I: The Facts of Heredity

The science of genetics has entered a new and revolutionary stage of its development. Knowledge of the principles has been achieved by the study of inheritance in garden peas, in the fruit fly, *Drosophila,* and in microorganisms, notably in the mold, *Neurospora,* and in various bacteria. As a result, the gene has become a central concept of biology. This fundamental knowledge is now being combined with modern techniques for culturing human cells outside the body and the recently acquired insights into the biochemistry of normal and abnormal metabolism to build a true science of human genetics. Cells of organisms as far apart as mouse and man are being hybridized in cultures. These methods are used to map the positions of genes on specific human chromosomes, an accomplishment which is prerequisite for any sound understanding of human heredity.

Not only is such knowledge of great theoretical interest, but it is also of great practical importance, especially as an aid to alleviating the great burden of human genetic disease. The National Foundation—March of Dimes organization estimates that the annual cost of caring for victims of Down's syndrome (mongolism), a condition caused by an extra chromosome, is far in excess of a billion dollars in the United States alone. No such figure can express the grief of parents nor the sheer disruption such tragedies inflict on families.

Happily the new interdisciplinary science of human genetics offers a real hope of significantly improving the human condition. However, before discussion of these issues, the science of genetics itself must be examined, for an understanding of it is essential not only to an understanding of human heredity, but also for a complete insight into any other biological problem from photosynthesis to ecology.

In the Beginning

The value of taking a hard look at some of man's ancient beliefs is twofold. Many of them, even though surely false, are still prevalent, and others raise basic and persistent problems. Moreover, genetics illustrates well the important but often forgotten fact that science is both cumulative and open-ended.

To begin with the ancient Greeks, Hippocrates (460(?)–377 B.C.), botanist and the "father of medicine," taught that the embryo is formed from a swarm of "seeds" or particles which come from all parts of the body of the two parents and are carried in the reproductive fluids. Thus heredity is particulate. The great philosopher-scientist, Aristotle, a younger contemporary of Hippocrates, attacked both parts of this theory and claimed instead that the embryo is formed not by particles but out of the reproductive fluids by the action of a vital formative force which he called "entelechy," a word revived in modern times by Driesch. These fluids, which form the link between parents and children, Aristotle claimed, are not formed all over the body but only in the reproductive organs.

Thus two questions came into clear focus. Is heredity due to particles or to fluids? In other words, is there such a thing as a "blood relative"? Does the hereditary material, whatever it may be, arise from all parts of the body or only from the reproductive organs? This second question obviously involves the question of whether or not the effects of use and disuse, or of anything which happens to the parts of the body, can influence heredity.

Enter Gregor Mendel

Modern genetics began with Mendel, not at the time he published his results, over a century ago, but after they lay completely neglected for nearly 40 years. Why this happened is a complex story, but it is important to remember that Mendel's discoveries, like all scientific advances, had antecedents. Mendel himself was widely and actively interested in many branches of science. One of the keys to his success was that he used in his basic experiments a plant, the garden pea, that had been used by many others in studies of heredity. As early as 1823, Thomas Knight had confirmed still earlier reports of dominance and recessiveness and the reappearance of ancestral types. Although neither Knight nor any of the others who crossbred peas noticed regular laws, the necessary basic information about the techniques of breeding pea plants had been obtained.

In addition, several workers, including Charles Darwin, who had studied hybrid pigeons, had stated the problem of heredity very clearly. In his epoch-making book, *Origin of Species,* Darwin wrote:

Fig. 4-1. Gregor Mendel, the monk who discovered the laws of heredity for all living things by the use of simple arithmetic and Isaac Newton's binomial theorem. (Courtesy of The Johns Hopkins University Institute of the History of Medicine.)

The offspring from the first cross between two pure breeds is tolerably and sometimes (as I have found with pigeons) quite uniform in character, and everything seems simple enough: but when these mongrels are crossed one with another for several generations, hardly two of them are alike, and then the difficulty of the task becomes manifest. . . . The slight variability of hybrids in the first generation, in contrast with that in succeeding generations, is a curious fact and deserves attention.

The final factor which may have given Mendel the clue to his discovery was his interest in beekeeping. Mendel was a contemporary and almost a neighbor of Johann Dzierzon, the most famous beekeeper of all time. Dzierzon crossed German with Italian bees and found that in the following generation half the drones were German, half Italian. Mendel was thus alerted to the possibility of finding definite ratios.

Why the importance of Mendel's analysis was not appreciated at once is also a complex story. Chromosomes were not taken very seriously by biologists, most of whom were primarily interested in evolution, and the facts of meiosis had not been worked out even though Weismann had predicted that meiosis must take place. Another factor was that Galton and his school had not yet shown the importance of applying statistics to the study of variation.

In 1900 three experimenters eventually rediscovered Mendel's basic laws. They were de Vries in Holland, Correns in Germany, and von Tschermak in Austria. Mendel's paper was found almost immediately afterwards. (It had been mentioned in a book by the renowned American botanist, Liberty Hyde Bailey). At about the same time, de Vries (1848–1935) discovered **mutations,** i.e. sudden changes in heredity. Once alerted to mutations, biologists began to find them in many animals and plants.

Mendel's Fundamental Discoveries

SEGREGATION: MENDEL'S FIRST LAW

Mendel's experiments, lasting 8 years, included the cross-fertilization and raising of many hundreds of plants and the counting of some 8,000 peas. He succeeded where others had failed first, because he simplified his problem and considered single pairs of characters, tall vs. short plants, or wrinkled vs. smooth seeds, instead of thinking about the whole complex organism. Second, he had the patience to use statistics.

To illustrate the principles discovered by Mendel, consider a cross between two special kinds of chickens, a so-called "splashed white" and a black. If two such chickens are crossed, 100 per cent of the first generation, the F_1 or **first filial generation,** will be neither splashed white nor black but a slaty blue, the "blue Andalusian" (Fig. 4-2). If one of these hybrid blue chickens, either a rooster or a hen, is mated with a black chicken, 50 per cent of the offspring will be black and the other 50 per cent blue. None will be splashed white. These are the facts.

What do these facts mean? Mendel's explanation was simplicity itself. It has since come to be called the **law of segregation.** Each individual produced by sexual reproduction possesses a double set of hereditary factors, one set from each parent. The hereditary make-up of a purebred black chicken can be represented as ●●, and the hereditary constitution of a purebred white chicken as ○○. The hereditary constitution of the first generation cross between the two is then ● ○. This produces the intermediate "blue" feathers.

What happens when a black rooster, ●●, is crossed with a blue hen, ● ○? Obviously the black rooster can contribute only a factor for black, represented here by ●, to his offspring. The blue hen, however, can contrib-

ute either a factor for white, ○, or a factor for black, ●. In other words, all the sperms will carry a factor for black, and one-half of the eggs will carry a factor for black and one-half will carry a factor for white. An egg carrying a factor for black, fertilized by a sperm carrying a similar factor, forms a pure black individual, ●●. Consequently, 50 per cent of the offspring will be black. The 50 per cent of the eggs which carry a factor for white will also be fertilized by "black" sperms. This 50 per cent of the eggs will thus form the intermediate blue chicks, ● ○.

This situation can be diagrammed by the construction of a grid in which the different kinds of eggs are represented along one side, and the different kinds of sperms (in this case only one kind) along the other side. Within the squares the possible crosses are then written. These represent zygotes or fertilized eggs (Fig. 4-3).

Suppose you attempt to breed a race of blue Andalusians by crossing two blue chickens. What will be the result? There will be two kinds of sperms as well as two kinds of eggs. Again the possible combinations can easily be seen by placing the two kinds of eggs along one side of a grid and the two kinds of sperms along the other. This is actually a graphic form of multiplication, to find all the kinds of products. Each kind of egg can then be united with each kind of sperm. It is at once evident that there are four possibilities. The egg and sperm may both carry a factor for black, or both may carry a factor for white

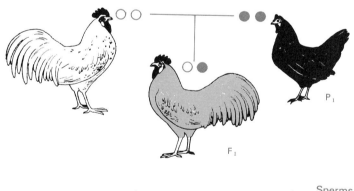

Fig. 4-2. The cross between a purebred "splashed white" and a purebred black chicken yields a "blue" Andalusian offspring. (From Moment, G. B., and H. M. Habermann. 1973. *Biology: A Full Spectrum.* The Williams & Wilkins Co., Baltimore.)

Fig. 4-3. Genetic grid representing the possible results of crossing a "blue" Andalusian hen and a purebred black rooster. (From Moment, G. B., and H. M. Habermann. 1973. *Biology: A Full Spectrum.* The Williams & Wilkins Co., Baltimore.)

In the third and fourth possibilities, the sperm may carry a factor for black and the egg a factor for white, or vice versa. You can thus predict that the result of crossing two blue fowl would be 25 per cent black, 50 per cent blue, and 25 per cent white. You could predict further that if the black offspring are mated to other black chickens, all of their offspring will be black, that the white segregated out of the cross of two blues will also breed true, but that the blues if mated together will again give offspring in a 1:2:1 ratio. This cross has been made repeatedly and the results always agree with prediction.

There are a number of other points to bear in mind. Fertilization takes place at random. In this case a sperm bearing a factor for black is just as apt to fertilize an egg bearing a factor for white as it is to fertilize an egg with a factor for black.

TERMINOLOGY OF GENETICS

At this point let us introduce some modern terminology to facilitate discussion. The hereditary factors are now called **genes.** An individual receiving similar genes for a given trait, e.g. white feathers, from each parent is said to be **homozygous** for that trait. An individual receiving different genes for a given trait is said to be **heterozygous** for that trait. All blue Andalusians are heterozygous for feather color. All black sheep must be homozygous for wool color because black is **recessive** to white in sheep. If a sheep bore the gene for white, the sheep would be white even though it also carried a gene for black. In such a case a gene for white is said to be **dominant.** Thus, white sheep may be either homozygous or heterozygous for color. A heterozygous individual is called a **carrier** because a gene for the recessive trait is present but does not show.

The different forms of a gene at the same position (locus) on a specific chromosome are called **alleles.** Thus, the genes for black and for white wool in sheep are alleles. In many known cases there is a series of alleles, as in some of the genes for eye color in *Drosophila.* Because alleles must be at the same position or locus on a specific chromosome (otherwise they would not be called alleles), any individual diploid animal or plant can carry a maximum of only two alleles of a given gene.

In symbolizing genes it is customary to capitalize the gene symbol for a dominant trait and put its recessive partner or allele in small

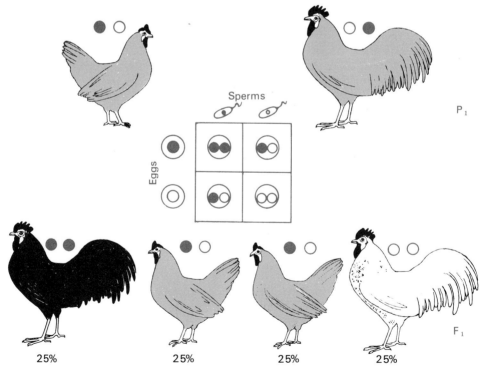

25%   25%   25%   25%

Fig. 4-4. Genetic grid showing the four possible results of a cross between two blue Andalusian chickens. In reality there would be an equal number of males and females in each of the three sorts of offspring. (From Moment, G. B., and H. M. Habermann. 1973. *Biology: A Full Spectrum.* The Williams & Wilkins Co., Baltimore.)

letters. For example, **a** symbolizes a recessive gene for albinism and **A**, the corresponding dominant allele, the gene for normal pigment. Thus an albino individual has the genetic constitution **aa** and is homozygous for albinism. A normal individual may be either homozygous, **AA,** or heterozygous, **Aa;** the latter would be a carrier.

Crosses of the kind just described, in which the two individuals differ with respect to a single pair of genes, are said to be **monohybrid** crosses.

Individuals that look alike are said to belong to the same **phenotype** *(phainein,* to show), whether they are genetically the same or not. If they are genetically the same, they are said to belong to the same **genotype** *(genos,* race). Thus all albinos belong to the same genotype **aa.** All individuals with normal pigmentation belong to the same phenotype but may be of either the homozygous genotype **AA** or the heterozygous genotype **Aa.** The complete **haploid (monoploid)** set of chromosomes characteristic of the cells of any individual constitutes that individual's **genome.** In other words, all the genes carried by a gamete are defined as a genome. However, this term is also commonly used to refer to the entire genetic complement of an organism, i.e. the genes in the diploid set of chromosomes of a particular individual.

PROBABILITY AND THE PRODUCT LAW

A moment's thought will show that in any cross involving Mendel's law of segregation, the results depend on chance, that is, on **probability.** If 50 per cent of the sperms carry a gene for albinism and 50 per cent carry a gene for normal pigmentation, then there is a 50-50 chance that any particular egg will be fertilized by a sperm carrying the gene for albinism. It will soon be seen that probability also applies to Mendel's second law (see below), as well as to sex determination, and to the genetics of populations in general.

Although it is not yet possible to control the kind of sperm which will fertilize a given egg, it is possible to make some useful predictions about the sex of unborn children, the likelihood of a given marriage producing an albino child, and many other matters. These predictions are based on very simple laws of probability and have an extremely wide application not only in theoretical genetics but also in human life.

Fractions are used to express probabilities. Thus, if the chance of an event happening is one in two, its probability is said to be $1/2$. For example, the probability of "heads" when a coin is tossed is $1/2$. Likewise the chance of getting "tails" is $1/2$. A dice cube has six sides; consequently, the chance that a "four" will land uppermost is $1/6$, and so for each of the other sides.

The law of probability which is most important for an understanding of the workings of heredity is the **product law.** It is very simple and can be simply stated. The probability that two independent events will coincide is the product of their individual probabilities. If two coins are tossed simultaneously, the chance that two "heads" will land uppermost is $1/2 \times 1/2$, or $1/4$. This means that in a sufficiently large series of such double throws, two "heads" will appear in 25 per cent of the cases. Any skeptic need only get two coins and try for himself. It always is true; but you must do it often enough to eliminate the vagaries of chance in small numbers.

Apply this now to a specific problem. A man and his wife are both normal but each is known to carry a gene for albinism because each had one albino parent. One-half of the man's sperms carry a gene for albinism and so do one-half of the woman's eggs. Thus, the chance that any particular child will be an albino is $1/2 \times 1/2$, that is $1/4$, or one in four. Every time they have a child there is one chance in four that an albino will be born. What is the chance for a homozygous child with normal pigmentation? Again $1/2 \times 1/2$, or a one-in-four chance. Thus 25 per cent of the children from such marriages will be homozygous albinos and 25 per cent will be homozygous normals.

What about the other 50 per cent of the children? The probability that a sperm carries an **A,** an egg carries an **a,** and the zygote (fertilized egg) **Aa** is therefore $1/2 \times 1/2$ or $1/4$; the reciprocal combination of an egg with **A** and a sperm with **a** and the resulting zygote **aA** is again $1/2 \times 1/2$ or $1/4$. Thus, 50 per cent of the offspring will be normal in pigmentation but carriers for albinism: 25 per cent having received the gene for that trait from their mother and 25 per cent from their father. Clearly this is the familiar 25:50:25 or 1:2:1 Mendelian ratio.

Suppose that the first child of a couple is an albino. What is the probability that their second child will also be an albino? Neither the second egg nor second sperm months or years later has any way of knowing what the first egg or sperm was like. Therefore the second child is an independent event and the chance of a second albino is again $1/4$. However, by applying the same product law to two independent events, each with a probability of only $^1/_4$, it is clear that the probability that both first and second children will be albinos is only one in sixteen: $^1/_4 \times ^1/_4 = ^1/_{16}$.

As Mendel himself pointed out, the results of a monohybrid cross can be predicted by use of the binomial theorem usually represented by the familiar equation $(p + q)^2 = p^2 + 2pq + q^2$. Here p represents the frequency of one gene and q the frequency of its allele. Thus, if 0.5 of the gametes in a given cross carry gene p and 0.5 carry its allele, gene q, then 0.25 of the progeny will be pp, 0.50 will receive both p and q, and 0.25 will be qq. The use of this formula is basic in population genetics, an important aspect of evolutionary theory.

INDEPENDENT ASSORTMENT: MENDEL'S SECOND LAW

Mendel's second law comes into play when the two individuals in a cross differ with respect to two, three, or more pairs of genes. Such crosses are called **dihybrid, trihybrid,** etc. For example, black in rabbits is due to a dominant gene, **B**, and white to its recessive allele or partner gene, **b**. Also, short hair is due to a dominant gene, **S**, and long hair to its allele, **s**. What happens when a homozygous black, short-haired rabbit is crossed with a homozygous white, long-haired one? All the offspring in the first generation will look alike, i.e. will be of the same phenotype, black and short-haired. Their genetic constitution, or genotype, will be heterozygous for both traits, **BbSs**.

When two of these heterozygous rabbits are crossed, the genes inherited from each parent separate in the germ cells of the offspring without any influence on each other. This is the normal segregation of Mendel's first law. Independent assortment means that the way one pair of genes segregates into the germ cells is independent of the way another pair does. In the present illustration one-half of the gametes, either eggs or sperms, will receive a gene for black, **B**, and one-half will receive its allele, the gene for white, **b**. But whether a given gamete gets gene **B** or **b** has nothing to do with whether or not it gets the gene for short or for long hair. This is determined simply by chance, so that one-half of the gametes that get the gene for black will get the gene for short hair and the other half will get the gene for long hair. This results in four kinds of gametes in equal numbers, symbolized as **BS, Bs, bS, bs**. To predict the results of a dihybrid cross, the four possible types of sperms are written along one side of a square, and the four possible types of eggs along the other. Within the squares appear the possible genotypes resulting from the cross.

It can be seen from the grid in Figure 4-5 that there are only four phenotypes present: black, short-haired; black, long-haired; white, short-haired; and white, long-haired; and that they occur in a ratio of 9:3:3:1. The ratio of the genotypes is very different. For instance, although $^9/_{16}$ of this generation are of the black, short-haired phenotype, only $^1/_{16}$ are of the genotype **BBSS**, which is homozygous for both traits. Notice also that only one out of 16 is the double recessive phenotype—long-haired and white—and that in this case there is only one genotype, **bbss**, that can give this particular phenotype.

It is important to remember in genetic prediction that fertilization is random. When it is said that one-half or $^1/_{16}$ of the offspring will be of a particular sort, what is really meant is that there is one chance in two or one chance in 16 that the offspring will be so. The genetic ratios predicted are actually realized only in large samples.

Chromosomes and Heredity

Mendel knew nothing of chromosomes, but after the rediscovery of his laws in 1900, several lines of evidence converged to prove that the unit factors Mendel had talked about, the genes, as we say today, are located on the chromosomes. This evidence lies in the precise and extensive parallelism between the behavior of the units of heredity and the behavior of the chromosomes. It is a parallelism so detailed and so extensive that there is no room for doubt that the chromosomes are the very stuff of heredity and hence provide the physical basis of evolution itself. The parallelism runs all through the cycle of fertilization and meiosis, as well as in particular aspects of it, e.g. sex determination and linkage. Furthermore, abnormalities of chromosome behavior are followed by abnormalities of inheritance.

FERTILIZATION AND MEIOSIS: THE CYCLE

The facts of fertilization in themselves provide some of the most cogent and obvious evidence for the chromosomal theory of heredity. The male contributes equally with the female to the heredity of the offspring, yet the only physical contribution of the male is the head of a sperm. What is the head of a

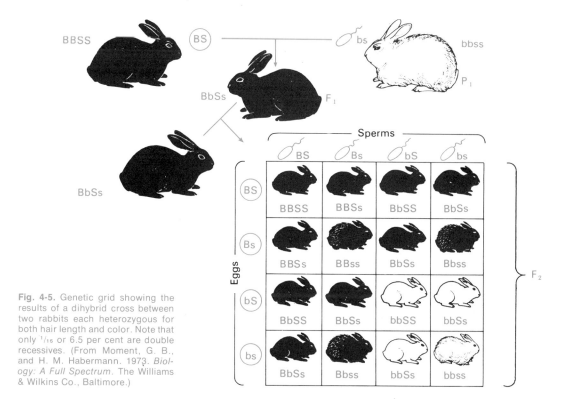

Fig. 4-5. Genetic grid showing the results of a dihybrid cross between two rabbits each heterozygous for both hair length and color. Note that only $\frac{1}{16}$ or 6.5 per cent are double recessives. (From Moment, G. B., and H. M. Habermann. 1973. *Biology: A Full Spectrum.* The Williams & Wilkins Co., Baltimore.)

sperm? Microscopic examination of sperm formation in the testis of any animal will reveal that the head of a sperm is little more than a condensed packet of chromosomes. Consequently, it follows that chromosomes are the physical bearers of inheritance. Once inside the egg, the sperm head gradually swells up into a nucleus which ultimately fuses with the egg nucleus.

It was discovered that the number of chromosomes in every cell of a given animal is always the same: 46 in humans, 8 in *Drosophila,* the fruit fly, and 36 in chickens. Except for some minor exceptions, this generalization is also true for plants, although a number of cultivated varieties are polyploid rather than diploid.

Each chromosome has its own permanent individuality as indicated by its size, shape, and the position of its **kinetochore** or centromere, the place where it is attached to the spindle during mitosis. The visible complex of chromosomes, their number, sizes, and shapes, characteristic of any species or individual, is called the **karyotype,** which means, literally, nuclear type.

Every cell in an animal has two sets of chromosomes, one set derived from the sperm and one set from the egg. Yet the number of chromosomes characteristic of any species remains the same, generation

after generation. How can this be? It was this question which led Weismann to predict that there must be some time when the number of chromosomes is reduced by half, for otherwise they would double in number with every generation. In animals, this reduction, known as **meiosis** *(meiosis,* to diminish), occurs during the formation of gametes, specifically in the two final cell divisions in the formation of sperms or eggs. In the flowering plants, meiosis takes place in the flower. Pollen grains have haploid nuclei as do the specialized structures within the female pistil, called embryo sacs. In fertilization, a sperm nucleus originating in the pollen grain fuses with the egg nucleus within the embryo sac. The resulting zygote develops into an embryo located within the seed, which is released from the parent plant.

When finally unraveled, the overall facts of fertilization and meiosis turned out to be rather simple. In meiosis the double set of chromosomes found in all the somatic or body cells is reduced to a single set in each sperm or egg. In fertilization the double or diploid set is restored by the fusion, into the nucleus of the fertilized egg, of a single or monoploid set of chromosomes from the sperm with a single set in the egg.

To understand heredity, it is a great advantage to understand in more detail than in the

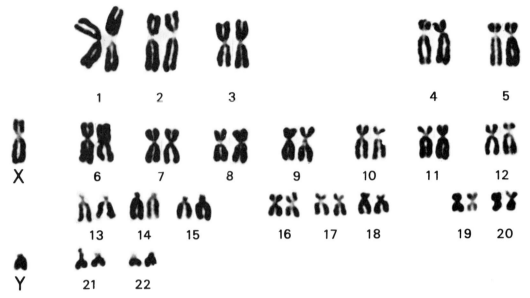

Fig. 4-6. A normal karyotype prepared from a human leukocyte dividing in culture. The chromosomes were fixed in acetic alcohol and stained with acetic orcein. When this preparation is viewed in the light microscope the metaphase chromosomes appear in a cluster on the slide. A photograph of these chromosomes is obtained and is then cut up so that the chromosomes may be grouped as shown here (× 1,500). (Courtesy of Dr. Orlando Miller and reproduced from Copenhaver, W. M., R. P. Bunge, and M. B. Bunge. 1971. *Bailey's Textbook of Histology*, 16th ed. The Williams & Wilkins Co., Baltimore.)

previous account exactly what happens during the process of meiosis. These details are very much the same in all animals and plants, which is of course the reason why the laws of heredity are the same throughout the living world. In all forms of life meiosis requires two successive divisions. It begins with a single diploid cell and ends with four monoploid (haploid) cells. (Fig. 4-7). Two divisions are required to segregate all the pairs of genes because the chromosomal DNA replicates before the first division begins. The occurrence of crossing over (see below) which is about as universal as meiosis itself permits genes located on the same chromosome pair to undergo recombination.

Like ordinary mitosis, meiosis begins with a **prophase,** during which the chromosomes gradually stain more darkly and become shorter and thicker. As in mitosis, a good microscope will show that each chromosome is really double along its entire length, and hence must have duplicated at some time before prophase began.

The first important difference between mitosis and meiosis occurs in late prophase. In mitosis the two sets of doubled chromosomes, one from the male and one from the female parent, pay no attention to each other. However, in meiosis the corresponding chromosomes of paternal and maternal origin come to lie side by side, closely aligned at every point from one end to the other. Thus, the largest chromosome in the paternal set lies alongside the largest in the maternal set, and so on down to the two smallest chromosomes. This pairing of corresponding or **homologous** chromosomes is called **synapsis.** Synapsis does not occur in mitosis.

Since in meiosis the chromosomes have doubled themselves during the previous interphase, there are really two maternal and two corresponding paternal chromosomes that come to lie side by side in synapsis. The two maternal chromosomes are still held together by their kinetochore or spindle fiber attachment point, and the same is true of the paternal pair. Such a synaptic group is called a **tetrad.** The four chromosomes still held together by their two kinetochores are usually called **chromatids** until they separate. In a species which has three chromosomes in a set, there will, of course, be three tetrads. In humans there are 23 chromosomes in a set, and 23 tetrads in the prophase and metaphase of the first meiotic division. In other words, there are four complete sets of chromosomes at the first metaphase of meiosis. This is precisely enough to provide each of the four resulting sperms (or four eggs) with one set of chromosomes. To do this, two cell divisions are required.

At the **first meiotic division** the two kinetochores for every tetrad separate and move to opposite poles of the spindle, each pulling its two chromosomes (more accurately, chromatids). This obviously affords a physical basis for Mendel's law of the segregation of hereditary factors in the formation of sperms or eggs. If, for example, a paternal chromosome carried a factor for red hair, and the homologous maternal chromosome carried the allele for black hair, one-half of the gametes would receive the chromosome carrying the gene for red, the other half the chromosome carrying the gene for black. Said in another way, Mendel's unit factors are present in pairs in the adult. So are chromosomes. Mendel's unit factors separate from each other during reproduction. So do the individual chromosomes in each pair during meiosis.

The way one pair of maternal and paternal chromosomes separate after synapsis is independent of the way any other pair separate. This means that almost no two eggs or no two sperms will have exactly the same assortment of maternal and paternal chromosomes.

At this point the chromosomes are still double and held together by their kinetochores, i.e. the points of spindle fiber attachment. The kinetochores now duplicate themselves. This second division separates the doubled

chromosomes so that each of the resulting cells has one chromosome from each of the original tetrads. There is now a total of four haploid cells, each with one complete set of chromosomes.

GENETIC VARIATION—ROLE OF MEIOSIS AND FERTILIZATION

The primary origin of genetic variation is **mutation**, that is, permanent change in a gene. In addition, all organisms which reproduce sexually possess a built-in mechanism to guarantee continual variation by the formation of ever new combinations of the genes via meiosis and fertilization. The way any particular pair of synapsing chromosomes segregates is entirely independent of the way any other pair does. This means that, on the average, every egg and every sperm receives a thoroughly mixed set of chromosomes partly of paternal and partly of maternal origin. What is the chance that a sperm (or an egg), in an organism with only three chromosomes in a set, might receive only paternal chromosomes? Using the product law for the coincidence of two or more independent events, you can easily calculate this

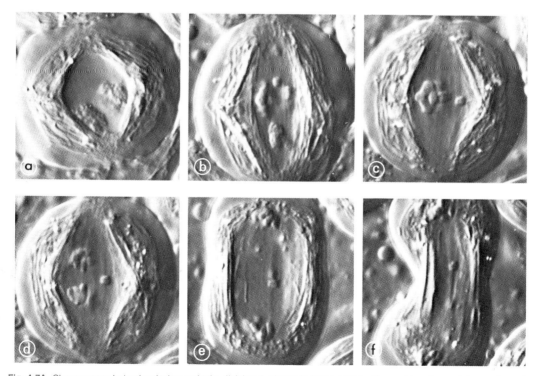

Fig. 4-7A. Chromosome behavior during meiosis, division 1, shown in living cells by Nomarski optics. a, late prophase; b, prometaphase; c, metaphase; d, anaphase; e, telophase; f, cytokinesis, i.e. separation of haploid daughter cells. The threadlike structures on each side of the cell are mitochondria. (Courtesy of J. R. LaFountain, Jr.; from LaFountain, J. R., Jr. 1972. Spindle shape changes as an indicator of force production in crane-fly spermatocytes. *J. Cell Sci.* 10:79.)

MEIOSIS: Division I

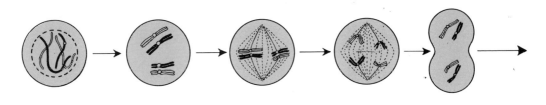

| Early Prophase I | Mid-Prophase I | Metaphase I | Anaphase I | Telophase I |

Fig. 4-7B. Diagram of meiosis. The colored cells are haploid (monoploid). (From Moment, G. B., and H. M. Habermann. 1973. *Biology: A Full Spectrum*. The Williams & Wilkins Co., Baltimore.)

chance. For each pair of synapsing chromosomes there is a probability of $1/2$ that the paternal member of the pair will enter a particular sperm. Consequently the chance that all three paternal chromosomes will enter the same sperm is $1/2 \times 1/2 \times 1/2 = (1/2)^3 = 1/8$. In other words, one in every eight sperms of such a species will contain only paternal chromosomes. By the same reasoning it follows that one in every eight sperms will receive only maternal chromosomes. And of course six of every eight sperms (or eggs) will contain a mixed set. The total number of possible kinds of sperms (or eggs) is 2^3, or 8.

The probability that a given human sperm will carry only paternal chromosomes is $(1/2)^{23}$ or one chance in 8,388,608. This is another way of saying that every man and every woman can produce over 8,000,000 kinds of gametes! In fact, crossing over, to be discussed later, increases this staggering number even more.

Fertilization then enters the picture to compound the amount of variation already guaranteed by meiosis. In a given mating, if there were 8,000,000 kinds of sperms and only two kinds of eggs, there would obviously be 16,000,000 possible kinds of zygotes. However, since any woman can theoretically produce over 8,000,000 kinds of eggs, the total possible kinds of zygotes any human couple can produce is over 8,000,000 times 8,000,000, without counting crossing over. Such is the biological basis of human individuality.

SEX DETERMINATION

Natural

In the early years of the present century American cytologists (specialists in the study of cells) made a peculiar discovery about chromosomes in certain insects where chromosomes are favorable for study. In females all the chromosome pairs match perfectly. In males, however, there is one pair (the **X** and **Y** **chromosomes)** which does not match even though they come together in synapsis. During meiosis it can be observed that females always possess two **X** chromosomes, the males an **X** and a **Y**. After meiosis every egg will carry an **X** chromosome, but half of the sperms will carry an **X** and half a **Y**. The two kinds of sperms will be produced in equal numbers because at synapsis for every **X** chromosome there is a **Y**.

This means that 50 per cent of the eggs will be fertilized by a **Y**-bearing and 50 per cent by an **X**-bearing sperm. The result is equal numbers of **XX** female-producing zygotes and **XY** male-producing zygotes. In most animals males and females are produced in approximately equal numbers. In some species, perhaps in most, a differential mortality begins before birth and continues long afterward, so that in different age groups the sex ratio varies somewhat to one or the other side of the 50:50 ratio. In humans and insects sex is determined by the nature of the sperm, and the sex of an individual is determined at the instant of fertilization.

Artificial Sex Determination

If some way could be found to separate **X**-bearing from **Y**-bearing sperms, it would have immediate practical applications in animal husbandry, where artificial insemination is widely practiced. There is some evidence that separation of **X**-bearing from **Y**-bearing sperms is possible by the methods of differential centrifugation and electrophoresis. The two kinds of sperms differ slightly in density and in their migration in an electric field, and some observers claim that **X**-bearing sperms appear under a phase contrast microscope to have slightly larger and more elongated heads. Some have expressed fears least the use of such knowledge would upset the sex ratio in human populations, but there

MEIOSIS: Division II

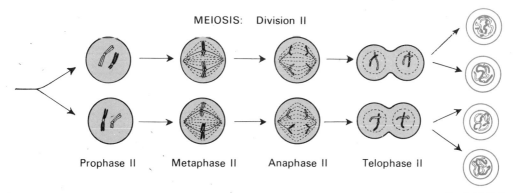

Prophase II   Metaphase II   Anaphase II   Telophase II

Haploid Cells

Fig. 4-7B (cont.)

is good evidence that most people desire both boys and girls.

Even though the sex of offspring cannot yet be controlled, it is possible to make some useful predictions about the sex of the unborn. This is done by the use of the product law already discussed. Since all human eggs carry an **X** chromosome, while 50 per cent of sperms carry an **X** and 50 per cent carry a **Y**, the chance of any particular child being a boy is $1/2$, and likewise the chance of a girl is $1/2$. In a family of three children, what is the probability that all three will be boys? Applying the product law, $1/2 \times 1/2 \times 1/2 = (1/2)^3 = 1/8$. This means that if a survey were made of families with three children, it should be found that in $1/8$ of the families all three children are boys. It also means that in $1/8$ of the families all three of the children should be girls and in $6/8$ of the cases there should be a mixture of boys and girls.

X-LINKED GENES

When a gene is located on the **X** chromosome, it is said to be **X**- or **sex-linked** because its inheritance follows the transmission of the **X** chromosome. The best known **X**-linked genes in humans are those for red-green color blindness and for hemophilia (a faulty clotting mechanism of the blood that results in excessive bleeding, even from a scratch). The gene for white eyes in *Drosophila* is similarly a sex-linked one.

A male has only one **X** chromosome. Therefore he can carry only one gene for such traits. His one **X** chromosome may carry a gene for red-green color blindness or a gene for normal vision. But his **Y** chromosome has no corresponding genes. Consequently, even though the genes for white eyes and for color blindness are recessive, they will produce their characteristic effects in a male. For example, if a man has a gene for hemophilia on

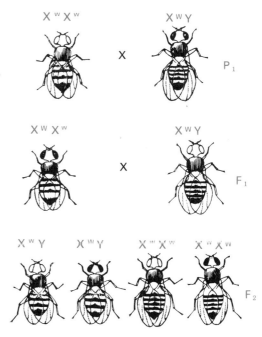

Fig. 4-8. Diagram illustrating the inheritance of genes linked to an X chromosome, often called sex-linked inheritance, in *Drosophila*. W represents a gene for red eyes and w a gene for white eyes. (From Moment, G. B., and H. M. Habermann. 1973. *Biology: A Full Spectrum.* The Williams & Wilkins Co., Baltimore.)

his **X** chromosome, he is bound to be hemophilic because there is no possibility of the recessive gene being counteracted by a dominant gene for normal blood clotting on a second **X** chromosome. This type of inheritance is illustrated in the case of white eyes in *Drosophila* (Fig. 4-8).

Of course a female with her two **X** chromosomes will carry two genes for each **X**-linked trait, one gene on each of the **X** chromosomes. She may be either homozygous or heterozygous. If she is homozygous, she may carry two genes for normal or two genes for white eyes (if she is a fruit fly) or two genes for color blindness (if she is a woman). If she is heterozygous she will carry, of course, one gene for normal eyes and one gene for the abnormal eye trait involved.

A female who is heterozygous for a recessive **X**-linked trait is known as a **carrier.** She will be normal, but will transmit a gene for the recessive trait in 50 per cent of her eggs. Whether or not the trait appears in her children will depend on whether they are boys or girls and, in the case of the girls, whether or not the father has a recessive or dominant gene on his **X** chromosome.

Autosomal Linkage

Just as Weismann predicted meiosis, so a young graduate student named Sutton at Columbia University predicted **linkage.** He did it on the eminently reasonable grounds that there are many more hereditary factors than chromosomes and that hence each chromosome must carry a group of genes. The group of genes on a single chromosome cannot assort independently at meiosis but must pass as a unit into the same gamete. In other words, the genes tied together in the same chromosome cannot follow Mendel's second law, the law of independent assortment. To assort quite independently, genes must be on different chromosomes of the set.

There are, consequently, as many linkage groups in any organism as there are chromosomes in its set, i.e. in the genome. So Sutton predicted, and so it has turned out. In *Drosophila* there are four chromosomes in a haploid set, one large, two middle-sized, and one small. Likewise there are four linkage groups of comparable sizes. After linkage was first established in *Drosophila* by T. H. Morgan and his students, it was found to exist in all animals and plants investigated. In humans there are 23 chromosomes in a set and hence 23 possible linkage groups. Twenty-two of the chromosomes are called **autosomes,** and one the sex chromosome, either **X** or **Y**. Over 1100 human genes have been studied and mapping information exists for at least 105 of these. One or more genes have been identified for all human autosomes except chromosome number 3.

Linkage Analysis by Somatic Cell Hybridization

One of the most important and exciting recent advances in genetics is the development of a new and powerful method of identifying human and other linkage groups. It grew out of several modern discoveries, notably that it is possible to hybridize somatic cells, i.e. body cells in contrast to gametes. For example, human and mouse cells can be cultured together in such a way that the cells fuse and produce human-mouse hybrid cells each with all the chromosomes of both species. Fortunately, human and mouse chromosomes differ enough in appearance to be distinguished when the cells are examined by the technique described for studying karyotypes (see Fig. 4-6).

Moreover, human cells share with those of the mouse, and of mammals in general, their basic biochemical capabilities. This means that if the human cells, or the mouse cells, that went into a hybrid came from a mutant individual lacking the gene essential for synthesis of some particular enzyme, the chromosome from the other partner (provided the second partner was normal) would carry the missing gene and thus compensate for its absence among the genes of the first partner. The result would be that the gene product would appear in the culture.

Probably because of a lack of perfect coordination of the chromosomes during mitosis in such hybrid cells, with the passage of time chromosomes get lost in a more or less random manner. Some cells will come to have a complete set of mouse chromosomes but lack some of the human, or vice versa. More frequently there will be some mixture of chromosomes from the two species. When such cells are separated and **cloned,** i.e. grown in pure cultures with but one type of cell, any clone lacking a particular chromosome is no longer able to synthesize the enzyme coded for on that chromosome. Therefore, the missing chromosome is the one that carries the gene for this enzyme and it is linked to all other genes on that chromosome. A knowledge of linkage groups is a first step to a truly scientific understanding of human genetics.

CROSSING OVER

It was learned early in the original work on linkage in *Drosophila* that a phenomenon called **crossing over** occurs. This takes place during meiosis between the two members of a pair of homologous chromosomes. In the words of T. H. Morgan: "Linkage and crossing over are correlative phenomena, and can be expressed by numerical laws that are as definite as those discovered by Mendel."

For example, on the long second chromosome of *Drosophila* are located the mutant genes for star eyes, black body, purple eye color, dachs (very short) legs, vestigial wings, plexus veins, and speck (black dot at base of wings). Suppose a fly had received a number

Fig. 4-9. Thomas Hunt Morgan, renowned geneticist, who established linkage groups, studied crossing over, and mapped chromosomes in *Drosophila*. (Courtesy of Condé Nast Publications, Inc., © 1945.)

2 chromosome with the normal alleles from one parent and a number 2 chromosome with these mutant alleles from the other. Then according to the usual events of linkage, the offspring of this fly will either get all these mutant genes and none of their normal alleles, or else all the normal alleles and none of the mutant genes.

However, in a predictable percentage of cases some offspring will get the genes for speck, plexus, vestigial, and normal alleles for the other mutants, while other offspring will get the normal alleles for the genes first mentioned and the mutant genes for dachs, purple, black, and star. The new combinations now remain as firmly linked as had the old.

Examine now a specific instance more closely. Let **b** symbolize a recessive gene for black body and **B** its normal allele, and let **v** symbolize a recessive gene for vestigial wings and **V** its normal allele. Then, if a fly, homozygous for the recessive genes, **bbvv**, is crossed with one homozygous for the normal alleles, **BBVV**, all the F_1 generation flies will be heterozygous **BbVv** and appear normal. When these heterozygous individuals are mated, crossing over shows itself. This can most easily be seen if one of the phenotypically normal heterozygotes, a female, is bred

to a homozygous recessive, vestigial winged, black male. In the heterozygous female parent, one number 2 chromosome carries both recessive genes **b** and **v**, while the other number 2 chromosome carries their normal alleles **B** and **V**. Thus, with complete linkage, 50 per cent of her eggs will be **bv** and 50 per cent **BV**. When mated with a homozygous recessive male, all of whose sperms must carry **b** and **v**, 50 per cent of her offspring would be expected to have normal wings and pigmentation, and 50 per cent to have black, vestigial wings. Breeding experiments, however, show that only 83 per cent of the offspring of such a cross belong to either of these two types. The other 17 per cent are recombinations in which either vestigial is combined with normal body pigmentation, or normal wings with black pigmentation. It is significant that these two new combinations appear in equal numbers (Fig. 4-10).

This result can be explained on the assumption that in the formation of 17 per cent of the gametes the homologous chromosomes in the heterozygous female have exchanged parts, so that gene **V** is now on the same chromosome as gene **b,** and gene **v** on the chromosome with **B**. Direct visible evidence of such crossing over can be seen in meiosis. During synapsis the tightly paired chromosomes become twisted; when separation occurs, the chromosomes have exchanged parts. Healing appears to be perfect, for once the new combination is formed the chromosomes are as stable as before.

Crossing over produces new combinations of traits. This is important both in natural evolution and in producing desirable new types of domestic animals and plants.

MAPPING CHROMOSOMES

Crossing over makes it possible to actually map the locations of genes on a chromosome. It is reasonable to assume that the farther apart two genes lie, the greater the likelihood of a break (with crossing over) between them. Assume also, and this appears to be true, that crossing over is equally probable at any point along the chromosome. Conversely, the closer together two genes are, the less likely it is that crossing over will occur between them.

Suppose that the percentage of recombinations, i.e. crossovers, between one pair of alleles, say **Aa**, and another pair, say **Bb**, is 5 per cent. It can be said that the two alleles are located 5 arbitrary units apart. Suppose now that crossing over is determined between the **Aa** alleles and a third pair of alleles, **Cc**, and this is found to occur in 15 per cent of the

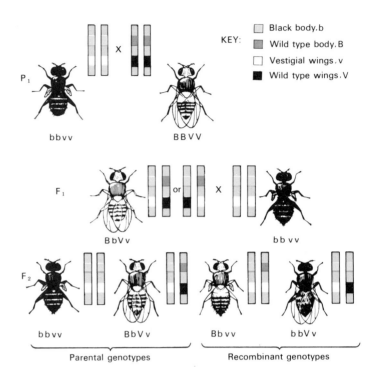

Fig. 4-10. Crossing over occurs between the genes for black body and vestigial wings and those for wild type body and wings in *Drosophila.* (From Moment, G. B., and H. M. Habermann. 1973. *Biology: A Full Spectrum.* The Williams & Wilkins Co., Baltimore.)

cases. Evidently locus **A** and locus **C** are 15 units apart (Fig. 4–11). There is no way of telling from these data whether locus **B** and locus **C** are on the same side of locus **A** or on different sides. This question can be answered by determining the percentage of crossing over between **B** and **C**. If they are on the same side of **A**, then the percentage of crossing over would be only 10 per cent between them. If however, **B** and **C** are on opposite sides of **A** then the crossing over should be the sum of the individual values of distances from **A**.

Very detailed chromosome maps have been constructed by this method, using a large number of genes, not only in the fruit fly and maize plant, but in the chicken, the mouse, the fungus *Neurospora,* and other organisms.

In addition, there is both logical and direct visual evidence of the linear order of genes on chromosomes. The fact that in cell division chromosomes duplicate themselves and pull apart longitudinally rather than breaking in half transversely argues that their important constituents are arranged in a linear series.

Chromosomes, as seen in most cells, are tiny irregular rods. However, about 75 years ago an Italian investigator, Balbiani, discovered that the chromosomes in the salivary glands of flies, gnats, and mosquitos are gi-

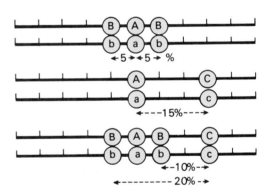

Fig. 4-11. Diagram illustrating the method of mapping the position of genes on a chromosome by comparing crossover frequencies. (From Moment, G. B., and H. M. Habermann. 1973. *Biology: A Full Spectrum.* The Williams & Wilkins Co., Baltimore.)

gantic in comparison with chromosomes in most cells. He also noticed that these **giant chromosomes** are banded. Beginning in 1930, various workers in this country and in Germany reinvestigated the bands which Balbiani had described long before the rediscovery of Mendel's laws. It soon became obvious that the bands were not haphazard but were constant in number, thickness, and position on any particular chromosome. From this fact it was possible to show that specific bands correspond to the position, or **locus,** to use a more technical term, of particular genes. An

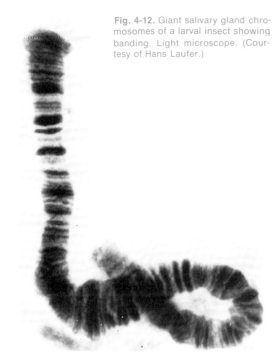

Fig. 4-12. Giant salivary gland chromosomes of a larval insect showing banding. Light microscope. (Courtesy of Hans Laufer.)

abnormality in any particular band is invariably correlated with a particular abnormality in the animal. Thus the chromosome maps constructed from crossover data have been validated by visual evidence.

Mutation

Mutations are sudden and relatively permanent changes in the hereditary material. These changes can sometimes be seen under the microscope as changes in the chromosomes; other mutations are changes in the genes themselves.

CHROMOSOMAL ABERRATIONS

At the chromosome level, mutations are of four general types. All are caused by the breakage and incorrect rejoining of chromosome fragments. They can be seen in human chromosomes after irradiation.

Deletions are losses of larger or smaller pieces of chromosomes. If too much of a chromosome is lost so that some essential enzyme is not formed, then the mutation will kill the organism or perhaps not permit development even to start. Such a mutation is called **lethal.** A well-known deletion is "notch" in *Drosophila;* this causes a small abnormality in the wings of fruitflies heterozygous for this trait and is lethal when homozygous. This abnormality can be identified cytologically as a small section missing from the third chromosome.

Duplications are repetitions of a portion of a chromosome. These also can be seen in the salivary gland chromosomes. A well-known example is "bar" eye in *Drosophila.*

Inversions are cases where part of a chromosome has been rotated 180° so that the genes, instead of running **ABCDEF,** run **ABEDCF.** Such events produce a wide variety of complications, such as loops, when homologous chromosomes attempt to pair during meiosis (Fig. 4–13). Some inversions are lethal.

Translocation is the term given to mutations in which part of one chromosome becomes permanently attached to the end of a nonhomologous chromosome. Translocation is sometimes called "illegitimate crossing over."

GENE MUTATIONS

The most interesting and important mutations are those which occur at the gene level and represent mistakes in gene duplication. As will be seen shortly, this means an error in the duplication of a sequence of purine and pyrimidine bases in the DNA (see Chapter 5).

Mutations have been found to affect just about every part of an organism. In *Drosophila* they change the size, shape, and color of the body; they alter the size, shape, color, and bristles of the eyes (Fig. 4–14). There is even a mutant gene which abolishes eyes. Other genes produce truly revolutionary changes, converting mouth parts into legs, legs into wings, wings into halteres, and halteres into wings, or abolishing wings com-

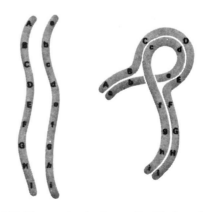

Fig. 4-13. Diagram showing the results of the inversion of a piece of a chromosome when pairing of homologous chromosomes takes place during meiosis. (From Moment, G. B., and H. M. Habermann. 1973. *Biology: A Full Spectrum.* The Williams & Wilkins Co., Baltimore.)

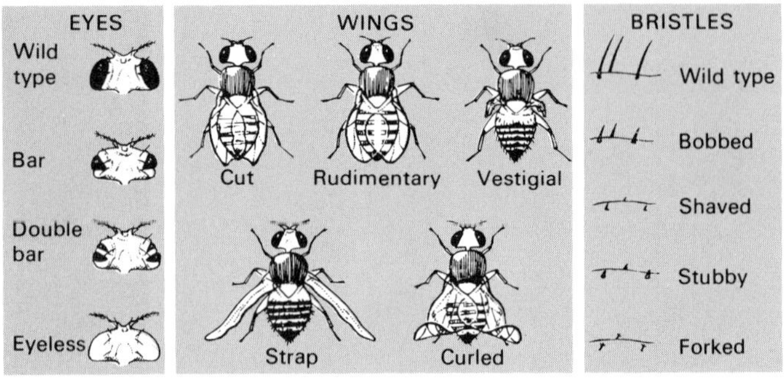

Fig. 4-14. Various mutations affecting eyes, wings, and bristles in *Drosophila*. (From Moment, G. B., and H. M. Habermann. 1973. *Biology: A Full Spectrum*. The Williams & Wilkins Co., Baltimore.)

pletely. Some mutations do not produce anatomical results but change physiological or biochemical abilities such as resistance to poisons or various nutritive capabilities. Such nonmorphological mutations are commonly called biochemical mutations. All mutations are, of course, really biochemical, because all the anatomical effects are the result of some change in the enzymatic activity of the cells during development.

RATE OF MUTATION

The spontaneous rate of mutation is very low and varies from one gene to another. In *Drosophila*, millions of which have been scrutinized in laboratories all over the world, the mutation rate at any one locus on a chromosome may be as high as 1 in every 1,000 gametes or as low as 1 in over 1,000,000 gametes, depending on the gene. The gene mutation that causes ancon (very short legged) sheep has been reported only twice in almost 200 years. Examination of 11,600 flies turned up a mutation which causes yellow body only three times, while a gene for echinoid eyes appeared 18 times.

Records of 94,075 babies born in Copenhagen hospitals showed 10 chondrodystrophic dwarfs. This is the kind of dwarf in which the head and body are normal in size but the arms and legs are abnormally short. The trait is inherited as a dominant. Two of these children had a dwarf parent but the other eight did not; that is, they represent new mutations. These facts show that the gene involved mutates to produce chondrodystrophy approximately once in 11,759 births. Since each child is the result of the fusion of two gametes, an egg and a sperm, and since this trait is a dominant, this means that one out of every 23,518 human gametes in Copenhagen carried this mutated gene.

ARTIFICIALLY INDUCED MUTATIONS

H. J. Muller, an old student of T. H. Morgan of *Drosophila* fame, then at the University of

Texas, and L. J. Stadler, of the University of Missouri, discovered that X rays will greatly increase the rate of mutation in animals and in plants. It has since been found that ultraviolet light and alpha, beta, and gamma radiation will also produce mutations. Because most mutations are deleterious, an increase in the number of mutations in the human population would be a disaster. Thus, clouds of radioactive dust drifting around the world could easily be ruinous for all mankind. Radiation has been used by plant breeders to obtain mutations in the hope of finding desirable ones from which to start improved varieties of domestic plants.

Many chemicals produce mutations. Such **mutagenic agents** include nitrogen mustards, various cancer-producing substances, alkylating agents, epoxides, caffeine, etc.

RANDOMNESS OF MUTATIONS

Mutations are random in two different senses: (1) they cannot be predicted except in a statistical way; and (2) they are not related to the needs of the organism. At least 99 per cent of mutations are harmful; this is not surprising when it is remembered that living species are well adapted for their modes of life. Consequently, any random change in the developmental blueprint is all but certain not to be an improvement. Whether or not a mutation is advantageous or disadvantageous depends on the environment. For example, mutations reducing wing size so that an insect cannot fly would be a very serious, perhaps fatal, disadvantage in most environments; however, on a tiny windswept oceanic island it would be a life saver. Many such islands support a variety of flightless insects. Similar mutations have occurred and persisted in laboratory cages where they do not affect survival.

There is a sense in which mutations are not random. The kinds of mutations which are possible depend upon the chemical properties of DNA.

Heredity versus Environment

Which is more important, heredity or environment? This question refuses to die. The answer depends on cases, even on your point of view, because the development of any organism is like the flight of an arrow. At every point the trajectory of the arrow is the result of two forces. One is the initial propulsion imparted by the spring of the bow. This may be compared to heredity, specifically the set of genes in the fertilized egg. The other force is the environmental force of gravity, modified by wind; and this may be compared to the action of regular and random environmental forces on the course of development.

In one sense the genes reign supreme. We are human and not starfish, or even anthropoid apes, precisely because of the kind of fertilized eggs we developed from and for no other reason. In the teeming waters of the ocean the fertilized eggs of thousands of kinds of animals and plants develop: fish, worms, sea urchins, medusae, clams, and the myriad kinds of algae. The environment may be the same over hundreds of cubic miles of sea, but the organisms that come from these different kinds of eggs are drastically different.

In some minor aspects as well, the genes are supreme. If a man is born without a gene for normal pigmentation he will be an albino regardless of environment. The blood groups to which a person belongs are determined by his genes and by them alone.

However, in another sense environmental forces are equally important. Genes do not function in a vacuum. Only within a narrow range of temperature found only on a planet within a certain distance from a sun is it at all possible for genes to form and function. In innumerable lesser ways the environment controls the actual as contrasted with the potential development of organisms.

In *Drosophila* it has been possible to produce copies of many of the well-known mutants by manipulation of the environment, treating the embryo with heat or various chemicals at a certain stage of its development. In wild flies, up to 75 per cent of the individuals can be converted into phenocopies of the "curly" mutant by heat. In humans, the persistent exposure to sunlight of an individual with a genotype producing a very light skin can cause conversion to the phenocopy of an individual of a genotype which in a relatively sunless environment produces an equally dark hue.

In summary, what is inherited by any organism is a capacity to respond in certain ways or a lack of ability to respond. As far as the external environment is concerned, there is a whole spectrum of cases, from genotypes that produce their characteristic effects in all known environments to genotypes that are very sensitive to environmental factors. Albinism is in the first class, melanin production during sun tanning in the second.

ARE THE EFFECTS OF ENVIRONMENT AND ACTIVITY INHERITED?

The belief that effects of use and disuse and of the environment are inherited is extremely ancient but still persists. Perhaps this is because it offers hope of being able to influence the hereditary nature of domestic animals and plants and even of people.

Shortly after the turn of the century an American zoologist, Castle, and his colleagues tried a new way to get at this problem. They removed the ovaries from a purebred white guinea pig and ingrafted ovaries from a black animal. Later this white female with ovaries from a black animal was mated to a white male. The offspring were all black. In no case was it possible to detect any modification of the black inheritance even though the ovaries from the black female were actually within the white female and were necessarily nourished by her bloodstream.

After the *Drosophila* work was well advanced, T. H. Morgan pointed out that some of the observations of his school really gave the old theory of the inheritance of acquired characteristics the *coup de grace.* If a fly with very small vestigial wings, which, of course, cannot be used in flight, from a long line of flies all with vestigial wings, is crossed with a normal fly, all the first generation will have normal wings, just as large and functional as ever even though half of their ancestors had never used their wings for many generations.

METHODS FOR GENETIC IMPROVEMENT

A method of heredity control that does produce improved breeds of animals and plants is the application of selection alone or, better, in conjunction with inbreeding and outcrossing. This is the method that has given such spectacular results with corn, now one of the most widely cultivated crops from the plains of Iowa to the Ukraine.

Hybridization

The development of improved varieties of grains, especially corn, wheat, and rice, and their impact on the methods of agriculture and the availability of food supplies, has been a phenomenon of the past half century. The development of hybrid corn caused revolutionary changes in U. S. agriculture. In the past decade, the introduction of new varieties of rice in Asia has had even more profound effects on the health and economy of nations.

Hybridization of corn is certainly not new. *Zea mays* is a hybrid in nature because cross-pollination ordinarily occurs. The American Indians traditionally planted different varieties of corn in the same fields because they were aware that this led to increased yields and greater vigor. The literature on corn hybridization goes back to the early days of the American colonies. In 1716, Cotton Mather, who is better known for his sermons and witch-hunting, wrote about his observations of the crossing of corn varieties in nature. It was not until the middle of the next century, however, that controlled experiments on the hybridization of corn were initiated by Charles Darwin. Darwin compared corn plants that were the progeny of self-pollinated vs. cross-pollinated individuals. He observed that crosses between unrelated varieties resulted in progeny that were often more vigorous and produced a higher yield than either parent strain, a phenomenon referred to as **hybrid vigor.**

Darwin corresponded about his experiments with the American botanist, Asa Gray. One of Gray's students, William Beal, working at Michigan State University, wrote the next chapter in this tale by attempting to obtain improved corn varieties through hybrid vigor. Although unsuccessful, Beal made a very significant contribution to the ultimate attainment of his objective: the method for obtaining hybrid plants. He grew two varieties in the same field and made certain that no other corn was grown nearby. All developing tassels (male flowers) were removed from one variety. Corn kernels developing in the ears of these emasculated plants all resulted from fertilization of the ovules by pollen from the second variety of corn planted in the field. Unfortunately, in Beal's attempts to obtain better varieties of corn, the yields from the resulting hybrids were not sufficiently improved to justify all the extra work involved.

The factor overlooked in Beal's experiments was revealed at the beginning of this century by the work of George H. Shull, of Princeton. Shull's contribution came from his attempts to obtain pure lines (i.e. inbred varieties) of corn. Shull's objective was to obtain better and more uniform corn strains by inbreeding, the so-called "pure line" approach.

His corn varieties did indeed become more uniform after several self-pollinated generations, and he developed, with continued inbreeding, a number of pure lines of remarkable uniformity. Some of these pure lines were greatly inferior in yield to the original stock; others were moderately inferior and some were about the same.

Shull's next research objective was a study of the inheritance of the number of rows of kernels, and he proceeded to make crosses between his pure lines. The resulting hybrids were indeed uniform but, most importantly, were greatly superior in vigor and yield to the parental strains. Thus, without intending to do so, Shull had developed a method for obtaining improved strains of corn with high yield. This approach involved two steps: (1) inbreeding to obtain the best pure lines and (2) crossing two pure lines to obtain a hybrid of vastly improved productivity.

By the 1920's one final step was added to Shull's classically simple approach to corn improvement by Edward M. East and Donald Jones of the Connecticut Agricultural Experiment Station. This was to obtain seeds from a double cross, thus combining desirable traits from four inbred strains and making large amounts of seeds for agricultural use from scarce single crossed seed. From a few bushels of single-cross seeds, several thousands of bushels of double cross seeds can be obtained (Fig. 4–15).

By the early 1930's commercial production of hybrid corn had begun and by 1950 more than 65 million acres of hybrid corn were grown in the United States (more than 75% of the total corn acreage). It was introduced in the U.S.S.R. by Nikita Kruschev. Research aimed at the improvement of corn strains still continues. Much of the current effort is to introduce characteristics such as stiff stalks (which make mechanical harvesting easier) or improved nutritional content of the kernels (for higher protein and vitamin content). The greatest impact of the story of hybrid corn has been, of course, on the yield per acre, which has enabled U. S. agriculture to consistently produce excess crops for export to parts of the world where critical food shortages exist.

Selection

In the past half century there has been wide application of genetic knowledge for improving strains of plants and animals that result in increased agricultural production. With a relatively unimproved group of cows, chickens, or other animals to begin with, rigorous **selection** of the best individuals results in he-

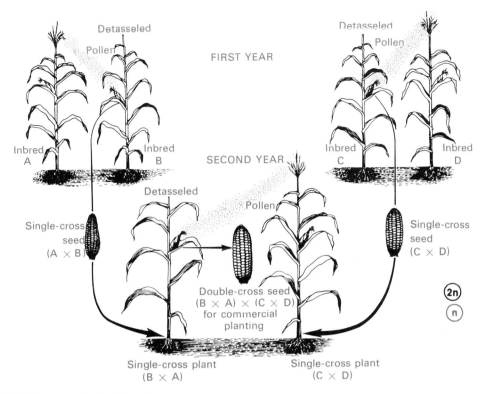

FIRST YEAR

SECOND YEAR

Detasseled
Pollen
Inbred A
Inbred B
Detasseled C
Pollen
Inbred C
Inbred D

Single-cross seed (A × B)
Single-cross seed (C × D)
Detasseled
Pollen
Double-cross seed (B × A) × (C × D) for commercial planting
Single-cross plant (B × A)
Single-cross plant (C × D)

2n
n

Fig. 4-15. Diagram of the Shull-East-Jones method of producing dramatic improvement in corn by double hybridization. (From Moment, G. B., and H. M. Habermann. 1973. *Biology: A Full Spectrum.* The Williams & Wilkins Co., Baltimore.)

reditary improvement. This is the way most of the current breeds of farm animals have been produced. By this method, improvement is usually rapid at first, but a plateau is reached after a variable number of generations.

Progeny Testing

To achieve further results, a method originated over a century ago by the great French plant breeder, Louis de Vilmorin, has been applied with success. This is the **progeny test.** It has been especially successful in dairy herds and in chickens. A bull obviously cannot be judged on his milk production. It is the milking record of his daughters that reveals whether or not he carries genes that make for high milk productivity.

Excellent results from using the progeny test on chickens have been achieved in agricultural experiment stations in several states. The older method of selecting hens for breeders on the basis of their own egg-laying records was followed until for several years no further improvement was obtained. Then the selection of the hens and roosters to be used as breeders was based on the egg production record of their offspring. This new method of

selection gave marked additional improvement.

Heterosis

Outcrossing (or **cross-breeding**) is an extremely valuable method for improving varieties of animals as well as plants. Outcrossing is used for two purposes. First, it may combine in one animal desirable traits from various breeds. A notable case is the Santa Gertrudis cattle of Texas, which have been produced by crossing Herefords and short-horns with Brahman cattle from India. The resulting hybrid has some of the superior beef qualities of the European breeds combined with the resistance to heat, drought and insects of the Indian cattle. The second generation after any such cross shows a considerable variability, as would be expected from the segregation and recombination of genes. Consequently, in the second and later generations, selection must play an important part in the elimination of undesirable traits. Some of the later generations might equally well combine the poor beef qualities of the Indian cattle with the susceptibility to drought and heat of the European breeds.

Cross-breeding, as shown by hybrid corn, not only produces new combinations of traits from which the desirable ones can be selected, it also produces hybrid vigor or **heterosis.** Heterosis was defined by Gowen as "the evident superiority of the hybrid over the better parent in any measurable character such as size, general vegetative vigor, or yield." The explanation of hybrid vigor is still obscure. Part of it may be due to heterozygosity of harmful recessive genes. Some think that for unknown reasons heterozygosity is in itself beneficial. Others believe that in certain strains there just happen to be genes for "vigor" which have a complementary reinforcing action when brought together.

Less is known about hybrid vigor in animals than in plants, although in *Drosophila,* heterosis, as defined above, has been shown to exist to a very marked degree in lifetime egg production and to lesser degrees in various other traits.

The hybrid mule is a much hardier animal and able to thrive on poorer food than either the donkey or the mare, its parents. The lack of fertility in mules is not due to any fault in reproductive anatomy or physiology but to a failure in chromosome behavior during meiosis. In synapsis, the horse chromosomes do not pair properly with the donkey chromosomes. The eggs or sperms produced do not have complete sets of chromosomes and hence do not give rise to viable offspring.

Inbreeding

The value of **inbreeding** is based on the fact that inbreeding produces homozygosity. This is a point made by Mendel in his original paper. He established what perhaps should be called Mendel's third law, namely that under conditions of self-fertilization or close inbreeding, the proportion of homozygous individuals becomes greater and greater and the proportion of heterozygotes becomes less and less. It is easy to understand why this should be so, because with close inbreeding the homozygous individuals will produce only homozygous offspring, whereas only half the offspring of the heterozygotes will be heterozygous. In the 10th generation of self-fertilized peas, Mendel calculated that the offspring from a monohybrid cross would be in the ratio of 1,023 homozygous dominants to 2 heterozygotes, to 1,023 homozygous recessives. In other words, homozygotes to heterozygotes would be in a ratio of 1,023:1. Of course the effect could not be so extreme with inbreeding in animals, but with brother-sister and father-daughter crossing, similar results follow.

The first great practical result of inbreeding is that the resulting homozygosity "stabilizes" the desirable traits so that animals breed true. This is the method that has been used in the production of most present-day breeds of domestic animals. The first step may or may not be a series of crosses, but there is always selection of the animals having the desirable genes, and this is usually followed by inbreeding to make the strain homozygous.

A second useful fact of inbreeding is that homozygosity reveals desirable and undesirable recessives. Such genes can then be either retained or eliminated. In corn breeding, which set the pattern, inbreeding produces a number of strains, some very poor. Only the best inbred strains are subsequently used in the crossing that gives spectacular hybrid vigor.

Concerning the Human Gene Pool

The grand total of all the genes, good, bad, and indifferent, possessed by the human race is unknown but certainly enormous. Scientific knowledge of this vast treasury of potentialities is only just beginning, but at long last a true information explosion has begun. The study of human heredity will always be hampered by great difficulties. Compared with *Drosophila* or mice, man is a slow-breeding animal. Moreover, a single female does not produce 75 to 100 offspring, nor can specific crosses be made to order for genetic purposes. Many of the most important and interesting human traits are behavioral and therefore governed by cultural as well as genetic influences. Added to these basic difficulties are several minor ones. Occasionally, individuals are not completely honest in reporting the facts. Furthermore, a number of human genes either vary in the degree to which they are expressed in different individuals or lack 100 per cent **penetrance.** In other words, the same gene may manifest itself more strongly in one person than in another, or in some cases it may not produce any detectable effect. The genes for hyperextension of the thumb and for lack of a complete set of permanent teeth fall into this category.

THE BLOOD GROUPS

It is now over half a century since Karl Landsteiner, in reinvestigating the old problem of blood reactions, discovered that the

Fig. 4-16. Karl Landsteiner, physician and protein chemist. (Courtesy of The Johns Hopkins University Institute of the History of Medicine.)

blood serum of some people would cause an agglutination, i.e. clumping, of the red cells of certain individuals but not of others. Following this discovery it became evident that all people fall into one of four blood groups, now called O, A, B, and AB. One who belongs to group O has blood serum (the fluid part of his blood minus the clotting fibrin) which carries proteins known as **antibodies** that will clump the red cells from any person belonging to group A, B, or AB. If a person belongs to group A, his serum carries antibodies that will agglutinate cells from a member of groups B or AB. If he belongs to group B, his blood will clump cells from members of groups A and AB. Finally, if a person is of group AB, his serum will not clump blood cells of A, B, or O type. A necessary corollary of the presence in group O serum of antibodies against both A and B cells is that the anti-B antibody in A serum and the anti-A antibody in B serum will not clump group O red blood cells.

In transfusions it is important to use blood of a type matching that of the recipient. In extreme emergencies group O blood can be transfused into persons belonging to the other three groups because the O cells cannot be agglutinated by the anti-A or anti-B antibodies present in the blood of the recipient, and at the same time the antibodies in the O serum are so diluted that they do not clump the recipient's cells.

The inheritance of the ABO factors is simple. Three alleles, all located at the same or nearly the same position, produce the four blood groups. Gene I^A (abbreviated from

isoagglutin A) yields A cells and anti-B antibody in the serum. Its allele I^B produces B cells and anti-A antibody in the serum. When a person receives gene I^A from one parent and gene I^B from the other, he is of the genotype $I^A I^B$ and belongs to group AB. Gene I^O, if homozygous, puts a man or woman in group O.

Subsequent studies have uncovered a dozen or more other blood groups by which people can be distinguished. The best known of these are the MN system and the Rh or rhesus factor blood group. In these systems there are normally no corresponding antibodies in the serum, but only antigens on the red cells. All are inherited according to Mendelian rules. The rhesus factor was discovered first in the blood of the macaque or rhesus monkey, *Macaca,* which is the short-tailed, brown monkey of Southern Asia used in medical research. The medical importance of the rhesus factor lies in the fact that if an Rh negative (Rh⁻) person receives a transfusion of Rh positive (Rh⁺) blood, the Rh⁻ person forms antibodies against the Rh⁺ erythrocytes or red cells. On a subsequent transfusion with Rh⁺ cells, the foreign Rh⁺ cells will be clumped, and that can lead to serious blocks in the vascular system. Furthermore, if an Rh⁻ woman becomes pregnant with an Rh⁺ child, some Rh positive factors may pass through the placenta into the blood of the mother. She will then produce antibodies against the Rh⁺ cells. These antibodies from the mother will pass through the placenta

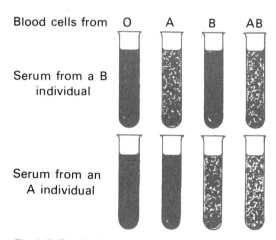

Fig. 4-17. Test for blood groups of the ABO series. Blood cells from O, A, B, and AB individuals are mixed with serum from a B individual which contains an antibody clumping cells carrying the A factor, and from an A individual whose serum will clump red blood cells carrying the B factor. (From Moment, G. B., and H. M. Habermann. 1973. *Biology: A Full Spectrum.* The Williams & Wilkins Co., Baltimore.)

into the fetal bloodstream. Consequently, in a subsequent pregnancy with an Rh⁺ child, this child may develop a hemolytic disease called **erythroblastosis fetalis** in which the red blood cells are destroyed.

Although about 15 per cent of people of European ancestry have Rh⁻ blood, which means that about 12 per cent of all marriages are between Rh⁻ woman and Rh⁺ men, fortunately very few of these marriages result in erythroblastic babies. This is partly because some women do not make Rh⁺ antibodies at all or fast enough to be very harmful, and partly because many of their husbands are heterozygous for the Rh factor. Rh⁺ is a simple dominant. One-half of the sperms of a heterozygous man carry a gene for Rh⁻. A child resulting from such a sperm would be homozygous for the Rh negative factor and would be unharmed by the mother's Rh⁺ antibodies, even if abundant. Consequently, instead of 12 per cent of marriages yielding one or more erythroblastic children, only about one-half of 1 per cent do so. Erythroblastic babies, if not too badly affected, can be tided over the crisis by transfusion with about 500 ml of Rh negative red cells which are immune to the antibodies the baby has received from its mother. It is now possible to prevent the mother from forming destructive antibodies by administering to her an Rh immunoglobulin (Rho-GAM) which suppresses the maternal immune system.

Blood group genes are especially useful to anthropologists interested in the historical interrelationships of the various races of mankind. This is because, unlike skull shape or tooth wear or skin color and many other traits, blood groups (1) are not changed by differences in food, climate, or other environmental influences; (2) are inherited according to very simple Mendelian laws; (3) are sharply defined, "all-or-none" characteristics in marked contrast to the blending nature of traits like skull shape or hair structure; and (4) can even be determined in prehistoric mummies and bones of other ancient human remains.

All three of the ABO blood group genes have been found in all racial groups, but their frequencies vary widely. Gene Iᴬ, for example, is almost completely absent among the Indians of Central and South America, but is very common among the Blackfeet and other tribes centering in Montana and among the natives of northern Norway.

As far as blood transfusions are concerned, the only thing that matters is that the two individuals concerned belong to the same blood group. Thus, natives of Sweden, China and Africa can safely exchange bloods if they are of the same blood group, while two brothers cannot if they belong to different blood groups.

SEX CHROMOSOMES AND HUMAN ABNORMALITIES

There are many genes now known to be located on the human **X** chromosome. In addition to the genes for red-green color blindness and hemophilia are the genes for childhood (Duchenne type) muscular dystrophy, glucose-6-phosphate dehydrogenase (G-6-PD) deficiency, favism, a severe illness due to eating beans (*Vicia fava*), total color blindness, a blood group termed Xg, a lack of gamma globulin essential as antibody against bacteria, and numerous others.

The Lyon Hypothesis

Male mammals have only one **X** chromosome while females have two. Since this is so, you would expect females to have twice as much of the products of all the genes on the **X** chromosome as do males. This is not the case. It has been established for some years that women do not have any more of the antihemophilia globulin, for example, than do men. The same is true for an abnormal hemoglobin due to a gene on the **X** chromosome.

Several mechanisms that would result in "dosage compensation" giving this result can be imagined. Several investigators hit upon what appears to be the correct explanation at about the same time, but the clearest statement and most convincing evidence was presented by Dr. Mary Lyon of Harwell, England, an explanation commonly known as the **Lyon hypothesis.** According to this theory, early in mammalian development (probably in the stage when the egg is becoming implanted on the wall of the uterus) one or the other of the **X** chromosomes in each cell of a female blastula becomes inactivated. These inactive **X** chromosomes are the **Barr bodies** which can be seen adhering to the nuclear membrane of many cells, whether from cheek epithelium, nerve ganglia or liver in females. They are the "drumsticks" attached to the irregular shaped, "polymorphic" nuclei of the white blood cells of a woman. Which **X** chromosomes becomes inactive is apparently a matter of chance, but once one of the two is committed to inactivity, all of its descendants follow suit and that **X** chromosome remains inactive. The evidence for the Lyon hypothesis is convincing and the results striking, if hardly surprising. Lyon based her theory on

Fig. 4-18. Mary F. Lyon of the MRC Radiobiology Unit, Harwell, England, originator of the Lyon hypothesis that half of the X chromosomes become inactivated during early development in female mammals. (Courtesy of Godfrey Argent, London.)

X-linked coat color in mice. A homozygous female has a single color but a heterozygous female is mottled with some patches of fur of one and some of the other color. A similar situation occurs in tortoise shell cats. Males fail to show such mosaics.

Another instance is the population of erythrocytes in the blood of a female heterozygous for G-6-PD deficiency. It is possible by a special staining technique to distinguish red cells which have this enzyme from those which do not. As expected, heterozygous women show both types of cells in approximately equal numbers.

Dosage compensation, however, is not always complete. If it were, **XXY** men would be normal rather than suffering from Klinefelter's syndrome.

Sex Anomalies

You will recall that in humans (and fruit flies) sex is determined by the sex chromosomes. **XX** individuals are females, while **XY** individuals are males. During meiosis mistakes can occur which result in gametes with missing or extra chromosomes. When such deletions or duplications occur, any resulting zygotes will have the wrong number of chromosomes. Abnormal numbers of **X** or **Y** chromosomes can influence not only the sex of the individual but also fertility. In an insect

such as *Drosophila* where all **XY** zygotes develop into males, **XO** zygotes, which have no **Y** chromosome because of its loss during early development of the egg, also develop into normal though sterile males. Any **XXY** zygote develops into a normal appearing and fertile female fruit fly.

The situation in humans is somewhat different. Human **XO** zygotes do not develop into males but into females. Such girls develop normally until the age when puberty should occur. But menstruation does not take place and there is neither development of breasts nor of axillary or pubic hair. Such a condition is called **Turner's syndrome.** At first it was thought that the trouble might be a failure of the pituitary gland to secrete its normal gonad-stimulating hormone, but it was soon shown that there is ample gonadotropin from the pituitary in the blood of afflicted individuals. Biopsy, however, revealed that the gonads, i.e. the ovaries, were virtually absent.

An egg which contains two **X** chromosomes plus a **Y** chromosome develops into a man with **Klinefelter's syndrome.** Such men appear more or less normal but are sterile because spermatogenesis does not occur.

The facts of Turner's and Klinefelter's syndromes make it evident that sex determination in human beings (and probably all other mammals), while basically determined by an **X** and **Y** chromosomal mechanism as in insects, nevertheless has some important differences. The **Y** chromosome, which appears to be without much function in insects except to exclude the presence of a second **X**, in humans carries genes which have a positive male-producing action. It is of interest here to point out that a very similar situation exists in certain plants, where a **Y** makes the plant a "male," i.e. a pollen producer only, whether one or two **X**'s are present.

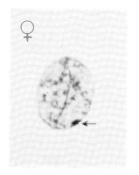
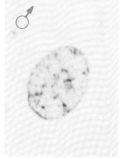

Fig 4-19. Human male and female epithelium cells from the lining of the cheeks, showing Barr body. (From Moore, K. L., and M. L. Barr. 1955. Smears from the oral mucosa in the detection of chromosomal sex. *Lancet* 2:57.)

Eggs with abnormal numbers of chromosomes can be produced by various agents — radiation, extremes of temperature, and certain chemicals. Any derangement of meiosis in which an **X** or **Y** chromosome is lost, or the **X** chromosome fails to separate properly, could lead to the sexual abnormalities just discussed.

Recently a number of men with two **Y**'s and an **X** have been found in penal institutions to which they had been committed because their extremely aggressive and even belligerent natures made them unacceptable as members of society. It is possible that the **XYY** condition of these unfortunate men has no causal relation to their behavior, but the evidence accumulated so far suggests such a relationship. Many more cases will have to be examined, especially among men outside of penal or mental institutions, before a firm answer can be given to the question of causality.

A Biological Look at Races

Racial differences clearly involve genetic differences, but they also involve so many cultural factors and so many unknowns that a scientific discussion of race is difficult. At the outset, value judgments must be set aside as irrelevant to a scientific study. That Swedes have big ears is a statistical fact which can be confirmed by going to Sweden and measuring people's ears. To say "big ears are beautiful" is to express a value judgment and is outside the sphere of the natural sciences.

Physical measurements are easiest to obtain and compare, and they form the classic criteria for racial distinctions — skin color, skull shape (whether long, medium, or round), hair type, and, more recently, the frequencies of various blood groups. An important conclusion can be drawn from these physical measurements, which is: "race" is a very fuzzy term. There are many genes shared in common by different races. Although it is useful to talk about yellow (which, when translated into Mongoloid, includes the American Indians), black, and white races, there are no sharp boundaries. Moreover, there are "races" which defy inclusion in even such a broad classification. The "hairy Ainu," the aborigines of Japan, in many traits resemble Asiatics but in others Europeans. The aboriginal Blackfellows of Australia are as black as the blackest Africans and resemble them in other respects, yet they possess traits which link them to the Caucasians.

There are different gene pools but how different the gene content of these great collections of genes is remains unknown.

It is certain that in many traits — height, physical strength, resistance to disease, hairiness, and others, including intellectual abilities — there is a great overlap among the three major races of man. Intelligence is extremely difficult, perhaps impossible, to measure within one group and far more so between individuals of diverse groups; but it does not require any elaborate psychological testing to know that the smart, the average, and the stupid are no monopoly of any one race. When individuals of all races vary among themselves in height, weight, body build, and a myriad of other traits, it is impossible to believe there is no variation in intellectual endowment. The only real question is averages, and there seems to be no satisfactory way to measure these.

One of the most persistent questions asked biologists concerns **racial crossing.** There are three classic biological objections that have been raised against racial crossing. The first is that harmonious adaptive patterns will be broken. A favorite example is the lightly pigmented skin and narrow nostrils of northern peoples compared with the heavy pigmentation and broad nostrils characteristic of many equatorial races. It has turned out to be harder to prove that these differences are truly adaptive than you might suppose, but it is a plausible idea that can be accepted for the sake of argument. Imagine a man with fair skin and a wide nose or one with dark skin and narrow nostrils. Such men would resemble Socrates and Gandhi and hosts of others who appear not to have suffered at all from their unusual combinations of traits.

The second objection, which is similar to the first, is that disharmonious combinations of traits will result. The classic example is in the work of C. B. Davenport on crosses between Europeans and Africans in the West Indies. The arms of these crossbreeds were said to be too long for their legs. The fact is that their arms were on the average 1.1 cm (not quite $1/2$ inch) longer than in the original white group and 0.6 cm longer than in the original blacks! Certainly a negligible effect and quite possibly due to errors of measurement, diet or even to a touch of hybrid vigor. Surely everyone has known mongrel dogs which showed neither physical nor behavioral deficiencies; in fact, many are smart, vigorous, and cooperative without being subservient.

The third objection raised against racial crossing is that if one race is superior, its qualities will be diluted. This objection depends on a value judgment. How can anyone decide which race is superior? One suggested criterion is similarity to the great apes; the more apelike, the more inferior. The apes have very thin lips and wavy hair, so by these traits blacks are clearly superior to whites. If body pigmentation is the test, then the Europeans are superior to Africans. If general hairiness is the test, then the Asiatics are first, the Negroes second, and the Europeans and the hairy Ainu last. Who can regard such traits as of any great importance?

Another approach is to consider genes for traits that are universally regarded as harmful. There is a list of hundreds: albinism, hemophilia, Down's syndrome, diabetes, Huntington's chorea, deaf-mutism, porphyria, etc. But the vast majority of these genes have been identified in all racial groups. Presumably the relative frequencies of these genes vary somewhat from race to race but that certainly provides no basis for judging one group as a whole superior to another.

Brain size is sometimes taken as a criterion. Within very broad limits, intelligence is correlated with size. Frogs have very small brains and no one is surprised that their intellectual abilites are limited. Great caution must be exercised, however, in extrapolating such observations to mankind. Microcephalic idiots exist in all races. But it must be remembered that much more than mere size is important. Cuvier had one of the largest human brains ever recorded and his contributions to biological science were great, but not nearly so important or original as those of the far smaller-brained Lamarck. Anatole France, a writer commonly regarded as the equal of Voltaire during his lifetime, possessed one of the smallest of recorded human brains. Remember that in size and complexity the brain of the porpoise is equal to man's. There is some evidence that average brain size is greatest in Central Asia and becomes less as you move out to the far corners of the world. Perhaps potential intellectual capacity is greatest in Central Asians and less in the rest of us. Perhaps, but no one knows. Quite possibly no one ever will.

Human racial crosses turn out well from the biological point of view even though there are sometimes sociological problems of a severe kind. Cross-cultural marriages often have rough going. One of the most dramatic racial crosses studied was between the Australian Blackfellows, who had one of the most primitive cultures ever seen, with no written language and no use of metals, and people from highly industrialized England. The offspring are healthy (no disharmonious gene combinations), capable, and do well at school and college. The most famous racial cross is probably that between the mutineers on H.M.S. Bounty and their Polynesian wives. Their descendants have lived in virtually complete isolation on Pitcairn and Norfolk Islands in remote parts of the Pacific. One human geneticist reported the results of this cross as deplorable, but earlier and several more recent visitors to Pitcairn have returned with moving pictures and other documentation showing that the Pitcairn Islanders are in actuality "robust, long-lived, smart, and ingenious," and in addition endowed with wit and friendliness. Certainly not inferior in anybody's language!

A specific question often put to biologists is how much gene exchange there has been between races in North America. By arbitrary convention, anyone with any sign of African genes is called Negro or black. From the biological point of view this becomes more and more absurd as the proportion of genes of European origin increases. Even when the number of African and European genes are present in equal numbers, to call a man a black is as arbitrary as calling a zebra a black horse with white strips or a white horse with black strips.

Casual observation makes it obvious that there has been racial crossing on a vast scale. Careful calculations based on the familiar Hardy-Weinberg principle indicate that approximately 20 per cent of the gene pool of American blacks is of non-African origin. This is a general average. Of course in some individuals it is much more and in others, less. Two individuals who happen to have the same proportion of European to African genes may have this proportion consist of rather different genes. The genes for skin color and hair type, for example, are not necessarily inherited together.

Finally, a biologist should warn against falling into the trap of thinking that cultural inheritance is transmitted via chromosomes. There are no genes which carry instructions for speaking English or Urdu, Swahili or Japanese. It is in no way necessary to possess DNA which originated in Greece to be an heir to the cultural heritage of Athens. That belongs to all mankind regardless of race. A biologist who thinks of the commonality of all living things based everywhere on the double helix of deoxyribonucleic acid finds it easy to remember the words which Socrates uttered so long ago: "As for me, it is not the Athenians, nor even the Greeks, that are my brothers, but all mankind."

USEFUL REFERENCES

Bodmer, W. F., and L. L. Cavalli-Sforza. 1976. *Genetics, Evolution* and *Man*. W. H. Freeman and Co., San Francisco.

Clarke, C. A. 1970. *Human Genetics and Medicine*. Institute of Biology Studies in Biology No. 20. Edward Arnold (Publishers) Ltd., London. Distributed in the U.S. by Crane, Russak and Co., Inc., New York. (Paperback.)

Lawrence, W. J. C. 1968/71. *Plant Breeding*. Institute of Biology Studies in Biology No. 12. Edward Arnold (Publishers) Ltd., London. Distributed in the U.S. by Crane, Russak and Co., Inc., New York. (Paperback.)

Mead, M., T. Dobzhansky, E. Tobach, and R. E. Light, eds. 1968. *Science and Concept of Race*. Columbia University Press, New York.

Mendel, G. (reprinted in 1967). *Experiments in Plant Hybridization*. Reprinted by Harvard University Press, Cambridge, Mass.

Suzuki, D. T., and A. J. F. Griffiths. 1976. *An Introduction to Genetic Analysis*. W. H. Freeman and Co., San Francisco.

5

Genetics II: Reading the Code

This chapter is about the "language" of genetics and the way in which it is read. In a sense, the information in the deoxyribonucleic acid of the chromosomes is comparable to ordinary written language, with structural and chemical analogies to paragraphs, sentences, words, and letters of the alphabet. There are four letters in the alphabet of life, the two purine and two pyrimidine bases in DNA. A sequence of three bases along the linear structure of DNA makes up a "word" or **codon** that specifies a single amino acid. A series of codons indicates the ordering of amino acids in a polypeptide chain. This information is analogous to a sentence in language and is the basic unit of genetic function, the **gene**. Genes are arranged in linear order on the chromosomes, and the several genes in a given segment, or **locus**, on a chromosome are comparable to paragraphs on a printed page.

This chapter is also about human genetic diseases and the possibility of gene therapy.

Genes

The word "gene" was first defined as a unit factor of heredity which followed Mendel's laws. It was coined without reference to chromosomes, frankly as a verbal tool to make it easier to talk and think about heredity. It is in this sense that we have been using the term. However, after T. H. Morgan and his students had mapped the positions of many genes on specific chromosomes by the use of crossover rates, it became fashionable to define a gene as the smallest unit of heredity not divi-

sible by crossing over. Others preferred to define a gene as the smallest unit of mutation. Because each gene controls the formation of an enzyme, or at least a polypeptide chain, others define a gene as the smallest functional unit of a chromosome: **one gene, one polypeptide.**

Obviously all these definitions are very similar, although none is precisely like any other. Each is useful under slightly different circumstances, so that none can be called wrong. This situation will doubtless continue until all the biochemical details of gene structure and action become known. It is to this topic that we now turn our attention.

THE CHEMICAL BASIS OF GENES

DNA, deoxyribonucleic acid, is formed something like a twisted ladder, i.e. a **double helix**. The sides of the ladder are composed of linked and alternating molecules of **phosphoric acid** and **deoxyribose**, a five-carbon sugar. The rungs connected to the sugars carry the genetic information. They must if the code really is written in the DNA, because only the rungs vary as you climb the twisted ladder.

These rungs are made up of only four kinds of nitrogenous bases. Two, **adenine** (A) and **guanine** (G), are purines, which are two-ringed compounds containing nitrogen; the other two, **cytosine** (C) and **thymine** (T), are pyrimidines, which are similar to the purines but have only one ring. **RNA,** found both in the nucleus and in the cytoplasm, resembles DNA except that the sugar is ribose instead of deoxyribose and **uracil** takes the place of thymine.

The structural formulas of the nitrogenous

bases of nucleic acids, i.e. the alphabet of the genetic code, are shown in Fig. 5-1.

Two molecules, one purine and one pyrimidine, make a rung. Since the length of the rungs does not vary, the pattern of a purine matched with a pyrimidine remains constant. Two pyrimidines would make a short rung that would not fit. Two purines would make a rung too long. So it comes about that, due to the size of the molecules, adenine is always matched with thymine by hydrogen bonds, and guanine with cytosine, also by hydrogen bonds. From this it follows that if the nitrogenous bases on one side of the ladder are (in order) adenine, guanine, guanine, thymine, thymine, thymine..., then the corresponding bases on the other side will be thymine, cytosine, cytosine, adenine, adenine, adenine....

There is a slightly different but also useful way of looking at DNA structure. A single phosphoric acid group, plus the sugar, plus the purine or pyrimidine, constitute a **nucleotide**. Thus, when the double helix, or twisted ladder, separates lengthwise into two separate strands of nucleic acid, each strand will consist of a series of nucleotides (a polynucleotide) just as a protein consists of a series of amino acids called a polypeptide.

THE GENETIC CODE

There is much evidence, some presented in an earlier chapter and some to come later, that genes preside over the synthesis of proteins, many of which are enzymes. This is not very surprising, since cells are what their enzymes make them, and genes control enzymes. Both nucleic acids and proteins are long chains of smaller units, nucleotides in one case, amino acids in the other. Consequently, it would seem probable that the sequence of bases in the DNA represents, or

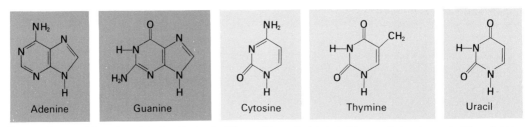

Fig. 5-1. Structural formulae of the five "letters" of the nucleic acid alphabet. (Purines dark color; pyrimidines light.)

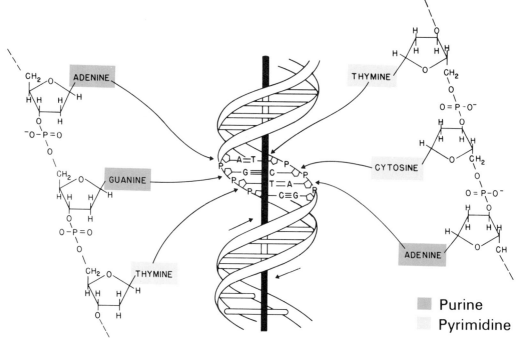

Fig. 5-2. A segment of DNA showing pairs of nucleotides. (From Moment, G. B., and H. M. Habermann. 1973. *Biology: A Full Spectrum*. The Williams & Wilkins Co., Baltimore.)

codes for, the sequence of amino acids in the proteins.

Since there are four bases, there is a four-letter alphabet. There are about 20 amino acids to specify. Consequently, one-letter words will not suffice because with four letters, A, T, G, and C, there could be only four one-letter words. How many two-letter words can be made from four letters? The first letter of each word might be A, T, G, or C. That makes four possibilities. The second letter might be any one of the four in each case, AA, AT, AG, AC, TA, TT, TG, etc. This makes 4 × 4, or 16 words. This is still not enough. With three-letter words the possibilities are 4 × 4 × 4, or 64. This is enough and perhaps too many, unless some words are merely nonsense or perhaps synonyms or even punctuation.

One of the first questions about a three-letter word system is whether the words can overlap. For example, if *ABC* spells a certain amino acid, and *BCD* spells another amino acid, would *ABCD* spell them both or would you have to write *ABCBCD*? How can this question be put to a test? One way is to note that if the code is an overlapping one, a change (i.e. a mutation) in one letter (one nucleotide) would affect two adjacent amino acids in the protein. The fact is that many cases are now known where a mutation changes a single amino acid but, so far, none has been found where two adjacent ones are changed—something which would happen with most mutations if the code were an overlapping one. Furthermore, if the code were overlapping, it would mean that a given amino acid in a protein would always have one of only four amino acids following it, which is not true. For example, if *ABC* were the code word for glycine, then glycine could only be followed by whatever four amino acids were coded by *BCA*, *BCB*, *BCC*, and *BCD*.

To summarize, the code consists of three-letter words made from a four-letter alphabet. It spells out the sequence of the amino acids in the proteins. These three-letter "words" are commonly referred to as **triplets** or **codons**.

REPLICATION OF THE CODE

In the life cycle of dividing cells, the chromosomes become duplicated and then in mitosis the duplicates separate, one passing to each pole of the spindle. The conditions and specific stimulus, if indeed there is one, which result in chromosome duplication and mitosis are unknown. However, Watson and Crick have proposed a theory, supported by

considerable evidence, as to the way in which DNA duplicates. The weak hydrogen bonds, which hold the two strands of the double helix together, break and the strands separate. Each then acts as a model or template for the formation of a new chain of nucleotides complementary to itself along its entire length. The result is two double helices, each identical to the original. In each, one strand will be from the old original double helix and one will be new (Fig. 5-3). In the words of Watson and Crick:

Now our model for deoxyribonucleic acid is, in effect, a pair of templates, each of which is complementary to the other. We imagine that prior to duplication the hydrogen bonds are broken and the two chains unwind and separate. Each chain then acts as a template for the formation on to itself of a new companion chain, so that eventually we shall have *two* pairs of chains, where we only had one before. Moreover, the sequence of the pairs of bases will have been duplicated exactly.

Much still remains to be learned. It is clear that this is an energy-using process and also that an enzyme, DNA polymerase, is essential. One of the ways of proving that chromosomes duplicate in this manner, i.e. by serving as models which induce a new structure to form alongside of themselves, is to feed

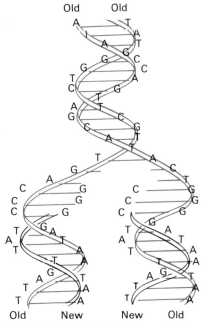

Fig. 5-3. The DNA double helix in the act of self-duplication. (From Moment, G. B., and H. M. Habermann. 1973. *Biology: A Full Spectrum*. The Williams & Wilkins Co., Baltimore.)

cells before division with radioactive thymine. A newer way is to grow cells in a medium in which bromodeoxyuridine (BrdU) is substituted for thymidine. When BrdU is built into a chromatid (as it is when there is no thymidine present) that chromatid no longer fluoresces brightly as normal chromatids do when stained with acridine orange or certain other dyes (see Fig. 5-5). After the formation and separation of the new chromosomes, one of each homologous pair has all the radioactivity or fluorescence and the other member of the pair has none.

TRANSLATION OF THE CODE INTO PROTEINS

It will be recalled that DNA not only makes more DNA but also, with the help of the proper enzymes, produces messenger RNA, which then moves out of the nucleus into the cytoplasm. The formation of messenger RNA from its template DNA is referred to as **transcription**. The formation of proteins from the code in the messenger RNA is called **translation**.

Three types of RNA can be identified in the cytoplasm by differential centrifugation and other methods. The first type is **messenger RNA**, usually written mRNA. The messenger RNA occurs in long strands or chains, as

Watson and Crick would say. It is the equivalent of a long computer tape carrying coded instructions. A second type is **ribosomal RNA** (rRNA). This makes up about 60 per cent of the structure of ribosomes; the rest is protein. A third type of RNA is soluble or **transfer RNA** (sRNA or tRNA). Transfer RNA consists of relatively short pieces of RNA.

Translation of the genetic code is merely a somewhat more informative way of saying **protein synthesis**. It occurs as follows. **Ribosomes** become attached to the starter ends of the long strands of mRNA and move along these strands, reading the code and producing proteins by adding one amino acid after another. Series of ribosomes lined up along mRNA can sometimes be seen under an electron microscope. Such a series of ribosomes is commonly referred to as a **polysome** or polyribosome (Fig. 5-6).

At the same time that the ribosomes are moving along the mRNA strand, specific enzymes in the cytoplasm activate amino acids. Adenosine triphosphate (ATP) complexes with an amino acid. This compound, specifically the amino acid part, then becomes attached to one end of the tRNA molecule.

For a time, transfer RNA was thought to exist in the form of an elongated twisted U or hairpin. More recent evidence indicates that

Fig. 5-4. J. D. Watson (left) and F. H. C. Crick with their model of the double helix. (Courtesy of The Bettmann Archive.)

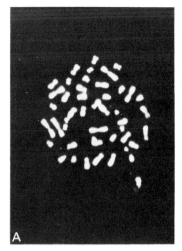
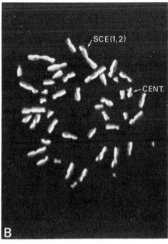
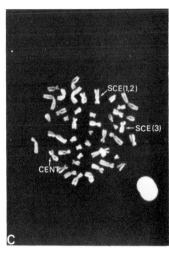

Fig. 5-5. Photograph of a human male set of metaphase chromosomes (i.e. male karyotype) treated to show the semiconservative way chromosomes reproduce. Each chromosome acts as a template or model along side of which a new sister chromosome is built. The original and the new "chromosome" are called sister chromatids until after they separate during metaphase. The cells, in this case lymphocytes, were grown in a medium in which bromodeoxyuridine (BrdU) was substituted for thymidine, a normal; component of DNA. When BrdU is built into a chromatid (as it is when there is no thymidine present) that chromatid no longer fluoresces brightly as do normal chromatids containing thymidine and stained with acridine orange. A, chromosomes before taking up BrdU showing both sister chromatids fluorescing; B, chromosomes after replication in BrdU showing that the new member of each pair lacks strong fluorescence; C, after another replication in BrdU only ¼ of all chromatid regions fluoresce. SCE (sister chromatid exchange, indicated by arrows and numbered); CENT (centromere, i.e. kineteochore). (Courtesy of R. Tice, J. Chailet, and E. L. Schneider.)

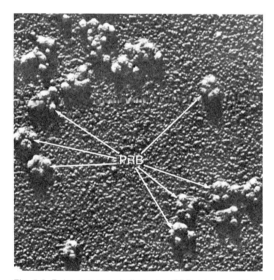

Fig. 5-6. Polysome or polyribosome (PRB) as seen with an electron microscope. (Courtesy of Dr. A. Rich; from Jensen, W. A., and R. B. Park. 1967. *Cell Ultrastructure.* Wadsworth Publishing Co., Inc., Belmont, Calif.)

the shape of the tRNA molecule is more like a cloverleaf (Fig. 5-7).

The tRNA molecules, each carrying an amino acid which corresponds to the code it bears, move through the cytoplasm, presumably by Brownian movement, and thus come into contact with ribosomes on the mRNA.

Each ribosome, as it passes along the series of triplets in the mRNA, accepts at each position only the appropriately coded tRNA. The tRNA molecule for methionine cannot get functionally tied up with a ribosome until the ribosome reaches a place along the mRNA which reads for methionine. When this happens the methionine-carrying tRNA moves into place on (or perhaps in) the ribosome and the tRNA previously there falls free minus its amino acid. Its amino acid, whatever it was, becomes attached to the methionine carried by the newly arrived tRNA. In this way a chain of amino acids linked by peptide bonds is built up, forming a polypeptide chain. Thus the code is translated into the **primary structure** of a protein, i.e. its characteristic sequence of amino acids.

It seems reasonable to suppose that the **secondary structure**—the twisted helical form of the protein molecule—would show itself as soon as the polypeptide chain forms. Likewise, there is no reason to suppose that the forces which produce the **tertiary structure**—the various foldings of the protein molecule—would wait to act until after the full length of the primary structure is complete. Consequently, when the finished protein finally drops free as its ribosome comes to the end of the mRNA ribbon, it would be held in its characteristic folds and bends by its sulfhydryl (−S−S−) and other bonds.

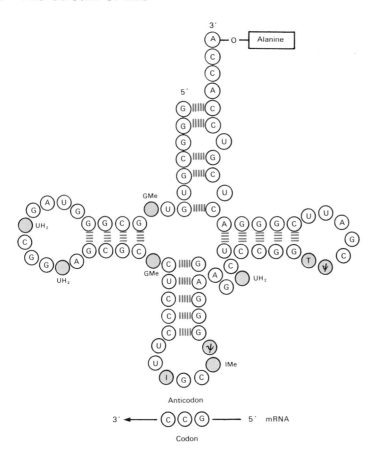

Fig. 5-7. Nucleotide sequence of the transfer RNA for alanine. Note position of the anticodon in tRNA and codon in mRNA. The amino acid alanine is attached to the 3′ end of the tRNA molecule. Note the presence of a number of unusual bases designated by color: inosine (I), dihydrouridine (UH$_2$), ribothymidine (T), methylguanosine (GMe), dimethylguanosine (GMe$_2$), methylinosine (IMe), pseudouridine (ψ). (Modified from Watson, J. D. 1970. *Molecular Biology of the Gene*, 2nd ed. W. A. Benjamin, Inc., Menlo Park, Calif.)

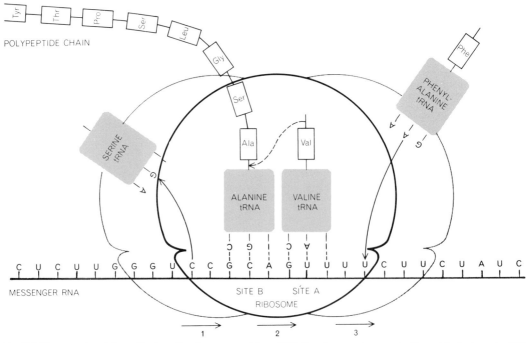

Fig. 5-8. Diagram of protein synthesis as the ribosome and the mRNA "tape" move along each other while the transfer RNA (tRNA) moves in with activated amino acids. (From Crick, F. H. C. 1966. The genetic code: III. *Sci. Am.* 215:55. Copyright © 1966 by Scientific American, Inc. All rights reserved.)

IS THE CODE UNIVERSAL?

Is the genetic code the same for microorganisms, plants, and animals? Since it has been firmly established that the proteins of all living things are composed of the same 20 amino acids, and that the nucleic acids of all living things contain the same four (five, counting uracil) purines and pyrimidines, it would be hard to believe that the code would be different. However, plausibility does not constitute proof. The best evidence supporting the idea that the code is the same in all organisms has been obtained by mixing amino acid-charged tRNA from one source (perhaps bacteria), with mRNA from a second source (e.g. from hemoglobin-producing cells of rabbits), and ribosomes from yet a third source; the results is always determined by the kind of mRNA present—in this case rabbit hemoglobin will be formed.

CRACKING THE CODE

One way of discovering what the code actually is, i.e. how to spell the word for any specific amino acid in the three-letter purine-pyrimidine words, is to make your own messenger RNA. This quite spectacular achievement was first accomplished by Marshall Nirenberg working at the National Institutes of Health. If the mRNA has no nucleotides except those carrying uracil, it will cause the synthesis of a polypeptide chain of repeating phenylalanine units. Thus, UUU codes for the amino acid, phenylalanine. Similarly, mRNA carrying only adenine will produce, in a test tube, when mixed with activated tRNA charged with all 20 naturally occurring amino acids, the necessary activating enzymes, and ribosomes, a polypeptide chain consisting of repeating units of only one amino acid, lysine. Thus, AAA codes for lysine. Three cytosines code for the amino acid proline. By changing the percentages of the different purines and pyrimidines in the mRNA, it has been possible to deduce the code words for all 20 of the naturally occurring amino acids. With different percentages of different mixtures of the four bases built into the mRNA, different amino acids are incorporated into the protein synthesized. For example, if there is no guanine in the mRNA, the amino acid glycine is not incorporated into the polypeptide formed. Therefore the triplet for glycine must contain at least one G (guanine). Messenger RNA with nothing but guanine will not call for glycine, hence there must be some code letter in addition to G. In fact, there has to be twice as much guanine as there is of some other base. Therefore the triplet for glycine has two G's plus one of the other bases. By this kind of analysis it has been found not only that all 20 of the amino acids can be spelled in triplet form, but that the code is redundant, i.e. there is more than one way to spell most of the amino acids. Thus UUU and CUU both correspond to phenylalanine (see Table 5-1).

Fig. 5-9. Marshall Nirenberg, whose investigations at the National Institutes of Health established the three letter purine-pyrimidine words of the genetic code. (Courtesy of M. Nirenberg).

TABLE 5-1
The genetic code (From Watson, J. D. 1970. *The Molecular Biology of the Gene,* 2nd ed. W. J. Benjamin, Inc., Menlo Park, Calif.)

First Position (5′ end)	Second Position				Third Position (3′ end)
	U	C	A	G	
U	Phe	Ser	Tyr	Cys	U
	Phe	Ser	Tyr	Cys	C
	Leu	Ser	Term[a]	Term	A
	Leu	Ser	Term	Trp	G
C	Leu	Pro	His	Arg	U
	Leu	Pro	His	Arg	C
	Leu	Pro	GluN	Arg	A
	Leu	Pro	GluN	Arg	G
A	Ileu	Thr	AspN	Ser	U
	Ileu	Thr	AspN	Ser	C
	Ileu	Thr	Lys	Arg	A
	Meth	Thr	Lys	Arg	G
G	Val	Ala	Asp	Gly	U
	Val	Ala	Asp	Gly	C
	Val	Ala	Glu	Gly	A
	Val	Ala	Glu	Gly	G

[a] Chain terminating (formerly called "nonsense").

Genes in Action: Genes and Enzymes

"One gene, one enzyme," or at least, "one gene, one polypeptide chain" (for there are cases known where two genes are required to produce the complete enzyme), is a central concept of modern biochemical genetics. The idea, however, that genes produce their effects by producing, or failing to produce, specific enzymes was proposed by a physician and biochemist at Oxford, Sir Archibald E. Garrod, during the first decade after the rediscovery of Mendel's laws of heredity. As happened with Mendel's discovery of the elementary laws of inheritance and MacMunn's discovery of the respiratory pigment cytochrome, no one paid much attention to Garrod's book, *Inborn Errors of Metabolism*, even though he was Regius Professor of Medicine at Oxford. The reasons for this complete neglect are complex but the basic one is surely that biologists were not intellectually ready to incorporate any of these ideas into the general body of their thinking. (After the battle, it is always much easier to see what the general should have done.)

Garrod studied a hereditary disease called **alkaptonuria**, which still serves as a model of how genes work and how the point at which they act can be discovered. The chief symptom of this rather benign disease is a blackening of the urine after exposure to air. This happens because the urine contains an abnormal constituent, alkapton or homogentisic acid. In normal individuals, the amino acid phenylalanine from food is incorporated into body proteins and also converted into a number of other products, including homogentisic acid. This is then converted into CO_2 and water (see Fig. 5-15).

In the families of people with alkaptonuria, Garrod showed that the disease behaves like one of Mendel's recessive factors (genes). He went on to argue that the accumulation of homogentisic acid which results in its being excreted by the kidneys is due to the absence of a specific enzyme. The missing enzyme is called homogentisic acid oxidase because it can oxidize homogentisic acid, not into a black pigment but into simpler products which in turn are converted into CO_2 and water.

The reasoning is simple. Within the body one compound, say some amino acid or amino acid precursor, is converted into some other compound via a series of steps. If a mutation changes a gene so that it can no longer form the enzyme responsible for some specific conversion of, say, A to B, then A will accumulate in the body and bloodstream and may be excreted in the urine.

This mode of thinking about how genes express themselves did not appear again until about 1935 when Boris Ephrussi and George Beadle undertook to investigate the biochemistry of the formation in *Drosophila* of various eye pigments which were known to be under the control of specific genes. The difficulties of this work led Beadle and E. L. Tatum to turn to *Neurospora*, the pink bread mold (Fig. 5-10). Actually it is more orange than pink, and is found growing wild on sugar cane. B. O. Dodge, of the New York Botanical Garden, had long urged T. H. Morgan to work with *Neurospora*, which he claimed was better than *Drosophila*. Morgan took some strains of *Neurospora* with him when he left Columbia University for the California Institute of Technology. It was there that Beadle and Tatum obtained this organism, and then the study of biochemical genetics really got under way.

The advantages of *Neurospora* are several. It can be grown readily in pure culture in test tubes and on media of known chemical composition. This means that the mold can synthesize for itself a large number of compounds from very few simple raw materials. The asexual spores, **conidia**, can be irradiated or otherwise treated to produce mutations. When germinated, the growth from these conidia can be crossed with wild strains. After crossing, fruiting bodies are produced which contain cigar-shaped capsules, each holding eight haploid spores. There are two adjacent spores for each of the usual four products of meiosis. Hundreds of these haploid cells can be planted singly in tubes with a complete medium containing a rich variety of vitamins, amino acids, and other substances. They can grow and produce conidia even if they carry mutations which deprive them of one of the enzymes necessary for some essential step in the synthesis of a necessary metabolite. After a good growth has been obtained, each strain can be tested in minimal medium. If it can no longer grow on it the way its ancestor did, then a mutation has occurred, depriving it of some particular enzyme (Fig. 5-11). Because *Neurospora* is a haploid organism, its phenotype reveals its entire genotype, and any change in its genetic constitution is immediately obvious. The next step is to test its ability to grow on media lacking one after another of the various vitamins, amino acids, and other growth substances to discover exactly where

ASEXUAL REPRODUCTION

SEXUAL REPRODUCTION

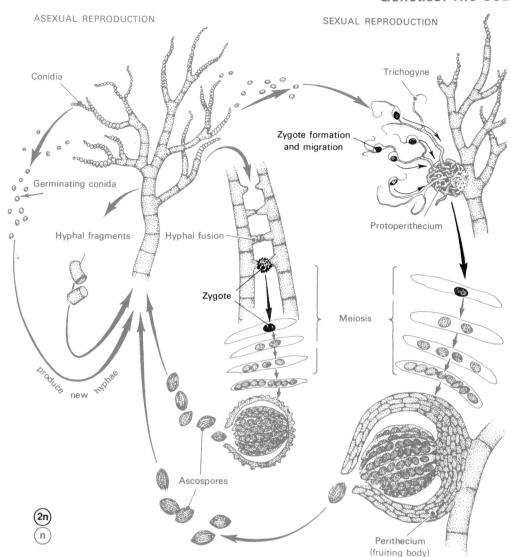

Fig. 5-10. Life cycle of *Neurospora*. The pink bread mold, *Neurospora* consists of a mass of branched hyphae called a mycelium. Asexual reproduction is by means of chains of pink spores (conidia) that form at the ends of hyphal branches. Sexual reproduction occurs only when two strains of different mating types are brought together. The hyphae can fuse and form a mycelium called a heterokaryon which contains nuclei of both strains. At times conidia can act as male gametes. When such conidia come in contact with a hypha serving as the female parent, they contribute their nuclei to heterokaryon formation and set into motion the formation of dark colored fruiting bodies called perithecia. The perithecia contain many asci. Within each ascus, eight ascospores develop. Young asci have two nuclei, one from each of the parental strains. These nuclei fuse and then undergo three divisions during which meiosis occurs. At maturity, there are eight haploid ascospores in each ascus. On germination, each ascospore can form a new mycelium. (From Moment, G. B., and H. M. Habermann. 1973. *Biology: A Full Spectrum*. The Williams & Wilkins Co., Baltimore.)

the biochemical lesion, i.e. damage, is located.

Clearly the "one gene, one enzyme" hypothesis fits the facts. It is the biochemical basis of mutational effects in organisms as widely separated as man and mold.

GENE INTERACTION

Genes themselves probably never interact with each other directly, but their products can interact in a variety of ways. A number of instances are known of what is called **epistasis**. In these cases a gene in one locus overrules the effect of a gene in a different locus, perhaps even on a different chromosome. Since this reaction is not between alleles, i.e. alternative forms of the same gene at a given locus on the same chromosome, it differs from ordinary dominance.

A well-known case of epistasis occurs in

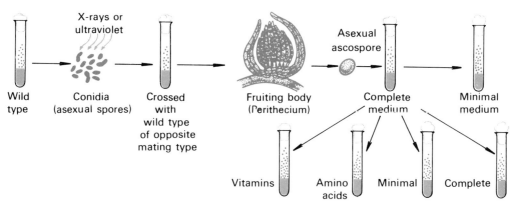

Fig. 5-11. Diagram of the basic experimental design for discovering biochemical mutants in *Neurospora*, the orange sugar cane mold commonly called pink bread mold. The two left media on the bottom row lack vitamins and amino acids. Growth under all these conditions would indicate wild type biochemical capability. (From Moment, G. B., and H. M. Habermann. 1973. *Biology: A Full Spectrum*. The Williams & Wilkins Co., Baltimore.)

dogs where the dominant gene **B** gives a solid black coat; the recessive allele **b** gives a brown coat when in the homozygous **bb** condition. However, there is another gene, **I** (for inhibitor), on a different chromosome which prevents pigment formation except in the eyes. To have fur of any color whatever, a dog must be homozygous **ii** for the recessive allele of this inhibitor gene. Thus, a dog may be homozygous for dominant black and yet be white because of the inhibitor gene **I** which masks the effect of the genes at the other locus. This **I** gene is different from the gene for albinism which prevents the formation of melanin, even in the eyes.

Another type of gene interaction occurs in all cases where many genes influence a single trait. Body size, intelligence, skin color, eye color, and numerous other general characteristics fall into this class.

HOW ARE GENES TURNED ON OR OFF?

One of the great paradoxes of modern biological science is the accepted fact that the cells of a multicellular animal or plant have the same diploid set of chromosomes and therefore the same genes, and yet these cells become differentiated in many very different directions. Clearly there must be some way in which certain genes are turned on in certain cells while other genes are activated in other cells. Or perhaps it sould be said the other way around: most of the genes are kept permanently inactive in most cells but the genes that are inactivated are different in different kinds of cells.

The control of the action of one gene by other genes has been known for many years. The cases of repressor genes in *Drosophila*

and position effects in corn (maize) also show that somehow genes can affect the action of each other. However, it was not until the work of F. Jacob and J. Monod in 1960 that a plausible model backed up with convincing evidence was proposed to explain how the activity of genes is regulated.

It should be noted first that enzymes are of two general types. There are **constitutive enzymes** which are formed by cells whether or not their substrate is present, at least in detectable amounts, and there are **inducible enzymes,** formed when and only when their substrate is present.

As early as the beginning of this century it was observed that certain enzymes of microorganisms are produced only in the presence of the specific substrates on which these enzymes act. For example, the colon bacterium, *Escherichia coli*, grows on rather simple media containing minerals and carbohydrate, and ferments maltose when this sugar is present as the carbon and energy source. If these bacteria are transferred to a medium where lactose is substituted for maltose, they do not begin to ferment lactose immediately, but are able to do so only after a lag period. During this lag period there is an **induced** production of the enzymes needed to break down this substrate. During such **enzyme induction**, the enzyme which initially is not present in appreciable amounts increases by as much as many thousandfold.

Escherichia coli exhibits yet another kind of regulated enzyme synthesis where a product of metabolism can control the amount of enzyme formed. If glucose and lactose are both present in the medium, glucose will be metabolized first and the fermentation of lactose will begin only after depletion of glu-

Fig. 5-12. Jacque Monod who with François Jacob proposed the operon theory of the control of bacterial enzyme synthesis. (Courtesy of J. Monod.)

cose. The reason for this is that the enzyme necessary for lactose fermentation (β-galactosidase) is an induced enzyme that is formed only in the presence of lactose. However, in addition, the presence of glucose suppresses the formation of this enzyme. Suppression can be caused by a nutrient or by a metabolic product near the end of a sequence of steps in which the enzyme is involved. The molecule responsible for suppression of enzyme synthesis is called the **repressor**, and this kind of suppression of enzyme synthesis is called **feedback repression**.

Induced synthesis of enzyme ensures that enzymes can be formed when specific substrates are available. Feedback repression guarantees that overproduction of enzyme can be avoided. Inducers and repressors can act antagonistically in regulating enzyme production. There are of course certain enzymes that are not subject to regulation by induction or feedback repression. Such constitutive enzymes are always present. It is worth noting that the kinds of enzymes which can be induced by substrates are limited by the hereditary constitution, i.e. the genome of the organism.

The Operon

In 1961, the French bacteriologists Jacob and Monod published a paper summarizing their ideas about bacterial synthesis of induc-

ible and repressible enzymes. This paper laid the groundwork for our understanding of the molecular basis of induction and repression, ideas which seem to be applicable not only to the bacteria, but also to higher plants and animals.

According to the generally accepted **operon** theory proposed by Jacob and Monod, information (i.e. structural genes) for the synthesis of one to several enzyme molecules comes under the control of a contiguous segment of the DNA molecule called the **operator**. The operator plus the closely linked sites under its control is called an **operon**. This segment of DNA consists of a **promoter gene**, a region of the DNA molecule where RNA polymerase binds to initiate synthesis of a strand of messenger RNA. Next is the **operator gene** which acts as an on-off switch. When "on" or open, the synthesis of messenger RNA can proceed along all segments of the DNA molecule linked to the operator. When "off" or closed, segments of the DNA molecule controlled by the operator are nonfunctional, i.e. no messenger RNA is formed and the polypeptides coded for by the structural genes linked to the operator are not synthesized. The operator itself is under the control of the product of a **regulator gene** that is located outside the operon. Two kinds of repressors are produced by regulator genes. One kind can shut off the operator but also can be neutralized by an **inducer** substance (a nutrient, hormone, etc.) so that it no longer represses the operator. Another kind of repressor must combine with a specific metabolite, the **corepressor**, in order to be effective, i.e. able to inactivate the operator gene. In the absence of the corepressor, the product of the regulator gene is not inhibitory and the operator remains in the open state.

There can be a change (or a mutation) of the regulator gene just as there can be mutations in portions of the DNA coding for specific proteins or enzymes. If the regulator gene becomes inactivated by mutation and no longer forms a product which can repress its operator, then the operator can no longer be repressed and remains permanently in its open state. Under these conditions one or several enzymes will always be produced and will appear to be constitutive. Also, a change in the operator can modify its susceptibility to repression. The information under the control of the operator can become permanently available and the associated structural genes would give rise to constitutive synthesis of enzyme. The general schemes by which genes can be regulated are summarized in Fig. 5-13.

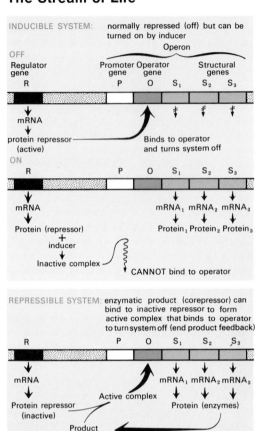

INDUCIBLE SYSTEM: normally repressed (off) but can be turned on by inducer

REPRESSIBLE SYSTEM: enzymatic product (corepressor) can bind to inactive repressor to form active complex that binds to operator to turn system off (end product feedback)

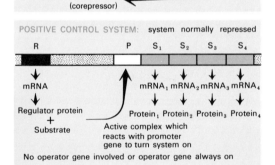

POSITIVE CONTROL SYSTEM: system normally repressed

No operator gene involved or operator gene always on

Fig. 5-13. Summary of mechanisms by which enzyme synthesis can be controlled.

Technology of Genetic Engineering

HEREDITARY MECHANISMS IN BACTERIA: WHAT THEY ARE AND WHAT THEY MEAN FOR MANKIND

It had always been assumed that nothing comparable to sex exists among bacteria and that mutation was either absent or extremely rare. Under these conditions a population of bacteria would remain virtually unchanged generation after generation. Consequently, when penicillin and other antibiotics were discovered during and after World War II, it was generally felt that the scourge of bacterial infections in disease and after surgery had been conquered at last. However, resistant strains of bacteria continued to appear, making it evident that mutations do play an important role among bacteria as they do among other organisms. This has meant important changes in the way antibiotics are administered to patients and also a continuing search for new kinds of antibiotics. In addition, bacterial mutations have furnished key information for the understanding of heredity in all living things.

Three different mechanisms exist for the tranfer of genetic information (DNA) from one bacterial cell to another. One mechanism, **conjugation**, is more or less comparable to sexuality in eukaryotic organisms. The other two, **transformation** and **transduction**, are not only remarkable in themselves but open up the possibility of developing methods for introducing new and desirable genes into adult plants and animals including man. This whole area of knowledge is in much the same undeveloped state that knowledge of atomic energy was in the 1920's when scientists believed that atomic forces might, perhaps, become available for human use at some time in the remote future. No one expects gene therapy, or genetic engineering as it is sometimes called, to become feasible within the next 10 years, but the technology to make them possible is being developed.

Transformation. Transformation was discovered in the 1920's when it was observed that sometimes one strain of bacteria, if grown in association with a second, can acquire properties characteristic of the second. Two American investigators, Frobisher and Brown, found that nonpathogenic streptococci from cream cheese grown in the presence of streptococci causing scarlet fever could acquire the capability of producing scarlet fever toxin. Other studies by Fred Griffith, an English medical bacteriologist, showed that bacteria could be transformed not only in the presence of other living cells but also by cell-free extracts of dead organisms. Extracts of dead capsule-forming pneumococci could transform a nonencapsulated strain so that cells could now form capsule substances. Later studies by O. T. Avery, Colin McCleod, and Maclyn McCarty at the then Rockefeller Institute in New York showed that the substance responsible for transformation is DNA. Once fragments of DNA have entered a recipient cell, they may or may not be incorporated. If incorporation occurs, the piece of transforming DNA becomes an integral part of the recipient's DNA and can be replicated. The fragment of DNA which has been replaced and fragments of transforming DNA which are not incorporated cannot be replicated and are eliminated. Normally, transformation occurs only between related species or strains having DNA's with structural similarities, that is, with similar nucleotide sequences. For altered phenotypes to be observable, the transformed DNA code must be read to form proteins. The resulting protein enzymes must produce products detectably different from those of the original recipient cell, and the transformed cells must multiply.

The penetration of DNA fragments into so-called "competent" cells is surprisingly rapid and can occur in a matter of seconds. The incorporation of transforming DNA into a cell's genetic material takes from minutes to hours. The **generation time** of bacterial cells (the time required for a single cell to undergo fission or for a population of cells to double in number) also ranges from minutes to hours. The most probable time for the insertion of fragments of transforming DNA is during DNA replication.

The phenomenon of transformation of bacterial cells thus raises the possibility that it may also be possible to transfer DNA from bacteria to plant or animal cells, perhaps even to insert specifically synthesized fragments of DNA into plants or animals to obtain improved varieties. There is evidence that higher plants, in the presence of certain bacteria, can produce an enzyme (choline sulfate permease) which enables their roots to take up a compound that is normally excluded (choline sulfate). A recent paper by the Norwegian, Per Nissen, showed that in order for the changed permeability to occur, bacteria able to produce this enzyme must be in contact with the plant (barley seedlings) and new protein must be synthesized by the plant. Nissen's work raises a number of questions. To what extent do soil bacteria influence the permeability of plant roots? To what extent are interactions of this kind important in bacterially caused plant (or animal) diseases? Can such interactions be used to enhance the yields of agricultural crops?

Conjugation. A second mechanism for changing the genetic potential of a bacterial cell involves direct contact between two cells of the same or closely related species. Following the formation of a **conjugation tube** between the two cells, DNA is transferred from the donor or "male" to the recipient or "female" cells (Fig. 5-14).

Transduction. The third mechanism for exchange of genetic information between bacteria was discovered in 1952 by Zinder and Lederberg. Norton Zinder, then a graduate student in Lederberg's laboratory, was assigned the problem of finding out whether strains of *Salmonella* exhibit the kind of exchange of genetic information by conjugation that had been demonstrated in *Escherichia coli*. When variant strains carrying appropriate genetic markers were mixed together and then plated out on media made up so that only recombinant types could grow, colonies appeared in numbers comparable to those observed in experiments with *E. coli*. It appeared that Zinder had found strains of *Salmonella* capable of exchanging genetic information by conjugation. When Zinder and Lederberg repeated these experiments, now separating the two apparently conjugating strains by a porous glass barrier that prevented the two strains of bacterial cells from coming in contact with each other, "recombination" still seemed to occur. Obviously, a mechanism other than conjugation was involved. This mystery was solved when Zinder demonstrated that one strain of *Salmonella* had been infected with a virus (bacteriophage) capable of passing through the glass filter and infecting the second strain.

Viruses can infect bacteria in the same way they attack plant or animal cells. The **bacteriophages** ("bacteria-eating" viruses) inject

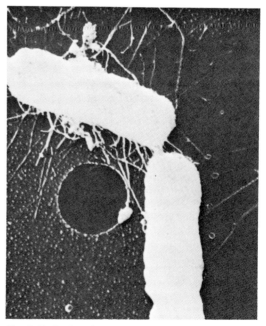

Fig. 5-14. Conjugation in *Escherichia coli*, × 20,000. Genetic material is transferred from donor to recipient cells through the conjugation tube. (From Anderson, T. F., and E. L. Wollman. 1957. *Ann. Inst. Pasteur* 93:450.)

their nucleic acid into bacterial cells and cause them to copy the DNA or RNA and protein of the virus rather than their own. Hundreds of new virus particles can be formed in a single bacterial cell. They are released when the bacterial cell ruptures (is **lysed**). Viruses behaving in this way are said to be **virulent**. Viruses do not always cause rapid destruction of the host's cells. They may remain **temperate** for generations of bacterial cells, and during this time there are no observable deleterious effects due to the presence of the virus. In this benign state, phage DNA becomes associated with the genetic material of the host, and pieces of the bacterial chromosome can be incorporated into the nucleic acid of the phage. If the phage becomes virulent, the bacterial cell lyses and the released phage particles carry with them portions of the nucleic acids of the bacterial cell. Subsequent infection of other cells introduces into the host not only phage nucleic acid but also nucleic acid from the previous host. The recipient cell thus acquires new genetic information as in transformation or conjugation. Thus, transduction, or virus-mediated recombination, which is now a well-documented phenomenon in bacterial systems, presents the possibility of comparable introduction of new genetic material into plants and animals. There is also evidence that virally caused cell transformation may be involved in tumors and cancers.

SOMATIC CELL HYBRIDIZATION IN PLANTS

Recently developed techniques for manipulating plant tissues in culture, isolating single cells, and enzymatically removing cell walls so that the resulting naked protoplasts can fuse, provide promising possibilities for developing new kinds of plants with astounding characteristics. It may now become possible to specifically engineer varieties that are resistant to cold, heat, salinity, or pathogens and thus produce plants able to survive under conditions where good harvests are not now feasible. Other possible improvements might involve grains with improved nutrient content, for example, rich in amino acids essential for human diets. The possibilities of introducing new biochemical capabilities such as a more efficient photosynthesis or association with nitrogen-fixing bacteria are now being regarded as serious goals rather than pipe dreams.

This all came about because of the convergence of several lines of investigation which made it possible to apply the techniques of microbial genetics to cells of higher plants.

Single cells can be isolated from plant tissue cultures and thereafter manipulated just as cultures of single-celled microorganisms. They can be exposed to mutagens such as X rays, cells having the desired biochemical characteristics can be isolated, and whole new plants can be regenerated from such isolated cells. The details of these techniques are discussed in the chapter on plant development (Chapter 7).

GENE THERAPY

Genetic investigators have recently opened up a highly promising, indeed revolutionary, new field exploring the possibilities of introducing desirable genes into individuals or species which lack them. For example, if an individual lacks the gene for pigmentation and consequently is an albino, or received a gene for defective hemoglobin and suffers from sickle cell anemia, it may be possible sometime in the future to introduce corrective genes. A new era in agriculture may be possible by the introduction of the bacterial genes for obtaining nitrogen from the air into crop plants like wheat, thus eliminating the need for the ever more expensive nitrogen fertilizers manufactured with the use of petroleum products.

Various agents appear to be able to carry genes into cells. Viruses apparently can do this. Another agent is an enzyme, **reverse transcriptase,** which reverses the usual flow of genetic information so that from RNA, instead of protein, the corresponding DNA is produced! It has long been known that in many of the tumor-producing viruses the genetic material is RNA. The puzzle was that once a normal cell has become infected and cancerous, its malignancy is passed on to all its descendants like other genetic traits. There are several ways this might happen. A young researcher, Howard Temin, suspected that RNA tumor viruses might possess an enzyme which worked backwards, so to speak, and produced DNA corresponding to its cancer-producing RNA. He looked for it and found it.

The possibilities for spectacular benefits to the human race from genetic engineering are enormous. There are also possibilities for unknown and unsuspected dangers. What might happen if a plant could obtain its own nitrogen? Would it become an uncontrollable weed overwhelming most other plants? Might new and terrifying diseases appear? To safeguard against these unknown and possibly even nonexistent dangers, several international conferences have been held and the

National Institutes of Health has issued a set of "Guidelines" for research in this area.

Human Genetic Diseases

Approximately 100 genetically caused human diseases have been identified including phenylketonuria, sickle cell anemia, Tay-Sachs disease, the self-mutilating Lesch-Nyhan syndrome, and royal porphyria. Some are common and some are rare, but all are debilitating to some extent and most of them include severe and permanent mental retardation. It is significant to note that a 1965 study by Reed and Reed showed that 5 out of 6 cases of mental retardation in the United States had either a retarded parent or a normal parent with a retarded sibling.

PHENYLKETONURIA (PKU)

Phenylketonuria (PKU) is caused by a recessive autosomal gene which blocks the conversion of phenylalanine into tyrosine.

This is a very simple reaction, merely the addition of an −OH group (hydroxylation) to the phenol ring (Fig. 5-15). Since tyrosine (p-hydroxyphenylalanine) is a normal precursor for adrenaline, melanin, and thyroxine, it might be thought that PKU sufferers would lack adrenaline, and be albinos with goiters. This is not the case because tyrosine is an amino acid found in a normal diet. In the disease, if the ingested phenylalanine is not hydroxylated into tyrosine, it and its derivatives, phenylpyruvic acid, etc., accumulate in the blood and cerebrospinal fluid, and appear in the urine.

Although some PKU cases are only slightly retarded, most are feebleminded or worse. Medical opinion holds that brain damage can be avoided by feeding only enough phenylalanine to provide for essential protein building. The results of this treatment are extremely hard to assess because of the variability of untreated cases. A rough diagnosis can be made with the "diaper test" because phenylpyruvic acid will turn $FeCl_3$ green. This procedure is being replaced by the more reliable Guthrie test.

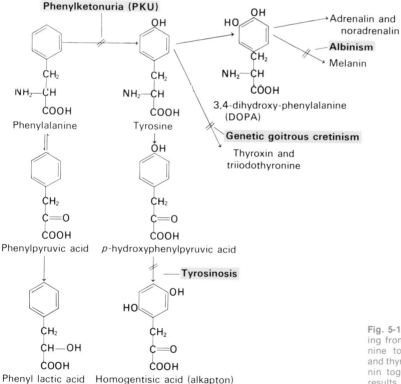

Fig. 5-15. Metabolic pathways leading from the amino acid phenylalanine to the hormones adrenaline and thyroxine and the pigment melanin together with the pathological results of enzyme deficiencies at specific points. (From Moment, G. B. 1967. *General Zoology*, 2nd ed. Houghton Mifflin Co., Boston.)

SICKLE CELL ANEMIA

Sickle cell anemia is so called because the red blood cells shrink into rough crescents when deprived of oxygen (as when placed on a microscopic slide, Fig. 5-16). In heterozygous individuals the disease may cause a more or less debilitating anemia, especially if the individual lives at high altitudes. It is characterized by occasional crises of swollen and very painful joints and spleen. The homozygous condition is invariably fatal. This gene is found among people of African descent and in certain groups in Greece and India. It appears to have been preserved by natural selection because heterozygous individuals have a degree of immunity to malaria. The sickle cell gene causes an error in one amino acid in the primary structure of the beta chain of hemoglobin. There is no known cure, but urea, an agent which tends to keep proteins in solution, is a palliative.

PORPHYRIA

Porphyria is a derangement of porphyrin metabolism in which the urine is the red color of Port wine and during acute attacks the patient suffers abdominal cramps, nausea, sweating, restlessness, and great mental confusion. The most famous case in history was George III of England, who is now thought to have suffered recurring attacks from 1765 until his death at the age of 81 in 1820. Not surprisingly this disease was rated as insanity. It may have been a factor in the mismanagement of the North American colonies which resulted in the American Revolution.

In recent years it has been found that various infections, even rather mild ones, may bring on an attack in a person with porphyria, and that various drugs such as barbiturates or sulfonimides always do. Many of the royal relatives of George III, including his remote cousin Frederick the Great and his ancestor Mary Queen of Scots, are now known to have had porphyria. One is reminded of the way the X-linked gene for hemophilia was spread around Europe among the descendants of another queen, Victoria.

LESCH-NYHAN SYNDROME

The Lesch-Nyhan syndrome is one of the most dramatic instances where a single gene, with a well understood mode of inheritance, produces a profound and tragic behavioral abnormality. This disease is X-linked, and so is inherited like hemophilia or red-green color blindness. So far it has been seen only in boys. These unfortunate children have an uncontrollable tendency for self-mutilation (Fig. 5-17). Unless their hands are kept bandaged, they chew off their fingers. Before their fingers are destroyed, if not prevented by wearing padded gloves, they will sometimes destroy their noses piece by piece. Physically these patients appear normal but their muscular coordination is very poor, often spastic; and they are mentally retarded. They are afflicted with a marked deficiency in a specific enzyme (hypoxanthine guanine phosphoribosyltransferase). The gene controlling this enzyme is located on the X chromosome. One of the consequences of the Lesch-Nyhan syndrome is excess purine synthesis. Exactly how this biochemical lesion is causally related to the behavioral abnormality and mental retardation is unknown.

DOWN'S SYNDROME

Down's syndrome or **mongolism** is now known to be associated with **trisomy** of chromosome 21. Why having three instead of the normal two of these chromosomes should produce the stolid, somewhat fat individuals with rounded face and head, stubby hands and body build, mental retardation, and (fortunately) a friendly disposition is quite unknown. The condition has nothing whatever to do with Mongolian ancestry and occurs in all human races. The chance of it occurring rises sharply in mothers over 35, although it can occur in the children born to women of any age.

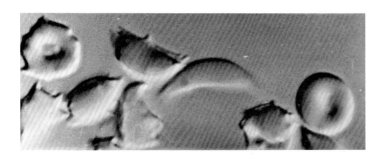

Fig. 5-16. Photomicrograph of sickled red blood cells. Normal cell is at far right (Courtesy of the National Foundation.)

TAY-SACHS DISEASE

Tay-Sachs disease is found mostly among people of Jewish ancestry, especially those

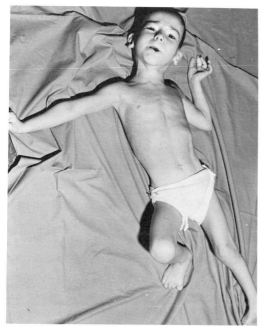

Fig. 5-17. Victim of the X-linked Lesch-Nyhan syndrome. Note the spastic posture and that part of the lower lip has been chewed away. (Courtesy of W. L. Nyhan; from Nyhan, W. L. 1967. Genetics of X-linked disorder of uric acid metabolism and cerebral function. *Pediat. Res.* 1:5.

originating in certain parts of Eastern Europe. It is a tragic disease that is always fatal after a child has developed normally for a year or so.

PREVENTION OF HEREDITARY DISEASES

All these hereditary diseases represent a great genetic load for the human race, more costly in personal tragedy and in money badly needed for other purposes than the vast majority of bacterial diseases. As such, genetic ailments are a public health problem of the first magnitude. Fortunately this situation can now be vastly alleviated by a new technique for early diagnosis of genetic defects, **amniocentesis,** followed by biochemical and cytological analysis of the cells so obtained (Fig. 5-18). The cells sloughed off by the fetus into the amniotic fluid can be grown in vitro and their chromosomes studied. In this way Down's syndrome can be detected long before birth. The metabolic defect causing Tay-Sachs disease can be detected, as can the enzyme deficiency in Lesch-Nyhan disease and others.

After amniocentesis and analysis of the fetal cells, parents can be informed as to whether they can expect a normal child. In cases like Lesch-Nyhan disease it is difficult to understand how any humane society would do anything but recommend the immediate termination of pregnancy. With some of

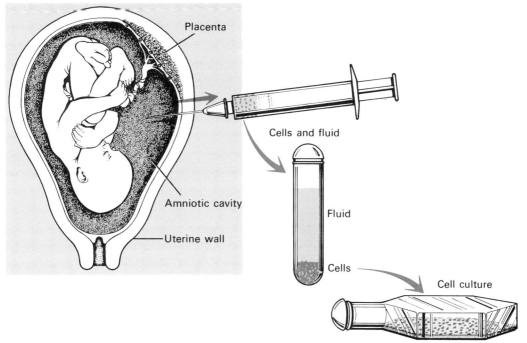

Fig. 5-18. Diagram showing technique of amniocentesis and subsequent plating out of fetal cells for growth *in vitro.*

the other diseases the situation may seen different but it is a noteworthy fact that few people who think parents should be forced to bring an abnormal child into the world are themselves willing to donate the large sums of money and the devotion required to support such tragic malformations of human life.

What of Eugenics?

Many people have been interested in **eugenics**, the science of "being well born." In fact, this concern extends far back into history. Plato and other Greeks of his age discussed it. The citizens of Sparta, that stern and bleak prototype of the dictator state, practiced a crude form of race improvement by exposing all weak or deformed infants to the weather to die.

Two basic issues are involved in eugenics. The first concerns what is biologically possible and what is impossible or extremely difficult to do. These are questions of fact and are a normal part of science. The second group of questions concerns value judgments. What traits are desirable in human beings? These questions begin in biology but extend far beyond the province of ordinary science.

What is biologically possible and what could be done without great social change depends to a large extent on whether or not a particular gene is dominant or recessive, and on whether the intent is to reduce the number of harmful genes to the lowest possible level or to increase the number of desirable ones. Consider first so-called **negative eugenics**, i.e. reduction in the human load of harmful genes. There are many which produce results so sad and so horrible that it is impossible to believe any sane person would wish them to be perpetuated. As noted 700 years ago by St. Thomas Aquinas, some of the most frightful kinds of idiocy are hereditary. Sickle cell anemia and Down's syndrome are mild ailments compared with the Lesch-Nyhan disease or dozens of others. The only real question is one of method. The ancient Spartans would have exposed afflicted infants on a hillside. The Nazis would have used gas chambers.

In our own times humane methods and potentially very effective ones are becoming available. If a deleterious gene is dominant, it can be completely eliminated, except for rare mutations, in a single generation by preventing afflicted individuals from reproducing. Minor surgery tying off the sperm ducts or oviducts makes the conception of offspring impossible. By the use of amniocentesis and appropriate testing of the cells obtained, even afflicted individuals, if heterozygous, can be assured of normal children.

The problem with completely recessive genes or dominant genes which are not expressed until after the age of childbearing is more complex. **Huntington's disease** (chorea) illustrates the latter type. It begins in middle age with involuntary twitching of face and limbs which slowly becomes more and more severe. In the end there is degeneration of the brain and complete dementia. In a freely intermarrying population, preventing the afflicted homozygous recessives from having children will do little to reduce the number of the recessive genes in the gene pool. For example, there are five albinos in every 100,000 people. To produce these five on the basis of random matings requires 1,420 heterozygous carriers. Thus, removing the 10 recessive genes present in the five albinos would reduce the number of genes for albinism in the population from 1,430 to 1,420, a negligible amount.

What are the possibilities for **positive eugenics?** The desire that children should come into the world sound in body and mind is universal. It is when increased numbers of superior children are demanded that eugenics runs into a difficulty far transcending the proper bounds of biology. Who is to judge what the really desirable traits are and in what proportions they are desirable?

Under a ruthless dictatorship, either of a single man or of a group of so-called philosopher-kings, it would be possible to produce as diverse and fantastic breeds of men as has already been done with dogs and pigeons. The methods are the same as those that have produced the various breeds of domestic animals—using mutations, crossbreeding, selection, inbreeding, and more selection. As explained in the chapter on animal development, it would be possible by the new methods of nuclear transplantation to produce hundreds and even thousands of individuals all with identical genotypes.

A world without Shakespeares, Pavlovas, Beethovens, Gandhis, Edisons, Einsteins, and opera singers would surely be a poor place. Yet who would care to face a world composed solely of such people? Or who would advocate producing a standard model human being so all men would be as alike as identical twins? There is even evidence that certain afflictions can serve as gadflies to achievements of great benefit to the human race. Homer was blind, Edison deaf, Steinmetz crippled, Byron club-footed. Clearly, our aim should be a golden mean of rich human diversity, perhaps not very different from what we now have but free of the present burden of genetic defects.

USEFUL REFERENCES

Clarke, C. A. 1970. *Human Genetics and Medicine.* Institute of Biology Studies in Biology No. 20. Edward Arnold (Publishers) Ltd., London. Distributed in the U. S. by Crane, Russak and Co. Inc, New York. (Paperback.)

Crick, F. H. C. 1966. The genetic code III. Sci. Am. 215:55–62. (Available in reprint: *Scientific American* offprint No. 1052.)

Hartman, P. E. and S. R. Suskind. 1969. *Gene Action*, 2nd ed., Prentice-Hall, Inc., Engelwood Cliffs, N. J. (Paperback.)

McKusick, V. A. 1975. *Mendelian Inheritance in Man. Catalogs of Autosomal Dominant, Autosomal Recessive and X-linked Phenotypes,* 4th ed. The Johns Hopkins University Press, Baltimore. (Frequently revised compendium listing all known human genetic diseases.)

Wade, N. 1975. Recombinant DNA: NIH sets strict rule to launch new technology. *Science* 190:1175–1179.

Watson, J. D. 1968. *The Double Helix.* Atheneum, New York. (Best selling book describing scientific intrigues involved in unravelling the structure of DNA.)

Watson, J. D. 1976. *Molecular Biology of the Gene,* 3rd ed. W. A. Benjamin, Inc., Menlo Park, Calif.

6

Genes in Action I: Development in Animals

The birth of a child is not only one of the most deeply moving of human events but it also forces on our minds our basic unity with the rest of the living world and leaves us wondering about the deep mysteries of the origin and continuity of life. What is the nature of the hereditary link between two parents and their offspring? How can a highly complex organism endowed with speech and conscious purpose possibly develop out of a minute speck—a fertilized egg? Anything like complete answers to these questions lies far in the future. Meanwhile, a significant amount of knowledge has been achieved. Within the fertilized egg is the code, written in sequences of purines and pyrimidines, which spells out the instructions for making a new human or a new starfish. The central modern problem in development is to discover how this information is translated into the adult; how the word, if you will, becomes flesh.

In the pursuit of this major problem, many satellite problems are also under investigation. Some carry important implications for mankind. The way is now open to produce hundreds or thousands of identical, i.e. monozygotic, men or women. Whether anything of the sort is desirable is a question extending far beyond the boundaries of the natural sciences. For other forms of animal life there are obviously instances where many individuals with the same genome would be desirable. The methods? It is now possible to remove a nucleus from a cell of the intestine and transplant it into an enucleated egg which can then develop into a second animal with a genetic constitution identical with the donor of the nucleus. Since, in any individual, there are thousands of gut cells which could provide nuclei and any number of eggs, very large numbers of individuals could be produced. The eggs with the donor nuclei would have to be placed in the uterus of a suitable female host, but this feat has been accomplished routinely in mice, rabbits, and other mammals.

The regeneration of new organs, which avoids the problems of organ transplants, has long been under investigation in Europe, North America, and Japan. Many of the lower vertebrates can regenerate new feet, new legs, or new lenses for the eye. Why are mammals unable to do this? Some of the new discoveries about development have been unexpected. Who would have anticipated, for example, that the development of the brain in mammals is permanently affected by male sex hormones soon after birth?

Removing the Roadblock to Modern Knowledge

The work of Fabricius at Padua on the chick and of his famous student, William Harvey, discoverer of the circulation of the blood, are often taken as the starting points of modern embryology. Harvey's work, published in the middle of the 17th century, illustrates the bewildering confusion then current. Like many others, the learned Harvey believed that mares could become pregnant not only by mating with stallions but also, according to ancient tradition, from breathing in the air on hilltops at certain seasons. He seriously discussed the comparison between the conception of an idea by the brain and the conception of an embryo in the uterus and argued for a basic similarity.

Amid all this confusion, Harvey made important advances. In a truly royal experiment on the king's deer, doubtless what we would today call "government-sponsored research," Harvey showed that the popular and age-old idea that the embryo is formed by coagulation could not be true. Harvey took some female deer very shortly after mating and divided them into two groups. One he left as controls. These gave birth to fawns at the end of the normal time, 8 months. Harvey sacrificed the does in the other group at intervals and examined their uteri. At no time, least of all soon after mating, was there any coagulating mass in the uterus. In Harvey's words, "After coition there is nothing at all to be found in the uterus, more than there was before." These experiments disposed of the time-honored theory that the embryo is formed by coagulation of blood and seminal fluid.

Gametogenesis and Gametes

The process by which gametes, whether sperms or eggs, are produced is called **gametogenesis.** It has two aspects. One is the formation of gametes, either motile sperms or yolk-laden eggs. The other concerns the reduction of the chromosomes from two sets to a single set, i.e. meiosis, in the sperm or egg.

The mechanisms underlying both meiosis and gametogenesis are very obscure. Both vitamin E (tocopherol) and vitamin A (a carotene derivative) are essential for gametogenesis, especially in males. A sufficiently high level of pituitary hormone is also essential. If the pituitary gland is removed, spermatogenesis stops in males, and in females all ova above a certain size degenerate.

SPERMATOGENESIS AND SPERMS

Spermatogenesis, the formation of sperms, takes place in the **testes.** The testes vary in number, position, and shape according to the kind of animal. In many invertebrates and in fish and salamanders among vertebrates, the testis is composed of boxlike compartments often called lobules. All the sperm cells in a given lobule develop more or

less simultaneously, so they will all be approximately in the same stage of development. Such a testis is relatively easy to study, and much of the pioneer research on spermatogenesis was done on the testes of insects for this reason. In the higher vertebrates, from frogs to men, sperms are produced in the **seminiferous tubules.** The stem cells, dividing by mitosis, lie around the periphery of such a tubule, and cells in the various stages of meiosis are crowded into the center.

The structure of mature sperm of either a vertebrate or an invertebrate varies little from species to species. These "little traveling libraries of genetic information" invariably consist of a head, a midpiece, and a tail.

The **head** of a sperm is the condensed nucleus of the spermatid. The tip of the head is covered by a caplike membrane called the **acrosome,** which plays an important role in the penetration of the egg.

The **midpiece** of the sperm is notable chiefly because it contains mitochondria, sometimes several wound in a spiral around a central axial filament. The midpiece may be a mere dot or longer than the head, depending on the species. Where the head and midpiece join, there is a pair of **centrioles** similar to those found at the base of any flagellum or cilium. Evidently, the function of the midpiece is to furnish the energy for the action of the flagellum. In many species of animals the midpiece enters the egg cytoplasm along with the head and contributes one or both of its centrioles which form the astral rays of the first cell division of the egg.

The **tail** has the same basic flagellar structure in all animals, from sponges to mammals. There is a pair of central **fibrils** surrounded by a circle of nine fibrils, the whole surrounded by a membrane. In some kinds of sperm a beautiful undulating membrane extends the length of the tail. This can readily be seen with a compound microscope in the living sperms of salamanders.

OOGENESIS AND EGGS

Oogenesis, the development of ova, takes place in **ovaries** which, like testes, vary in number, shape, and position according to the kind of animal. In frogs, the two ovaries are hollow sacs each made of a thin double membrane; in higher vertebrates the ovary is solid. Within it are blood vessels, nerves, and developing eggs. Each egg is surrounded by **follicle cells,** which are important in supplying the egg cytoplasm with yolk. Surrounding the egg cell, and separating it from the follicle cells, is a yolk (or **vitelline**) **membrane.** In

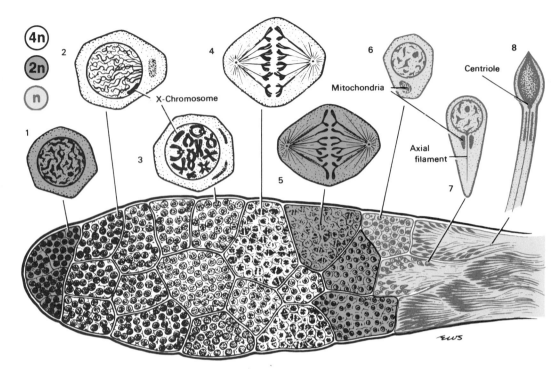

Fig. 6-1. Longisection through the easily studied testis of a grasshopper. Note how all the cells in a compartment are in the same stage. 1, spermatogonium; 2, early primary spermatocyte; 3, primary spermatocyte with tetrads (synapsis); 4, metaphase of first meiotic division; 5, metaphase of second meiotic division (separation of kinetochores and production of haploid cells; 6, spermatid; 7 and 8, developing sperms. (From Moment, G. B., and H. M. Habermann. 1973. *Biology: A Full Spectrum*. The Williams & Wilkins Co., Baltimore.)

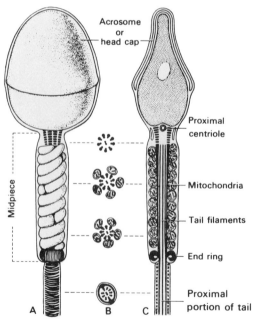

Fig. 6-2. A typical mammalian sperm as seen in the electron microscope. A, side view; B, cross sections; C, edge view in section. (After D. W. Fawcett; from Moment, G. B., and H. M. Habermann. 1973. *Biology: A Full Spectrum*. The Williams & Wilkins Co., Baltimore.)

mammalian eggs this is a thick conspicuous structure. As a developing frog's egg increases in size, the follicle cells stretch out flat and thin. In mammals the follicle cells remain with the egg at ovulation and are not dispersed until fertilization. The smallest, least developed egg cells are known as **oogonia.** The larger ova undergoing meiosis are called **oocytes.** A notable feature of the mammalian ovary is that all the oocytes that will every be present are present at birth. In the human ovary the number is estimated to lie between 200,000 and 400,000.

During their gradual growth in the ovary, eggs acquire a polarity. The yolk accumulates more densely around one pole of the egg, which is known as the lower or **vegetal pole,** because the old naturalists believed it was primarily vegetative and plantlike in function. The opposite pole of the egg contains the nucleus and much more cytoplasm. This pole floats uppermost and is called the **animal pole.** The follicle cells not only provide yolk but also nucleoproteins, which are absorbed by the ovum and are ready to be built into new chromosomes during the rapid series of mitoses in the cleavage stages which follow fertilization.

When the primary oocyte with its giant nucleus, i.e. the **germinal vesicle**, undergoes the first meiotic division, the division of the cytoplasm is so unequal that the egg seems to be extruding a tiny sphere. This minute cell is called the **first polar body**. It is, of course, a secondary oocyte, the egg itself being the other secondary oocyte. The first polar body seldom, if ever, divides. The egg undergoes a second meiotic division, extruding a **second polar body** containing a single set of chromosomes. The egg, now an **ootid**, also contains a single set of chromosomes.

Types of Eggs

Eggs differ in two important respects. The most far-reaching distinction is between the **spirally cleaving, mosaic eggs** characteristic of flatworms, annelids, mollusks, and many other groups of animals and the **radially cleaving, regulative eggs** found chiefly in echinoderms and vertebrates. This distinction will be discussed more completely in the section on cleavage of eggs.

Eggs also differ in the amount and disposition of yolk, a fact which makes a big difference in how development proceeds. Those in which the yolk is fairly uniformly distributed throughout the cytoplasm, as in the human or starfish egg, are known as **isolecithal** (*isos,* equal, + *lecithos,* yolk). These eggs are all small. Egg cells in which the yolk is concentrated toward one pole (or end), as in frogs and birds, are called **telolecithal** (*telos,* end, + *lecithos,* yolk). Frogs and salamanders have moderately telolecithal eggs; reptiles, birds, and some very primitive egg-laying mammals of Australia, such as the platypus, have extremely telolecithal eggs. In these cases the actual ovum or egg cell is commonly called the yolk of the egg. Only this yolk is formed in the ovary of the bird or reptile. It is a single cell and is the true ovum. The white of the egg and the shell are se-

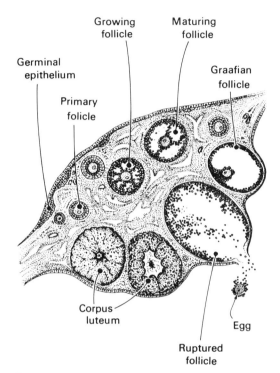

Fig. 6-3. Mammalian ovary in section showing enlarging Graafian follicles and ovulation. (From Moment, G. B., and H. M. Habermann. 1973. *Biology: A Full Spectrum.* The Williams & Wilkins Co., Baltimore.)

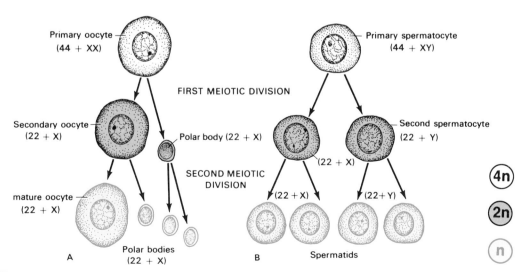

Fig. 6-4. Diagram showing meiosis during egg formation, A, and sperm formation, B. (Modified from Langman, J. 1975. *Medical Embryology,* 3rd ed. The Williams & Wilkins Co., Baltimore.)

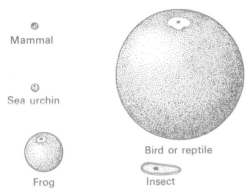

Mammal

Sea urchin

Bird or reptile

Frog

Insect

Fig. 6-5. Types of eggs, according to the amount and distribution of yolk. Mammalian and sea urchin eggs are isolecithal, frog and bird telolecithal, and the insect egg centrolecithal. (From Moment, G. B., and H. M. Habermann. 1973. *Biology: A Full Spectrum*. The Williams & Wilkins Co., Baltimore.)

creted by the oviduct around the true ovum as it passes to the exterior. This enormous ovum is inert except for a small disk of living cytoplasm containing the nucleus and lying on the top of the yolk. The third type of egg is the **centrolecithal** found in insects and other arthropods. It may be round, oval, or even sausage-shaped, but the living cytoplasm always forms a thin layer over a centrally placed yolk.

Ovulation

Ovulation, the process of extrusion of the egg from the ovary, is, in vertebrates at least, under the control of the pituitary gland which is located on the underside of the brain. Vertebrates in general can be induced to ovulate by injecting anterior pituitary glands or their extracts. This method is used to cure certain types of sterility in women but has to be administered with great precision; too little extract has no effect, and too much produces multiple ovulations and therefore multiple births of as many as 10 infants which have almost no chance of survival.

In many animals, especially birds but also various mammals such as deer, ovulation is also controlled by the number of hours of light per day to which the animal has been exposed. In most rodents and in the primates, including man, ovulation follows an internal rhythm, with a 4- or 5-day cycle in the mouse, a 28-day cycle in the human, and a 32-day cycle in the baboon. In a very few mammals, such as cats and rabbits, ovulation occurs a short and definite time after the stimulus of mating. Rabbits, in fact, can be made to ovu-

late by stamping near their cages. In all these cases, the pituitary is believed to be involved, stimulated in some way by light, mating, or internal rhythms.

Fertilization

The location of the egg when it is fertilized and becomes a **zygote** depends largely on the environment in which the adults live. Most, though by no means all, aquatic animals have external fertilization in which both eggs and sperms are simply poured out into the water to meet there. Many species of sharks and other live-bearing fish are, of course, exceptions. Terrestrial animals such as insects and mammals are forced to have internal fertilization by the simple fact that neither their sperms nor their eggs can live or move on dry land. In reptiles and birds fertilization takes place at or near the upper end of the **oviduct,** before the albumen and shell are secreted around the yolk, the actual ovum. In mammals, likewise, the sperms are carried to the upper ends of the oviducts, or **Fallopian tubes** as they are known in human beings, by muscular contraction of the ducts.

RESULTS OF FERTILIZATION

There are three major results of fertilization. First, fertilization stimulates the egg to begin its development. The second major result of fertilization is the restoration of diploidy, since the sperm brings its set of chromosomes into the egg. Diploidy in itself seems to be a source of vigor, but the main advantage of introducing a second set of chromosomes is an evolutionary one, the production of variation, of new kinds of individuals which result from new combinations of the genes. This is one of the main driving forces of evolution, and what sex is really all about. Third, in mammals and many other animals but not all, the sex of the new individual is determined by whether the sperm carries an X or a Y chromosome.

In addition to the three primary results of fertilization there are a number of lesser ones having to do with the activation of the egg. The most obvious is the almost immediate appearance of a **fertilization membrane** within which the egg is free to rotate. This membrane prevents other sperms from entering the egg. The fertilization membrane also serves to hold together and protect the **blastomeres** as the first cells into which the egg cleaves are called. In the case of the human

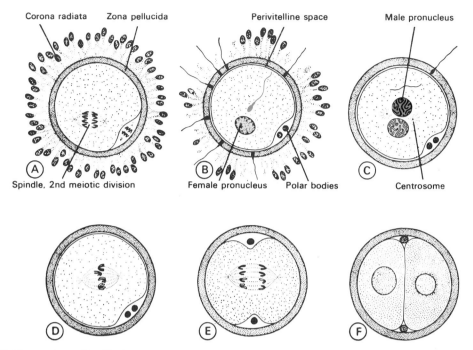

Fig. 6-6. Microanatomy of fertilization, completion of meiosis, and first cleavage in a mammal. (From Langman, J. 1975. *Medical Embryology,* 3rd ed. The Williams & Wilkins Co., Baltimore.)

egg, where the first polar body is given off before and the second after fertilization, the first polar body lies outside and the second inside the fertilization membrane. The fertilization membrane appears to be formed directly from the vitelline membrane, which tightly surrounds the egg during its ovarian development, by the absorption of water and its collection immediately under the membrane. This lifts the membrane away from the cytoplasm.

ARTIFICIAL FERTILIZATION

This term refers to the introduction of sperms into the female reproductive tract by mechanical means. It has been used successfully in cases of human sterility. It is a common practice in certain branches of animal husbandry. Pedigreed lambs have been produced in Idaho from sires living on the U.S. Department of Agriculture farms in Beltsville, Maryland. Cows are frequently bred by this method because it saves the expense, trouble, and danger of keeping a bull, and allows wide use of sires that are known to transmit desirable milking qualities to their daughters.

The longevity of functional sperms has been much studied, especially in connection with artificial insemination. If chilled and kept in thermos jars, mammalian sperms can be kept about a week. If kept anaerobic at ap-

proximately −70°C. in "deep deep-freeze," sperms seem capable of indefinite existence. This is the method of "sperm banks" where human sperms are deposited.

MICROANATOMY OF FERTILIZATION

Many of the details of fertilization were first clearly seen during the summer of 1875, which Oscar and Richard Hertwig spent on the shores of the Mediterranean Sea at Naples. There they discovered that the eggs of starfish and sea urchins are so transparent that the major events of fertilization can be seen in the living egg. In these animals the head and midpiece of the sperm enter the cytoplasm of the ovum. The midpiece gives rise to a star-shaped aster, or division center, which soon becomes double. The head of the sperm moves toward the egg nucleus and gradually becomes transformed into a nucleus itself, lying pressed against the egg nucleus.

There is no compelling evidence that animal sperms are attracted toward eggs. Swimming appears to be random. The first visible event in the actual fertilization process is the extrusion of the acrosome into a longer or shorter tubular thread or vesicle which penetrates the jelly around the egg and makes contact with the cytoplasmic surface of the ovum. The **acrosome reaction** is almost certainly in response to a glycoprotein, **fertilizin,** which diffuses from the egg. The response of the egg is the formation of a cytoplasmic fertilization cone which engulfs the head and midpiece of the sperm. In some species, including the human, the second polar body is given off and in most the egg membrane swells away from the egg.

PHYSIOLOGY AND BIOCHEMISTRY OF FERTILIZATION

Despite nearly a century of hard work in many marine laboratories since that summer when the Hertwigs discovered what favorable material for research echinoderm eggs are, we still cannot answer the question of what fertilization does that breaks the block to cleavage and subsequent development of an egg. We do not know the nature of the block. Nor do we know the similarity or dissimilarity between the stimulus of fertilization and the normal stimulus for ordinary cell division. There are at least two main possibilities. Entrance of the sperm might somehow inactivate or remove an inhibitor. Or it might trigger the activity of some enzyme or cofactor or even add a factor essential in minute amounts.

In addition to the elevation of the fertilization membrane and the disintegration of granules at the egg surface, other events are known to occur very soon after fertilization. The **permeability** of the eggs changes. The egg becomes impermeable to other sperms but much more permeable to certain ions. Potassium and calcium exchange between the egg cytoplasm and the surrounding water increases about 15-fold. Radioactive phosphorus, ^{32}P, is taken up by the egg over 100 times faster after fertilization.

In some species the rate of **respiration** greatly increases immediately after fertilization. This is true of sea urchin eggs and was at first supposed to be the key aspect in activating development. Then is was discovered that in some species, such as *Chaetopterus,* a marine annelid, and in various mollusks, the rate of respiration falls after fertilization. Still more puzzling, if sea urchin eggs are fertilized immediately after they are shed and while they are still healthy, they show no appreciable change in respiration.

There is also a marked rise in **protein synthesis**. This can be demonstrated by various methods such as measuring the rapid increase in uptake of labeled amino acids. It would be logical to suppose that immediately after fertilization the zygote nucleus begins to pour out messenger RNA into the cytoplasm. However, Ethel Harvey showed that enucleated fragments of sea urchin eggs can be stimulated to cleave parthenogenetically without any nucleus whatever. They never develop beyond a ball of cells. Moreover, mRNA can be extracted by the usual methods from unfertilized eggs. Transfer RNA and ribosomes are also present in unfertilized eggs, along with amino acids.

PARTHENOGENESIS

Many investigators have thought that **parthenogenesis,** the development of an egg without fertilization, would reveal how fertilization removes the block to development. Parthenogenesis was first extensively studied years ago by Jacques Loeb and others at the laboratory at Woods Hole on Cape Cod. They and their successors have found that a large variety of treatments will stimulate unfertilized eggs to develop. Adding weak organic acids such as butyric acid or various fat solvents such as ether, alcohol, or benzene, temperature shock by exposure to either heat or cold, osmotic changes which can be produced by sugar or urea, ultraviolet light, and the prick of a needle (especially if previously dipped in blood plasma), all will stimulate the development of unfertilized eggs. Among the most interesting of the more recent experiments are those of John Shaver, who showed that injection of small amounts of the granular, presumably mitochondrial and ribosomal, fraction obtained by centrifuging homogenated adult tissue will produce parthenogenesis, while the clear supernatant fluid will not. The trick in all these experiments is to have the concentrations or dosage and the timing precisely right.

Parthenogenesis has been artificially induced in many organisms, including sea urchins, marine annelids, frogs, and rabbits. The resulting animals become normal looking adults. Natural parthenogenesis has long been known to occur in rotifers, certain small crustaceans, aphids, and honeybees (where it gives rise to males).

So many different procedures will result in artificial parthenogenesis that one can only conclude that an egg is set up to begin cleavage and embryo formation much as a muscle cell is set up to contract. A stimulus can elicit the only response of which the cell is capable. So far, the study of various possible stimuli has not furnished the clue to the old problem of how a sperm breaks the block to development.

Cleavage of the Zygote

Cleavage is the name given to the series of cell divisions which divide the fertilized ovum, or zygote, into successively smaller and smaller cells until first a mulberry-like group of cells, the **morula** and then a hollow ball called the **blastula** is formed. The chromosomes replicate by ordinary mitosis, so that each new cell is diploid. The cell divisions follow each other rapidly; hence cleavage must be a period of rapid synthesis of nucleic acids and proteins to form new chromosomes and new protoplasm.

Anatomically, depending on the amount of yolk, cleavage differs markedly in different kinds of animals. In eggs with little yolk, cleavage is total, or **holoblastic,** and completely divides the egg successively into 2, 4, 8, 16, etc., separate cells. Isolecithal eggs, such as those of echinoderms and mammals, have this type of cleavage. So also do the moderately telolecithal eggs of frogs. In the extremely telolecithal eggs, such as those of reptiles and birds where there is only a tiny

disk of protoplasm on top of an enormous yolk, cleavage does not divide the entire ovum into separate cells. Cleavage here involves only the thin disk on top of the egg, and is termed partial or **meroblastic** cleavage (*meros,* part).

In mammals, cleavage takes place as the egg passes down the oviduct. This requires about 4½ or 5 days in all mammals regardless of size. On the 5th day, whether in mice or women, the egg is a hollow blastula, called the **blastocyst** in mammals, which enters the uterus and becomes more or less deeply implanted on the uterine lining.

There are two patterns of cleavage in the animal kingdom. In echinoderms and vertebrates cleavage is **radial** and also **indeterminate.** It is called indeterminate because there is only a very general relation between the position of any particular cell formed during cleavage and the specific tissues it will form in the embryo.

In contrast, in marine annelids and mollusks, etc., the first two cleavages result in four equal cells, but the third time the cells divide they each give off a small cell in a clockwise direction, as seen from above. The

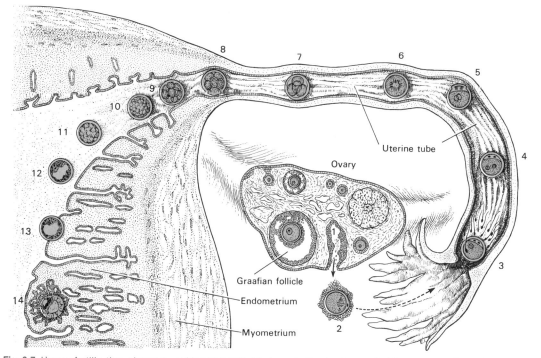

Fig. 6-7. Human fertilization, cleavage, and implantation. All stages in the development of the ovum are shown in color. 1, empty Graafian follicle from which an egg has just been released. Note the solid round mass just above and to the right, a corpus luteum. 2, the egg still surrounded by a mass of follicle cells and with the nucleus in metaphase of the second meiotic division. 3, fertilization. 4–10, cleavage stages. 11–12, blastula stages. 13–14, Implantation into the wall of the uterus. (From Moment, G. B. 1967. *General Zoology,* 2nd ed. Houghton Mifflin Co., Boston.)

large cells later give off a second quartet of cells, this time in a counterclockwise direction. At about the same time the first quartet of small cells divides in a spiral direction. In this way an elaborate and extremely precise jewel-like design of cells is produced which is exactly the same in every blastula, and in which the ultimate destiny of every cell can be precisely foretold. Consequently, such cleavage is termed **spiral** and also **determinate.**

Nucleus or Cytoplasm?
The Paradox of Differentiation

IS IT THE NUCLEUS?

As soon as it became clear that every cell, whether a brain or liver cell, carries exactly the same double set of chromosomes, a very awkward but fundamental question arose. If the chromosomes really carry the hereditary factors which determine all the physical traits of the body, why are not all the cells exactly alike? How can they become differentiated into so many very different sorts of cells? Despite great advances in knowledge there is still no satisfactory answer to this question.

August Weismann, the blind biologist and founder of the germ-plasm theory that an organism is the expression of its hereditary material (germ-plasm), proposed that the answer to this question is that the nuclei of different kinds of cells carry different hereditary determinants (we would say genes). Specifically he held that during the cleavage of the egg the hereditary factors for the different parts of the body are sorted out into different cells. At the end of cleavage certain cells contain only genes for the right arm or for skin, or red blood cells, and that is why they become a right arm or skin or blood cells.

Other investigators found that if you separate the first two blastomeres (as the cells of a cleaving egg are called) of the egg of an animal which has determinate cleavage, then two half-embryos develop. Each of the first four cells, if separated, forms a quarter-embryo. This seemed to support Weismann's differential chromosomal segregation theory.

But at about this point in the argument a speculative zoologist, Hans Driesch, showed that if you separate the first two cells of a starfish egg by shaking in calcium-free sea water, two whole embryos will form. In his flamboyant phrase, each cell is a **"harmonic equipotential system."** His experiment certainly argues against the idea that nuclei become different from each other in genetic

content during cleavage so that all the genes for the left half of the body get into the left of the first two blastomeres, and the genes for the right half into the right blastomere. Hans Driesch himself became so excited by his experiment that he left his laboratory forever and became a philosopher. He proposed that the development of an embryo is due to an "entelechy," which is something like Henri Bergson's vital force, and which "carries its purpose within itself." No mechanical or truly scientific explanation of embryonic development is possible, he claimed. After all, one cannot cut a typewriter or other machine in two and expect each half to behave as a harmonic equipotential system and regulate into two perfect small typewriters. Therefore, at the very least, these early nuclei must all be equivalents and cannot be responsible for differentiation.

IS IT THE CYTOPLASM?

If all the nuclei are equivalent, then it would seem that cytoplasmic differences must somehow be responsible for differentiation.

The cytoplasm of uncleaved eggs commonly shows its heterogenous character by differences in pigmentation and cytoplasmic constituents in different regions of the ovum. Perhaps these cytoplasmic substances are organ-forming, acting either directly or in some roundabout way by influencing genes. To test such a hypothesis, eggs can be centrifuged. If this is done in a solution of approximately the same density as the eggs, the eggs will be gradually pulled apart. The lighter components of the cytoplasm will move to the upper (centripetal) pole of the egg and the heavier to the lower (centrifugal) pole. By repeating this process with each of the separated halves of the egg, sea urchin eggs can be neatly separated into four quarters. Each quarter can be fertilized and thus tested for developmental potentialities. This experiment was carried out by Ethel B. Harvey.

The lightest, uppermost quarter contains mostly a clear fluid plus a few oil droplets, together with the egg nucleus. The second is a mitochondrial quarter which contains all, or at least all the detectable, mitochondria. The third and fourth are, respectively, a yolk-filled quarter and a pigment quarter. Upon fertilization all four quarters will develop into ciliated swimming larvae, although, of course, all but those derived from the clear first quarters are haploid. More remarkable, the clear quarters, which lack mitochondria, yolk, pigment, and whatever other substances are localized in the other three quarters, develop into approx-

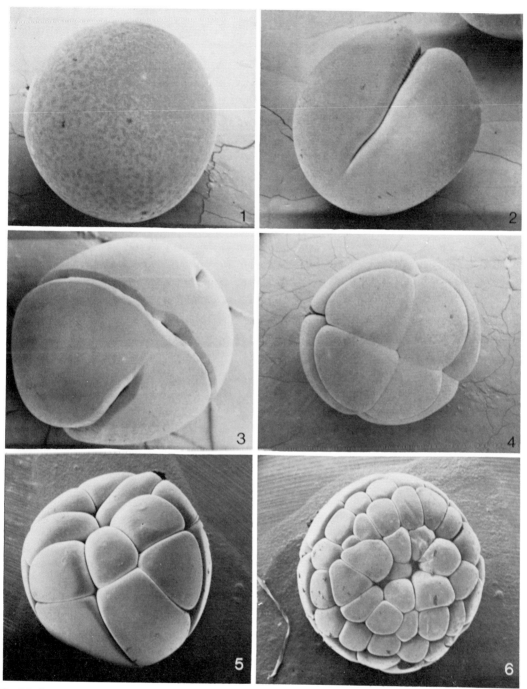

Fig. 6-8. Cleavage of a frog's egg viewed under very low power of a scanning microscope. 1, uncleaved egg; 2, side view of first cleavage furrow cutting down from upper pole of the egg toward the yolky lower pole; 3, egg viewed from the lower pole showing the completed first cleavage with the ends of the second cleavage coming in at right angles to the first; 4, the eight-cell stage showing four upper and four lower cells; 5, 16-cell stage; 6, many-cell stage. (From R. G. Kessel and C. Y. Shih. 1974. *Scanning Electron Microscope in Biology—A Student's Atlas of Biological Organization.* Springer-Verlag, New York.)

imately normal larvae with skeleton, gut, and pigment spots. The yolk quarters also form larvae with skeleton, gut, and pigment, but with a less normal body shape. Other centrifugation experiments on the eggs of this and other species of echinoderms make it certain that the visible substances in the egg of sea urchins are not a direct cause of differentiation.

Thus, we are left with the paradoxical conclusion that neither the nucleus nor the cytoplasm is responsible for initiating differentiation. Yet differentiation obviously does occur and chromosomes can be shown to control the development of the specific traits of an animal.

MODERN ANSWERS

Several avenues of escape from this dilemma of development are open. Theodor Boveri showed that in a nematode worm chromosomal material is lost during cleavage. The amount of loss depends on the cytoplasm in which the nuclei come to lie. Then E. G. Conklin, of Princeton and Woods Hole, showed in a long series of largely overlooked experiments with the eggs of marine relatives of vertebrates called ascidians that abnormalities can be produced by centrifuging the cytoplasm to abnormal positions, provided that the centrifugal force is sufficiently strong and applied at the right time. Conklin's line of work is now being pursued by T. C. Tung in The People's Republic of China but the results are not available. It may turn out after all that there are substances in the cytoplasm which derepress or activate specific genes.

Recently Briggs and King and others in the United States and abroad have been investigating the problem of whether or not the nuclei become different during cleavage or are interchangeable, a problem that Pflüger had investigated in 1884 by squeezing frog eggs between glass plates. The new method, a very difficult one, is the **transplantation of nuclei** from various parts of frog embryos and even tadpoles in different stages of development into activated but enucleated uncleaved eggs. With proper care, the host eggs with foreign nuclei will develop in as many as 80 per cent of the cases. The advantage of the new method is that it allows the testing of nuclei from much more differentiated stages than did the old Pflüger technique. The first results of the new method quickly confirmed Pflüger's conclusion that all the early cleavage nuclei are interchangeable. However, when older and older donors are used, the resulting embryos show more and more abnormalities. Nuclei taken from such malde-

veloping embryos and retransplanted into second and third generations of enucleated eggs have given clones of embryos showing similar abnormal development. The actual meaning of this is very puzzling because more than a few cases have been found where nuclei from older embryos gave entirely normal development. Furthermore, work by Gurdon at Oxford with the African toad *Xenopus* has given large numbers of normal tadpoles with nuclei from differentiated epithelial cells of the tadpole intestine. By this method large numbers of tadpoles can be obtained, all with the same genome because all were produced by nuclei from the same donor intestine. It is this technique combined with intrauterine implantation into host females that would make possible large numbers of identical mammals, including humans.

Experimentation during later stages of development has revealed much about how differentiation is induced, but before discussing these discoveries, we will consider normal embryo formation.

Building the Embryo

GASTRULATION AND THE GERM LAYERS

Gastrulation is the process that converts a ball of cells, the blastula, into an embryo. After fertilization, it is the most crucial stage in development. A gastrula stage occurs in most animals, but the following account will deal only with sea urchins and vertebrates. The details of gastrulation are very different in different kinds of animals, although the basic process is similar in all, as are the results.

After gastrulation three layers of cells, called germ layers, can be distinguished. The **ectoderm** covers the embryo externally and will form the external layers of the skin and skin derivatives like hair, hoofs, and nails, plus the entire nervous system. The **endoderm** forms the gut and will become the epithelial lining of the entire alimentary canal and also all the structures derived from the gut in the course of development, such as the lungs, liver, and pancreas. The **mesoderm** lies between ectoderm and endoderm and is itself split into two layers. The **parietal** or somatic layer presses against the ectoderm. The **visceral** layer of mesoderm presses against the endoderm of the gut. The cavity between parietal and visceral mesoderm is the **coelom.** From the mesoderm are formed muscle, skeleton, vascular system, and the connective tissues which hold the body together.

In the eggs of starfish and sea urchins, where there is very little yolk, gastrulation begins with the inpocketing of the cells around the lower or vegetal pole. It looks as though a thumb had pushed in one side of an old hollow rubber ball. The cavity of the blastula, i.e, the **blastocoel,** is slowly obliterated. The new cavity formed (after the hypothetical thumb is removed) is the primitive gut or enteron and is therefore called the **archenteron** (*archaios,* ancient, + *enteron,* gut). The opening into the archenteron is called the **blastopore.** The archenteron is the origin of the endoderm. The mesoderm is formed as a pair of hollow buds from the inner end of the archenteron.

In a frog's egg there is so much yolk that the cavity of the blastula is very small, and consequently gastrulation of the type just described is impossible. The egg "cheats," as it were. The archenteron pushes into the blastula cavity from one side. Cells move down from the animal hemisphere and roll in over the edge or **dorsal lip** of the **blastopore** to

add themselves to the wall of the archenteron. The dorsal lip is the "organizer" of the embryo.

In the telolecithal eggs of reptiles and birds, where the yolk is enormous, gastrulation takes place by the inpushing of cells along a line called the **primitive streak.** At the anterior end of the primitive streak is a pit beside a hillock of cells. This is Hensen's node and corresponds to the dorsal lip of the blastopore. At the end of gastrulation the embryo consists of three layers of cells, a layer of endoderm against the yolk, a layer of mesoderm on top of it, and a layer of ectoderm over the mesoderm. The three are commonly adherent along the **notochord,** which extends anteriorly from Hensen's node and lies in the mesoderm. It is the forerunner of the backbone.

In mammals gastrulation occurs in the same way as in birds and reptiles, clearly an inheritance from our reptilian ancestors. The blastula of placental mammals, such as man, is a hollow sphere called the **blastocyst.** It is the blastocyst which becomes implanted on or, in higher mammals, into the lining of the uterus, an event which occurs on the 5th or 6th day of development. A solid mass of cells

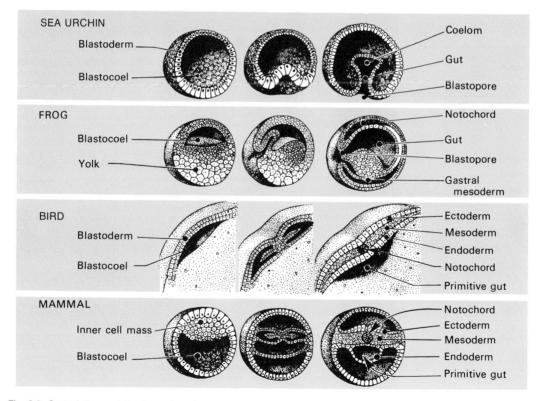

Fig. 6-9. Gastrulation and the formation of germ layers in four types of animals. Note the basic similarity of end result despite striking differences. (From Moment, G. B., and H. M. Habermann. 1973. *Biology: A Full Spectrum.* The Williams & Wilkins Co., Baltimore.)

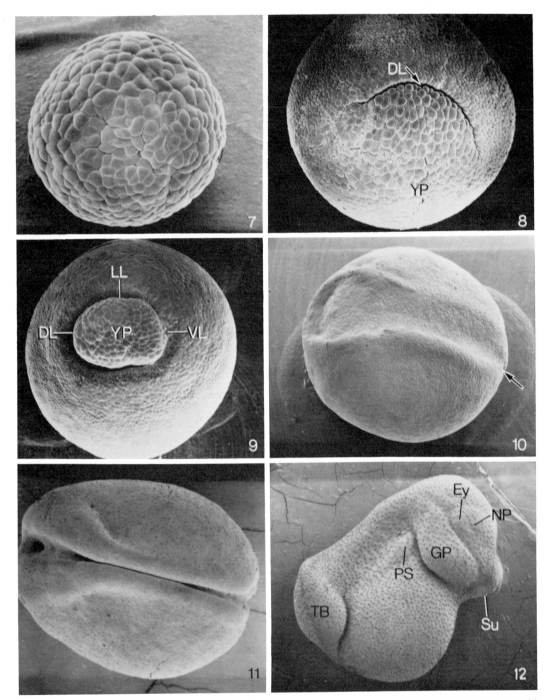

Fig. 6-10. Later development of a frog's egg viewed under a scanning electron microscope. 7, blastula stage (compare with Fig. 6-9). 8, posteroventral view showing the dorsal lip of the blastopore, DL, and the yolk-ladened ventral cells, YP. 9, late gastrula. DL, VL, and LL, dorsal, ventral, and lateral lips of the blastopore; YP, yolk plug. 10, early neural folds. The arrow indicates the last remnant of the yolk plug. 11, closure of the neural folds. The primordium of the brain is at the left. 12, early embryo. Ey, eye; NP, nasal placode which will form the olfactory epithelium; Su, "suckers," i.e. adhesive glands; GP, future gills; PS, pronephros hillock, i.e. embryonic kidney; TB, tail bud. (From R. G. Kessel and C. Y. Shih, 1974 *Scanning Electron Microscope in Biology—A Student's Atlas of Biological Organization.* Springer-Verlag, New York.)

lies at one pole of the blastocyst. Two cavities form in this mass, and in the plate of cells separating the two cavities, the primitive streak is formed and gastrulation takes place. The upper cavity becomes the **amnion,** a membrane which surrounds the embryo until birth. A third cavity, the **primitive gut,** pushes into the lower cavity, which is obliterated.

NEURULATION

Neurulation is the process by which the primitive nervous system is formed. Immediately after gastrulation a broad strip of ectoderm overlying the notochord and the mesoderm adjacent to it begins to thicken. This strip of ectoderm, the **neural plate,** forms the brain and spinal cord in a most remarkable way. The edges roll up to form a groove, and by the meeting and fusion of the edges of the groove, a tube is formed. The anterior part of this tube enlarges into the brain, and the posterior part becomes the spinal cord. In these early stages the embryo is called a **neurula.**

The motor (efferent) nerves grow out from the brain and the spinal cord and make connections with their appropriate muscles and

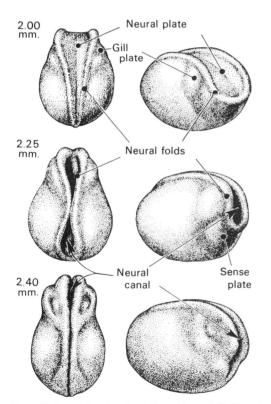

Fig. 6-11. Neurulation in a frog. (From Moment, G. B., and H. M. Habermann. 1973. *Biology: A Full Spectrum.* The Williams & Wilkins Co., Baltimore.)

glands. This remarkable fact was deduced from the study of sectioned embryos by two rather isolated investigators, Santiago Ramon y Cajal in Spain and Wilhelm His in Switzerland. However, the scientific world continued to hold to the prevailing theory that nerves are formed by the coalescence of long chains of cells until Ross G. Harrison discovered that cells can be grown outside the animal's body and was able to demonstrate the outgrowth of nerve fibers from isolated pieces of embryonic spinal cord. Harrison's famous experiment at Johns Hopkins was the beginning of animal tissue culture in vitro as a usable technique.

Along the edges of the neural folds is a narrow strip of tissue called the **neural crest.** The cells of the neural crest separate from both the surface ectoderm and the neural folds as they are closing, and migrate down over each side of the body between the ectoderm and the underlying mesoderm. The neural crest cells develop into at least four different important tissues; although this has been hard for anatomists and embryologists to believe, it has been abundantly proved by microsurgical studies in which parts of the neural crest have been removed and transplanted to other sites in the embryo. The neural crest cells form the dorsal sensory ganglia and hence the sensory (afferent) nerves to both brain and spinal cord. They become the pigment cells, both the melanocytes of a fish or a frog and the pigment-forming cells in the feathers of birds and the hair of mammals. They form the cartilage and bone of the so-called visceral skeleton which comprises the jaws and gill arch skeleton of fish and the jaws and of the throat (Fig. 19–3) bones of higher vertebrates. They form the medulla, i.e. central portion, of the suprarenal (adrenal) gland and related endocrine glands. In connection with the nerves, some of the neural crest cells become Schwann cells which line up along the nerve fibers and, by winding around them, myelinate them, i.e. cover them with a special kind of protective sheath which insulates one fiber from another and increases the speed of the nerve impulses.

GILL SLITS

In the most anterior part of the gut in all vertebrates, **gill slits** develop. These are paired openings on the side of the throat from the exterior into the pharnyx or throat itself. At the most anterior and ventral part of the gut, the mouth breaks through. This stage, in which pharyngeal gill slits appear, is called the **pharyngula.** As pharyngulas, vertebrates

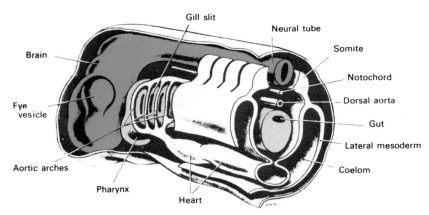

Fig. 6-12. Diagram of basic structure of a vertebrate embryo in the gill slit or pharyngeal cleft stage. This corresponds to a chick of 72-hr incubation or a human embryo at the end of the 1st month. (From Moment, G. B., and H. M. Habermann. 1973. *Biology: A Full Spectrum.* The Williams & Wilkins Co., Baltimore.)

of widely separate groups—birds, mammals, fish—resemble each other more closely than in any other stage, for the eggs differ as markedly as do the adults. In this stage a simple diagram will do for them all. The heart is located forward, not far behind the mouth. Blood is pumped forward and then up to the dorsal side of the gut, through six pairs of aortic arches. In fish and in larval frogs and salamanders, the aortic arches send blood vessels out into gills, where the blood vessels become thin-walled capillaries through the walls of which oxygen and carbon dioxide are exchanged with the surrounding water. The capillary circulation in the gills makes a handsome sight when brilliantly illuminated under a binocular dissecting microscope. In mammals all the gill slits grow closed except the first pair, which form the Eustachian tubes of mammalian ears. The hands and feet grow out as paddle-shaped structures which then develop fingers or toes.

The relationship between a frog embryo and a chick in the neurula or pharyngula stage can be visualized by cutting a chick embryo off from the underlying yolk and pulling the edges together ventrally. Since the days of Fabricius and Harvey, the chick embryo has been a favorite object of study because of its ready availability and ease of handling. A 72-hour chick embryo corresponds to a human embryo at about the end of the 1st month of intrauterine life. Both are in the gill slit stage. It is in the early stages during the first 3 months that human embryos are most susceptible to deleterious influences, such as the virus of rubella (German measles) or the products of thalidomide. However, there is no stage in which the unborn cannot be damaged.

Embryonic Induction—Turning on Genes

Investigators seeking the forces that produce gastrulation have made little progress, but in the effort important facts have been learned about the causes of differentiation, which visibly begins during gastrulation. The pioneer work was carried out on amphibian embryos by Warren Lewis in Baltimore and Hans Spemann in Freiburg. Briefly, they found that the dorsal lip of the blastopore is in some unknown way the **"organizer,"** or better, the **evocator**, which calls forth the development of an embryo. If a frog blastula is constricted with a fine hair into two separate halves, two embryos, identical twins form, but only if the constriction cuts through the part of the blastula that will form the dorsal lip of the blastopore. If the blastula is so constricted that one-half gets all the dorsal lip material, that half, and that half only, will form an embryo. The other half remains a mere ball of cells.

More dramatically, if the dorsal lip is cut out and implanted in another embryo by injecting it into the cavity of the blastula with a fine pipette, the transplanted dorsal lip will be pushed down against the ventral belly wall during gastrulation and will there induce or evocate a secondary embryo in the host tissue (Fig. 6-15). Specifically, the cells which lie over the notochord and its adjacent mesoderm become induced to form neural plate. These inducing tissues are found as the dorsal lip of the blastopore in the early stages of gastrulation. In birds and mammals, where gastrulation occurs along the primitive streak which thus corresponds to the blastopore, no one was surprised when it was discovered that Hensen's node is the essential evocator or organizer of bird and mammalian embryos.

The dorsal lip is the primary evocator without which no embryo forms. There are also **secondary evocators.** For example, the eye is formed as an outgrowth from the brain. Where this outgrowing **optic vesicle**, as it is called, comes into contact with the skin of the embryonic face, a lens forms in the embryonic facial ectoderm. If the optic vesicle is removed before it touches the skin, in most species of animals, no lens forms. The optic vesicle is therefore a secondary evocator.

How does the evocator work? Does it have to be living? The answer to the second question is no. Subsequent work has shown that a wide variety of substances, when soaked up in agar jelly and imbedded in a blastula, will evocate a secondary embryo. These include: extracts of adult brain or other tissues; various pure chemicals (especially phenan-

threnes, which are chemicals resembling both the sex hormones and adrenal cortical hormones); and a number of cancer-inducing agents. Hypotonic salt solution will also cause ectoderm explanted into a glass dish to form neural tissue. The situation is comparable to that in fertilization and artificial parthenogenesis. So many agents will produce the effect that all one can say is that, as with muscular contraction, many agents can trigger the reaction.

Experiments suggest that evocators are diffusible chemicals. Clifford Grobstein separated the inducing tissue from the inducible tissue by cellulose ester membranes of different thickness and porosity. Even if the pores in such membranes are small enough to prevent any cytoplasmic contact (about 0.1 μm in diameter), induction can still take place. For example, kidney tubules are induced in mesoderm by substances diffusing from a region of adjacent neural tube. Twitty and Niu have shown that tissues grown in vitro exude

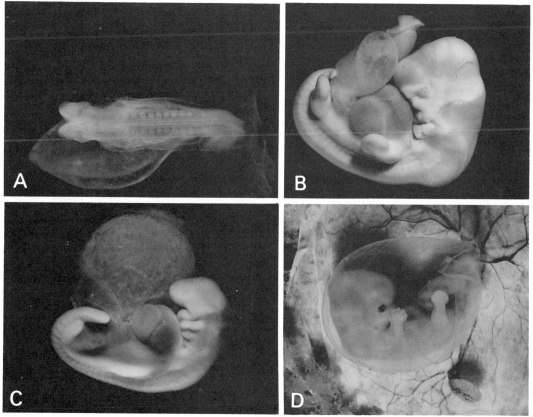

Fig. 6-13. Human embryos at approximately 3, 4, 5, and 6 weeks. At 3 weeks note the neural folds at the left and the somites on either side of the midline. Gill slits (pharyngeal clefts) are visible at 4 and 5 weeks and the rudiments of fingers at 6. An amnion surrounds the 6-week embryo. (Courtesy of the Carnegie Institution of Washington, Department of Embryology, Davis Division.)

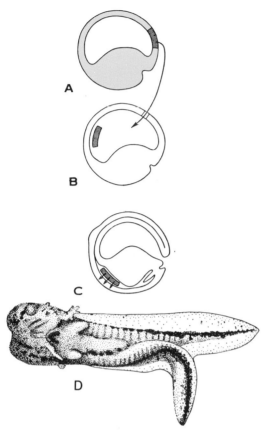

Fig. 6-14. Inductive action of dorsal lip of the blastopore using a salamander gastrula. The dorsal lip is placed inside the blastocoel of the host and is pushed against the lower wall of the egg during gastrulation. (From Kühn, A. (Translated by R. Milkman) 1971. *Lectures on Developmental Physiology*, 2nd ed. Springer-Verlag, New York.)

inducing molecules into the medium. The work of Yamada in Japan strongly suggests that these inducer substances are proteins; other experiments suggest that they are mRNA.

POSSIBLE BIOCHEMICAL MECHANISMS

It will be recalled that the functional genetic unit consists of an operator gene and one or more structural genes which the operator controls. The structural genes are the templates for the formation of messenger RNA. Also, the operator gene itself is under the control of a regulator gene which keeps the operator gene repressed by a repressor substance, probably a protein. According to this Jacob and Monod model of gene action and its control, genes become activated, i.e. derepressed, by specific derepressor, i.e. inducer, substances. These may be proteins or other things that combine with or otherwise counteract the repressor substance. If embryonic inducers act in this way, they are acting essentially at the gene level.

It is also possible that embryonic inducers act at the level of proteins by what is called **end product feedback inhibition.** There are a number of well-worked-out cases in microorganisms where the entire biosynthetic pathway is known from substance A to B to C to D to E, each step controlled by its own enzyme.

In such cases it has been found that if the end product E is added to the medium in which the organisms are growing, the cells no longer form E. Further investigation has revealed that it is not the final step from D to E that is blocked by the presence of excess E but the enzyme governing the first step, A to B, that is blocked. Here clearly is another mechanism that could function during induction of embryonic differentiation. Does it?

Meryl Rose, an investigator with wide experience with animals from different phyla, has proposed that as each tissue is differentiated it exudes some kind of substance which inhibits the differentiation of additional cells of the same kind—a "don't-make-any-more-of-me" substance. Such a substance would explain the limitation of a given tissue, although it does not explain how the genes controlling the next tissue to differentiate become derepressed (activated).

A number of striking cases are known where a continuously acting influence of one structure on a second is essential to maintain a differentiated condition. If the retina is removed from the eye of a salamander, the black pigmented cells just back of the retina lose their pigment and become embryonic. The stages in this process can be readily seen if salamanders are sacrificed at intervals after removal of the retina. The formerly pigmented cells undergo mitosis and form a new layer of cells lining the big vitreous chamber of the eye. The cells on the inner surface of this layer differentiate into a new retina, and those on the outer surface form a new pigmented layer. This demonstrates that the state of differentiation of at least the pigmented cells requires the continuous presence of some influence from the retina. It also shows that differentiation is not an irreversible process.

SEQUENTIAL GENE ACTION

Recent discoveries about the sequence in which different specific proteins appear during development have brought some of the problems just discussed into a sharper focus.

Each protein, or each polypeptide part of a protein composed of more than one chain, is the expression of a single gene. Therefore, since the proteins found in later stages of development are different from those found earlier, the genes responsible must be different. Among the best understood cases are hemoglobin, the visual pigments, and liver enzymes of a metamorphosing frog tadpole.

Fetal and Adult Hemoglobin

It has been known for nearly a century that fetal hemoglobin is different from the hemoglobin of an adult. Adult hemoglobin is easily denatured by acid or alkalis but fetal hemoglobin is not. Modern analyses show that the normal adult hemoglobin molecule is composed of four polypeptide chains, two of alpha hemoglobin and two of beta. The normal fetal hemoglobin molecule is composed of two alpha chains and two gamma chains. During human development the total production of hemoglobin continues at a steady rate, but the synthesis of the gamma chains falls while that of the beta chains increases.

What is the adaptive meaning of this change? Fetal hemoglobin has a greater affinity for oxygen than does adult hemoglobin. Consequently, it is able to acquire oxygen from maternal hemoglobin on the other side of the placental barrier. This is true of the monkey, sheep, cow, and mouse, as well as man.

Eventually the structural gene for gamma hemoglobin becomes completely repressed. The interesting question is: how does the gene for the gamma polypeptide get repressed while the gene for the beta chain becomes derepressed? There are at least three possibilities. The lower oxygen concentration in the fetus may be responsible. Severe anemias due to various diseases and to bleeding, all of which would lead to oxygen deprivation, also lead to the appearance of fetal hemoglobin even in fully mature adults.

It may be that hormones play an important role. It is known that some hormones act at the gene level. Furthermore, women with certain types of ovarian tumors produce small amounts of fetal hemoglobin. Another suggestion is that the precursor cells for erythrocytes begin with some cytoplasmic repressor for the beta gene, which becomes diluted or disappears as the precursor cells proliferate. At that point the structural gene for beta hemoglobin begins producing its mRNA. Such mRNA not only furnishes the instructions for the synthesis of beta chains but also may act as a repressor for the gamma gene.

Biochemical Changes during Frog Development

Changes in gene activity are also known to occur during the life history of the frog. The visual pigment in a tadpole is porphyropsin (as in fish). In the adult it becomes rhodopsin as in terrestrial vertebrates. In the tadpole, waste nitrogen simply diffuses out through the gills as ammonia, while in the frog it is converted into urea by enzymes in the liver. Clearly genes are active in the adult frog which were repressed in the tadpole. Again it is easy to believe that hormones are responsible. The "turning on" of specific genes by the activity of known hormones can be seen in certain insects; we will now turn to this topic.

VISIBLE GENE ACTIVATION

It has been known for some time that in the nuclei of salivary gland cells and of cells in various other tissues of the larvae of flies and gnats are **giant-sized** and **banded chromosomes**. Moreover, the bands correspond to specific gene loci. Several years ago W. Beerman discovered that various bands show puffs or swellings, and that specific bands become puffed in a definite sequence during the development of the larva. In a cross between two species of gnats *(Chironomus)*, one species forms special granules in certain cells of the salivary gland near its duct; the other does not. Conventional genetic experiments showed that the ability to secrete these granules is inherited as a simple Mendelian

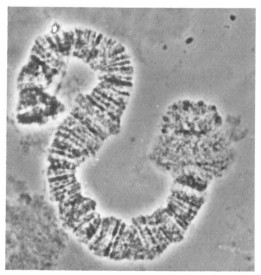

Fig. 6-15. Chromosome in action. Bands and a puff in a salivary chromosome of a gnat, *Rhynchosciara*. "Squash" preparation, × 765. (Courtesy of Mary T. Handel and the Oak Ridge National Laboratory.)

dominant and that the gene responsible is located adjacent to the kinetochore of one of the chromosomes. Cytological studies showed that in the cells which had these granules — and only in those cells — there is a conspicuous puff adjacent to the kinetochore. Most significantly, Beerman then showed that in heterozygous individuals only the chromosome carrying the dominant allele shows a puff in that locus. What more convincing evidence could there be that puffing represents gene activity?

More convincing evidence has been presented by Ulrich Clever. Also using *Chironomus,* Clever showed that injections of the insect steroid hormone ecdysone into the larva will result in pupation much sooner than would otherwise occur. Normal pupation in this gnat is preceded by some very characteristic puffs in specific chromosomes. The same puffs appear within 2 hours after ecdysone injection, and a long time before any other signs of impending metamorphosis can be detected. Histochemical stains and radioactive labeling reveal that the puffs are sites of messenger RNA synthesis. Moreover, actinomycin D, which is known to inhibit the synthesis of mRNA, inhibits the formation of these puffs. Hence the puffs apparently correspond to the loops of the lampbrush chromosomes found in the germinal vesicles of oocytes. Precisely what this newly formed material does has not been established, but it is clear that steroid hormones can act directly at the gene level.

In higher organisms it may be that whole "batteries of genes" are derepressed together and that gene derepression, which is another name for differentiation, may be a much more complex process in higher organisms than in bacteria on which the current Jacob-Monod model of gene activation is based. Alkaline proteins known as histones, which adhere to chromosomes, may play a key role in the repression and derepression of genes.

Sex Differentiation

Both male and female vertebrate embryos begin development with the same set of reproductive primordia. In mammals a pair of **gonads** is located on either side of the midline near the kidneys, and two pairs of ducts, the **mesonephric** or **Wolffian ducts** and the **Mullerian ducts,** extend from near the kidneys posteriorly to connect with the exterior via the common urogenital opening. In addi-

tion there are the external genitalia. These consist of a median **genital tubercle** immediately anterior to the urogenital opening, on either side of which is a **genital fold.** If the zygote from which an embryo develops contains one X and one Y chromosome, i.e. if it is a male, the gonads will develop into testes and descend from the abdomen into a scrotum, the genital tubercle will grow into a penis, each Wolffian duct will develop into a vas deferens which will conduct sperms to the penis, the Mullerian ducts will degenerate and virtually disappear, and the genital folds will form the scrotum.

If the egg carries two X chromosomes and develops into a female, the gonads become ovaries and remain abdominal although they move much farther down toward the pelvis than they were originally placed, the Wolffian ducts degenerate and virtually disappear, the Mullerian ducts develop into the oviducts (Fallopian tubes of human anatomy), the uterus, and vagina. The genital tubercle becomes the clitoris and the genital folds become the labia majora.

The differentiation of sex is a very complex affair with considerable medical importance and much variation in mechanisms from one kind of animal to another. Sex determination as distinguished from the process of sex differentiation during development has been discussed in Chapter 4.

In some vertebrates and invertebrates, notably insects and crustaceans, many sexual differences are controlled by local gene action within the tissues rather than by a circulating hormone. The classical and best studied case is **gynandromorphism** in *Drosophila,* where one-half of the body is male and the other half is female. Cytological study of the chromosomes of such animals shows that the female side has the normal XX composition but the male side has lost one of the X's and is XO, which is the same thing as an XY in this fly. In the gynandromorph *Drosophila* shown in Figure 6-17, the remaining X on the male side carried recessive genes for white eyes and for miniature wings. The normal female side must have had an X carrying the normal dominant alleles for these characteristics.

In placental mammals present evidence indicates that **Wiesner's one-hormone theory,** first proposed in 1934, is correct. According to this theory female structures develop autonomously, and male structures develop under the influence of male sex hormone. If male genital primordia from a mouse or rat fetus are cultured in vitro with embryonic testes or with testosterone (male sex hormone), the Wolffian ducts continue to grow

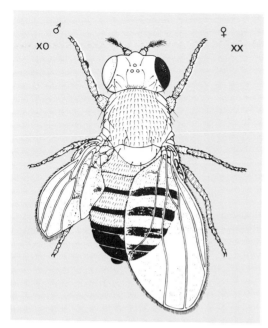

Fig. 6-16. Gynandromorph in *Drosophila*. The X chromosome on the male carried the recessive genes for white eyes and miniature wings while the other X carried their normal dominant alleles. Note the male sex comb on the left front leg and the male type abdomen on that side only. (From Moment, G. B., and H. M. Habermann. 1973. *Biology: A Full Spectrum*. The Williams & Wilkins Co., Baltimore.)

confirmed existence of true gynandromorphs in certain strains of mice where the sex ducts and other structures are male on one side of the body and female on the other presents some interesting problems. It is perhaps possible that mammalian gynandromorphs can be explained by the loss of a chromosome as in insects, plus the operation of Wiesner's theory.

One recently discovered and important fact is that the **hypothalamus** (the portion of the brainstem immediately above the pituitary gland) behaves differently in adult males and females and that this difference is imprinted into the hyopthalamus of the newborn mammal by male sex hormones very soon after birth. A single injection of as little as 1.25 mg of testosterone into a female rat 2 or 3 days old produces permanent sterility, apparently by producing a continuum of hormone production instead of the cyclical one characteristic of the female. Since it is the hypothalamus which signals the pituitary to release gonad-stimulating hormones at the appropriate intervals, the investigators, Barraclough and Gorski at U.C.L.A., tested the idea that the lack of rhythms of activity in the hypothalamus was responsible for lack of ovulation by stimulating the hypothalamus electrically. Such stimulation produced ovulation.

Membranes

The membranes which surround an egg, it will be recalled, are of two different origins. The vitelline membrane is secreted by the follicle cells and probably also by the cytoplasm of the egg. It is formed while the egg is still in the ovary and becomes transformed into the fertilization membrane immediately after fertilization. Various other membranes are secreted around the egg by the oviduct during its passage to the exterior of the body. In a frog these are proteinous coats which swell to form a jelly as soon as the eggs reach the water. In the chick the first membranes secreted by the oviduct are several layers of protein, commonly called the white of the egg. The layer next to the yolk itself is drawn out into two whitish strings called **chalazas** which help keep the yolk properly oriented within the shell. Around the albumenous layers, the lower portion of the oviduct secretes two very thin but somewhat leathery shell membranes, and finally, a calcareous shell.

With reptiles, birds, and mammals, the embryo itself grows four membranes of living tissue which surround and protect the em-

while the Mullerian ducts degenerate. If the testes or testosterone is omitted, the Wolffian ducts degenerate. If female genital primordia are explanted in vitro, the female ducts continue to develop with or without ovaries. In other experiments normal and castrated female fetuses and castrated males all developed essentially alike, furnishing further support for Wiesner's nonhormonal theory of female sexual differentiation.

Finally, the fact that female sex hormone passes through the placenta and reaches the developing fetus from its mother virtually necessitates a system whereby the male develops under the influence of its own hormones while the female develops without hormones. If female sex hormone were necessary to ensure that a genetic female developed in that direction, then, since female sex hormone is always present in placental transmission from the mother, only females would be possible.

The development of the secondary sexual characteristics that appear at sexual maturity, such as differences in hair, voice, etc., are clearly under hormonal control in both sexes of mammals. Whether various other sex differences, such as general bodily conformation, are or not is an open question. The well-

bryo as well as mediate its nutrition, respiration, and excretion. These four embryonic membranes make it possible for reptiles and birds to lay their eggs on dry land, and for mammals to be effectively viviparous, that is, to bear living young.

The membrane closest to the embryo is the **amnion,** which completely encloses the embryo within a fluid-filled space—its own private pond in which it passes through the gill slit stage, although the gill slits are never used for respiration. Occasionally a baby is born with a bit of the amnion on its head. Termed a "caul," it is counted good luck in folklore. It will be noted later that it is the amnion which gives the name **amniotes** to the three groups of vertebrates possessing it.

The second membrane, the **yolk sac**, grows out from the belly side of the embryo and encloses the yolk from which it absorbs food. A yolk sac membrane forms in the higher mammals also, even though there is no yolk to enclose.

The third membrane is called the **allantois** (*allas,* sausage), because in many animals it becomes an elongate cylindrical structure. The allantois grows out from the hind gut, so that the cavity of the hind gut is continuous with the cavity of the allantois. In reptiles and birds it is highly vascular and lies under the shell, serving as a lung. The allantois stores fetal urine in mammals with poorly developed placentas, and plays a major part in the formation of the placenta in other mammals.

The **chorion** is the most external of the four membranes. In reptiles and birds it presses against the shell membrane. In mammals it develops directly from the blastocyst wall and presses against the lining of the uterus. Where the allantois fuses with the overlying chorion, the blood vessels sprout out into rootlike extensions which fit tightly into the lining of the uterus. The region where these chorionic extensions grow out forms the **placenta,** which is the only place where actual exchanges of materials between mother and fetus occur.

In carnivores, like cats and dogs, the chorion is an elongate sac. The chorionic villi develop a placenta that encircles the sausage-shaped chorion like a cigar band. In deer and cows, many small rounded placentas are scattered over the chorion. In primates there is a single disk-shaped placenta; the lining of the uterus proliferates, forming part of the structure.

It is important to note that even in cases, as in man, where the maternal tissue lining the uterus is extensively digested away, forming crypts into which maternal blood empties, there is no direct exchange of blood between fetus and mother. The fetal blood is enclosed

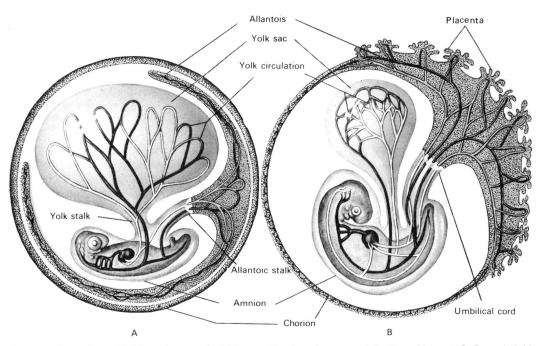

Fig. 6-17. Comparison of fetal membranes of a bird or reptile; A, and a mammal, B. (From Moment, G. B., and H. M. Habermann. 1973. *Biology: A Full Spectrum.* The Williams & Wilkins Co., Baltimore.)

within capillaries which are themselves enclosed within the various tissues of the villi. Only by some breakage, which rarely occurs, is there direct mixing of bloods. Nevertheless, a host of materials are continually exchanged via the placenta between fetal and maternal bloodstreams. Indeed, that is the whole point of the placenta. Carbon dioxide, oxygen, digested food, and metabolic wastes all pass through, as do antibodies. The odor of garlic, though not of onions, passes through, so that if a woman chews garlic shortly before her baby is born, it will come into the world with garlic on its breath, i.e. excreting it from its bloodstream via the lungs.

FETAL MEMBRANES AND TWINS

Monozygotic, or genetically idential, twins bear a special relationship to the fetal membranes. One-egg twins develop from a single ovum in which there are two centers of development in the blastocyst. Then, since the chorion develops directly from the blastocyst, one-egg twins are always enclosed within a single chorion and have more or less fused placentas. Each twin is provided — or rather provides itself — with its own amnion and usually with its own yolk sac.

Fraternal twins, whether boys, girls, or a boy and a girl, result from simultaneous fertilization of two separate eggs. Each is enclosed in its own chorion and provided with a separate placenta. Unfortunately from the diagnostic point of view, occasionally a pair of fraternal twins become implanted so close together on the wall of the uterus that their placentas and even their chorions tend to fuse.

The cause of fraternal twins, fraternal triplets, or any higher number of fraternal sibs is multiple ovulation. To produce a pair of fraternal twins, two eggs must be ovulated at approximately the same time. There is evidence that fraternal twinning tends to be inherited. A woman can inherit this tendency to ovulate two or even more eggs at a time from either her mother or her father. On the other hand, one-egg twins are produced by certain factors within the fertilized egg. Consequently, the male as well as the female parent might contribute the factor which makes for the appearance of two centers of development in a single egg. What factors actually cause this are unknown, but it has been found in many organisms from coelenterates up that if the original growth center is inhibited, then one or more secondary growth centers will arise. If the inhibition is removed, multiple growth buds will result

Development of the Nervous System

BRAIN AND NEURAL CIRCUITRY

The development of the vertebrate nervous system can be conveniently divided into three phases. First is the induction of the neural plate by the underlying chordamesoderm. The chordamesoderm is derived directly from the primary evocator or organizer of the vertebrate embryo, the dorsal lip of the blastopore. Immediately following this induction comes formation of the neural groove and neural tube.

The second phase is the conversion of the neural tube into the five basic subdivisions of the brain, plus the spinal cord. The five brain subdivisions are the same in all vertebrates from fish to man and form comparable structures in the adults. As soon as it is formed, the anterior end of the neural tube develops three hollow swellings — **forebrain, midbrain,** and **hindbrain.** The cavities become the various ventricles of the adult brain and remain open into each other and into the lumen of the spinal cord. The forebrain develops into the right and left cerebral hemispheres and, immediately posterior to them, the **diencephalon** from which the optic vesicles grow out on either side. Dorsally the diencephalon forms the pineal gland or third eye in many reptiles. The sides become the thalamus and hypothalamus which contain centers controlling hunger, thirst, pleasure, sleep, etc. From the midventral surface, the posterior lobe of the pituitary gland grows out.

The midbrain becomes a visual center and is where the optic lobes of the frog brain are located. In mammals this visual center to which nerve fibers go which come directly from the eye is called the **optic tectum.** The hindbrain forms the cerebellum and the medulla oblongata which narrows into the spinal cord.

NEURAL CIRCUITRY AND THE DEVELOPMENT OF BEHAVIOR

During the later stages in the development of the nervous system, a neural circuitry of enormous complexity and incredible precision is spun out between the hundreds of millions of cells in the system. The nerve fibers which compose this system are longer or shorter extensions of single nerve cells. A group of nerve fibers running close together and parallel within the brain or spinal cord constitutes a **fiber tract.** When a group of nerve fibers run together outside the central nervous system, they form a **nerve.** Ever since the descriptive investigations of Ramon y Cajal and His and experiments of Ross Harrison, it has been clear that nerve fibers grow out from their cell bodies and make connections with their end organs—muscle, gland, sense organ, or (within the nervous system) other nerve cells.

The relationship of the development of behavior to the development of this neural circuitry is a highly important and also highly controversial area where biology and psychology overlap. One persistent question is the relative importance of cytological maturation of the nervous system compared with learning in the appearance of any specific behavior.

G. E. Coghill and L. Carmichael early showed that the various swimming movements in a developing salamander larva appear in a fixed sequence, the same in every individual, and that these motions are correlated with the development of the appropriate connections within the nervous system. The larva first bends only into a C, then later into an S. Swimming becomes more and more perfect as the neural connections become more completely formed. The growth of the nerve fibers concerned was traced in histological sections stained with silver, which blackens nerve fibers. Coghill also showed that if an embryo salamander is kept under an anesthetic until control animals have developed into swimming larvae, the anesthetized animal swims immediately on being removed from the anesthetic and does so without passing through the usual awkward stages which look like learning. Flying in many species of birds—pigeons, swallows, and cliff-nesting sea birds, among others—is not learned but appears fully developed the first time the bird attempts it, even if the birds are raised with their wings bound to their bodies.

THE CENTRAL PROBLEM

A central problem remains which presents many unanswered questions. Does the developing nervous system grow as a diffuse, equipotential nervous network on which function, i.e. conditioned reflexes or some other type of learning, imposes reaction patterns? Or does the growth of nerve fibers within the central nervous system establish genetically programmed neural circuits which, in turn, result in preordained behavior patterns such as the precise kind of web a spider will spin or the elaborate coordination of muscular action essential to produce skillful flight? It should be noted that the second alternative does not preclude the possibility that the same nervous system could also grow circuits which provide the physical basis for learning. How do the outgrowing nerve fibers know where to grow and what connections to make? How does the brain know which muscle or gland a certain nerve fiber innervates or from what spot on the retina, for example, a certain nerve fiber comes?

Here are some of the facts. They give only partial answers but they are a beginning. In embryonic and even larval salamanders and frogs, muscles can be transplanted to unnatural sites on the body. A whole leg can be grafted into the socket from which an eye has been removed. In such cases nerves which would normally never see a leg will make functional connections with the strange muscles. Very commonly such transplanted muscles will contract synchronously with the corresponding muscles in the normally placed leg. Facts such as these have led to the conclusion that nerves merely innervate the first muscle they meet and are guided in their outgrowth purely by physical topography. The explanation of the synchronous contraction with the corresponding muscles in the normally placed limb is very obscure. In the higher vertebrates, especially in mammals, scrambling nerves and muscles leads to serious malfunction. The meaning of these facts remains to be clarified.

Highly specific behavior in the growth of nerve fibers from specific parts of the retina to specific parts of the brain (in the optic tectum, of course) has been shown by R. W. Sperry at Cal Tech in fish and by Leon Stone at Yale in salamanders. In the common pond salamander, *Diemictylus (Triturus) viridescens* it is possible to transplant the eyes with a return of vision because the low oxygen requirements of a cold-blooded animal permit revascularization before the death of the eye. The new nerve fibers grow into the brain from their cell bodies located in the retina. It is necessary only to make sure that the cut ends of the optic nerve are held together; otherwise the regenerating fibers grow out of the cut end and merely form an irregular mass.

If a salamander is anesthetized and one eye cut out, rotated 180°, and replaced, when vision returns, that eye will see everything upside down and backwards. If both eyes are thus rotated the animal will live indefinitely but must be hand-fed because it never learns not to reach up for food placed below it or reach down in trying to grab food which is, in fact, dangled above it. If the left eye is grafted into the place of the right eye but not rotated, the salamander will see backwards but not upside down with that eye. If that left eye is rotated 180°, then the salamander will see upside down but not backwards with that eye. In other words, it will jump up to get food presented under its nose but lurch forward to get a piece of meat held a short distance in front of it. These facts indicate that nerve fibers coming from each particular spot on the retina either carry some kind of label (which the brain can recognize) as to their place of origin in the retina, or else that in regeneration each nerve fiber somehow grows back to its proper location in the brain.

In fish, which are favorable material for such investigations, Sperry showed that nerve fibers from the eye actually do grow back very precisely to their original terminal sites in the brain. How do they do this? Perhaps they are guided by some kind of chemical gradients.

Aging

The fact of aging is obvious to all, yet nothing certain is known about its causes. It may very well be that it is controlled in different kinds of animals by entirely different mechanisms. What determines the time when teeth will erupt, or when the thymus gland undergoes atrophy, or sexual maturity occurs, or menopause takes place? Why should a rat have a potential life span of three years, being the equivalent of a man in his 90's at 36 months, while a flying squirrel having basically the same size, anatomy, and physiology easily lives 6 or 7 years? The life span of *Drosophila* can be lengthened by a factor of three by low temperature, or cut in half by raising the temperature. However, despite some intensive studies, no one knows how temperature affects the mechanisms that control aging.

There are many theories about senescence, most of them easily disposed of. Metschnikoff held that aging was due to poisons absorbed from intestinal and other bacteria. But animals raised in a sterile environment grow and age like others — or even die sooner.

Steinach in Germany proposed that aging is caused by changes in the reproductive glands. But castrated animals grow and age at almost the same rate as do normal ones. Others have suggested dietary factors and change in the calcium or water content of the cells.

Various enzymes change during a life span. But whether this is merely a concomitant of aging, or is a cause of aging, remains unknown.

A now popular theory, which has been under discussion for over a decade, holds that senescence is due to the accumulation of random errors in the replication of DNA during cell division until so much "noise" is built up in the genetic information system that an adequate number of functional cells cannot be maintained.

Perhaps the clue to the problems of aging will be found by studying the programmed death of blocks of cells which occurs during development. Even when transplanted to a presumably favorable environment, these cells die, although surrounding cells remain healthy. J. Saunders in the United States and L. Amprino in Italy have been investigating this strange phenomenon in chicks. Recently, others have found the same thing in mammals but no one can even guess what the "death clock" is. Ciliates and very probably metazoan cells have a limited clonal life span when cultured outside the body. If this is a fact then it will influence the maximum attainable age.

USEFUL REFERENCES

Arey, L. 1965. *Developmental Anatomy of Man.* 7th ed. W. B. Saunders Co., Philadelphia.

Ebert, J. D. 1965. *Interacting Systems in Development.* Holt, Rinehart and Winston, Inc., New York.

Hamilton, H. L. 1952. *Lillie's Development of the Chick.* Henry Holt & Co., Inc., New York.

Leeper, E. M. 1975. Fetal Research: Commission Sets Guidelines for Experimentation. *BioScience* 25:357–399.

Markert, C. L., and H. Ursprung. 1971. *Developmental Genetics.* Prentice-Hall, Inc., Englewood Cliffs, N.J.

Moment, G. B. 1975. The Ponce de Leon Trail Today. *BioScience* 25:623–628.

Patten, B. M. 1951. *Early Embryology of the Chick,* 4th ed. Blakiston, Philadelphia.

Rugh, R. 1951. *The Frog: Its Reproduction and Development.* P. Blakiston's Sons & Co., Philadelphia.

7

Genes in Action II: Development in Plants

The question, "How does a plant grow?" can be asked, and answered, in many ways. When the question is asked in the sense of "How well?" it assumes vital importance to mankind, because all animals, including ourselves, are completely dependent on plants to provide the base for food chains, to supply the essential raw materials for products as diverse as lumber and drugs, and to maintain adequate levels of oxygen in the air we breathe.

When we ask "What happens?" or "What is formed?" we refer to the developmental patterns that have evolved since the origin of plants on this planet. Here we will limit the consideration of these questions to the key events common to all plants and then concentrate on the dominant plants in our terrestrial environment, the seed plants.

Until relatively recently, studies of development were limited to the questions "What happens?" and "What is formed?" It was not until the last decade, when an understanding was achieved of how DNA in the nucleus of a cell controls the synthesis of specific proteins in the cytoplasm, that questions about development could be posed in a more sophisticated way: "How is it that one kind of DNA can produce an alga while another can give rise to an oak tree?"

Development of a plant (or an animal) is an orderly sequence of events, predictable for any species. It can be thought of as the product of reading the genetic code. The basic information needed to make a new plant is present in the nucleus of each cell. In some ways, the genetic code and the plant which is formed from its set of instructions are analogous to the written musical score of a symphony and the performance of it. In both cases a number of variables can influence the quality of the product. The performance of a symphony orchestra is influenced by factors such as the expertise of the players, the quality of the instruments, the acoustical properties of the hall, and the correct proportion and variety of instruments and musicians. In other words, the product is variable and dependent on factors other than the musical score. Plants can develop in ways that fall far short of their genetic potential, just as the performance of a musical score may leave much to be desired. A number of external factors must be present and available in the correct proportions for the vigorous growth and development of plants. These factors include appropriate temperature, moisture, minerals from the soil, oxygen and carbon dioxide from the atmosphere, and light. The influence of these external factors will be discussed in detail in the chapters on plant metabolism (Chapters 10, 12, 15, and 18).

Life Cycles: Readouts of the Genetic Code

The life cycles of plants present what appears to be a bewildering diversity. What should be recognized is that, in spite of the apparent complexity of the living things around us, there is a common underlying theme. The life cycle of any plant from the simplest alga or fungus to a large and complex conifer or flowering plant can be sum-

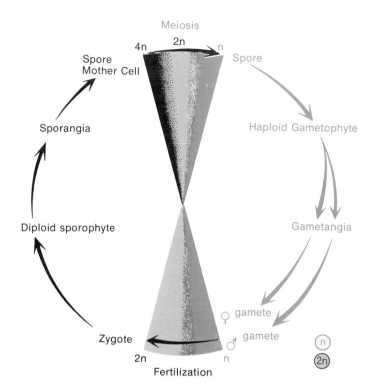

Meiosis

4n   2n   n

Spore
Mother Cell

Spore

Sporangia

Haploid Gametophyte

Diploid sporophyte

Gametangia

♀ gamete

♂ gamete

Zygote

2n   n

Fertilization

ⓝ

②ⓝ

Fig. 7-1. Generalized plant life cycle. The sequence of events in the
life of any plant can be represented
by appropriate variation of this basic
scheme.

marized by means of a diagram (Fig. 7-1).
Once a few key events are recognized, it becomes relatively simple to understand all the
diverse life cycles of plants.

The two most significant events that occur
are **meiosis,** a specialized pair of cell divisions in which haploid cells are formed, and
fertilization, the fusion of nuclei which restores diploidy. A widespread phenomenon
in plants is the separation of these key steps
in the life cycle. In many plants they occur in
separate and independent plant bodies.

The plant body composed of haploid cells
is called the **gametophyte.** As it reaches maturity, the gametophyte plant produces gametes in specialized structures called **gametangia.** In most plants, two kinds of gametes
are formed which differ in size and structure.
The smaller, frequently motile, gametes are
called **sperms** and are produced in male gametangia or **antheridia** (singular, antheridium). Larger, usually nonmotile, gametes
called **eggs** are produced in the female gametangia or **archegonia** (singular, archegon-
ium). Archegonia and antheridia can be
formed on the same gametophyte plant or on
separate plants. Fusion of haploid sperm and
egg nuclei in the process of fertilization results, as in animals, in a single diploid cell
called the **zygote.** This cell is the beginning of
the diploid phase of the life cycle. From it, a

diploid organism called the **sporophyte** develops. When the sporophyte matures, structures called **sporangia** are formed. In the
sporangia, special cells called **spore mother
cells** are produced. These cells undergo
meiosis and the resulting haploid cells,
spores, are released. Spores, on development, form new haploid gametophyte plants.

Among several aspects of the life cycles of
plants that are different from the circumstances found in animals is the already noted
phenomenon of separate haploid and diploid
phases with morphologically different and independent gametophyte and sporophyte
generations. This life style, which is apparent
in the mosses and ferns, has been given the
name **alternation of generations,** an unfortunate choice of terminology because the same
term is used by zoologists when referring to
the morphologically distinct, but both diploid, phases in certain jellyfish life cycles.

The alternation of haploid and diploid
phases in the life cycles of plants ensures a
separation in time of the key steps of meiosis
and fertilization. There is a general evolutionary tendency in plants for the gametophyte
phase of the life cycle to become reduced in
size and duration. In the most highly evolved
plant group, the familiar seed plants, the haploid portion of the life cycle is reduced to a
few specialized cells.

Life Patterns in the Seed Plants

The seed plants dominate our environment with a vast host of species—the grasses and maple trees, sunflowers, pines, and orchids. They form the basis of our horticulture and forestry. The plants that surround us are in the diploid (sporophyte) phase of their life cycles. The haploid or gametophyte stages have been reduced to such an extent that haploid cells occur only as the immediate products of meiosis in parts of specialized reproductive structures, **flowers** and **cones.** The key events in the life cycle of a flower-forming seed plant (an angiosperm) are summarized in Figure 7-2. New diploid plants develop from **seeds.** In each seed there is an **embryo** plus food reserves, frequently in a specialized tissue called the **endosperm.** The entire structure is covered with a protective **seed coat** originating from tissue of the sporophyte flower. After seeds are formed, there is often a period of arrested development **(dormancy)** before further growth of the new sporophyte plant takes place.

GERMINATION

Once dormancy has ended and appropriate conditions of temperature, moisture, oxygen, and light prevail, **germination** begins. Factors such as gravity and light assure the proper orientation of the young diploid plant during its early development from embryo to seedling. Vegetative growth continues until internal and external factors are appropriate for the onset of reproductive development, or flowering (see Chapter 9).

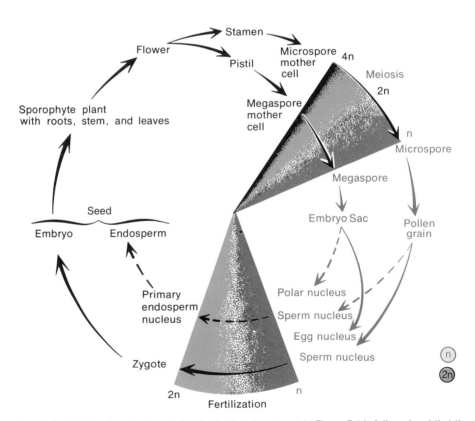

Fig. 7-2. Life cycle of a flowering plant. Note that the basic scheme seen in Figure 7-1 is followed and that there is an alternation of haploid and diploid stages.

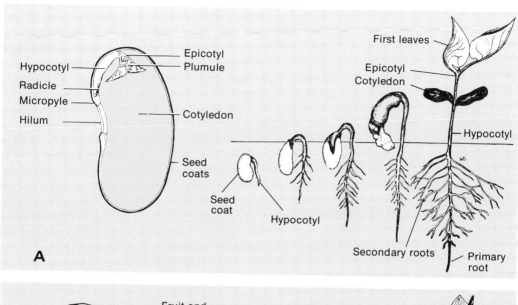

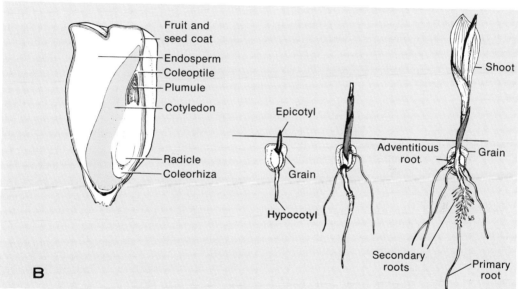

Fig. 7-3. Seed structure and germination. A, bean; B, corn. (Modified from Greulach, V. A., and J. E. Adams. 1967. *Plants, An Introduction to Modern Botany*. John Wiley & Sons, Inc., New York; and Fuller, H. J., and O. Tippo. 1954. *College Botany*. Henry Holt & Co., Inc., New York.)

The Structure of Seed Plants — the Product of Development

A basic difference between plants and animals is that animals tend to grow "all over" while in plants new parts are added only in restricted regions called **meristems.** However, before the organization and function of meristems are discussed, we will have to ex-amine the basic structural patterns of the seed plants.

It is to the distinct advantage of the botanist that plants are relatively simple in organization with far fewer organs to be named and recognized than in animals. There are only six major parts of a flowering plant: **roots, stems,** and **leaves** (vegetative organs); and **flowers, seeds,** and **fruits** (reproductive organs). The simplest way to describe a seed plant is to say that it is made up of a vertical axis plus appendages. That portion of a plant normally above the ground is called the **shoot,** and that below the soil level called the **root.** The vertical axis of the shoot is the **stem** and the appendages found on the stem are the **leaves.** The usually broad, flat portion of the leaf, the **blade,** is attached to the stem by a stalk termed the **petiole.** A region of the stem where leaves are attached is called a **node.** For every species there is a definite pattern of leaf attachment or **phyllotaxy.** Portions of a stem between nodes are called **internodes.** The angle between the petiole of a leaf and the stem is called an **axil.** This is where **buds** are formed. A bud can develop into a branch of the stem with its own attached leaves, or can develop into a flower, or both.

Stems have three functions: support, conduction, and storage. The rigidity and strength of stems is the result of thickened and reinforced walls of the cells found in stem tissues.

Cells of the **vascular** (or conducting) tissues are arranged in very definite quantities and ways within stems. Herbaceous plants, which frequently complete an entire life cycle in one growing season, contain relatively less vascular tissue in their stems than do trees living for many years. There are two kinds of conducting tissues: the **xylem,** making up the woody portion of a tree trunk; and **phloem,** found in the bark region of trees. In herbaceous plants, xylem and phloem are located together in **vascular bundles** embedded in the soft, relatively undifferentiated inner tissues of the stem. Two patterns of arrangement of vascular bundles characteristic of many herbaceous species of flowering plants are illustrated in Figure 7-5. In the **monocots,** such as corn and lilies, vascular bundles are scattered at random through the central **ground tissue.** In the **dicots,** such as sunflowers and beans, strands of conducting tissue are usually found in a ring around a central **pith.** Within each strand, xylem cells are located on the inside and phloem cells toward the outside of the stem. Beyond the ring of vascular bundles is a second layer of relatively thin-walled, undifferentiated cells called the **cortex,** and at the outer surface of the stem we find an **epidermal layer.** Frequently the epidermal cells of both stems and leaves secrete a waxy coating (cutin) to form the **cuticle,** which helps to retard water loss.

The arrangement of tissues in woody stems is somewhat more complicated than in herbaceous stems. Here the xylem and phloem are generally laid down as concentric layers that constitute hollow cylinders running lengthwise within the stem (Fig. 7-6).

Primary growth results in the formation of

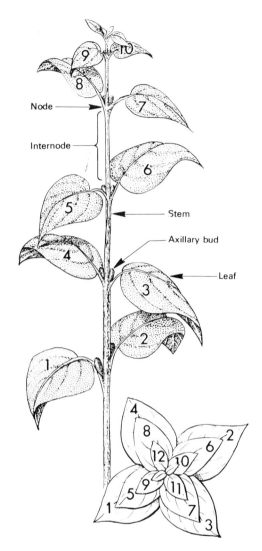

Fig. 7-4. Structural organization of a shoot showing stem, leaves, nodes and internodes, and axillary buds. Top view shows phyllotaxy with leaves numbered to indicate order of formation. (From Moment, G. B., and H. M. Habermann. 1973. *Biology: A Full Spectrum.* The Williams & Wilkins Co., Baltimore.)

new stems or lengths added to old stems or roots. The tissues laid down during the addition of such new plant parts are referred to as **primary tissues.** Primary growth is responsible for establishing branching patterns of the shoot and root and is the product of terminal growing regions called **apical meristems.** **Secondary growth,** on the other hand, brings about a thickening of already formed stems or roots. The regions of actively dividing cells responsible for secondary growth are layers of cells called the **vascular cambium** and the **cork cambium.**

The vascular cambium is a single layer of

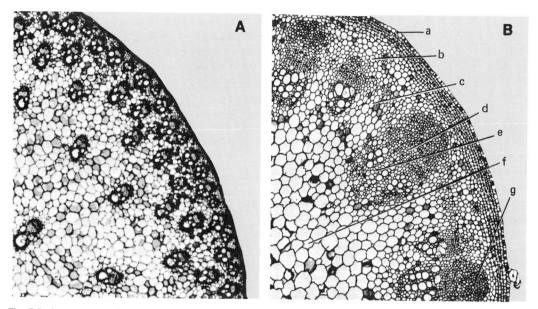

Fig. 7-5. Arrangement of vascular bundles in herbaceous stems. Left, monocot pattern: cross section of a corn stem. (Courtesy of Carolina Biological Supply Co.) Right, dicot pattern: cross section of a sunflower stem. a, epidermis; b, cortex; c, pith ray; d, primary phloem; e, primary xylem; f, pith; g, cortical fibers. (Courtesy of Turtox/Cambosco, Macmillan Science Co., Inc; Chicago.)

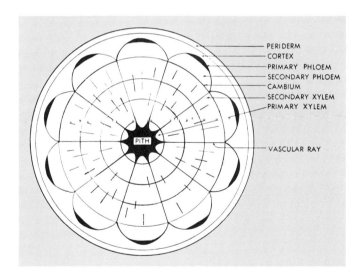

Fig. 7-6. Diagrammatic cross section of a young woody stem at the end of 3 yr of growth. Note three annual rings. (From Fuller, H. J., and O. Tippo. 1954. *College Botany.* Henry Holt & Co., Inc., New York.)

cells located at the interface between the xylem and the phloem. The products of mitosis in this type of cambial cell are a new cambial cell plus either a secondary xylem or a secondary phloem cell. The accumulation of a new ring of secondary xylem cells inside the vascular cambium pushes the cambium and all tissues beyond it toward the outside. Thus, a stem or root increases in diameter by adding layers of secondary xylem and phloem. New secondary tissues are formed each growing season. Particularly in the secondary xylem, which makes up the bulk of the wood

in tree trunks, the seasonal increments are readily seen in cross section. Relatively larger cells are formed early in the growing season while cell size later in the season is smaller because environmental factors are less favorable for growth. These seasonally formed layers, or **annual rings,** are visible in most wood products finished so that the grain is visible.

The outermost layers of the stem, the phloem, cortex, and epidermis, are obviously stretched during secondary growth as new layers of cells are laid down on the inside. Eventually the epidermis and cortex can actually crack open. This stimulates some of the exposed cells of the cortex to differentiate into cork cambium. Cork formed by division of cells in the cork cambium provides a waterproof protective layer for the outside of the stem. Phloem cells are located outside of the vascular cambium and, like all other cells in the bark region, are subject to great stress and the pressure of outward movement. Only the innermost portion of the secondary phloem forms a continuous ring. Older secondary and primary phloem form discontinuous patches in the bark region.

There are many structural similarities between roots and stems. Both can exhibit secondary as well as primary growth. Products of the vascular and cork cambium in roots are comparable to those described in stems. Several fundamental differences exist, however, including an absence of pith (instead there is a core of primary xylem at the center of roots)

and the absence of leaves (and nodes). The mode of origin of secondary roots also differs from that of stems. In contrast to the new branches of the shoot which arise from buds formed in the axils of leaves, lateral roots arise internally from the **pericycle,** a layer just inside the cortex. Developing lateral roots actually penetrate outward through the cortex and epidermis.

Roots not only anchor plants in the ground, but also serve as the port of entry for water and dissolved minerals from the soil. These necessary raw materials enter the root system through the thin-walled extensions of epidermal cells of young roots called **root hairs** (Fig. 7-7). Water and dissolved minerals diffuse from cell to cell from the epidermis through the cortex and **endodermis** (a layer of cells just inside the cortex) to the cells of the vascular tissue which serves to carry water and dissolved materials to all parts of the plant.

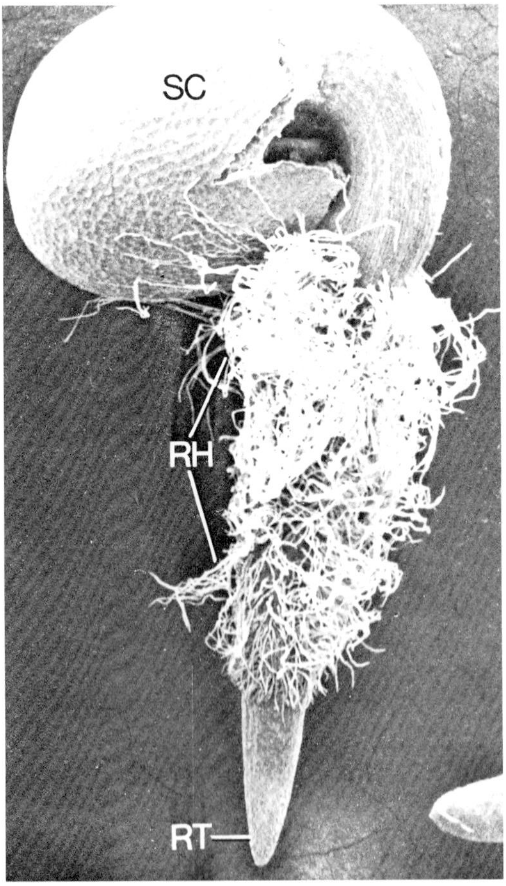

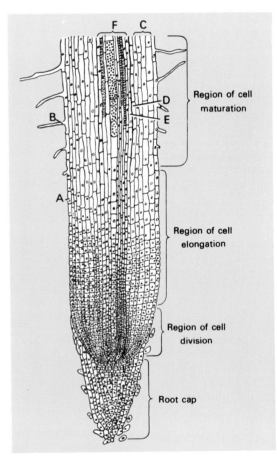

Fig. 7-7. Germinating radish seedling as viewed in the scanning electron microscope, × 25. The hypocotyl has emerged through the seed coat (SC) and the root tip (RT) has grown downward. Root hairs (RH) are readily visible. (From Kessel, R. G., and C. Y. Shih. 1974. *Scanning Electron Microscopy in Biology: A Student's Atlas.* Springer-Verlag, New York.)

Fig. 7-8. Longitudinal section of a young root of barley. A, epidermis; B, root hair; C, cortex; D, endodermis; E, pericycle; F, differentiating conducting tissues. (From Fuller, H. J., and O. Tippo. 1954. *College Botany.* Henry Holt & Co., Inc., New York.)

Apical Meristems: the Regions of Growth

As we have already mentioned, primary growth (or increasing length) of stems and roots results from the activity of the apical meristems. The apical meristem of the root is somewhat simpler in organization than its counterpart in the shoot (Fig. 7-9). It is covered by a protective mass of cells called the **root cap.** New cells are formed in a region of cell division located just behind the tip of the root. Once new cells are formed, they undergo elongation and finally differentiation. The elongating, newly formed cells provide the impetus to push the tip of the root through the soil. The growth of a root is analogous to the unrolling of a ball of yarn. The ball of yarn moves forward as it unwinds, leaving the unwound yarn stationary behind it. The older portions of a root system are in fixed position in the soil with only the tip portion pushing forward as a result of new growth.

The products of shoot growth are new stem and new leaves. The apical meristem of the shoot is surrounded by young developing leaves, the youngest located closest to the apex with the larger expanding primordia attached beneath and surrounding the younger, more apical leaves (Fig. 7-9).

In most seed plants the shoot apex consists of two regions, an outer mantle of one to several layers of cells (called **tunica**) plus an inner core of central tissue made up of larger dividing cells (the **corpus**). While the terms tunica and corpus describe a structural pattern common to apical meristems of many flowering plants, they are of little help in explaining the workings of these structures or the manner in which new plant parts are formed. There is some evidence that the summit area of the shoot apex is meristematically inactive. It is the subterminal and peripheral zone that produces the leaf primordia and gives rise to cortex and vascular tissue of the stem. A third region, located below the apical zone and inside the peripheral region where leaf primordia are formed, is thought to give rise to cells which mature into pith.

Although the structural patterns of apical meristems have been identified and named, there are still many unanswered questions about their functioning. What factors control the orientation and frequency of cell divisions? What controls the pattern of initiation of leaf primordia? Speculative answers to some of these questions are available. To a great extent the location of cells determines

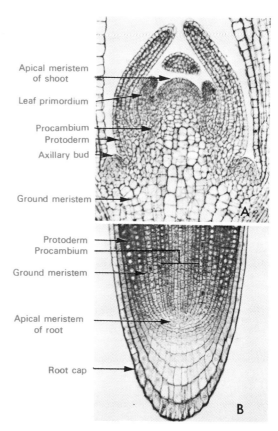

Fig. 7-9. Apical meristems of shoot and root, longitudinal sections. A, flax seedling shoot tip, × 120. (From Sass, J. E. 1958. *Botanical Microtechnique,* 3rd ed. © Iowa State University Press, Ames.) B, root tip, × 130. (From Esau, K. 1960. *Anatomy of Seed Plants.* John Wiley & Sons, Inc., New York.)

their potential. For instance, internal cells are under very different physical restrictions and pressures from those in the epidermal, or surface, layers. There is some evidence that the distribution of metabolites and regulators of growth becomes nonuniform in the apex, and that their patterns of accumulation may control the inception of leaf primordia. Existing leaf primordia may also influence the positioning of additional primordia. Certainly the genetic code is a major factor in determining the patterns of leaf attachment.

Growth vs. Differentiation

We are all familiar with the changes in morphology (form) that occur during the development of most organisms. Just as tadpoles are very different from frogs, the early developmental stages of most plants differ in structure and function from the adult. While **growth** increases the size of an individual, the process of **differentiation** results in the internal and external modifications of the organism which make a maple tree different from a fern. We can consider differentiation at the levels of cells, tissues, and organs. At every level, form is closely related to function. For example, chloroplasts, the green cellular organelles that are the site of photosynthesis (a process requiring light), are found in the parts of plants located above ground, primarily in the cells of the leaves but also in the outer layers of young stems (the regions most easily exposed to light). The form of leaves is well adapted to their primary function as the food-synthesizing organs of the plant. The broad, flat surface is appropriate for absorption of radiant energy from the sun, and the relative thinness allows quanta of incident light to penetrate to the inner layers of cells where the chloroplasts are found. As a further aid to function, there are openings called **stomates** in the leaf's surface layers through which the gases involved in photosynthesis and respiration can be exchanged with the atmosphere. Leaf veins provide the means for importing to the leaves the water and minerals absorbed by the roots of the plant and for exporting the products of photosynthesis. The veins of the leaves are made up of cells and tissues specialized for transport of water and dissolved substances, and they are connected to the conducting tissues of the stem and root.

On the cellular level, differentiation can result in formation of specialized subcellular organelles, such as the chloroplasts, or in structural modifications such as the thicken-

ings found in cell walls of supporting and vascular tissues. In tissues, differentiation can result in the association of various kinds of specialized cells as in the xylem and phloem, conducting tissues found in the veins of a leaf and in the vascular strands of the stem.

On the level of organs, the unequal rates of cell division in different directions result in the characteristic form of the organ. The characteristic shapes of a maple leaf, of flower petals, or of fruits such as an apple or a squash are results of differences in rates of cell division (along their various axes) during development.

An important fact to keep in mind about growth and differentiation is that they occur simultaneously in practically all organisms. Thus it is difficult to study just one of these two aspects of development. There is, however, a group of organisms in which the processes of growth and differentiation are separated in time. These organisms, the slime molds, are fungi. They have served as the experimental material for many classical studies of development. With these organisms, cell multiplication can be studied without the complication of cell differentiation and vice versa.

The slim molds are not the only experimental material available to botanists who wish to study differentiation in the absence of growth. The advent of the nuclear age has made available new tools for the investigation of biological phenomena. At the Oak Ridge National Laboratory, biologists are studying the effects of radiation on living organisms. Haber, Foard, and their co-workers have studied the development of seedlings from

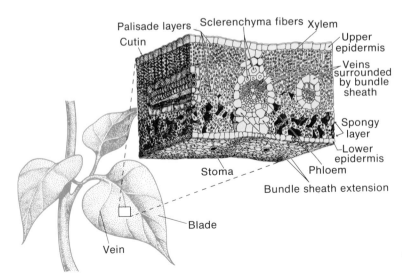

Palisade layers — Sclerenchyma fibers — Xylem

Cutin

Upper epidermis

Veins surrounded by bundle sheath

Spongy layer

Lower epidermis

Stoma

Phloem

Bundle sheath extension

Blade

Vein

Fig. 7-10. Internal structure of a leaf. (From Moment, G. B., and H. M. Habermann. 1973. *Biology: A Full Spectrum*. The Williams & Wilkins Co., Baltimore.)

Fig. 7-11. Wheat seedlings 9 days after the beginning of germination. Left, unirradiated control; right, "gamma plantlet" from grain that had been exposed to 800 kiloroentgens of ^{60}Co gamma radiation. The gamma plantlet has grown by expansion of cells within the embryo without cell division and DNA synthesis. (From Haber, A. H., W. L. Carrier, and D. E. Foard. 1961. Metabolic studies of gamma-irradiated wheat growing without cell division. *Am. J. Bot.* 48:431–438, 1961.)

seeds receiving sufficient gamma irradiation to completely inhibit DNA synthesis, mitosis, and cell division. Grains of wheat can still germinate following exposure to 800 kiloroentgens of gamma radiation. The resulting seedlings (called "**gamma plantlets**," Fig. 7-11) can grow in the sense that cells present in the embryo at the time of irradiation can enlarge and differentiate. Thus, botanists have utilized radiation as a tool and have taken advantage of the fact that cell division is relatively radiation-sensitive, while cell expansion and differentiation are relatively radiation-resistant. The gamma plantlets developing from gamma-irradiated wheat have provided experimental materials for the study of many aspects of the differentiation and metabolism of young seedlings in the absence of cell multiplication.

Environmental Control of Plant Growth and Development

All aspects of the environment, from climate to man-made smog, affect not only the development but also the very survival of plants. Climate, a collective term for many factors including rainfall and temperature extremes, is influenced by latitude, altitude, and geographical features such as lakes and oceans. Climate has a great influence on the kinds of plants that are native to or can be introduced into any given area. We acknowledge this when we associate palm trees with tropic islands or hemlock and spruce trees with northern forests. Climate is the prime factor determining whether agriculture is feasible. Regions of limited rainfall or seasonal extremes of temperature are usually unproductive and not used for farming or grazing. Unexpected fluctuations in climate (too much or too little rainfall, abnormally cold summer weather, etc.) can affect agricultural yields and ultimately the cost of commodities in the local supermarket.

TEMPERATURE

All plants respond to temperature and usually grow best within a limited temperature range. Frequently plants will grow more rapidly under conditions of warm days and cool nights than they will under conditions of constant warm temperature. This response is called **thermoperiodism.** An example of such effects of temperature on growth is shown in Figure 7-12. Under conditions of constant temperature, growth rates for the stems of tomato plants were highest at 26.5°C. However, when tomatoes were grown at this optimum temperature during the day but at lower temperatures during the night, stems elongated even more rapidly. Vegetative growth is not the only developmental response exhibiting thermoperiodism. Fruit set in many garden plants from peppers to squash is improved under conditions of lower night temperature, and thermoperiodism can influence such varied responses as leaf shape and flowering.

Seasonal fluctuations in temperature also play an important role in plant life cycles. Buds and seeds of many species remain dormant until subjected to a period of cold temperature. In some kinds of grains normally planted in the fall (e.g. the so-called "winter" varieties of rye and wheat), a cold period during vegetative growth is required for flower-

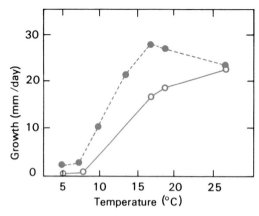

Fig. 7-12. Effects of temperature on growth of tomato plants. Greatest growth occurs when night temperatures are lower than day temperatures. ●---●, day temperature of 26 C alternating with night temperature indicated; ○——○, constant temperature (day and night) as indicated. (After Leopold, A. C. 1964. *Plant Growth and Development.* McGraw-Hill Book Co., New York.)

ing. These varieties are usually planted in the fall and harvested the following year. If such varieties are planted in the spring, they will grow but not flower. After seeds of winter varieties are soaked in water and then exposed to low temperature (0–10°C) for a period of a few hours to several days, they can be planted in the spring and will flower in the same season. This process is called **vernalization** and is an agricultural practice used extensively in Russia. Many biennials, such as carnations, which normally form flowers only in their second season of growth, can be made to bloom in their first season if subjected to a period of low temperature. The requirement for cold prior to reproduction in many plants is comparable to the situation in some insects which will not emerge from their pupal stage unless subjected to a period of cold.

Basically, temperature affects plants through its control of the rates of individual biochemical reactions. Within the physiological range of temperatures that a species can tolerate, an increase in temperature will generally increase the rates of cellular metabolic processes just as rates of all chemical reactions are increased by increased temperature. The adverse effects of temperatures which are too low (at the freezing point of water or below) or too high (in excess of about 40°C) generally can be explained on the basis of their damaging effects on cellular components. Freezing temperatures can result in the formation of ice crystals which rupture cell membranes. Elevated temperatures can denature proteins and thus destroy the capacity of enzymes and membranes to function normally.

MOISTURE

The prevailing conditions of rainfall and soil moisture determine which species can survive in any given area. Plants exhibit wide differences in their tolerance for unusually dry or wet conditions, and the selection of appropriate varieties plays an important role in determining the success or failure of agriculture.

All plants require water and are dependent on their environment for its adequate supply. In many respects, plants act like wicks in their removal of moisture from the soil. The movement of water is a one-way process from the soil, through plants, to the atmosphere (see Chapter 15). Although relatively little of the water moving through a plant becomes incorporated into cellular components, water is necessary for a great many metabolic processes, from the first steps in the germination of seeds and spores (which involve an imbibition of water) to hydrolysis of stored foodstuffs (a splitting of complex molecules through addition of water) and synthesis of nutrients in the process of photosynthesis (for which water is a raw material).

ATMOSPHERE

The gaseous mantle which now surrounds the earth is a product of many billions of years of chemical and biological evolution. Three component gases in air (nitrogen, 78.1 per cent; oxygen, 20.9 per cent; and carbon dioxide, 0.03 per cent) play major roles in plant metabolism and are therefore essential for growth and differentiation.

Except for the nitrogen-fixing bacteria, and a few blue-green algae, plants are not able to utilize atmospheric nitrogen. Thus, the nitrogen required for synthesis of all amino acids, nucleic acids, porphyrins, and a wide range of other essential constituents of cells must be absorbed from the soil solution in the form of soluble nitrate or ammonium ions.

The maintenance of a constant level of 20.9 per cent oxygen in our atmosphere is a consequence of the balance between utilization in all oxidation reactions requiring oxygen and the single process which adds significant amounts of molecular oxygen, namely green plant photosynthesis (see Chapter 10). With the exception of anaerobic forms such as some of the bacteria, plants require oxygen to remain alive. Early in the process of seed germination, after imbibition of water has initiated the hydrolysis of food reserves, there is a sharp increase in the rate of oxygen uptake. This occurs because respiratory metabolism requires oxygen for the complete oxidation of foodstuffs and the release of chemical energy. Under conditions where oxygen is restricted (this can happen in waterlogged soils following unusually heavy rains), seeds sometimes fail to germinate. Roots are the plant organ most likely to be subjected to anaerobic conditions in nature. In flooded soil, air spaces are filled with water and the root systems can no longer function normally. Most of us have observed trees killed by the damming of streams by beavers or the creation of man-made lakes.

Carbon dioxide, even though a minor component of air, is constantly being added to our atmosphere as a product of the combustion of fuels and food. It is an essential raw material for the process of photosynthesis, the basis for all food chains, and under normal conditions is the most likely factor to be limiting in this process.

Within the past several decades we have become increasingly aware of the presence of man-made atmospheric pollutants and their effects on the biosphere. We are only beginning to understand the nature of damage to plants and animals caused by automobile exhaust and industrial fumes. It is becoming clear, however, that plants can be harmed by automobile exhausts just as humans are (Fig. 7-13). In many forested areas around the Los Angeles basin, some pine species are being killed by ever spreading automobile-generated smog. In many ways, plants are a good indicator of the general "health" of our atmosphere.

SOIL FACTORS

Soil provides a medium in which the root system anchors a plant and absorbs water and essential minerals. Soils vary greatly in their composition, consistency, and their capacity to support plant growth. Agricultural practices such as crop rotation, contour plowing, addition of chemical fertilizers, irrigation, and the like help to maintain fertility and productivity of agricultural lands and make possible the sustained high yields necessary to feed our growing world population. The essential mineral elements which plants must obtain from the soil are discussed in Chapter 12. For now it will suffice to point out that, if the soil is deficient in any of the essential minerals, plant growth will be impaired and productivity will be reduced.

Even when essential minerals are present in adequate amounts, other factors, such as

Fig. 7-13. A, effects of air pollution. Young grafted white pine plants. Left, after 7 months of exposure to polluted air; right, after 7 months in pollution-free air. (Courtesy of U.S. Department of Agriculture.) B, plant varieties differ in resistance to air pollution. Left, Greenpod 407 bean, a variety resistant to air pollution. Note uninjured foliage and good yield. Right, Tempo bean, a variety which is very sensitive to air pollution. Note damaged foliage and poor yield. Both of the plants in this photograph were grown under the same conditions in unfiltered air. (Courtesy of Robert K. Howell, U.S. Department of Agriculture.)

pH of the soil, can control their availability and thus influence plant growth. Soils may be acid, neutral, or alkaline. Some plants that grow under a variety of pH conditions actually can indicate, through responses such as flower color, the approximate pH of the soil. The familiar hydrangeas respond in this way with red flowers in alkaline soils and blue flowers in acid soils. The latter response results from formation of a blue complex between petal pigment and aluminum salts under acid conditions. In nature, acid-loving plants (azaleas, rhododendrons, and the like) often flourish in forest communities, while entirely different species such as grasses populate soils which are neutral or alkaline.

Although the preferences of various plant species for a particular soil pH are not readily explainable, the reasons for the deleterious effects of extremes of pH are quite obvious. Among other effects, pH can control the solubility (and hence the availability) of essential minerals. Iron, which is needed for chlorophyll synthesis and is a necessary component of a number of enzymes, including cytochromes and catalase, tends to form insoluble salts at pH levels above neutrality. Extremes of pH can also denature proteins and thus influence not only membrane permeability but also the status of cellular proteins. Fortunately, soil pH can be modified by appropriate use of fertilizers. Lime (predominately calcium carbonate) is frequently applied to make soils less acid. Calcium ions in limestone serve not only as an essential mineral but also as a factor in the release of other essential elements which may be bound to soil particles.

Another factor influencing soil fertility is its consistency. Clay soils contain much fine-grained material; they drain poorly and tend to become very hard when dry. The addition of organic matter (leaves, decomposing plant materials, peat, etc.), inorganic materials such as sand or vermiculite, or synthetic soil conditioners can make clay soils loose, crumbly, and porous. These conditions are necessary for proper aeration, percolation (drainage) of water, and penetration by roots.

LIGHT

Light is an essential factor for the sustained growth and normal differentiation of most plants. In addition to its essential role as an energy source in photosynthesis (see Chapter 10), it has profound effects on such varied aspects of development as leaf and stem morphology, pigment synthesis, direction of growth, and time of flowering.

Some Facts about Light

Before we discuss any specific aspects of light-dependent development (**photomorphogenesis**), let us review some basic facts about its physical nature and how light can promote chemical change in living systems. Without sunlight no life would exist on earth. Thus, a knowledge of light is basic to any complete understanding of the total ecology of our planet. **Light** is usually defined as the visible portion of the electromagnetic spectrum of radiation. Thus, the term light, by definition, involves human perception of one part of a continuum including cosmic rays, gamma and X rays, ultraviolet and visible light, infrared (or heat) waves, and radio waves. These types of electromagnetic radiation differ in wavelength (Fig. 7-14). Those wavelengths which influence plant development extend beyond what our eyes can see to longer wavelengths in the far red and near infrared and to shorter wavelengths in the ultraviolet region. The units most frequently used for measuring wavelengths of light are the Angstrom unit (1 Å = 10^{-8} cm), and the nanometer (1 nm = 10^{-9} M, 10^{-7} cm, or 10 Å). Light from the sun or from an incandescent lamp contains a mixture of wavelengths which we perceive as white light. When light

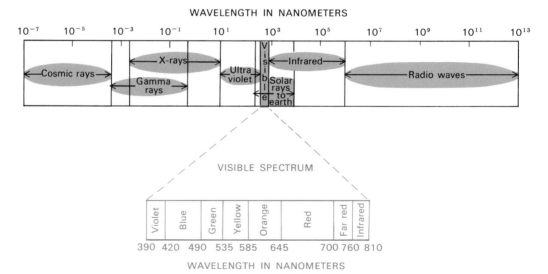

Fig. 7-14. The electromagnetic spectrum. Note that light, the visible portion of the spectrum, is a very narrow band in this continuum. (From Moment, G. B., and H. M. Habermann. 1973. *Biology: A Full Spectrum.* The Williams & Wilkins Co., Baltimore.)

hits an object, the component wavelengths can be absorbed, reflected, or transmitted.

Colored materials absorb some wavelengths while transmitting or reflecting the rest. In order for light to be absorbed, a pigment must be present. This light absorption by a pigment (called the **photoreceptor**) is always the initiating step in photochemical or photobiological processes. The pattern of light absorption by each pigment is unique and this **absorption spectrum** can be measured with a spectrophotometer. Colored substances can be identified and distinguished from similarly colored compounds on the basis of their absorption spectra. Absorbance increases with increasing concentration. This relationship has been formulated as the **Beer-Lambert law** and can be utilized to measure the concentration of colored compounds.

In addition to wave properties which allow us to describe light in terms of wavelength and **frequency** (number of repeating waves per unit time), electromagnetic radiations also have the properties of particles. In an apparently paradoxical way, light, and also ultraviolet, gamma and X radiation, etc., is absorbed or emitted in discrete packets called **quanta** (singular, quantum). A quantum of visible radiation is called a **photon.** The energy per quantum is inversely proportional to its wavelength.

A fixed amount of energy is associated with each quantum of electromagnetic radiation, and the amount of energy can be calculated from the following relationship formulated by Planck early in this century:

$$\text{Energy per quantum} = h\nu = hc/\lambda$$

where h = Planck's constant, and

ν = frequency = velocity (c)/wavelength (λ)

Both h (Planck's constant = 6.625×10^{-27} erg-sec) and c (the velocity of light = 2.99×10^{10} cm sec^{-1} are constants. The only variable is the frequency.

Thus, short wavelength (high frequency) radiations such as ultraviolet or gamma rays contain more energy per quantum than longer wavelengths such as visible or infrared radiations. This relationship has important implications because the absorption of a high energy quantum can result in the input of considerably more energy, with greater potential for cell damage or useful work, than a low energy quantum.

In order for light to affect a chemical or biological system, it must be absorbed. This obvious fact, recognized since the early 1800's, is known as the first law of photo-chemistry or the **Grotthus-Draper law.** The absorption spectrum of the photoreceptor tells us something about the relative probability that a given wavelength will be absorbed. Obviously only those wavelengths likely to be absorbed will be effective in promoting a light-dependent process. The relative response to light plotted as a function of wavelength is called an **action spectrum.** Usually the action spectrum of a light-dependent process corresponds closely to the absorption spectrum of the photoreceptor. This correspondence is used to determine what pigment is the photoreceptor of a light-dependent reaction.

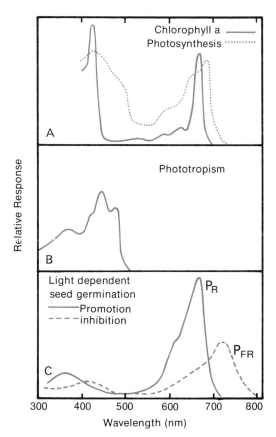

Fig. 7-15. Action spectra of several important light-dependent plant processes. A, chlorophyll synthesis and photosynthesis. Chlorophyll, the photoreceptor of photosynthesis, is a green pigment absorbing in the blue and red portions of the spectrum. These portions of the spectrum are most effective in promoting both chlorophyll synthesis and photosynthesis. B, phototropism, or the bending growth toward light, is promoted by blue wavelengths. The photoreceptor for this response is not known. C, light-dependent seed germination is one of many plant responses mediated by phytochrome. Red light promotes, while far red light inhibits germination. (Modified from Withrow, R. B., ed. 1959. *Photoperiodism and Related Phenomena in Plants and Animals.* American Association for the Advancement of Science, Washington, D.C.)

Some Examples of Light-controlled Plant Development

Many kinds of seeds can germinate in the dark and can grow for a limited time at the expense of their food reserves, but such dark-grown seedlings are abnormal in form and pigmentation, with long, weak stems and small, unexpanded, yellow leaves. Plants exhibiting such symptoms are said to be **etiolated**. We can observe etiolation to some extent in houseplants or in garden plants grown in deeply shaded areas. At least moderate intensities of light are needed for normal chlorophyll synthesis and leaf expansion, while moderate to high intensities are necessary to inhibit stem elongation. Plants grown in adequate light have shorter and stronger stems than those grown at lower light intensities.

Plants exhibit many other readily observable light-dependent developmental responses. When houseplants are kept near a window, growth of the shoot tends to be toward the source of light. Such a growth response to unilateral illumination is called **phototropism** and is caused by an asymmetric distribution of the plant hormone auxin (see Chapter 18).

Other developmental responses are controlled by the periodicity of illumination, i.e. the relative length of day and night in each diurnal cycle. A well known example of **photoperiodism** is the control of flowering (see Chapter 9).

There are numerous light-dependent phenomena involved in the germination of seeds and the early development of seedlings. Some varieties of seeds (such as lettuce and certain grasses) require light for germination to occur (Fig. 7-16). Later in seedling development when a part of the seedling emerges from the soil, light is required for its hypoco-

tyl to straighten, a change which moves the developing shoot into a vertical position. In both of these responses the quality of light (i.e. its wavelength or color) is critical. Wavelengths in the red portion of the spectrum are most effective in promoting, whereas longer wavelengths (in the far red region) can inhibit these responses or even counteract the effects of an earlier exposure to red light. Promotion by red and inhibition by far red light are characteristic of a wide variety of responses in plants mediated by the **phytochrome** pigment system. The action spectra for these responses indicate that the photoreceptor can exist in either of two interconvertible forms: one, known as P_R (the red absorbing form of phytochrome), is a blue pigment with an absorption peak at approximately 660 nm. It can be converted by red illumination to P_{FR} (the far red absorbing form of phytochrome) with an absorption maximum at 710-730 nm. The far red absorbing form is converted back to P_R by slow dark reactions or, more rapidly, by exposure to far red light.

The recognition that many light-dependent developmental phenomena in plants are mediated by the phytochrome pigment system came from studies in numerous laboratories in many parts of the world. However, the pioneering studies on action spectra and the first successful isolation of phytochrome were accomplishments of scientists at the U.S. Department of Agriculture Plant Industry Station in Beltsville, Maryland. The pioneer work on photoperiodic induction of flowering was begun by Garner and Allard. Careful studies on action spectra were contributed by Borthwick and Hendricks, and finally the phytochrome pigment was isolated by Siegelman and Butler. Among the responses now known to be under control of phytochrome are day length-dependent flowering, light-dependent seed germination, production of anthocyanin

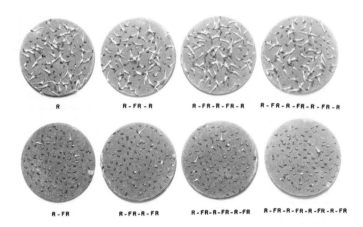

R　　R-FR-R　　R-FR-R-FR-R　　R-FR-R-FR-R-FR-R

R-FR　　R-FR-R-FR　　R-FR-R-FR-R-FR　　R-FR-R-FR-R-FR-R-FR

Fig. 7-16. Effects of light on the germination of lettuce seeds. Red light (R) promotes, while far red light (FR) inhibits germination. When exposed to red and far red light in sequence, it is the quality of the light used last that determines the response. (Courtesy of Harry A. Borthwick, U.S. Department of Agriculture.)

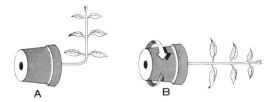

Fig. 7-17. A plant placed in a horizontal position soon turns upward (A), but if rotated in a horizontal plane, such a plant will continue to grow normally (B). (From Moment, G. B., and H. M. Habermann. 1973. *Biology: A Full Spectrum.* The Williams & Wilkins Co., Baltimore.)

pigments in fruits and leaves, stem elongation, leaf expansion, sleep movements of leaves, and the straightening of seedlings as they emerge from the soil.

GRAVITY

Everyone knows that plant shoots grow up, and roots grow down. The environmental factor controlling this phenomenon is the earth's gravitational field; the growth response to gravity is called **geotropism.** The mechanism of geotropism, like that of phototropism, involves the asymmetric distribution of the plant hormone, auxin (see Chapter 18). Plant physiologists have long questioned the effects on plant morphology of an absence of gravity. The original ingenious approach was to tie potted plants to a rapidly turning water wheel. This was done by Thomas Knight, an 18th century horticulturalist and forerunner of Mendel in crossbreeding peas. The later 19th century approach was to use a simple instrument called **a clinostat.** Such experiments established that a plant can grow quite normally in a horizontal position if it is rotated continually around the horizontal axis (Fig. 7-17). Recently, with the advent of the space age, it has become possible to test the effects of much reduced gravitational fields on developing seedlings and excised plant shoots by monitoring their response while in orbiting satellites. After retrieval, orbited materials can be compared to controls maintained under normal conditions on earth. An unexpected result of space experiments has been evidence that weightlessness does not appear to impair the development of seedlings. If anything, a slight stimulation of growth has been observed.

Internal Control of Development

So far we have considered the influence of external, i.e. environmental, controls over plant development: light, soil, water, temper-ature, and the like. Plants respond to all these factors, but the manner of any plant's response depends on its genetic makeup. Obviously no amount of nutrients or water will convert a daisy into a sunflower. Often the responses of a plant are mediated by hormones the plant synthesizes. There is another and, in a sense, more fundamental way in which the nature of plant development is subject to internal controls. This is the master plan, the information in the genetic code which dictates the patterns of development and the sequence of events in the life cycle. In the nucleus of each cell there is a complete set of instructions for the plant.

EVIDENCE THAT THE NUCLEUS CONTROLS CELLULAR DIFFERENTIATION

One of the most convincing and informative organisms, either plant or animal, in which to investigate the role of the nucleus is the remarkable *Acetabularia,* a single-celled marine alga. This plant has cells 10–25 cm long that differentiate to form a long stalk anchored by a rhizoidal (rootlike) base and an umbrella-shaped cap. Because of their unusual capacity for regeneration and the fact that *Acetabularia* can be cultured in the laboratory, these single-celled plants have been utilized for many studies of regeneration, the influence of the nucleus on cell development, and the interaction between the nucleus and the cytoplasm. If the cap of an *Acetabularia* cell is cut off, the basal segment can regenerate a new cap.

Apical and basal segments of *Acetabularia* can be grafted together by pushing together the cut ends of the two segments and thus bringing their cytoplasms into contact.

It is possible to graft together cell segments from two different species of *Acetabularia* that differ in cap morphology (Fig. 7-18). When a capless stalk from *A. mediterranea* is grafted onto a short basal piece of *A. crenulata* stalk with its nucleus-containing rhizoid, a new cap will regenerate. The shape of the new cap will be intermediate between the two species, indicating an interaction between cytoplasmic substances in the *A. mediterranea* stalk and substances produced by the *A. crenulata* nucleus in the rhizoid. However, if the new cap is removed, the second regenerated cap is of the *A. crenulata* type. Apparently the cytoplasmic cap-forming substance originally present is used up during formation of the first cap; so, when the second cap is formed, its morphology is determined by cap-forming substance of the *A. crenulata* type. This indicates that synthesis is now under the total control of the *A. crenulata* nucleus in the

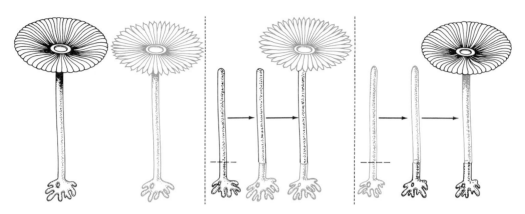

Fig. 7-18. Grafting experiments with different species of *Acetabularia* demonstrate the effect of the nucleus on development. *A. mediterranea* is shown in black, *A. crenulata* in color. When a young *A. mediterranea* is cut in half and the top is grafted onto an *A. crenulata* rhizoid, the grafted cell will develop a cap characteristic of *A. crenulata*. A young *A. crenulata* stalk grafted onto an *A. mediterranea* rhizoid forms an *A. mediterranea* cap. The type of cap formed corresponds to the kind of nucleus in the rhizoid. (From Moment, G. B., and H. M. Habermann. 1973. *Biology: A Full Spectrum.* The Williams & Wilkins Co., Baltimore.)

rhizoid. The reciprocal graft (*A. crenulata* stalk on *A. mediterranea* base) similarly produces an intermediate type cap at first; afterwards all new caps are of the type corresponding to the nucleus present in the base.

Although the nature of the cap-forming substance (or substances) is not known, it is probable that messenger RNA's synthesized in the nucleus are involved as the templates for synthesis of the enzymes and structural proteins utilized in the formation of new cap material.

EVIDENCE OF TOTIPOTENCY FROM TISSUE CULTURE

In multicellular organisms only a fraction of the information in the DNA of each cell nucleus is utilized. Cells and tissues differ not only with respect to the details of their structure but also in their complement of enzymes and chemical products. It has long puzzled students of development how an organism's various cells can develop so differently from the same set of instructions. The position of any cell within the plant appears to have a large effect on its developmental pattern because the position of a cell determines the conditions under which it develops. In plants, however, individual cells appear to retain great developmental potential. This can readily be demonstrated when stem cuttings form new roots, or when new plants form from leaves.

The term **totipotency** has been used by both zoologists and botanists to express the idea that the nucleus of any cell contains a complete set of genetic information for the organism. Therefore, an entire plant or ani-

mal can, at least in theory, be regenerated from a single cell provided that conditions inducing multiplication and differentiation can be provided.

Evidence supporting the concept of totipotency comes from studies of the growth of plant tissues in culture. Many plant scientists have worked to develop media in which isolated plant parts can be cultured. F. C. Steward and his collaborators at Cornell University studied the conditions necessary for proliferation of pieces of mature phloem tissue from carrot roots. Small explants of carrot phloem were removed aseptically and transferred to media containing minerals plus a carbohydrate source and necessary growth factors. Under appropriate conditions of temperature and aeration, cell division occurs in these explants; the tissue increases in size and weight. New cells differ from those removed from the carrot root, however. The undifferentiated and unorganized mass of new cells is called a **callus**. Steward found that for prolonged growth it is necessary to add special growth factors to the medium. The liquid endosperm of coconuts was found to be a convenient source of these factors, now recognized to be a class of plant hormones called the cytokinins (see Chapter 18). In time, differentiation can be observed within the callus. Growth centers of lignified elements that resemble the cells of vascular tissues form. Further growth and organization leads to the formation of roots. If a rooted callus is transferred from the original liquid medium to agar and added plant hormones are carefully controlled, buds can be induced which give rise to shoots. Thus, in culture, a sequence of events including proliferation of cells, dedifferentiation of the cells of mature tissues, and redifferentiation of callus cells with initiation of new roots, shoots, and finally entire new plants, is possible (Figs. 7-19 and 7-20).

Even more intriguing, however, are the events following isolation of single cells from the mass of carrot callus. Such single-cell isolates divide and follow a pattern of development which closely parallels that of an embryo developing from the zygote. Thus, removed from the normal restraints imposed by the surrounding cells within a tissue, and

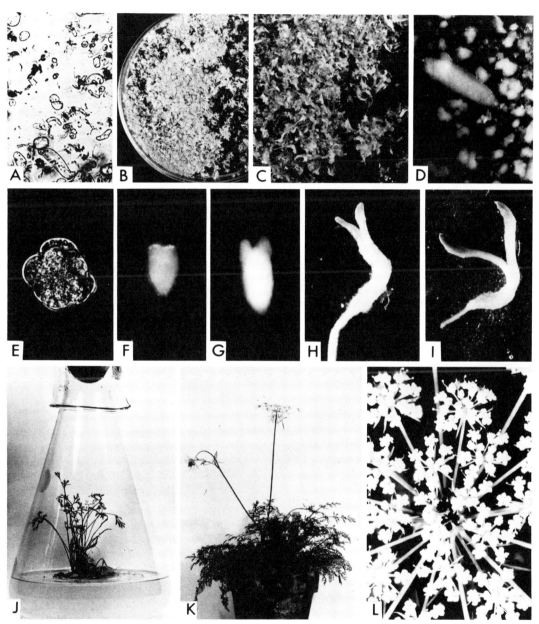

Fig. 7-19. Development of carrot plants from free cell suspensions. A, carrot cells grown in liquid medium; B, growth of embryoids from carrot cell suspension transferred to agar plate; C, higher magnification of embryoids shown in B; D, growth of globular masses in liquid medium. These were derived from a cell suspension similar to that shown in A. One torpedo stage of embryo is seen. E, globular mass of cells developed from free cell suspension; F, heart-shaped embryoid; G, torpedo-shaped embryoid; H, young plantlet developed from embryoid; I, older plantlet; J, carrot plant developed from embryoid growing in agar medium; K, carrot plant in flower after about 6 mo of growth starting with free cell suspension; L, close-up of inflorescence of *Daucus carota* plant grown from cells of embryo origin. (From Steward, F. C. 1964. Growth and development of cultured plant cells. *Science* 143:3601. Copyright © 1964 by the American Association for the Advancement of Science.)

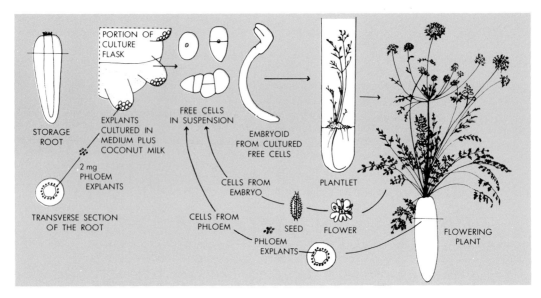

Fig. 7-20. Carrot cells to new carrot plants. Diagram showing sequence from mature phloem explants or isolated embryos to proliferating cultures, free cells in suspension, embryoids, plantlets, and mature carrot plants. (From Steward, F. C. 1964. Growth and development of cultured plant cells. *Science* 143:3601. Copyright © 1964 by the American Association for the Advancement of Science.)

provided with nutrients normally available to the zygote, developmental events occur which ordinarily only follow fertilization. First, polarity is established, and then the growing cell mass goes through a filamentous stage. Later, recognizable parts of a plant embryo (cotyledons, epicotyl, and hypocotyl) are formed, and ultimately an entire plant can be regenerated from a single diploid cell.

Other recent studies have indicated unusual regenerative powers in tissue cultures derived from parts of the carrot plant other than mature phloem in the root. Wetherell and his associates at the University of Connecticut have shown that immature embryos of the wild carrot can be grown in tissue culture. If such developing embryos are disintegrated, entire new embryos can be regenerated from individual cells or group cells. Thus, from a single embryo it is possible to obtain literally thousands of genetically identical plants.

Support for the Concept of Totipotency from Molecular Biology

Except for cells which are the products of meiosis, special polyploid cells in the endosperm, and interphase cells about to divide, every cell of a seed plant contains the same amount and kind of DNA. Evidence that the complete genetic complement is retained in differentiated cells and can be utilized under

appropriate conditions has been reviewed in the description of tissue culture experiments in the preceding section. However, differentiated cells differ greatly in the amounts and kinds of proteins and other cellular constituents normally present. Thus, it is obvious that much (if not most) of the genetic information available to an individual cell is ordinarily inert or **repressed**. Clearly, there must be some mechanism that determines in which cells and at what times during development a particular gene will be active (**derepressed**) and can be used as a template for synthesis of its characteristic messenger RNA. As the plant embryo develops from the zygote, cells begin to differ from one another and to acquire the characteristics of the specialized cells of the adult organism. Within the nucleus of each cell there is a program which determines the proper sequence of repression and derepression of genes necessary for orderly development. This program for the sequence of gene activity must itself be a part of the genetic information, since the course of development and final form are inheritable.

We must turn to molecular biology for insights into the mechanisms which control the repression and derepression of genes. It is now possible to break open cells and to separate the nuclear material from other cell components. The isolated chromatin thus obtained can be used as a template for synthesis of messenger RNA. The product of this DNA-dependent RNA synthesis can then be used in an in vitro protein-synthesizing system as a template for protein synthesis. The proteins formed can be compared with those normally synthesized by cells from which the chromosomal DNA was isolated. These procedures have been followed by James Bonner and Ru Chih Huang in their studies of the protein globulin which is synthesized and stored in the cotyledons of germinating peas. This globulin is not found in other tissues of the pea plant.

Once purified chromatin is obtained, it can be used as a template for messenger RNA synthesis. The messenger RNA synthesized in this step can then be used as a template for protein synthesis. The extent of amino acid incorporation into protein can most readily be followed by labeling the amino acids with radioactive carbon (^{14}C).

Following the procedures just outlined, Bonner and Huang prepared chromatin from pea seedling cotyledons and also from apical bud tissues. They utilized the chromatin from these two sources as the templates for RNA synthesis, used the mRNA's thus formed as the templates for protein synthesis, and finally compared the kinds of proteins formed.

Their experimental results indicate that isolated chromatin can function in the same manner as it does in the nuclei of intact cells. Genetic information that is inert in the intact nuclei of cells of the apical meristem of the pea seedling (i.e. information needed for globulin synthesis) remains repressed in the chromatin isolated from this source. On the other hand, the significant amounts of globulin that are synthesized from the genetic information available in the chromatin isolated from cotyledonary tissues correspond to the amounts synthesized in the corresponding intact tissue.

The obvious question raised by these findings is: What is the mechanism by which part of the genetic information remains inert (or repressed)? Bonner and his associates were aware of the fact that, in isolated chromatin, large amounts of protein are bound to DNA. A major portion of this bound protein consists of histones. It was found that bound proteins can be separated from DNA by dispersing isolated chromatin in 4 molar cesium chloride. The high ionic strength of the solution disrupts the ionic bonds linking the protein to DNA. DNA can then be separated from protein by centrifugation. The resulting deproteinized DNA can still support messenger RNA synthesis.

After removal of histones, the DNA's isolated from two different pea tissues have equal capacity to support globulin synthesis. However, the percentage of globulins produced from deproteinized cotyledonary chromatin is reduced because the genes for synthesis of many other proteins have been derepressed. While these experiments support the hypothesis that the histones bound to DNA can act as repressors, they do not prove that all repressors of gene activity are histones. Furthermore, we are left with the question of how repressors (histones, other bound proteins, or whatever) are selectively removed so that appropriate bits of genetic information become available at appropriate times in development. Certainly the removal of histones by subjecting isolated chromatin to high salt concentrations is a technique that cannot be used by living cells.

HORMONAL REGULATION OF DEVELOPMENT

There is increasing evidence that hormones act as derepressors of genes. The classic definition of a **hormone** is that it is a specific chemical synthesized in one organ and affecting another organ (the target organ). Generally, very small amounts of hormones have very large effects. How are such amazingly amplified effects achieved?

It is now generally agreed that the application of a hormone to its target organ derepresses previously repressed genes which control the production of specific enzymes. Thus, hormones cause the production of specific new species of messenger RNA. Once again, it is appropriate to ask how such derepression is brought about. One possible mechanism is that a hormone binds with a specific protein repressor (presumably histone), changing the conformation of that protein so that the resulting hormone-protein complex no longer can act as a repressor.

In addition to genes which can be derepressed by the presence of specific small molecules, i.e. hormones, there are also genes which are known to be regulated by specific metabolic products. Such genes can be turned on (or off) by the presence or absence of a product of enzymatic activity. For instance, in higher plants the enzyme nitrate reductase is not formed so long as an adequate level of nitrogen in the form of ammonium ions is available. This is clearly an example of the repression of genetic information by the product (NH_4^+) of an enzyme (nitrate reductase).

PHYTOCHROME AND DIFFERENTIAL GENE ACTIVATION

Although the mode of action of phytochrome is still not completely understood, there is considerable evidence that P_{FR} (the far red absorbing form) is the effector molecule which can trigger many varied responses. Although at first it was difficult to imagine how the small changes in a few pigment molecules could be amplified to produce such large and obvious changes in the whole plant, there is now good evidence that two basic mechanisms exist which make amplification possible. The first operates through **differential gene activation,** a mechanism proposed by Hans Mohr. His studies of phytochrome-induced anthocyanin synthesis in mustard seedlings indicate that P_{FR} promotes the synthesis of mRNA specific for enzymes involved in formation of these blue or red pigments. Phytochrome-mediated pigment synthesis can be blocked either by actinomycin D (which inhibits DNA-dependent RNA synthesis) or by puromycin (which inhibits mRNA-dependent protein synthesis). An amplification of such hormonal response can readily be understood if we consider that a number of mRNA's can be synthesized from the information in one activated gene, a number of enzyme molecules can be synthesized per mRNA, and each enzyme molecule can catalyze a given reaction many times.

A second mechanism of phytochrome action involves the control of **membrane permeability.** Some responses are too rapid to be explained by the multiple steps in gene activation and protein synthesis which require a minimum of several hours to be observed. Changes such as the movement of leaflets of *Mimosa pudica* (the sensitive plant) are evident within times as short as 5 sec. When these plants are irradiated with far red

light and then transferred to the dark, their leaflets remain open for many hours; but the leaflets close rapidly when exposed to red light prior to the dark period (Fig. 7-21). Such immediate responses are most readily explained in terms of changes in membrane permeability which are known to occur with great speed.

HAPLOID PLANTS

In discussing the topic of totipotency we emphasized that any somatic cell of a diploid plant (and apparently any animal, also) contains all the information needed to produce the entire organism. But one can raise serious questions about the necessity for all this DNA. Do diploid organisms really require duplicate sets of information? Recent work indicates that somatic cells of plants may possess more information than they actually need. Several years ago Nitsch and Nitsch at Gif-sur-Yvette in France succeeded in growing haploid (monoploid) tobacco plants from single pollen grains. The ability to produce a haploid plant at will has certain practical advantages. As you recall, the pink mold, *Neurospora crassa,* has been widely used in genetic studies precisely because it is haploid: induced mutations are readily visible; genotype and phenotype are the same. Thus, if haploid higher plants could be produced at will, mutations could be detected immediately and the desired changes in their genotypes preserved. The means for doubling chromosome numbers are available, for example, applying colchicine or by the process

FR   FR R FR   FR R FR R FR

FR R   FR R FR R   FR R FR R FR R

Fig. 7-21. Effects of light quality on sleep movements of *Mimosa pudica* leaves. Leaflets normally open in the light and close in the dark. Exposure to red light (R) prior to darkness promotes closure while far red light (FR) prior to darkness delays closure. Effects of red and far red in sequence are comparable to their effects on lettuce seed germination. The quality of the last light period governs the response. (From Fondeville, J. C., H. A. Borthwick, and S. B. Hendrick. 1966. Leaflet movement of *Mimosa pudica* L. Indicative of phytochrome action. Planta (Berlin), 69:357–364.)

of **endomitosis,** a doubling of chromosomes without nuclear division. Callus cultures tend to undergo endomitosis when grown on special media. Thus, completely homozygous individuals, a rarity among plants, can be obtained (Fig. 7-22).

The secret of growing monoploid tobacco plants from pollen grains is to remove stamens at a very early stage of flower development and to culture them on a suitable medium. After 3–4 weeks, embryos and plantlets emerge (Fig. 7-23) which can be transferred to a simpler medium. Once a root system has formed, the young plants can be transferred to pots where they will grow to maturity and flower. One strange (and perhaps reassuring) aspect of the development of embryos from pollen grains is that their pattern of development follows precisely the same steps seen in the normal development of the embryo plant from a zygote (Fig. 7-24).

Fig. 7-23. Haploid (monoploid) tobacco plantlets emerging from anther cultured on a synthetic agar medium containing salts and sucrose. (From Nitsch, J. 1972. Haploid plants from pollen. *Z. Pflanzenzücht.* 67:3–18.)

SOMATIC HYBRIDIZATION—NEW APPROACHES TO GENETIC ENGINEERING OF PLANTS

Recent advances in the technology of plant tissue culture which make possible the regeneration of whole new plants from single cells have laid the groundwork for exciting new approaches to the genetic engineering of plants, particularly for the improvement of crop plants. It is now possible to remove cell walls enzymatically. The resulting isolated **protoplasts** can be cultivated on a defined medium and will eventually regenerate new cell walls. Ultimately whole new plants can be grown from isolated protoplasts.

The significant thing about wall-less plant cells is that they can fuse—sometimes spontaneously, sometimes with persuasion. Fusion is enhanced by a variety of conditions such as alkaline pH and added calcium ions, or the presence of polyethylene glycol. Thus there is a means for combining two different genotypes, cytoplasms and nuclei from two plant species, even two species so different that they can never be hybridized by ordinary sexual crossing. The resulting fused protoplasts are called **heterokaryocytes.** Such heterokaryocytes have recently been reported for barley and soybean, pea and soybean, and for corn and soybean combinations. A fusion of nuclei to form true hybrids has been observed during the first mitotic division after fusion of the protoplasts. Ways to persuade such biologically (and commercially) interesting somatic cell hybrids to continue to divide and ultimately to grow into viable whole plants have yet to be developed.

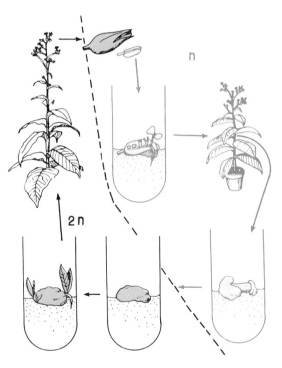

Fig. 7-22. Homozygous plants from cultured anthers. Flower buds excised from a diploid plant are grown in culture. The resulting haploid plants are transferred to pots. A piece of haploid stem or petiole is transferred to a synthetic medium stimulating callus growth. Doubling of chromosomes occurs during callus development. Diploid callus cultures are transferred to a medium that stimulates bud formation and give rise to diploid plants which are homozygous. (From Nitsch, J. P. 1969. Plant propagation at the cellular level, a basis for future developments. Proceedings of the International Plant Propagator's Society, vol. 19, pp. 123–132.)

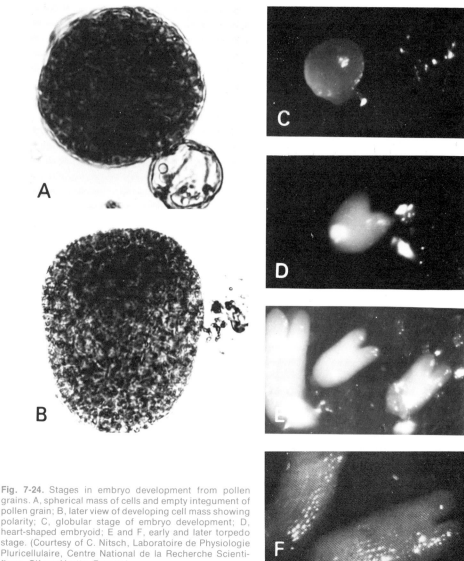

Fig. 7-24. Stages in embryo development from pollen grains. A, spherical mass of cells and empty integument of pollen grain; B, later view of developing cell mass showing polarity; C, globular stage of embryo development; D, heart-shaped embryoid; E and F, early and later torpedo stage. (Courtesy of C. Nitsch, Laboratoire de Physiologie Pluricellulaire, Centre National de la Recherche Scientifique, Gif-sur-Yvette, France.)

However, it has been established that these procedures are at least theoretically workable by utilizing biochemically distinguishable strains of tobacco (Fig. 7-25). The possibilities of a corn-soybean hybrid that can obtain its nitrogen from the air through the cooperation of symbiotic nitrogen fixing bacteria (a characteristic of its soybean "parent") and simultaneously has the biochemical potential for doing the very efficient C_4 type of photosynthesis (as does corn) with the added nutritional potential of both soybeans and corn are indeed astounding.

Naked plant protoplasts present still further possibilities for introducing desirable new characteristics into crop plants. Cellulose cell walls are an insurmountable barrier for macromolecules and organelles. Naked protoplasts, on the other hand, can readily engulf rather large particles from bits of DNA to viruses or even whole organelles such as chloroplasts and mitochondria. Thus it appears possible that certain desirable genes from microorganisms (such as the instructions for fixation of atmospheric nitrogen) might eventually become incorporated into the genomes of higher plants. Some experiments of this kind have already been performed successfully. Genes carrying information for the utilization of the sugars lactose and galactose have been transferred by means of bacteriophages (bacterial viruses) into naked proto-

NICOTIANA LANGSDORFFII   NICOTIANA GLAUCA

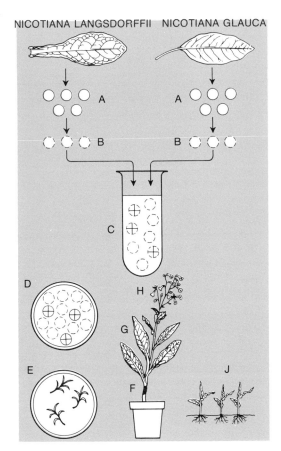

Fig. 7-25. Procedures for somatic hybridization by fusion of protoplasts. Leaf mesophyll cells (A) of two "parental" species are separated by enzymes that digest away cell walls leaving "naked" protoplasts (B). Protoplasts are suspended in NaNO₃ solution (C) and then plated on an agar medium (D). Fused hybrid cells grow and differentiate (E) into leaves and stem that are grafted (F) onto an appropriate stock. The hybrid scion matures (G) producing fertile flowers (H) and seeds. The seeds germinate (J). Seedlings are similar to sexual hybrids of the two parental species. (Redrawn after Smith, H. H. 1974. Model systems for somatic cell genetics. *Bioscience* 24:269–280.)

plasts prepared from cultured cells of a variety of plants including tomato and maple (plants that do not have these specific genes). The bacteriophages used in these experiments had acquired the genes from bacteria through the process of transduction (see Chapter 5).

Aging, Senescence, and Death

Old age is a predictable phase in the life of any multicellular organism, except those meeting accidental death. Associated with

aging and senescence is a loss of vigor and deterioration of cells and tissues. In annual plants, senescence usually follows the development of the reproductive structures (flowers, fruits, and seeds), and death of the organism is precipitated by changes in climate, such as the first frost. The underlying causes of aging in either animals or plants are unknown. Consequently, an intensive search for clues is underway in laboratories throughout the world.

JUVENILE vs. ADULT MORPHOLOGY

Although senescence and death are the final phases in the life cycle, they are not the only changes associated with the aging process. There is a constant turnover in cellular constituents from the time of fertilization in animals, and from the time of germination in seed plants. Furthermore, plants undergo changes in morphology as they grow that are comparable in many ways to the morphological changes so familiar in animals. We can thus recognize certain shapes and proportions as characteristic of juvenile or adult forms. Such differences in shape are readily observable in leaves formed at different stages of development. Some examples of these differences in leaf shape in ivy are shown in Figure 7-26. As leaves are products of the apical meristem of the shoot, it follows that there must be changes in the apical meristem preceding changes in leaf morphology. An even more drastic change occurs in this region when vegetative growth stops and the onset of reproductive growth begins. With the onset of flowering, massive changes in the shape and organization of the apical meristem itself can be observed. Can we then conclude that the meristem itself ages?

There are two views concerning this question. According to the first view, meristems do not age. Juvenile or adult leaves, or the initiation of flowering, result from a balance of factors such as supply of mineral nutrients, carbohydrates, nitrogenous substances, and growth factors. This view implies that the di-

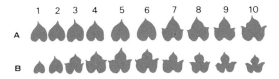

Fig. 7-26. Changing morphology of ivy leaves formed at successive nodes. A, leaves formed at first 10 nodes of a plant grown in the shade; B, leaves formed at first 10 nodes of a plant grown at a higher light intensity. (From Briggs, D., and S. M. Walters. 1969. *Plant Variation and Evolution.* McGraw-Hill Book Co., New York.)

rection and extent of meristematic activity are controlled by chemical and/or hormonal signals originating elsewhere in the plant and can be significantly influenced by environmental factors. According to the second hypothesis, meristems do age during the development of plants, so that an apical meristem can be classified as juvenile, adult, or senescent. An extreme interpretation of the latter view would predict an unalterable sequence of development set by the genetic code of the organism. Plant development actually follows a middle road, with the pattern set by heredity but modified by internal and external factors. If apical meristems are isolated in tissue cultures or as cuttings, they retain their characteristic organization and continue to form new appendages (leaves) characteristic of the species. When isolated surgically from adjacent tissues, but left in contact with the plant, an apical meristem can proceed to form a normal shoot even though connected with the rest of the plant only through undifferentiated pith cells.

As we have already stated, changes in apical meristems can be detected by changes in the products of their activity. Leaves formed at various stages of development can differ in their morphology. At the end of the last century, the German botanist Goebel used the term **heteroblastic development** to describe the situation where juvenile and adult leaves are strikingly different in shape. Such changes in leaf shape from node to node have been used as a measure of physiological age. Krenke, another German botanist whose works were published early in the 1900's, regarded successive nodes as units of physiological time. Leaves formed at successive nodes often differ in shape, and flowering normally follows a predictable succession of changes in leaf morphology.

The ideas of Goebel and Krenke have been used in studies of aging in the sunflower (*Helianthus annuus*). This species lends itself very well to studies of plant aging because its development is not only heteroblastic but also determinate, i.e. it develops a definite and predictable number of leaves. The latter trait is an unusual feature in plants, which are normally **indeterminate** in their growth patterns (i.e. most plants form an unpredictable number of appendages during vegetative growth). In the sunflower, however, the extent of vegetative growth, i.e. the number of leaves formed, is controlled by photoperiod. Thus, by maintaining a constant day length (13 hours), it is possible to control the number of leaves formed (14–16 pairs) prior to a shift in the activity of the single apical meristem to reproductive development.

Thus, a sunflower's apical meristem can be assigned a physiological age based on the number of leaves it has formed. Under conditions of controlled photoperiod, the remaining vegetative life expectancy of the meristem is predictable because its total life expectancy is fixed, and the number of leaves and primordia already formed can be counted.

Experiments can then be designed to answer the specific question: What effect would a changed biological environment have on the life expectancy of an apical meristem? The experimental approach to such a question is relatively simple because sunflowers can be grafted. Using this very old horticultural technique, the apical meristems of seedlings have been grafted onto physiologically older plants. By counting the number of leaves formed by a **scion** (the grafted seedling), it is possible to determine whether the older stock has influenced the rate of aging of the scion. The results clearly show that transfer to a physiologically older environment can accelerate the aging of an apical meristem.

The sum of the leaves on the stock plus scion always adds up to approximately the same total. A seedling grafted onto a physiologically old stock already in bud will form only two or three pairs of leaves before flowering (compared to 14 or 16 pairs normally formed by a sunflower plant). Accompanying such early flowering is a compression of the pattern of changes in leaf shape: morphological changes occur in fewer nodes (or units of physiological time). It is therefore obvious that the rest of the plant can influence the functioning of an apical meristem.

Although grafting provides a means for transferring young apical meristems to older environments, the nature of the chemical signal transferred from old tissues that triggers changes in the developmental patterns of physiologically young scions is as yet unknown. When such experiments are done in reverse and the old apical meristems are grafted onto seedlings, there is practically no influence of the stock on the scion. In other words, while exposure to older tissues can accelerate aging of young meristems, young tissues apparently are unable to influence the rate of aging of older meristems.

SENESCENCE

At any point in its life cycle, a higher plant is a mixture of young and old, living and dead components. In perennial plants such as the trees, life spans of several hundred years are not uncommon, yet there is an annual production of new tissues and organs. In deciduous species, new leaves are formed at the beginning and shed at the end of each growing season. In evergreen species such as the conifers, leaves persist for two to several seasons, but some leaves are replaced each year. If we examine the stems of older trees, we find that large amounts of the central xylem and also of the outer bark tissues are dead, so that only relatively narrow regions on either side of the vascular cambium are made up of living cells.

In annual plants, even while the entire bulk of the organism is composed of living tissues, we find a mixture of young, newly formed tissues, older organs, and aging parts. Leaves at the base, formed early in vegetative growth, are obviously older physiologically than recently formed leaves near the apical meristem. We can expect to find differences in morphology, composition, and metabolism related to plant age and leaf position.

Most studies of aging in plants have been concentrated on changes in single tissues or organs after removal from the plant. Such isolated plant parts, or **explants,** do deteriorate with time and undergo chemical and physical changes comparable to those occurring in intact plants. The most obvious of these changes are the decomposition of chlorophyll and hydrolysis of proteins.

Delay of Senescence by Cytokinins

There is a whole class of naturally occurring plant hormones, the **cytokinins** (see Chapter 18), that have the general effect of slowing the changes in plant tissues characteristic of senescence. A knowledge of the effects of these hormones has already been applied in the preservation of vegetables such as spinach and broccoli after harvesting. You will recall that the cytokinins are now known to be the mysterious components in the coconut milk used by Steward and others to supply the factors needed for continued cell division in plant tissue cultures. The cytokinins are essential throughout the life of a plant beginning with the earliest cell divisions in the developing embryo.

LEAF ABSCISSION

Among the most interesting studies of aging phenomena using explants have been the investigations of factors controlling **leaf abscission,** the separation of leaves from the stem in autumn. The loss of leaves in the fall is usually preceded by the formation of an abscission layer across the base of the petiole. The weak, thin-walled cells of the abscission layer soon disintegrate, leaving only the strands of conducting tissues in the petiole to support the leaf. Many factors are known to influence the formation of the abscission layer. Abeles, Rubenstein, and their co-workers at the former U.S. Army Biological Laboratories in Frederick, Maryland, have shown that all factors promoting leaf abscission have one effect in common: they induce the production of ethylene. Furthermore, they have shown that the introduction of ethylene gas during incubation (Fig. 7-27) of petiole explants can accelerate abscission. The mechanism by which ethylene promotes abscission appears to be comparable to the way in which hormones control development. Abeles and Holm have shown that ethylene promotes the synthesis of new kinds of messenger RNA and that inhibitors of RNA and protein synthesis can counteract the promoting effects of ethylene on abscission. Thus, ethylene can be classed with other hormones which act as specific derepressors of genetic information.

A Look to the Future

In the development of plants, there is a constant interplay of internal factors (genes and hormones) and the essential components of environment such as water, soil, light, at-

Fig. 7-27. A, freshly cut explant of cotton includes section of stem and bases of two petioles. B, explant of cotton after formation of abscission layer and drop of petioles. Ethylene promotes changes associated with abscission. (Courtesy of F. B. Abeles.)

mospheric gases, and climate. The earliest studies of plant development were descriptive, dealing primarily with establishing the sequence of events in the life history of a given species. Once the predictability of development and life cycles was recognized, more penetrating questions could be asked: What is the basis for this predictability and for the orderliness of development in living things? How can the patterns of development be modified? In recent decades the ideas and tools of molecular biology have brought new insights to studies of plant development, and of animal development as well. In the decades ahead we can look for a better understanding of the ways in which the blueprint for an organism provided by its genetic code is read and carried out.

USEFUL REFERENCES

Bonner, J. 1965. *The Molecular Biology of Development.* Oxford University Press, London. (A classic work on the developmental aspects of molecular biology.)

Cutter, E. G. 1969/71. *Plant Anatomy: Experiment and Interpretation.* Part I: Cells and Tissues; Part II: Organs. Addison-Wesley Publishing Co., Reading, Mass, (Paperback.)

Galston, A., and P. J. Davies. 1970. *Control Mechanisms in Plant Development.* Prentice-Hall Inc., Engelwood Cliffs, N.J. (Paperback.)

Gemmell, A. R. 1969/71. *Developmental Plant Anatomy.* Institute of Biology Studies in Biology No. 15. Edward Arnold (Publishers), Ltd., London. Distributed in U.S. by Crane, Russak & Co., Inc., New York (Paperback.)

Hess, D. 1975. *Plant Physiology. Molecular, Biochemical and Physiological Fundamentals of Metabolism and Development.* Springer-Verlag, New York. (Textbook of plant physiology emphasizing molecular aspects of development; paperback.)

O'Brien, T. P., and M. E. McCully. 1969. *Plant Structure and Development: A Pictorial and Physiological Approach.* The Macmillan Co., New York, and Crollier-Macmillan Canada, Ltd., Toronto. (Paperback.)

Smith, H. H. 1974. Model systems for somatic cell plant genetics. *BioScience* 24:269–276. (Research paper summarizing new approaches to plant breeding.)

Steward, F. C., ed. 1972. *Plant Physiology: A Treatise.* Vol. VI C: Physiology of Development: From Seeds to Sexuality. Academic Press, Inc., New York.

Torrey, J. G. 1967. *Development in Flowering Plants.* The Macmillan Co., New York, and Crollier-Macmillan Canada, Ltd., Toronto. (Paperback.)

8

Reproduction in Animals

Biological Meaning of Reproduction: Sexual and Asexual

Reproduction involves two conflicting requirements if life is to continue and if evolution is to take place. First is the basic requirement of self-duplication. Second is the production of novelty.

Without the first, obviously, life would not continue to exist. On the molecular level reproduction means replication of DNA. For cells it means mitosis. On the level of multicellular organisms reproduction means either some kind of asexual division in which fairly large replicates can be produced directly, or some kind of sexual reproduction in which the organism is reduced to the level of single cells. Why this second method?

The answer is the production of novelty, or variation. In a very real sense this is the opposite of reproduction. At the basic molecular level novelty is the result of two factors: errors in the replication of deoxyribonucleic acid, i.e. gene mutations, and the inherent properties of the DNA, for the kinds of errors that can occur reflect the chemical nature of nucleic acids.

The basic novelty produced on the molecular level is compounded on the level of cells by sexuality, that is, by the making of new combinations and permutations of chromosomes in the cycle of meiosis and fertilization. The variation produced by meiosis alone may be very great. With 23 pairs of chromosomes, each human may produce over 8,000,000 kinds of eggs or sperms. Fertilization multiplies these two possibilities by each

other in the production of the zygote, 8×10^6 multiplied by 8×10^6 equals 64×10^{12}, the number of possible kinds of zygotes from any marriage, and what a very large number it is — 64,000,000,000,000! Equally important, sexual reproduction makes it possible to bring together in a single individual and his or her descendents beneficial mutations which arose in different individuals. Clearly then, sexual reproduction is a highly important part of the machinery of evolution. This is, in fact, what sex is all about from the biological point of view. From the evolutionary point of view, asexual reproduction can advance only at a snail's pace.

This does not mean that asexual reproduction is unimportant among living things. Once a well-adapted animal has been produced as a result of mutation, sexual combinations, and natural selection, asexual reproduction can produce standardized duplicates in enormous numbers. Along side of sexual reproduction, asexual reproduction is frequent among protozoans, coelenterates, the parasitic flatworms, tapeworms, some annelids, and even in certain insects.

Adaptations to Ensure the Meeting of Gametes

TEMPERATURE AND LIGHT

If eggs and sperms are to meet, the spawning of animals which merely discharge their gametes into the sea water must be synchronous. In animals which mate, there must be mechanisms which bring both sexes into a state of sexual readiness and together at the

151

same time of the year. Two of the earliest and most widespread of such adaptive mechanisms are responsiveness to temperature or light.

Temperature is said to be the chief factor in triggering the shedding of gametes into the water for oysters and many other aquatic organisms. However, some animals such as the hydroids among the coelenterates have been very clearly shown to shed eggs and sperms at very definite times after the stimulus of light.

Throughout most of the vertebrates, and perhaps in all, photoperiod, namely the relative number of hours of light and dark in every 24, sets the timing for annual breeding periods. This discovery was made by William Rowan in Edmonton in northwestern Canada. Like many people before him, he was impressed by the astonishingly precise timing of migration and breeding of several species of birds. Unlike anyone preceding him in human history, he discerned the explanation. The temperature, food supply, and weather conditions varied greatly from year to year. The one factor that was precisely constant was the increasing number of hours of daylight beginning after the winter solstice on December 22. By artificially increasing the length of daylight each day, Rowan was able to prove that his theory that gradually increasing day length is the stimulus which leads to the annual breeding and migration. Juncos in outdoor cages in the bitter cold of the Alberta winter responded to increasing periods of daylight by a dramatic increase in the size of their gonads. When released, such birds flew northward in what resembled a migratory flight. This phenomenon resembles the control of flowering which Garner and Allard of the U.S. Department of Agriculture had demonstrated. Many plants blossom not under the stimulus of nutritional or thermal conditions but when the photoperiod is right for them.

Since Rowan's epoch-making discovery, photoperiod has been shown to affect reproduction in many animals, especially birds and insects. In the crayfish, the ovarian cycle and molting can be manipulated by controlling the photoperiod. On the practical side, in all industrialized nations, hens are stimulated to lay more eggs by artificially increasing day length.

The mechanisms of photoperiodic control of reproduction are very imperfectly understood. Apparently light perceived with the eyes stimulates the neurosecretory cells of the hypothalamus to send their secretion down into the anterior hypophysis via the pituitary portal system described in Chapter 20.

The role of light, acting through the pineal gland, has been discovered in the case of rats and a few other mammals. It is interesting to note that there are long-day and short-day plants (see Chapter 9.) The sheep and goat are short-day animals which come into estrus or "heat" with the decreasing daylight hours of the fall. The raccoon and horse are long-day animals which come into estrus with the lengthening days of spring. In all cases, the reproductive activity of the individuals of the species is coordinated. Moreover, it is coordinated at such a time that the young will hatch or be born at a favorable time of the year.

SOUNDS, SIGHTS AND SMELLS

The mating calls of animals are familiar, although many of them serve equally as a warning to rival males. This is especially true of bird song which is produced almost exclusively by males and is an invitation to a female and a warning to other males.

Visual stimuli which attract one sex to the other are common. The flashing of fireflies is a conspicuous case. The two common species in the eastern United States illustrate this nicely. The low-flying *Photinus pyralis* emits a yellowish-orange flash while flying upward an inch or two. Individual flashes may be several minutes apart. The higher flying *Photuris pennsylvanica* emits a yellowish-green light in a series of three or four closely spaced flashes, giving a twinkling effect. Females answer the flashing of males by flashing themselves about a second after the flash of the male.

Teleost fishes, among the animals with color vision, recognize the sex of members of their species by sight. Among the sticklebacks, males have a scarlet belly in the breeding season. In birds, of course, color and pattern play a predominant role in sex identification and attraction. The fantastically gorgeous and exotic plumage of male birds of paradise, peacocks, and turkey gobblers illustrate how important visual stimuli are.

The overwhelming majority of mammals are color-blind. Bulls can no more see red than they can see blue or yellow. Typically, mammals find members of the opposite sex by odor. An odoriferous material secreted by one animal which affects a second is called a **pheromone,** a class of substances discussed in Chapter 1. The most familiar, although undetectable by the human olfactory epithelium is the pheromone secreted by a female dog in estrus. Similar substances are produced by mares, cows, does, and other mammals. Primates like ourselves are an exception. In this order of mammals, vision, not olfaction, is the dominant sense, for primates are primarily arboreal. In most primates the distinction between the sexes rests on differences in bodily form and behavior, although in male baboons secondary sexual traits include bright blue buttocks and fierce bright blue and red faces. Pheromones which are sex attractants play vital roles in the lives of many insects. Knowledge of them has proven useful in agriculture, making it possible to attract males to poisons or into traps.

COURTSHIP AND MATING

Animals present a wide range of courtship and mating behavior ranging from some of the flatworms at one extreme where no known equivalent of courtship exists and

Fig. 8-1. Male Adelie penguin in "ecstatic display" with bill pointed skyward and flippers (wings) extended. This is the male territorial display which attracts females and repels other males. Thus it is comparable to song in many other birds. (Courtesy of W. J. L. Sladen, The Johns Hopkins University.)

sperms are merely injected into the body of the receiving individual through the epidermis at almost any point. In most species of flatworms, however, the penis is inserted into a special passageway, the vagina.

The spiders represent an opposite extreme. The sex of a spider is extremely difficult to discover until after the final molt, when a male spider is conspicuous because of a complex bulbous apparatus on the end of its palps. Any spider that looks as though it were wearing boxing gloves is a mature male. Before mating, the male spins a small flat web on which he deposits a drop of semen. He then carefully fills the bulbs on the ends of his palps with this sperm-laden fluid. Thus equipped, he begins to seek a female. Male spiders have a number of signs which act as stimuli for the female, conveying the message that "here is a potential mate, not a meal." Some species signal the female by fantastic posturings with the legs. Other males begin by striking the female with their legs, while in still other species the males commence by tentatively and gingerly touching the tip of one of the female's legs. Many orb weavers begin by tweaking the web of the female. Males are very likely to be eaten by the females, if not before mating, then afterwards. In the act of copulation, if in spiders it can be called that, the male places the seminal bulb at the end of his palp against the genital opening of the female. Here the transfer of sperms takes place.

In the vertebrates courtship behavior often involves elaborate and highly ritualized displays. These species-specific patterns have been carefully studied in birds, particularly in different kinds of wild ducks. The males of each species show stereotyped hereditary sequences of head bobbing, wing spreading, tail waggings, preening, and swimming in circles. By contrast, in members of the parrot family known as "love birds," initial courtship seems to consist of merely looking at a member of the opposite sex. This happens well before the onset of sexual maturity in either member of the pair. After such mutual inspection for a period as short as 4 minutes, a pair of love birds are firmly mated for the rest of their lives. The prolonged and irreversible nature of this response suggests another example of very early and long-lasting learning, the attachment of offspring to parent known as **imprinting** which occurs in many birds and probably in some mammals.

NEUROLOGICAL BASIS

Built into the nervous system are behavioral patterns which are essential for the tim-

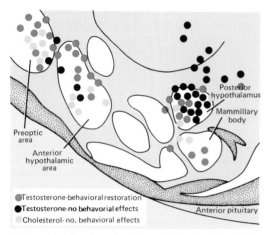

Posterior
hypothalamus

Mammillary
body

Preoptic
area

Anterior
hypothalamic
area

Anterior pituitary

- Testosterone-behavioral restoration
- Testosterone-no behavioral effects
- Cholesterol- no. behavioral effects

Fig. 8-2. Section through the hypothalamic region of the rat brain showing the method of locating the site where male sex hormone acts to influence reproductive behavior. Note the double control, testosterone injection over a wide area and the injection of a neutral substance, cholesterol. (Redrawn after J. M. Davidson.)

ing of reproduction by photoperiod or otherwise, and for bringing together and subsequent mating of the sexes. Any male mosquito with a defective nervous system that did not respond correctly to the "song" of the female or any male dog which could not smell the pheromone of a bitch in heat would contribute no genes to the following generation.

The neural mechanisms responsible are still very obscure. It has been found, however, that in both rats and cats male and female sex hormones act directly on specific centers within the hypothalamus of the brain. Minute amounts of testosterone, a male sex hormone, injected into precise areas of the anterior and posterior parts of the hypothalamus will restore sexual activity in castrated rats. A short distance outside these centers, testosterone is without effect while within the sensitive areas a neutral substance such as cholesterol is ineffective (Figs. 8-2 and 20-12).

Parental Care

Care of the young is an essential feature of reproduction in birds and mammals, in marked contrast to the rest of the animal kingdom, where parental care is extremely rare. Aristotle claimed that the male catfish guards the eggs and newly hatched fry. So unusual is such action that European zoolo-

gists regarded this tale as just one of the myths Aristotle accepted until Louis Agassiz in modern times discovered that male catfish in North America behave as Aristotle had described 2,000 years before! Female wolf spiders can sometimes be seen carrying spiderlings on their backs.

The necessity for parental care arises partly because of the evolution of intelligence, i.e. the ability to learn. If an animal is born with a full set of instincts and innate responses, i.e. prefabricated answers to the problems of life, the young are equipped to fend for themselves very early. If learning is to occur, then some protected period is required during which learning can take place. This is not the only factor in the evolution of parental care but it is an important one.

The kind of parental care and whether or not the male plays a part is adaptively correlated with the number of offspring, infant mortality, and, in birds, with the number of eggs laid and the stage of development of the young at birth. In penguins, one of the most intensively studied of birds, both sexes cooperate in incubating the single egg and feeding the young. Among ducks and chickens the male has a minor role after the eggs are fertilized. In such species the female lays a large number of eggs and the so-called **precocial** chicks are able to move about and feed themselves within a few hours after hatching. Infant mortality is high. Among song birds both sexes are deeply involved in incubating the eggs and in the care and feeding of the young. Such species lay only a small number of eggs, the helpless young, known as **altricial** *(altrix,* nurse), receive intensive parental care, and infant mortality is much lower.

In mammals the requirement for milk enforces at least a measure of maternal care which is usually both intensive and extensive. Paternal care varies widely. Male baboons show a great interest in very young baboons and solicitude for them. The males of the larger hoofed mammals commonly ignore their offspring except in defending them as members of the herd. Male foxes, wolves, and lions cooperate with their females not only in the care and education of the young but in the business of living generally. Male tigers live alone and are thought to eat newborn kits if the opportunity arises. How all these very different patterns of reproductive behavior are coordinated with all the other anatomical, physiological, hormonal, and behavioral adaptations in the lives of these animals is complex and very imperfectly understood — a challenge for future study.

The Anatomy of Reproduction

THE MALE REPRODUCTIVE SYSTEM

The male gonads or testes vary greatly in shape and number throughout the animal kingdom. Meiosis is, however, always the same. In nematode worms the testis has been much used to study meiosis because it is a continuous tube with primordial diploid germ cells at the distal end and mature haploid sperm at the proximal end. In between, the various stages of meiosis can be read in sequence like a printed sentence.

The testes of arthropods and of lower vertebrates, fish, and salamanders are much alike. Each testis is elongate and composed of a series of squarish compartments. Within each compartment all the cells are in approximately the same stage. At one end of the testis are primordial germ cells, and at the other are mature sperms, so that the observation of meiosis is not too difficult. Sex chromosomes were first discovered in the testis of the squash bug when it was noticed that in females there are two complete matching sets of chromosomes while in males there is one pair which does not match.

The testes of reptiles, birds, and mammals are all very similar. Each testis is composed of a mass of **seminiferous tubules** in which meiosis and sperm development take place. The diploid **spermatogonia,** which multiply by mitosis, lie around the periphery of the tubule; the **primary and secondary spermatocytes** lie closer to the lumen of the tubule; and finally the haploid **spermatids** lie closest to the center. As in other animals, the spermatids develop into mature sperms while embedded in the cytoplasm of nurse cells, usually called **Sertoli's cells.**

Between the seminiferous tubules are masses of **interstitial** or **Leydig's cells** which secrete **testosterone,** the male sex hormone.

Mature sperms pass through the lumens of the seminiferous tubules into the **epididymis,** a clump of coiled tubules lying against the testes, and finally into the **vas deferens.**

The **accessory sex organs** in different mammals are much the same. Most species possess a **seminal vesicle,** probably more accurately called a seminal gland because it is known to pour a secretion into the vas deferens, but whether it stores sperms is uncertain. Surrounding the junction of the vas deferens with the urethra from the urinary bladder is the **prostate gland.** Its secretion forms a major part of the **seminal fluid.** The position is unfortunate because in older males it has a tendency to enlarge and obstruct the flow of urine. A very small pair of glands, **Cowper's glands,** open into the urethra just below the prostate. Their secretion

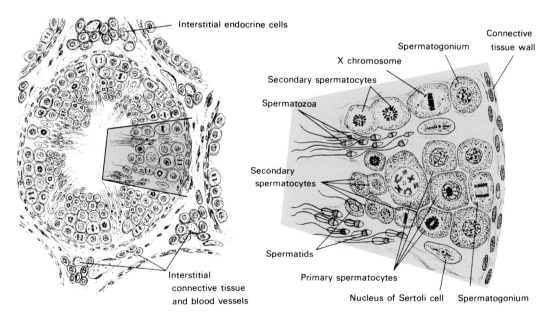

Fig. 8-3. Portion from cross section of a human testis, typical of all higher vertebrates. (After Arey, L. B. 1965. *Developmental Anatomy.* W. B. Saunders Co., Philadelphia.)

has a lubricating and acid-neutralizing function.

The seminal fluid is ejaculated into the vagina of the female from the **penis,** which is brought into a state of erection when specialized cavernous tissue within it becomes engorged with blood. In some mammals, notably the carnivores such as dogs and seals, the penis is further strengthened by a special bone.

The puzzling fact about the male reproductive system in mammals is the existence of the **scrotum.** This is essentially a muscular sac which acts as a thermoregulator, keeping the temperature of the testes several degrees below body temperature. If the temperature of the scrotum of an experimental animal is kept at body temperature either by insulated wrapping or by surgically placing it within the abdomen, spermatogenesis ceases, and the seminiferous tubules degenerate. This is probably why **cryptorchid** individuals, whose testes lie within the abdomen, are sterile. Yet the curious fact remains that in elephants, seals, and porpoises, the testes are normally and continuously abdominal. They are, also, in birds where the body temperature averages about 40–43°C. (104–108°F.), a dangerous fever for a mammal! In all females the ovaries are abdominal.

Male Secondary Sexual Characteristics

The **secondary sexual characteristics** of male mammals differ greatly among the different orders. The lion's mane and the moose's antlers are familiar examples. In porpoises and rodents secondary sexual characteristics are virtually nonexistent, except in behavior. In mammals, secondary sexual characteristics develop under control of gonadal hormones. In other animals they are independent of sex hormones or they are due to a combination of hormones and more local genetic action. In the common English house sparrow, for example, the color of the bill is a very sensitive indicator of the concentration of sex hormone in the blood, while the color pattern of feathers is independent of sex hormones and can even be seen in birds castrated while still in their nestling down. In other species of birds—chickens, for example—feather type and pigmentation are controlled by sex hormones.

THE FEMALE REPRODUCTIVE SYSTEM

The ovaries of animals vary considerably from one group to another. In flatworms and echinoderms, they are irregular in shape. In the much studied nematode worms, each of the two ovaries consists of a long tube in which meiosis occurs in a neat order as one looks along the ovary, continuous with the oviduct and vagina. In insects the ovary is usually a cluster of tubules, each with a single row of developing egg cells.

In lower vertebrates, fish and amphibians, the ovary is a hollow sac, although, as anyone who likes to eat fish roe knows, in the breeding season each of the two ovaries becomes solid with eggs with no central cavity apparent. In the higher vertebrates, the ovaries are solid and covered with a continuation of the peritoneal epithelium which lines the coelom. A peculiarity of birds is that only the left ovary and left oviduct develop.

The peritoneal epithelium over the surface of the ovary is known as the **germinal epithelium.** From it groups of cells containing descendants of the primordial germ cells migrate into the interior of the ovary. Here the germ cells begin meiosis. Each egg cell grows enormously in size and is surrounded by layers of cells between which a cavity forms. This cavity enlarges and becomes filled with **follicular fluid** rich in **estrogen,** the female sex hormone secreted by the cells surrounding the cavity. Outside these cells is a thin layer of smooth muscle fibers. The entire capsule is known as a **Graafian follicle.** The number of ova which mature at any one time depends on the species: normally one in women (the right and left ovaries are thought to alternate), three or four in vixen, seven or eight in mice. Some very delicate hormonal balance apparently controls this number.

In mammals, the **vagina,** or birth canal, located between the openings of the urethra and the alimentary tract, is merely a tube lined with mucoid epithelium which leads into the **uterus** or womb. At the opening of the uterus into the vagina is a thick muscular region, the **cervix.** In most mammals the uterus is more or less double. The inner end of each horn of the uterus is continued as an **oviduct,** known in human anatomy as a **Fallopian tube.** The oviduct on each side ends in a thin, funnel-shaped membrane that more or less encloses the ovary.

Fertilization takes place at the upper end of the oviduct. The stimulus of mating induces the release of a hormone, **oxytocin,** which causes uterine and oviductal contractions that carry the sperms up the oviduct in remarkably quick time; in some species only a few minutes are required to reach the upper end. The function of the swimming of sperms is chiefly to prevent them from becoming stuck on the inner walls of the reproductive

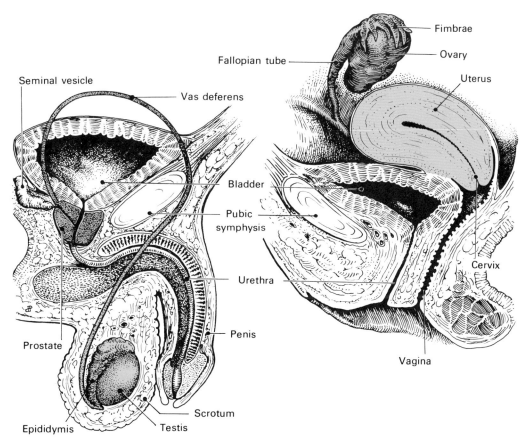

Fig. 8-4. Human male and female reproductive systems. (From Moment, G. B. 1967. *General Zoology,* 2nd ed. Houghton Mifflin Co., Boston.)

tract and perhaps to assist in the actual process of fertilization (see Fig. 6-7).

If pregnancy occurs, the developing egg and its surrounding cells signal the ovary by hormones. The result is that the cells of the ruptured follicles do not degenerate but instead proliferate and differentiate into a new structure, called from its yellowish color a **corpus luteum.** There are as many corpora lutea as there were ruptured follicles. They secrete **progesterone,** a hormone essential to maintain pregnancy.

Reproductive Hormones

It will be recalled that both male and female sex hormones are **steroids,** differing only slightly in the chemical groups that are attached to the basic sterol nucleus. Male sex hormones are termed **androgens,** and female sex hormones **estrogens. Prolactin,** a pitui-

tary hormone, stimulates the secretion of milk. **Oxytocin,** another proteinaceous pituitary hormone, causes the flow of milk by producing contraction of smooth mammary muscles. Oxytocin also stimulates the contractions of the female reproductive tract which carry sperms up the uterus and oviducts.

ESTROUS CYCLES

Ovulation in mammals occurs at more or less regular intervals: once a year for deer, every 4–6 months for dogs and cats, approximately every 28 days for women, every 21 days for mares and cows, and every 4 or 5 days for rodents. The period of ovulation, or **estrus** in many species, notably cats and cows, is accompanied by a period of restlessness and sexual receptivity known as "heat." In the higher primates the lining (**endometrium**) of the uterus undergoes growth, which prepares it for implantation of the blastocyst, culminating at about the time of ovulation. If no implantation and pregnancy take place, the hormones which have stimulated the growth of the endometrium recede, and the lining of the uterus is shed along with a certain amount of blood. This phenomenon is called **menstruation** from *mensis*, meaning month. Menstruation will be discussed later in this chapter. A species with but one estrous period a year is termed monestrous; a species with several, polyestrous.

The timing of estrus is under complex control. As mentioned above, in a very few cases—the rabbit and cat, for example—although the estrus condition is due to other factors, the actual shedding of the eggs from the ovaries is due to the stimulus of copulation. Psychic or at least neural factors are important in the mouse. A group of female mice showing a great deal of asynchrony and variation in their estrous cycles will become regulated by the presence of a male, even though he is separated from the females by a

fine mesh cage. Since merely the used bedding from a male will produce this effect, it is clear that the actual stimulus is odor. The African chacma baboon has an extremely regularized sexual cycle of 32 days, but this will be interrupted if the female is permitted to watch a fight between two other baboons. (This would result in an automatic limitation on population growth when numbers become great enough to make fighting frequent.)

Photoperiod, that is, the number of hours of daylight and of darkness, also has a profound influence on reproduction. In the rabbit light stimulates the retinal cells and sends impulses back to the brain. Artificially increasing the hours of daylight will bring wild rabbits into estrus even in the dead of winter. It has recently been shown that continuous 24-hour illumination results in an increase in the weight of the rat ovary and a decrease in the pineal gland, as discussed in chapter 6.

The anterior lobe of the pituitary secretes

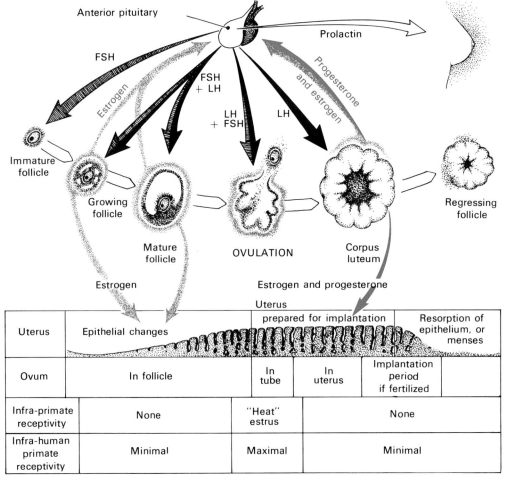

Fig. 8-5. Pituitary-ovary-uterus relationship. The two lower lines indicate behavioral correlates in infraprimate and in nonhuman primate species. (After Moment, G. B. 1967. *General Zoology*, 2nd ed. Houghton Mifflin Co., Boston.)

gonadotropin, the **FSH** (follicle-stimulating hormone) mentioned earlier, which stimulates the Graafian follicles to secrete estrone, the **estrogen** produced by the ovary. It has five main effects. (1) It stimulates the growth of the lining of the vagina and uterus. (2) It stimulates the growth of the milk-secreting ducts of the mammary glands and stimulates the development of female secondary sex characteristics in general. (3) It activates female reproductive behavior. (4) It stimulates the anterior pituitary to secret a luteinizing hormone, **LH,** which triggers the release of eggs by the follicles and stimulates the growth of the corpora lutea. (5) It inhibits the secretion of the follicle-stimulating hormone of the pituitary. Thus, it saws off the limb that supports it, for once the follicle-stimulating hormone of the pituitary is no longer present in the bloodstream, the follicles in the ovary regress, and the concentration of estrogen falls drastically. At this point the pituitary is again free to form FSH, the follicle-stimulating hormone, and so the cycle is repeated. There is no better example of biological feedback.

From these facts it might be concluded that the anterior pituitary and ovary constitute a self-sustaining oscillating system. Perhaps they do, and the role of the eyes and the pineal gland is to turn the oscillator on or off in such a way that it will be correlated either with the seasons or the time of day.

UTERINE CYCLES

In most mammals the **endometrium,** that is, the epithelial lining of the uterus, undergoes rhythmic cycles along with ovulatory and estrous cycles. The effect of estrogen on the endometrium of an ovariectomized rat can be seen in Figure 17-3.

As now understood, the full sequence of events in the menstrual cycle runs as follows. Estogen, which stimulates the proliferation of the lining of the uterus, also stimulates the pituitary to secrete the LH. The LH not only triggers ovulation, but also stimulates the growth of the corpus luteum, which as already described secretes progesterone which further stimulates the final stage of growth of the uterine lining, making it ready for implantation and the nourishment of an embryo. If pregnancy does not take place, the corpus luteum degenerates, the concentration of progesterone consequently falls, the lining of the uterus loses its hormonal support and sloughs off in the characteristic bleeding. Menstruation can therefore be inhibited by suitable injections of progesterone. Ovulation usually occurs on or about the 14th day after the onset of menstrual bleeding, when the uterine lining is prepared to receive the early embryo, but again there is considerable variation.

If pregnancy occurs, the placenta itself acts as an endocrine gland, for it secretes both estrogen and the luteinizing hormone. This additional supply of LH prevents the corpus luteum from degenerating. Consequently the supply of progesterone is maintained and the lining of the uterus is not sloughed off.

Hormonal Control of Mammalian Reproduction

One of the most important events of modern history was the discovery that hormones can be used to increase or decrease fertility. Barren women who desire children can sometimes be helped by precisely the correct dosage of pituitary gonadotropin to stimulate the development of a Graafian follicle and ovulation.

Of an even greater human import is the discovery that for reasons as yet imperfectly understood, oral administration of various combinations of estrogen and progesterone inhibit either ovulation or the implantation and development of the embryo. Some hormone preparations cause a marked thinning of the uterine lining. Others are thought to block the response of the ovary to gonadotropin from the pituitary. There is much talk about *the* Pill, but actually there are many different pills available, each with somewhat different hormones and hormone combinations and of different degrees of purity. It is not surprising that there are sometimes unwanted side effects.

There is much concern about the number of users of the Pill who develop blood clots (thromboses) in the veins or lungs. Present data show that this complication is almost exclusively limited to patients belonging to blood group A. Why this should be is a puzzle. In any specific case the patient and physician must decide on the wisest course of action, weighing the dangers of side effects against the dangers of excessive pregnancies. Recent tests suggest that if may be possible to develop long-lasting agents. A single dose of the progesterone derivative depo-medroxyprogesterone acetate, seems to be effective for 6 months, but unfortunately it has highly undesirable side effects, including frequent uterine bleeding.

Various contraceptive intrauterine devices, IUD's, have been tested, with considerable though not complete success. These plastic rings or coils may be expelled or cause bleeding or cramps. In a study group of over 10,000 women in the United States and Puerto Rico, the frequency of pregnancy was only 3 per cent but 20 per cent experienced some kind of difficulty. Immunological techniques have been tried but offer little promise of success because of the great difficulty in producing antibodies specifically against one particular tissue.

Anti-androgens are also available. Most are derivatives of progesterone. They have been used to control unwanted facial hair in women with deranged adrenal glands and to control prostatic cancer in men.

It should be remembered in connection with the problem of conception control that simple surgical procedures have long been available. Either the vas deferens in the male or the oviducts in the female can be tied off. The fact that these surgical procedures result in permanent sterility, although leaving the individual's sexuality unaffected, may be a large part of the reason why they have never been widely accepted.

The biological as well as sociological basis for the importance of human birth control rests primarily on the great advances in modern medicine which began with the establishment of the germ theory of disease. These advances have not extended the potential life span—the traditional "three score and ten" with occasionally a little more—but they have greatly increased life expectancy. In the 17th century the mean life expectancy at birth was about 25 years. Today it is about 67 years for men and 74 for women in Western industrialized countries.

Over the long course of history, human birth rates have more than compensated for high mortality rates. Save for times of pestilence, famine, and war with their accompanying toll of human lives, there has been a persistent trend of population growth. So long as vast areas of the earth's surface were still sparsely occupied, these ever increasing numbers of humans could readily be accommodated. We have now reached the point where population densities in many areas exceed what can be adequately supported by available natural resources. Clearly birth rates must ultimately be adjusted so that the frightful consequences of overpopulating this planet with our own species can be avoided. A valid analogy can be made with the reproductive situation in birds. Mankind is passing from a condition comparable to the precocial species, where there are many offspring and a high mortality, to a state more like that of the songbirds, where there are very few offspring but where the parents give their young

prolonged care with consequent lower mortality.

Gestation and Birth

The period of growth of the embryo in the uterus is known as the period of **gestation** or pregnancy. Ever since the time of Aristotle, zoologists have tried to find some firm correlations between the size of the animal, the number of young in a litter, and the length of the gestation period. In a general way, the larger the animal, the longer the period of gestation (see Table 8-1).

The physiological factors are still very obscure which set the length of pregnancy and which initiate and maintain the muscular contractions of the uterus, known collectively as labor, which expel the offspring. That the baby itself plays an important role in determining the length of gestation is well demonstrated by crosses between horses which have a gestation period of about 340 days and donkeys which have one of 365 days. In these crosses the male parent and therefore the foal influences the length of gestation.

The pain involved in this process varies greatly, depending on many factors both physical and psychological. The first stage of labor is the period during which the os, or mouth, of the uterus becomes dilated. The second stage is the descent and expulsion of the infant, and the third stage is the period following the birth of the infant and before

TABLE 8-1
Gestation and litter data (From Moment, G. B., and H. M. Habermann. 1973. *Biology: A Full Spectrum*. The Williams & Wilkins Co., Baltimore.)

Animal	Gestation Period	Litter Size	Life Span (yr)
Elephant	21 to 22 mo	1	65
Whale	10 to 12 mo	1	75
Horse	11 mo	1 or 2	21
Cow	9 mo	1 or 2	13
Human	9 mo	1 to 5	70
Deer	8 mo	1 or 2	12
Bear, black	7 mo	1–3	22
Sheep	5 mo	1 or 2	12
Pig	4 mo	8–25	
Lion	108 days	2–4	20
Guinea pig	68 days		
Dog	63 days	3–12	20
Cat	56 days	3–12	20
Rabbit	32 days	4–7	8
Rat	20–21 days	4–10	3
Mouse	20–21 days	5–15	3
Opossum	13 days in utero, 50 in pouch, and 30 free nursing		

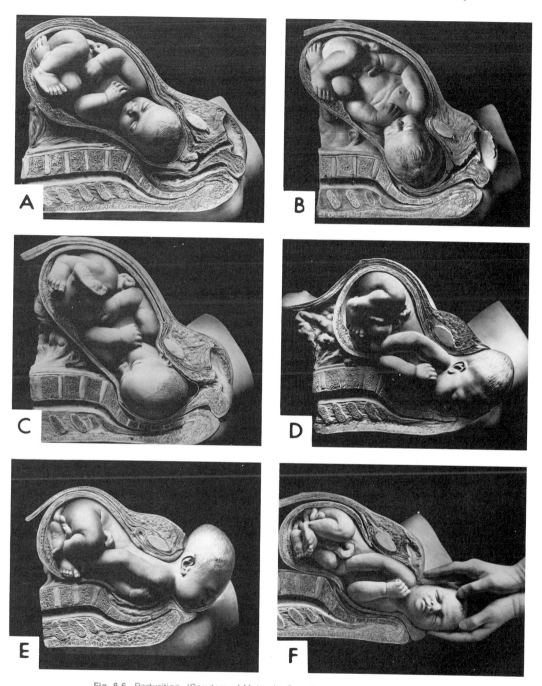

Fig. 8-6. Parturition. (Courtesy of Maternity Care Association, New York, N.Y.)

the expulsion of the placenta and membranes. The later stages of parturition are always times of potential danger, but many highly civilized people from Mme. Curie to an increasing number of present-day women have found no anesthetic necessary or desirable. Apparently in the final phases there is hypnotic numbness. Like other branches of

medicine, obstetrics is a blend of art and science requiring skilled judgment.

Lactation

The process of lactation, or milk production and secretion, is under endocrine con-

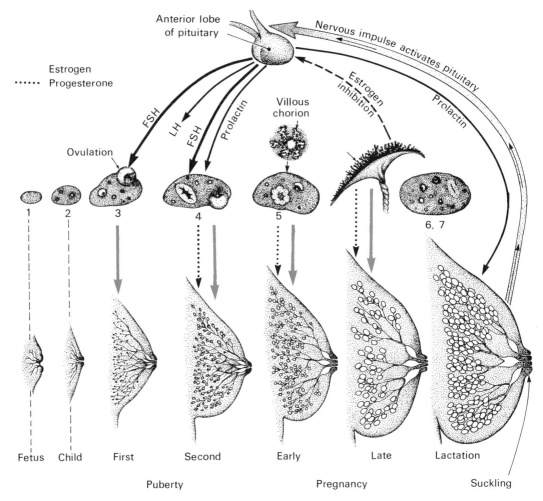

Fig. 8-7. Hormonal and neural control of lactation. (From Moment, G. B. 1967. *General Zoology*, 2nd ed. Houghton Mifflin Co., Boston.)

trol. The rudiments of the mammary glands are laid down in embryos of both sexes. In females at puberty the estrogen from the ovarian follicle cells stimulates the growth of the milk ducts and hence the general enlargement of the mammary glands. During maturity, and especially during pregnancy, progesterone from the corpus luteum stimulates the growth of alveoli, which are rounded glands at the tips of the branches of the milk ducts (Fig. 8-7). After the birth, the maternal pituitary secretes a new hormone, **prolactin.** This lactogenic hormone, first discovered in pigeons, causes the actual secretion of milk. Pigeons, it may be recalled, secrete true milk from esophageal glands. Its production seems to be due to the sudden fall in level of estrogen caused by the loss of the placenta, which has been secreting large quantities of both estrogen and progesterone. Estrogen has been used clinically to inhibit milk secretion. Mammals are not all alike in this respect, for small amounts of estrone increase milk production in cows and ewes.

Many secondary factors also govern lactation. Lack of the adrenal cortex or of thyroxin or insulin inhibits the mammary glands. Lack of an abundant supply of water in the diet also greatly diminishes milk secretion. Last, the mechanical stimulus of sucking stimulates the hypothalamic region of the brain which, in turn, results in greater production of oxytocin by the pituitary. This hormone stimulates contraction of muscles within the mammary gland which squeeze milk to the outside. This fact can be effectively demonstrated by injecting oxytocin into an anesthetized lactating animal. Hormones from the adrenals, thyroid, and posterior lobe of the pituitary also influence this long familiar but still incompletely understood process.

USEFUL REFERENCES

Arey, L. B. 1965. *Developmental Anatomy,* 7th ed. W. B. Saunders Co., Philadelphia. (Vertebrates with emphasis on human development.)

Karmel, M. 1959. *Thank You, Dr. Lamaze: A Mother's Experience in Painless Childbirth.* J. B. Lippincott Co., Philadelphia.

Millen, J. W. 1962. *Nutritional Basis of Reproduction.* Charles C Thomas, Publisher, Springfield, Ill.

Nalbandov, A. V. 1965. *Reproductive Physiology; Comparative Reproductive Physiology of Domestic Animals, Laboratory Animals, and Man.* W. H. Freeman and Co., San Francisco.

Pincus, G. 1965. *The Control of Fertility.* Academic Press, Inc., New York.

Money, J. and A. A. Ehrhardt. 1972. *Man & Woman, Boy & Girl.* The Johns Hopkins University Press, Baltimore.

Stonehouse, B., ed. 1975. *The Biology of Penguins.* University Park Press, Baltimore. (The most complete and best illustrated recent volume on these birds by 25 experts.)

9

Reproduction in Plants

Most plants (and a great many animals) have the capability of forming new individuals both asexually, that is, by vegetative propagation (either naturally or with human assistance), and by sexual reproduction. Within each mode of procreation, there are countless variations. Yet, despite such diversity, all asexual means of reproduction share the advantage of producing offspring identical to the original organism in genetic endowment, while sex, whether in plants or animals, provides for the recombination of inherited characteristics, the basis for evolutionary change in populations.

Consider the case of the common potato, which is a highly heterozygous organism. If potatoes are grown from seed, the resulting plants show enormous variability. Some even produce bright red or bright purple potatoes, most of them very small in size or number and highly unsatisfactory for farmer and consumer alike. If new plants are propagated asexually by cutting out and planting the buds ("eyes") from the tubers, then all the offspring produce potatoes like the parent, whether good or bad.

Clearly, if the objective of a horticulturalist is to obtain new plants from a desirable but heterozygous strain, to plant a new orchard of apple or grapefruit trees, or to propagate a newly discovered variety of flowering shrub, seeds from that genetically mixed strain cannot be used. The techniques of vegetative propagation are necessary to obtain new plants with precisely the characteristics of the original. On the other hand, a plant breeder wishing to obtain new types of plants will use sexual reproduction. To obtain a new strain of disease-resistant wheat, for example, thousands of plants are grown from seed and tested for disease resistance. To obtain a new kind of corn combining uniform height for easy mechanical harvesting with high yield, the plant breeder would begin by crossing plants having each of the desired traits (see Chapter 4).

Genetic engineering (see Chapter 7) is the most modern development in a long history of agriculture beginning with the first domestication of plants in Neolithic times. Over thousands of years, desirable crop varieties have been selected. The exploration of new continents has led to the importation of new species of agriculturally useful plants. The development of new and better varieties of crop plants is undoubtedly the most promising means available for increasing agricultural productivity.

Asexual Reproduction in Plants

One of the strangest generalizations that can be made about plants is that unstable or changing environments (whether shifting sand dunes, a strip of land cleared for a new road, or a puddle of rainwater) tend to be inhabited by species capable of reproducing asexually. This is borne out by common experience. Everyone has noticed that in places where water drains slowly after a rain, "blooms" of algae develop rapidly, only to disappear when the surface of the soil again dries. The shifting sand dunes along our coasts are inhabited by grasses and other seed plants that spread by means of **stolons** or **rhizomes,** modified stems that spread horizontally at, or just below, the surface of the soil. Vegetative propagation of these plants permits them to spread out around established individuals. In doing so, they help to stabilize the environment and make possible the invasion of other species having more specialized requirements for the germination of seeds. Thus, the asexually reproducing plants play a major role in ecological succession, a topic that will be discussed in detail in Chapter 24.

SOME EXAMPLES OF ASEXUAL REPRODUCTION

Plants in Which Cell Division Equals Reproduction

When unicellular organisms divide, two new individuals are formed. Here the replication of the nuclear DNA and division of the cytoplasm give rise to daughter cells genetically identical to the parent cell. Procreation of the prokaryotic bacteria and blue-green algae, of the unicellular eukaryotic algae and fungi among plants, or of the protozoa in the animal kingdom can be just this simple: in single-celled organisms cell division is a means of asexual reproduction.

In the bacteria, this kind of multiplication is commonly termed **binary fission,** a division into two equal parts. In many yeasts new cells are produced by **budding,** the process by which mature cells produce one or more daughter cells (Fig. 9-1). The small daughter cells, or buds, often remain attached to the parent cell for a time, sometimes forming clumps or chains of cells. Some species of yeasts multiply by fission as do the bacteria.

Spore Formation

In many of the unicellular plants (both prokaryotes and eukaryotes), cells repond to unfavorable environmental conditions, by forming spores. In the process there is usually a condensation of the protoplasm with a reduction in the water content and the formation of a new cell wall. Such changes make it possible for spore-forming species to survive conditions that would probably be lethal for physiologically active cells. Spores can usually withstand high temperatures, ultraviolet light, and many chemical agents, precisely the factors normally used in maintaining asepsis. Although asexual spore formation provides a mechanism for surviving unfavorable conditions, spores are not necessarily formed only in response to hostile environments; they can form under favorable conditions as well. In fact, the mechanism of asexual spore formation is a poorly understood phenomenon. The advantages of spores are

Fig. 9-1. Asexual reproduction in yeast by budding (bud, Bu) as seen by the scanning electron microscope, × 10,000. (From Kessel, R. G., and C. Y. Shih. 1974. *Scanning Electron Microscopy in Biology: A Student's Atlas of Biological Organization.* Springer-Verlag, New York.)

obvious, however. When conditions are again favorable, spores absorb water, germinate, and are transformed into physiologically active cells. In a strict sense, when a single cell forms a single spore, reproduction has not occurred. However, such a spore may be carried to a new location and there found a new colony of cells. It is important to remember that these spores are very different from the spores formed in groups of four as a result of meiosis that are part of sexual reproduction.

Asexual Reproduction in Multicellular Plants

With the association of cells to form filaments, colonies, or more complex plants, cell division results in an increase in the size of the organism rather than producing new individuals. Asexual means of reproduction were not abandoned by multicellular organisms, however. They only become more varied.

Filaments or colonies of the kinds found in the many groups of algae can form new individuals by fragmentation. In the fungi we find a variety of forms of vegetative reproduction. Pieces of mycelium, when removed, will continue to grow. Asexual spores are borne on special branches (Fig. 9-2). In most of the fungi, asexual means of reproduction predominate and sexual reproduction is a rare occurrence — in some cases unknown.

The more complex plants have also evolved specialized means for vegetative propagation. In the simplest of terrestrial forms, the mosses and liverworts, new individuals are produced by continually dying behind and growing ahead. New and separate plants are formed when disintegration of the posterior end reaches a branch point. There are, in addition, more elaborate means of asexual reproduction in these simple plants. Multicellular structures called **gemmae** can form on the surfaces of the vegetative plants (Fig. 9-3). Once detached from the parent plant, they germinate and form new plants. The mosses and liverworts share with their more complex evolutionary descendents the ability to regenerate entire new plants from practically any piece, be it leaf, stem, or rhizoid.

Adaptations for Vegetative Reproduction in Higher Plants

The spread of populations of many of the seed plants ranging from beach grasses to

Fig. 9-2. Asexual reproduction by conidia in *Aspergillus nidulans* as seen by scanning electron microscopy, × 1000. Spherical conidia are borne on a stalk (conidiophore, Cp) which develops by enlargement of certain cells of the mycelium (My). (From Kessel, R. G., and C. Y. Shih. 1974. *Scanning Electron Microscopy in Biology: A Student's Atlas of Biological Organization.* Springer-Verlag, New York.)

strawberries occurs in nature by rather minor modifications of ordinary plant parts. Strawberry plants have special stems, called **stolons** or **runners,** that grow above the soil surface. If they touch the soil they develop roots and new shoots that can form a new plant. The standard method for propagating strawberry varieties in agriculture is by this vegetative means. Many other plants can form roots on branches covered with soil. Blackberries, raspberries, and many other plants do this in nature even though they do not form specialized stems like the stolons of strawberries. The horticultural practice of **layering,** where a branch is purposefully buried so that roots will develop, is based on the capacity of stems to form such adventitious roots when brought into contact with the soil.

Other plants, including the irises and a great many grasses, have specialized underground stems (**rhizomes**), that send up aerial shoots. Many varieties of lawn grass have been selected precisely for this tendency to spread rapidly by means of rhizomes. Some trees, including aspens, and many weeds, such as goldenrod, morning glory, and crabgrass, proliferate by this means.

Tubers are another kind of specialized underground stem. The best known example of a tuber is the familiar potato. This bulbous underground stem, like all stems, has nodes and internodes but its leaves are reduced to small scales. Buds develop in the axils of these nonfunctional leaves. When planting potatoes in the spring, farmers cut the tubers into pieces having one or more such buds or "eyes." In the soil, the buds sprout to form new shoots and roots. Functionally, tubers serve two purposes: food storage and reproduction.

Two further types of modified stems that are important structures of vegetative propagation in the higher plants are **bulbs** and **corms.** A **bulb** is really nothing more than a bud. There is a short stem at the base with many fleshy, scalelike leaves growing from its upper surface. Adventitious roots develop from the base of a bulb. The bulk of any bulb is comprised of the fleshy storage leaves. Buds developing in the axils of these fleshy leaves can be separated from the parent bulb and used to propagate the species. Many familiar spring flowers including hyacinths, tulips, and daffodils are grown from bulbs. The readily available food reserves stored in the over-wintering bulbs are a major factor in the early flowering of many of these species. Animals, too, sometimes take advantage of such food reserves: the onion is a bulb which appears regularly on our dinner tables. **Corms**

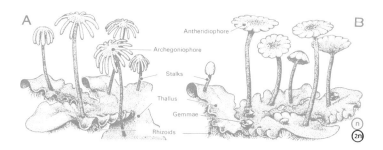

A

Antheridiophore

Archegoniophore

Stalks

Thallus

Gemmae

Rhizoids

B

n
2n

Fig. 9-3. Gametophyte plants of the liverwort, *Marchantia*. A, female plant with archegoniophores, × 2/3; B, male plant with antheridiophores, × 2/3. Note reproductive gemmae. (From Moment, G. B., and H. M. Habermann. 1973. *Biology: A Full Spectrum*. The Williams & Wilkins Co., Baltimore.)

are sometimes referred to as "bulbs" by the uninitiated, but differ from bulbs in consisting primarily of stem tissue. The outer surfaces of corms are covered with leaves that are usually thin and scaly. Adventitious roots form at the base as they do in bulbs. The crocuses and gladioli are familiar garden flowers propagated by means of corms.

In still other common garden varieties, reproductive structures are formed on the shoot. The tiger lilies form aerial **bulbils** about the size of peas in their leaf axis. These, too, are nothing more than modified stems surrounded by fleshy leaves. When they abscise and fall from the parent plant, they sprout to form new individuals.

Even leaves may be the means for vegetative propagation. The leaves of African violets and begonias readily form adventitious buds and roots when removed from the plant and placed in contact with the soil. In the walking fern, new plants are formed when the tips of the leaves touch the soil. In bryophyllum tiny plantlets develop in notches along the edges of the fleshy leaves. When shed from the parent plant, they form roots and develop into new plants. In some species long photoperiods promote plantlet formation while short days promote flowering. The tiny species of *Lemna* (duckweed), flowering plants found floating on the surfaces of ponds and streams, usually reproduce asexually by forming new individuals from leaves.

Thus, in nature all of the vegetative organs of higher plants (roots, stems and leaves) have been modified by one or another species as a means for asexual reproduction. Vegetative propagation is not restricted to the vegetative organs of plants by any means. Some types of aerial bulbs develop from flower buds. In species such as the common dandelion, seeds can be produced by **parthenogensis** (i.e. without fertilization). Recent tissue culture experiments indicate that embryo-like development is possible from almost any diploid cell, if appropriate nutrients and hormones are provided. Even some haploid cells such as developing pollen grains

can give rise to morphologically normal plants (see Chapter 7).

ROOTING AND GRAFTING— HORTICULTURALLY USEFUL METHODS FOR ASEXUAL PROPAGATION

Two common horticultural means of vegetative propagation—**stem cuttings** and **grafting**—are feasible because of the ability of plants to replace missing parts and to repair wounds. The many commercially important plants that are propagated by means of stem cuttings include chrysanthemums, roses, sugar cane, grapes, many kinds of fruit trees, and most ornamental shrubs. Although cuttings from many of these plants readily form adventitious roots, a knowledge of the effectiveness of plant hormones in promoting root development (see Chapter 18) has made it possible to propagate even varieties that do not readily form adventitious roots. Adequate moisture is absolutely necessary during rooting. The bases of cuttings are usually stuck into moist sand, vermiculite (modified mica), or peat. The tops are kept moist by means of sprays, mists, or plastic covers. Often the formation of roots is promoted if slightly higher temperatures are maintained around the bases of cuttings.

Grafting is no more than an artificial means of providing roots for a cutting. The cutting, which may be a piece of stem or only a bud, is called the **scion.** The plant or the root system onto which it is grafted is called the **stock.** Many kinds of woody and herbaceous dicots from apple trees to sunflowers have been grafted. The probability of sucess is greatest when scions and stocks are the same or different varieties of a single species. However, both interspecific and intergeneric grafts are possible. Such mixed grafts are not uncommon in fruit trees where flowers and fruits of numerous varieties and species can be produced on a single composite tree. Grafting is important in horticulture for two purposes. The first is to make possible the asexual prop-

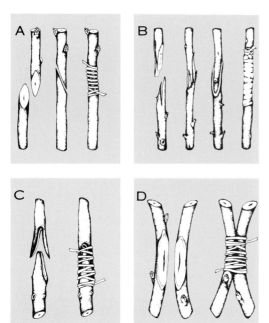

Fig. 9-4. Some widely used grafting techniques. A, "splice" graft in which obliquely cut stock and scion are fitted together to match cambium layers. B, "whip and tongue" graft. Tongue cleft aids in holding stock and scion together. C, "saddle" graft. Obliquely cut stock forms a blunt point or saddle. Scion is cut to fit over tapered end of stock. D, "approach" graft. Similar shields are cut from the two components. Whatever the type of graft, stock and scion are securely tied together with waxed string, nursery tape, etc. Often waterproof coverings such as wax or plastic film are used to retard water loss. (From Moment, G. B., and H. M. Habermann. 1973. *Biology: A Full Spectrum*. The Williams & Wilkins Co., Baltimore.)

agation of a highly desirable variety, e.g. a fruit tree, which is heterozygous and so will not breed true. Many varieties of apples, pears, grapes, and other kinds of fruit-bearing plants are propagated in this way from a single fortunate mutation. The second purpose is to make it possible to use a more vigorous or disease-free stock to propagate a scion that is difficult to grow on its own root system. Ordinarily ornamental roses are grafted onto wild rose stocks; many fruit varieties including most grapes are grown on less productive or desirable but more disease-resistant root systems.

Sexual Reproduction in Plants

Vegetative propagation, whether by means of modified stems or roots, or even leaves that are shed and then sprout to form new plants, provides a means of making new indi-

viduals with genotypes that are well adapted to a given environment. However, any change, be it unseasonable weather, the arrival of a new insect pest or disease-causing microorganism, or slight changes in soil or in average annual rainfall, threaten the survival of asexually reproducing species. A fact of evolution (and of life) is that change within a population is the only effective way to combat change in the environment. Survival in the evolutionary sense of adapting to change and of exploiting new environments is dependent upon sex because only in sexual reproduction is variability assured.

Meiosis and fertilization are the two mechanisms that assure the variability of progeny. During meiosis there is an independent segregation of genes on separate chromosomes. Fertilization not only multiplies the amount of variation but makes possible the combination of favorable mutations which have appeared in different strains, something impossible in asexual reproduction.

It is important to remember that not all variability is genetic. Variability remains even in plants that have been inbred for a number of generations and have become homozygous. A classic demonstration of such variability which is dependent on environment can be found in the work of the Danish biologist Johannsen. Early in this century Johannsen inbred the progeny of a single bean plant for several generations. He gathered seeds varying in size and weight from this population of plants grown under the most uniform conditions possible. He then selected the largest and smallest seeds and grew another generation of plants from them. When the seeds from these two sets of plants were compared, they had the same average size as the batch of seeds from which large and small ones had been selected for planting. This demonstrated that selection for a particular characteristic (large seeds) is effective only when there is genetic variability for this characteristic in the population. In plants, where development is profoundly influenced by environmental factors, it is more obvious than in animals that it is not final form but rather the potential for development that is inherited.

FACTORS THAT PROMOTE SEXUAL REPRODUCTION IN PLANTS

The annual succession of blooms—the crocuses and violets of spring to the chrysanthemums and asters of the fall—has long stirred human interest. Moreover, an understanding of this phenomenon is of great practical importance to florists the world over. As a result of modern investigations by plant physiologists, important new information is available to answer some of the ancient puzzles about why higher plants blossom when they do and even why algae and fungi go sexual.

To the long recognized but only vaguely understood factors like temperature and nutrition, and the mysterious "ripeness-to-flower," modern work has added various chemical agents. Equally important is the discovery of the role of photoperiod. We have already mentioned how plants such as bryophyllum reproduce vegetatively under conditions of long days and short nights but form flowers and thus reproduce sexually when the days are short and the nights long.

The development of the structures involved in sexual reproduction is usually triggered by environmental factors such as temperature and light that change with the seasons. Other internal and external conditions, including chemical growth regulators and mineral nutrition, also influence the onset of flowering. Most plants, as they become older, have an increased tendency for the changes associated with sexual reproduction to be initiated. For the flowering plants, this tendency has been referred to as "**ripeness-to-flower**" and some species will form flowers regardless of environmental conditions once an appropriate stage of maturity has been reached.

Environmental changes serve the useful

Fig. 9-5. Day length and chemicals are used to control form and flowering of ornamental plants. U.S. Department of Agriculture plant scientist, H. M. Cathey, has used short days to keep the petunia plant at left short and vegetative. Tall and flowering plant at center was grown under long days. Plant at right was sprayed with a growth retardant (which inhibited elongation) and grown under long days to promote early flowering. (Courtesy of U.S. Department of Agriculture.)

function of synchronizing reproduction, thus promoting cross-fertilization and variability. A further advantage of seasonal flowering is that it permits a plant to utilize all of its available resources for vegetative growth during part of the growing season. This enables it to compete more effectively with other species.

Temperature

Both diurnal and seasonal fluctuations in temperature can affect plant development. In Chapter 7 we noted that some plants require a period of cold following a time of vegetative development in order for sexual reproduction to occur. The cold treatment of germinating seeds (a process called **vernalization**) is an important agricultural practice (particularly in Russia) that makes it possible to plant and harvest certain cereal grains in the same growing season. Without such treatment, these varieties of rye and wheat would have to be planted in the fall and then harvested the following summer, and they could not be grown at all in climates too cold for fall plantings to survive the winter.

Seasonal fluctuations in temperature are also important in the breaking of dormancy, whether in buds or in underground storage organs such as the bulbs, corms, and tubers that form in response to decreasing day length. A period of days to months at temperatures below 10°C is usually required for the breaking of dormancy in buds. In certain bulbs and corms, the formation of flower primordia requires specific temperatures. In tulips, for example, bulbs must be maintained near 20°C for flower primordia to form. Their further development requires a drop in temperature to 9°C (about 48°F) or less for about 14 weeks. Such thermoperiodic requirements correspond to seasonal fluctuations of temperature in regions where these plants are native. Because of these temperature requirements, tulips cannot be grown successfully in warm climates, even though bulbs imported from colder regions may bloom during the first year.

Diurnal thermoperiodicity can influence not only the rate of vegetative growth of plants, but also certain aspects of their reproductive development. For unknown reasons, most plants grow best when temperatures at night are lower than during the day. In certain species such as squash and cucumbers, where both male (staminate) and female (pistillate) flowers are formed on the same plant, the relative numbers of male and female flowers are influenced both by photoperiod and by diurnal fluctuations in temperature. Long days and warm nights promote the formation of male flowers; while cool nights decrease, and warm nights with short days increase the production of female flowers.

Chemical Growth Regulators

Certain of the naturally occurring plant hormones and chemical plant growth regulators can substitute for the effects of light and temperature on dormancy and flowering. Gibberellins, 2-chloroethanol, and thiourea are effective in breaking dormancy of potato tubers; while naphthaleneacetic acid (a synthetic auxin) and maleic hydrazide have been used to prolong dormancy and prevent the sprouting of potato tubers during storage. The flowering of many cold-requiring species is promoted in the absence of cold by gibberellins. Thus, this plant hormone is potentially useful as an alternative to vernalization. It is particularly effective in those species where flowering is accompanied by the shift from a rosette growth pattern, i.e. one in which hardly any stem elongation occurs, to one with greater internode elongation.

Many examples of the chemical control of sexuality in plants have been discovered. In the more primitive plant groups such as the algae and the fungi there is evidence that the production of gametangia, the maturation of the sex organs, the release of gametes, and the attraction of sperms to eggs are influenced by specific chemicals produced by these organisms at particular stages of the life cycle. The investigation of specific chemical action in the production of gametangia has been studied in considerable detail in the ferns. The haploid, gametophyte prothallus of the fern can be cultured in the laboratory on synthetic media. Grown singly in test tubes, the prothallia develop normally except for the fact that they do not form male gametangia (**antheridia**). If grown in the presence of other gametophyte plants, or if some medium from cultures of **old** gametophyte plants is added, then antheridia are formed. When sufficient quantities of antheridium-inducing factor are added to the culture medium, practically every cell of the prothallus differentiates and forms an antheridium. This factor appears to be produced in old prothalli but only younger gametophyte plants can respond to it (hence an isolated prothallus can form no antheridia). Female gametangia (**archegonia**) can be formed on old gametophyte fern plants even though no antheridia have

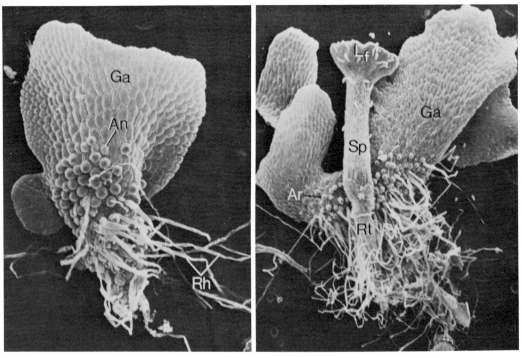

Fig. 9-6. Fern gametophytes as seen in the scanning electron microscope, × 32. These gametophytes (Ga) are heart-shaped and only 2–3 mm across. Rhizoids (Rh) on the lower surface absorb water and minerals from the soil. Antheridia (An) and archegonia (Ar) are also found on the lower surface. After fertilization, the diploid sporophyte (Sp) develops rapidly; a young leaf (Lf) and root (Rt) can be seen in the plate on the right. (From Kessel, R. G., and C. Y. Shih. 1974. *Scanning Electron Microscopy in Biology: A Student's Atlas of Biological Organization.* Springer-Verlag, New York.)

been formed. The production of male gametangia in neighboring plants through the production of a specific chemical stimulus Insures that sperms will be available to fertilize the eggs formed in the archegonia. The antheridium factor appears to be specific to each family of ferns.

Mineral Nutrition

We must not forget that mineral nutrition can influence the vegetative and reproductive development of plants. An excess of the essential element nitrogen promotes vegetative development. We take advantage of this effect of excess nitrogen to grow lush, dark green lawns. However, a comparable zeal in applying fertilizer to tomatoes or other species grown for their fruits has disappointing consequences. Overfed plants respond in a way that is somewhat comparable to the situation in overfed animals. Instead of getting fat, however, their excess nutrients are utilized for a superabundance of vegetative growth, while there is very little flowering and fruit set. Under circumstances of nitrogen deficiency, plants respond in the opposite way and flower after very limited vegetative growth. Obviously this quickened reproduction of a starving plant can have important survival value for the species.

PHOTOPERIODIC CONTROL OF FLOWERING—THE REASON FOR GIANT TOBACCO AND SEPTEMBER SOYBEANS

One of the amazing aspects of the photoperiodic control of flowering is that such a widespread and, one would think, obvious phenomenon was not discovered by plant scientists until the present century. Not until the classic publications of W. W. Garner and H. A. Allard appeared in the early 1920's was it realized that relative lengths of day and night are controlling factors in the onset of flowering. These two workers and their associates at the U.S. Department of Agriculture Plant Industry Station in Beltsville, Maryland, were puzzled by the strange behavior of two kinds of plants.

One plant, a mutant variety of tobacco (*Nicotiana tabacum,* var. *Maryland mammouth*), when grown in the field during the summer, attained a height of over 3 meters but persistently remained vegetative. This was frustrating to those who wanted to obtain seeds for commercial use of the mutant or wished to use it in plant breeding experiments. When cuttings of this plant were grown in the greenhouse during the winter months, they flowered profusely and set seed after only limited vegetative growth. Plants grown from seeds produced by winter-grown plants again grew to unusually large size and failed to flower in the field during the following summer. Another plant that had a puzzling behavior was the Biloxi variety of soybean (*Glycine max*). No matter when these soybeans were planted, whether in May, June, or July, they all flowered in September. This meant that those planted early in the season grew vegetatively for 4 months while those planted in July flowered when much smaller after only 2 months of growth. It appeared that all of the plantings responded in unison to the same environmental signal.

After considering a number of possible factors such as temperature, light intensity, and mineral nutrition, Garner and Allard realized

Fig. 9-7. H. A. Allard (right) collaborated with W. W. Garner in pioneering studies of the photoperiodic control of flowering in plants. In this photograph, taken in 1960, Dr. Allard is discussing more recent aspects of light-controlled growth with Dr. H. A. Borthwick. (Courtesy of U.S. Department of Agriculture.)

that relative length of day and night was the condition controlling the flowering response in both the Maryland mammoth tobacco and the Biloxi soybean. Garner and Allard were able to induce or prevent flowering in tobacco and soybean experimentally by shortening or lengthening the day. Shortening was accomplished by placing plants in dark chambers, lengthening by turning on artificial light at night. This knowledge has proved to be of great practical importance, especially to florists who want plants to blossom at particular times.

Classification of Plants According to Their Photoperiodic Requirement for Flowering

Garner and Allard found that flowering plants can be classified according to their responses to photoperiod into the following three categories: (1) **short-day** plants, (2) **long-day** plants, and (3) **day-neutral** plants. **Short-day** plants flower only when subjected to day lengths shorter than a particular number of hours during each 24 hour cycle of light and dark. The Maryland mammoth variety of tobacco, the Biloxi soybean, and the

plantlet-forming bryophyllum belong in this category. Other examples of short-day plants are the cocklebur, chrysanthemum, aster, ragweed, and poinsettia. In **long-day** plants, flowering is initiated under conditions where day length exceeds a critical number of hours. Examples of long-day plants include spinach, black henbane, and some varieties of barley. **Day-neutral** plants are indifferent to photoperiodic conditions in that after reaching a critical age or size, they flower regardless of day length. Many common plants, including tomatoes, peas, and sunflowers, are day-neutral.

One strange thing about the terms "long day" and "short day" is that they do not necessarily imply anything about the absolute length of day needed to induce flowering in a particular species. For example, many kinds of plants flower in response to a 14-hour photoperiod; i.e. 14 hour of light and 10 hours of darkness. This happens to be shorter than the critical photoperiod necessary for the short-day cocklebur and longer than the critical number of hours of light needed by the long-day black henbane.

Patterns of Seasonal Changes in Day Lengths and Their Effects

Following Garner and Allard's discovery of photoperiodism in plants, zoologists found comparable control of reproductive cycles in animals such as aphids and birds. In addition, there are in animals other seasonal responses related to photoperiod such as changes in fur color and the migration of birds. Seasonal change in day length thus appears to be a signal used universally as a cue for changing environmental conditions. The amount of annual change in day length increases with distance from the equator. In the tropics, there is relatively little fluctuation in day length or temperature. Seasonal changes in these two climatic factors increase with increasing latitude. Tolerance of temperature extremes and capacity to respond to the lengthening or shortening photoperiods that signal climatic

Fig. 9-8. S. B. Hendricks, U.S. Department of Agriculture scientist, whose studies of the action spectra of phytochrome-mediated plant processes led to the prediction of many of the properties of this important photoreceptor. He is shown here measuring soybeans grown under controlled photoperiod. (Courtesy of U.S. Department of Agriculture.)

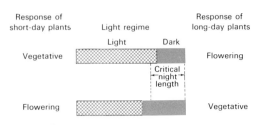

Fig. 9-9. Summary of responses of short-day and long-day plants grown under controlled photoperiods. (From Moment, G. B., and H. M. Habermann. 1973. *Biology: A Full Spectrum*. The Williams & Wilkins Co., Baltimore.)

changes determine whether a given species is able to survive in a particular geographic region. It should come as no surprise that plants native to the tropics, where seasonal changes in photoperiod and temperature are small, tend to be day-neutral. On the other hand, in temperate climates, where the need to respond to seasonal changes in temperature is greater, short-day and long-day plants are common.

Other aspects of plant development besides the onset of sexual reproduction are influenced by photoperiod. Long days promote onion bulb development and are favorable for the vegetative growth of most plants. In some species growth patterns are influenced by photoperiod. Internodes remain very short during vegetative growth in spinach and mullein; i.e. in these plants the leaves form a **rosette** close to the ground when photoperiods are short. However, when their critical day lengths are exceeded, they "bolt" as the onset of flowering is accompanied by rapid internode elongation. Short days generally promote tuber formation, leaf abscission, and the dormancy of buds. All of these modifications facilitate survival during the cold winter period that follows.

The Mechanism of Photoperiodic Control of Flowering

For a time after the discovery of the photoperiodic control of flowering, the length of the light period was thought to be the single critical factor. However, it was soon discovered that in short-day plants, flowering could be inhibited not only by artifically prolonging day length, but also by interrupting the dark period. Such interruptions can be surprisingly short—usually light of only a few minutes' to 1 hour duration is sufficient provided that the interruption comes at about the middle of the dark period. This discovery indicated that slow biochemical reactions taking place in the dark are required for flowering to occur in short-day plants and that they perhaps should have been termed **"long-night" plants**. In long-day plants such dark reactions probably proceed more rapidly and they could be referred to as **"short-night" plants.**

A clue about the mechanism of photoperiodism came from asking a logical question about light interruption of the dark period: "What wavelength is effective?" It turned out that red light is as effective as white light. Furthermore, the effects of red light can be canceled out by subsequent exposure to far red light, provided that the exposure to far red light follows within about $1/2$ hour. The effectiveness of red light and the antagonistic

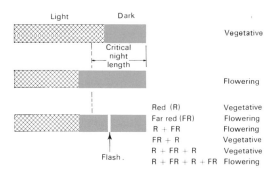

Fig. 9-10. Short-day plants grown under conditions of short days and long nights can be kept vegetative by a short flash of light during the dark period provided that the interrupting light is of the appropriate color. A red flash keeps plants vegetative, while a far red flash has the same effect as a continuous dark period. When red and far red flashes are given in sequence, the response is determined by the color of the last flash. (After Galston, A. W. 1961. *The Life of the Green Plant,* 2nd ed. Prentice-Hall, Inc., Englewood Cliffs, N.J.)

effect of far red light are characteristic of responses mediated by **phytochrome**. We encounter many such light-regulated responses in plant development, including light-dependent seed germination, the straightening of seedlings as they emerge from the soil, leaf expansion, and synthesis of the anthocyanin pigments (see Chapter 7). All of these responses have the same action spectrum and are triggered by light-dependent changes in phytochrome. You will recall that phytochrome can exist in either of two interconvertible forms. P_R, the blue form with an absorption maximum at about 660 nm is converted on absorption of red light to P_{FR}, having an absorption maximum in the far red at about 730 nm. P_{FR} is thought to be the enzymatically active form of phytochrome. In the flowering response, it appears that several events must occur during the critical dark period. First, P_{FR} formed during the preceding illumination must be converted back to the red absorbing form. This pigment conversion is thought to be only a preliminary event followed by further dark reactions which are necessary to produce a chemical which can be translocated to the apical meristems where the changes associated with the induction of flowering occur.

There is an abundance of experimental evidence supporting the hypothesis that the diffusible chemical stimulus which is responsible for flower induction is formed in the leaves. It is then translocated to apical meristems where it causes a shift from development of leaf primordia to development of flower primordia. In some species, such as the cocklebur, it is necessary to expose only a single leaf or branch to the appropriate photoperiod to induce flowering in the entire plant. When photoperiodically induced branches are grafted onto stocks maintained in light regimes that inhibit flowering, the uninduced stocks will flower. Neither the nature of this diffusible flowering hormone (sometimes called **florigen**) nor the sequence of metabolic changes that it induces in the apical meristems are known. Nevertheless, there have been many practical applications of our knowledge of the photoperiodic control of flowering. Plants grown in greenhouses are commonly brought into flower at times when the market demand is high. During winter months, day length can be extended by means of ordinary incandescent lamps. During the summer, days can be shortened by covering plants with special kinds of black cloth manufactured specifically for use by florists. These practices are now so widespread that we have come to expect that poinsettias will be blooming at Christmas time and that chrysanthemums will be available at any time from midsummer until late fall. There has even been limited use of photoperiodic controls for field crops. It is possible to obtain greater yields from sugar cane when flowering is prevented by brief exposure during the night to flood lights or flares.

Annuals, Biennials, and Perennials

Plants may be classified according to their flowering behavior as annuals, biennials, or perennials. **Annual plants** usually grow from seeds, flower, and die within one growing season. **Biennials** grow vegetatively during the 1st year and flower and die during the following year. **Perennials** persist for many years and usually flower repeatedly. Although such classifications are consistent with what occurs in nature, many species fall into one group or the other depending on the conditions that they encounter. After all, a biennial is probably nothing more than an annual that needs a period of cold before flowers can develop. Many annuals turn out to be perennials when planted in a different climate. This is particularly true for a number of plants that are annuals in temperate climates but continue to flourish as perennials when grown in the tropics. For example, common geraniums grow to almost treelike proportions after several years in California.

It is perhaps more meaningful to classify plants according to the number of times they flower regardless of environmental factors. **Monocarpic plants**, or those flowering only once followed by death, include such species as the common garden pea (*Pisum sativum*, a true annual in every sense), the century plant (*Agave*) that grows for 5–20 years before flowering, and certain bamboos that develop vegetatively for 2–50 years. True perennials, **polycarpic plants**, on the other hand, flower and fruit repeatedly over a period of many years and include apple, pear, and orange trees.

Flower Development

The onset of sexual reproduction, be it flowering in the seed plants or formation of gametangia in the gametophytes of more primitive plants, often means drastic changes in the nature and products of the meristems, or growing regions. In the flowering plants, instead of new leaves and internodes, the apical meristem forms grossly modified shoots called flowers. Flowering signals an end to the juvenile stage of plant development; it is also frequently associated with the onset of senescence and the ultimate death of the plant.

Once the internal and external stimuli necessary for the induction of floral primordia have been received by a plant, buds are formed that are destined to develop into the specialized shoots that we recognize as flowers. The working parts of a flower are the male **stamens** and the female **pistils** (Fig. 9-11). The showy, often very colorful **petals** and the green leaflike **sepals** are not essential, except to attract insects or birds that serve the useful purpose of disseminating pollen.

Stamens usually have an enlarged part called the **anther**, borne on a stalk called the **filament**. Within the anther, specialized cells called **microsporocytes** undergo meiosis to form haploid **microspores**. The nuclei of these haploid cells usually divide before they are released from the anther. The resulting cells are called **pollen grains**. These cells often have elaborately sculptured cell walls. Their very beautiful surface features are a factor in their distribution by animal vectors because they contribute a certain "stickiness."

Pistils are frequently vase-shaped with an enlarged base (**ovary**) connected by a tubular **style** to the **stigma**. Within the ovary there are one to many **ovules** containing the **megaspo-**

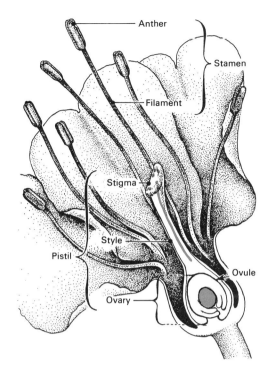

to the **endosperm tissue,** which is polyploid in many species. Fertilization stimulates further development of the ovule which at maturity forms the **seed**. The surrounding tissues of the ovary can enlarge. Mature ovaries, with their one to many embryo-containing seeds, are called **fruits**.

Flowers may be bisexual (containing both stamens and pistils) or unisexual (**staminate** or **pistillate**), depending on the plant species. Those plants with unisexual flowers may have both on the same plant and such plants (for example, corn) are called **monoecious**. An alternate arrangement is for male and female flowers to be borne on separate plants, and such species as hollies and the ginkgo are termed **dioecious**. With separate staminate and pistillate plants, cross-pollination and the resulting recombination of genetic traits and variability of progeny are assured. Cross-fertilization is also guaranteed in many monoecious plants (even those with bisexual flowers) by mechanisms such as self-sterility or the maturation of stamens and pistils at different times. Still other species have a flower structure that guarantees self-pollination. One classic example of such a naturally inbreeding plant is the common garden pea, *Pisum sativum*, the object of Gregor Mendel's well-known early studies of genetics.

Agents of Pollen Dispersal

Some of the most impressive examples of interaction between plants and animals are found in the adaptations of many flowers that facilitate pollen dispersal by specific animal carriers. Various plant species have flowers with colors, structures, odors, or nectar that attract particular pollinating agents, including bees, moths, butterflies, flies, beetles, birds, and bats. Often the shapes of flowers are especially adapted for pollination by a particular animal species. The elaborate structures of orchids provide fascinating examples of modifications appropriate for insect pollination that were first described over a century ago by Charles Darwin. One such modification not recognized by Darwin is a very clever trick whereby a male insect mistakes the pattern on an orchid flower for the female of his species and tries to copulate with it. Examples of this phenomenon of **pseudocopulation** are found in an Australian orchid species pollinated by a fly and in two European and North African genera pollinated by bees (Fig. 9-12). The specificity of such a device for pollination insures that pollen will not be distributed indiscriminately to other species of flowers. There are examples

Fig. 9-11. Structure of flower parts essential for reproduction, the stamens and pistils. (From Moment, G. B., and H. M. Habermann. 1973. *Biology: A Full Spectrum*. The Williams & Wilkins Co., Baltimore.)

rocytes. These cells undergo meiosis to form **megaspores**. Usually only one of the four haploid megaspores formed from each megasporocyte develops further. Through successive divisions of the megaspore nucleus, an **embryo sac** containing as many as eight or more haploid nuclei develops.

A unique feature of sexual reproduction in the seed plants is the **double fertilization** that occurs. After landing on the sticky surface of the stigma, pollen grains germinate. By this time a second nuclear division has occurred and three haploid nuclei, one **tube nucleus** and two **sperm nuclei**, enter the developing pollen tube. Extensive growth of the pollen tube (up to many centimeters) is necessary before the two sperm nuclei can be released into the embryo sac. Growth of the pollen tube appears to be guided by the tube nucleus which disintegrates when this job is completed; nutrients to sustain such phenomenal growth appear to come from the tissues of the style and ovary. Once released into the embryo sac, one sperm nucleus fuses with the egg nucleus to form the diploid **zygote**; the second fuses with the polar nuclei to form the **primary endosperm nucleus**. This second product of fertilization gives rise

of plants restricted in their geographical distribution to the areas where their pollen-carrying agents can survive. For example, the natural range of monkshood in the Northern Hemisphere corresponds to that of the bumble bee. Alfalfa tends to be infertile in California unless appropriate kinds of bees are imported to ensure pollination. Wind-pollinated species predominate at high altitudes or extreme latitudes where few insects and birds can survive.

Recent studies have indicated that some insect-pollinated flowers have patterns on their petals that humans cannot see but that are visible to insects with eyes sensitive to ultraviolet light. Ultraviolet-absorbing com-

Fig. 9-12. Flower of the bee orchid (*Ophrys apifera*). Note resemblance of central part of the flower to a bee. (Photograph by R. H. Noailles, from Jaeger, P. 1961. *The Wonderful Life of Flowers.* E. P. Dutton and Co., Inc., New York.)

Fig. 9-13. Nectar-collecting honeybee on alfalfa blossom also picks up pollen, which is the white mass packed on hind leg. (Courtesy of the U.S. Department of Agriculture.)

Fig. 9-14. The marsh marigold *(Caltha palustris)* is one of many flowers that have an ultraviolet reflection pattern visible to insects but not seen by human eyes. A, flower as we see it appears evenly yellow; B, ultraviolet reflection pattern seen by insects is observable with a television camera that has an ultraviolet transmitting lens. (Courtesy of T. Eisner.)

pounds in these flower petals are distributed in patterns that can be seen in photographs taken with ultraviolet sensitive films or with television cameras equipped with ultraviolet transmitting lenses (Fig. 9-14).

There are many examples of flower structures specifically adapted for pollination by a particular agent. Bee-pollinated flowers often have petals that serve as landing platforms where bees alight. Such flowers secrete nectar from glands at the base of a tube of petals. As a bee reaches for the nectar with its long, slender tongue, its body hairs pick up pollen from the flower's stamens. Bees tend to feed on one species of flower at a time and thus tend to distribute pollen to other flowers of the same species where it will do the most good. In bee-pollinated flowers, stamens and pistils are grouped together so that a bee simultaneously picks up pollen and delivers enough pollen from other flowers to fertilize many ovules.

Plant Dispersal

Whether reproduction is by asexual or sexual means, some mechanism for the dispersal of progeny is essential to the survival of plants. After all, a plant is destined to spend its entire life in the spot where the spore or seed from which it develops happens to germinate. Unlike animals that can migrate as the seasons change or search for favorable habitats, plants flourish, exist, or perish in response to the environment that they encounter by chance. The immobile plants are dependent on factors such as wind, water, and the freely moving animals for the dispersal of their progeny. Animals frequently are rewarded for this service just as they are for their important role in pollination. Many kinds of fruits are consumed, seeds and all. Because seeds often have tough, indigestible protective layers, they are excreted in the feces, usually at considerable distances from the spot where they were eaten. Some seeds contain abundant food reserves in the form of starch or oil which are as useful to an animal consumer as they are for the plant embryo. Such seeds are frequently transported over long distances by ants or squirrels and tend to be deposited in locations where they can germinate. Other kinds of seeds are marvelously adapted for being attached to the fur of passing animals. The amazing variety of hooks, spines, and bristles that facilitate this kind of hitch-hiking is known to all who have walked through fields and woods in the fall of the year.

The wind, which is the agent of pollination of many seed plants, also serves to transport seeds. Many kinds of devices enable seeds to be carried great distances by air currents. We have all seen the parachute-like structures of dandelion and milkweed seeds carrying their tiny packets of DNA through the air. Those of some trees such as the maples, elms, and lindens have winglike structures that enable these relatively heavy seeds to glide with the wind. These rather large seeds are exceptions to the general rule that structures dispersed by air currents must be small and light. The microscopic spores formed by bacteria, fungi, and other primitive plant groups are natural components of the dust carried everywhere by air currents.

There are some rather remarkable ways of catapulting seeds and spores from the structures in which they are formed. As some fruits dry, their walls split open, forcibly ejecting the seeds. In the mosses and liverworts there are many examples of structures that expand and contract in response to changes in atmospheric moisture content and therby expel spores at times when the conditions for their germination are optimal. *Pilobilus*, an unusual fungus that can easily be cultured on fresh horse dung, expels its spores with considerable force, sending them over distances as great as 2 meters and always aimed, with amazing accuracy, at a source of light.

HUMAN ROLE IN PLANT DISPERSAL

No consideration of plant dispersal would be complete without acknowledging the involvement of mankind. Since the beginnings of agriculture, humans have gathered and preserved the seeds of plant varieties having unusually good yields, desirable flavors, useful fibers, and the like. With the movement of peoples and especially since the exploration of the Western Hemisphere beginning in the 15th century, mankind has transported innumerable plant species from their places of origin to new areas. Many tales can be told about the ways in which newly introduced plant species have changed the course of history. The potato, a native of the Western Hemisphere, was probably introduced in Europe by the Spaniards. Sir Walter Raleigh is credited with the first cultivation of the potato for food in Britain in 1586. For the next 250 years, both potatoes and people prospered; in some areas such as Ireland, potato tubers became the mainstay of human diets. The widespread failure of potato crops in the mid-1800's caused by a fungus disease brought about famines and was the precipitating cause of a large migration to the New World.

The Western Hemisphere was the original

home of many other plant species that have been transported by man to Europe, Africa, and Asia to be perpetuated and modified for agriculture. Tobacco, corn (maize), tomatoes, the sugar beet, and peanuts are good examples of such New World plants that have become widespread in their distribution. In fact, the agriculture of this century is truly cosmopolitan in that crops of very diverse origins are grown in most of the temperate regions. The production of soybeans, native to China, is now rapidly increasing because of their extensive use as a raw material for industries manufacturing oils, soaps, and meat substitutes. Among crops that were already cultivated by ancient civilizations, barley and coffee originated in Ethiopia, oats and wheat in Iran and Afganistan, oranges and peaches in China, rubber along the Amazon, and cotton in India.

Our knowledge about the regions where plants originated comes from the detective work of plant geographers and is based on the premise that the greatest numbers of varieties of a given plant are found in its original home. With increasing distance from this center, fewer kinds are found. Information about the early use of a plant species in agriculture comes from the remains of early civilizations (excavated tombs, villages and cities), and even from ancient paintings and sculpture.

With the eras of exploration and settlement of the earth now at an end, the worldwide dispersal of edible plant species is fairly complete. Plant products are exported from those regions where they are most easily grown to wherever a market demand exists. For the future we can look to the botanical sciences for improvement of many plant species to produce higher yields and enhanced nutritional value. As remote areas become more accessible, and there is increased contact with formerly isolated peoples, new knowledge may become available of medicinally useful plants that are now restricted in their distribution. Perhaps we will see in the next century a worldwide dispersal of pharmacologically valuable plant species comparable in extent to the dissemination of edible ones in the past.

USEFUL REFERENCES

Baker, H. G. 1970. *Plants and Civilization*. Wadsworth Publishing Co., Inc., Belmont, Calif. (Paperback.)

Cook, S. A. 1964. *Reproduction, Heredity and Sexuality*. (Fundamentals of Botany Series). Wadsworth Publishing Co., Inc., Belmont, Calif. (Paperback.)

Heiser, C. B., Jr. 1973. *Seed to Civilization*. W. H. Freeman and Co., San Francisco. (Paperback.)

Heiser, C. B., Jr. 1969 *Nightshades, the paradoxical plants*. W. H. Freeman and Co., San Francisco. (Popular book written for nonscientists about a group of related plants including tomatoes, potatoes, and tobacco.)

Hillman, W. S. 1962. *The Physiology of Flowering*. Holt, Rinehart and Winston, New York.

Laetsch, W. M., and R. E. Cleland. 1967. *Papers on Plant Growth and Development*. Little, Brown and Co., Boston. (A collection of classic research papers including the report of W. W. Garner and H. A. Allard on day length control of flowering; paperback.)

Lawrence, W. J. C., 1968/71. *Plant Breeding*. Institute of Biology Studies in Biology No. 12. Edward Arnold (Publishers) Ltd., London. Distributed in U.S. by Crane, Russak & Co., Inc., New York (Paperback.)

Schery, R. W. 1972. *Plants for Man*, 2nd ed. Prentice-Hall, Inc., Engelwood Cliffs, N.J.

THE STRATEGIES OF LIFE:
How to Cope

Hermit crab living in abandoned snail
shell.
[Courtesy of William H. Amos]

STRATEGIES:
Part A

CAPTURE AND RELEASE OF ENERGY

[Photograph by R. H. Noailles.
Reproduced from Jaeger, P. 1961.
The Wonderful Life of Flowers.
E. P. Dutton & Co., New York]

10

Photosynthesis: Harvest of the Sun

Photosynthesis is unique among all the metabolic processes that occur in living cells, because it provides the only means for an input of energy from the sun into the earth's biosphere. All living things except for a few obscure chemosynthetic bacteria are ultimately dependent on this process, a complex set of reactions that result in the conversion of light energy into the chemical bond energy of organic compounds.

Photosynthesis occurs only in those cells containing the green pigment chlorophyll. From the substrates carbon dioxide and water, new complex organic molecules are formed and molecular oxygen is released. Accompanying this, the energy of sunlight is stored in the energy-rich bonds of ATP. This compound can be used later as a source of energy, for making more cells or reserve foodstuffs.

As intellectually curious animals, we should be appropriately impressed by the biochemical versatility and productivity of green plants. Not only our present well-being, but also much of our history is directly dependent on past photosynthetic productivity. For example, the industrial revolution was possible only because of the availability of vast reserves of fossil fuels, such as coal and oil, which are the partially decomposed products of plants living in past ages. If we look to the future, we realize that long voyages into space will be feasible only if appropriate kinds and numbers of photosynthetic organisms are taken along to recycle wastes and to provide a continuing source of oxygen and food. Photosynthetic organisms are the primary producers that form the basis of all food chains. Furthermore, these organisms are responsible for maintaining our aerobic atmos-

181

phere. Photosynthesis is the atmospheric source of molecular oxygen and all aerobic organisms are dependent on green plants for its continued availability.

The photosynthetic process has certainly influenced the evolution of all the organisms found on this planet. Aerobic respiration could not have evolved before photosynthetic organisms appeared. Furthermore, atmospheric oxygen made possible the accumulation of an ozone layer which acts as a shield to screen out much of the potentially harmful ultraviolet radiation in sunlight. Without such a screen, plants and animals could not live on the land.

Considering the amount of oxygen utilized in respiration and in the ceaseless oxidation of materials in our environment, it is evident that the continuing production of oxygen by photosynthesis is essential to the maintenance of life on earth. There have been many estimates of the total amount of photosynthesis occurring annually. No two estimates agree but all are fantastically large. One such estimate, expressing the worldwide rate of photosynthesis in terms of the weight of *carbon* incorporated into new organic compounds, is given as 16×10^9 *tons* per year. Most of this 16 billion tons of carbon is captured by photosynthesis occurring in the oceans that cover approximately four-fifths of the earth's surface.

Chloroplasts: The Photosynthetic Machines

Chlorophyll, the molecule that captures light energy, is located in a cellular organelle called the **chloroplast.** There is great variety in chloroplast shape and size, especially in algal cells, where chloroplasts range from stars to spiral ribbons in general appearance. Most chloroplasts resemble a football, i.e. are ellipsoidal. The number of chloroplasts per cell can vary from one to 50 or more.

Viewed with a light microscope under low power, chloroplasts appear to be uniformly green. Under oil immersion it is evident that chlorophyll is not uniformly dispersed within the chloroplasts, but is instead limited to certain regions called **grana,** which appear to be scattered through an optically clear colorless matrix called the **stroma.**

As revealed by the electron microscope, the basic structural pattern of a chloroplast is much like a many-layered sandwich (Fig. 10-1). This organelle is bounded by a membrane that, like the membranes of the nucleus, mitochondria, endoplasmic reticulum, etc., is made up of protein and lipid. Internally, there are many membranes called **lamellae.** Some of these lamellae appear to extend through the entire length of the plastid, and at intervals there are discrete stacks of membranes much like collapsed balloons. The electron microscope reveals that these stacks of lamellae correspond to the grana visible in the light microscope. These chlorophyll-containing layers are also called **grana lamellae;** the layers extending though the colorless matrix are called **stroma lamellae.**

EVIDENCE THAT CHLOROPHYLL IS THE LIGHT RECEPTOR FOR PHOTOSYNTHESIS

An appreciation of the uniqueness of green plant tissues—their ability to remove CO_2 from air and to generate oxygen in the light—originates with the earliest chemical studies of gases and the composition of air. Priestley observed that an animal placed in a closed container would soon suffocate, but if a plant were added to the container, the animal could survive. Therefore, he correctly concluded that plants remove a product of the animal's respiration from the air and so, in a sense, purify that air. However, Priestley did not realize why, on later occasions, this kind of experiment was not completely successful. In the 1790's Jan Ingen-Housz, the personal physician to Maria Theresa, Empress of Austria, and pioneer in the use of smallpox immunization, correctly interpreted Priestley's work in terms of the newly discovered gases, carbon dioxide and oxygen, and explained Priestley's later failures by showing that plants require light in order to remove CO_2 from the air and generate oxygen. Ingen-Housz also noted that only the green parts of plants can do this. He was the first to point out the importance of plants in the cycling of inorganic and organic material in nature: organic compounds synthesized from CO_2 and H_2O in green plant photosynthesis are consumed by herbivores; herbivores are in turn consumed by carnivores. CO_2 and H_2O, the products of respiration and of the decay of dead plants and animals, are released into the atmosphere. The input into this cycle of carbon compounds is photosynthesis.

Although the importance of green plants was recognized by earlier investigators, it was not until the late 19th century that experimental proof was obtained for the participation of chlorophyll and chloroplasts in photosynthesis. This was the achievement in the 1880's of the German investigator, Theodor

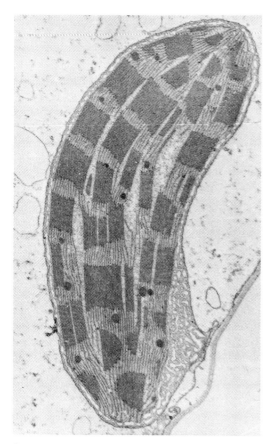

Fig. 10-1. Electron photomicrograph of chloroplast from maize (Indian corn), × 17,000. (From Shumway, L. K., and I. E. Weirer. 1967. The chloroplast structure of iojap maize. *Am. J. Bot.* 54:773.)

cluster when parts of the cell other than the chloroplast received light. Engelmann therefore concluded that photosynthesis must occur within the chloroplasts.

The most apparent colored component of chloroplasts is the green pigment chlorophyll. How was it possible for Engelmann to demonstrate that this is the **photoreceptor,** i.e. the light-absorbing component, of the photosynthetic apparatus?

As you will recall from the discussion of light effects on plant development in Chapter 7, white light is a mixture of all colors in the visible spectrum. Component wavelengths can be separated by passing a beam of white light through a prism. Projected on a white screen or a piece of paper, the separated wavelengths appear as bands of color blending gradually through violet, blue, green, yellow, orange, and red. The appearance of any colored object is determined by which wavelengths of the visible spectrum are absorbed by the pigments present and which are reflected or transmitted. Chlorophyll appears green because it absorbs poorly the green portion of the spectrum, whereas it readily absorbs the blues, oranges, and reds. In order to determine whether a given pigment, such as chlorophyll, can be the photoreceptor for a light-dependent process, it is necessary to demonstrate that the wavelengths of light absorbed by the pigment in question correspond to those portions of the spectrum that are effective in promoting the light-dependent process. If, for example, we find a

Engelmann, a physiologist also known for his work on the striations of voluntary muscle fibers. He initiated modern experimental studies of photosynthesis using very simple tools. An old-fashioned light microscope and the response of motile bacteria, which move toward regions of greater oxygen concentration, enabled him to detect the site of photosynthetic oxygen evolution. By cleaver manipulation, he was able to focus a narrow beam of white light on filamentous green algae and diatoms, thereby illuminating only one portion of a cell at a time: a chloroplast, nucleus, or an area of the cytoplasm, etc. By placing the motile bacteria in the aqueous medium surrounding the algal cells and observing their response, it was possible to determine under what circumstances oxygen was being evolved. Engelmann observed that the bacteria clustered around the illuminated cell as close as possible to a chloroplast when it received light. He observed a random distribution of bacteria with no tendency to

Fig. 10-2. Theodor Engelmann, the German physiologist, who proved that chrorophyll is the photoreceptor of photosynthesis and is also known for his studies of muscle fibers. (Courtesy of The Johns Hopkins University Institute of the History of Medicine.)

pigment that absorbs only blue light but we subsequently find that it is not blue but red light that makes the reaction go, then clearly the blue-absorbing pigment is not the photoreceptor we are looking for.

The relative effectiveness of various parts of the visible spectrum in promoting a process like photosynthesis can be ascertained in many ways. Usually some measure of rate such as the amount of oxygen evolved or CO_2 absorbed per unit time (in the case of photosynthesis) is measured at wavelengths in all parts of the spectrum with a constant intensity of light quanta used for each measurement. **Quanta** (singular, quantum) are the smallest particles of light energy. (The term **photon** is often used interchangeably with the term "quantum" for visible light.) These rate data are then plotted against wavelength to provide an **action spectrum** for the light-dependent process. Such an action spectrum should closely resemble the **absorption spectrum** of the **photoreceptor,** a pigment, in this case chlorophyll. This is because the absorbing material is in fact the photoreceptor for the process measured. The absorption spectrum of a pure compound is a unique property related to its chemical composition and molecular structure. It is one of the best ways to identify and describe the compound. An absorption spectrum can be determined on an instrument called a **spectophotometer,** which measures the amount of light absorbed by a dissolved pigment relative to that absorbed by its solvent. The units of **optical density** or **absorbance** are used as a measure of the light absorbed.

The basic concepts concerning the relationship between color and absorbed light were known to Engelmann even though sophisticated spectrophotometers were not yet developed. Even without this instrument, however, Engelmann was able to measure the action spectrum of photosynthesis in an elegantly simple way and to compare it with the absorption spectrum of chlorophyll, thereby proving conclusively that chlorophyll is, in fact, the photoreceptor for the photosynthetic process—a major accomplishment, indeed. To measure the action spectrum, he placed a prism in the path of the microbeam of white light used in previous experiments and focused the resulting visible spectrum across the chloroplasts of a single algal cell or along a filament (chain) of cells. The motile bacteria in his system clustered in greatest numbers around those cells where rates of oxygen production were highest. Thus, the distribution of bacteria provided a picture of the relative effectiveness of various wavelengths of light in promoting oxygen produc-

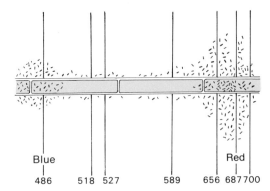

Blue   Red
486   518 527   589   656 687 700

Fig. 10-3. Portion of an algal filament (*Cladophora*) in a microspectrum of light. Swarming bacteria cluster around those regions receiving red and blue light. Numbered lines indicate wavelength in nanometers. This was the first action spectrum of photosynthesis. (After Engelmann, T. W. 1882. On the production of oxygen by plant cells in a microspectrum. *Bot. Zeit.* 40:419–426. Translation of this paper is in Gabriel, M. L., and S. Fogel, eds. 1955. *Great Experiments in Biology.* Prentice-Hall Inc., Englewood Cliffs, N. J.)

tion and depicted the action spectrum of photosynthesis. He found that the most effective wavelengths of light are in the blue and red regions of the spectrum. This is to be expected because these are the wavelengths of light most readily absorbed by the photoreceptor, a fact that is easily demonstrated by measuring the absorption spectrum of chlorophyll in a modern spectrophotometer.

QUANTASOMES: THE MACHINERY FOR CONVERSION OF LIGHT TO CHEMICAL ENERGY

From the work of Engelmann and subsequent investigators, we know conclusively that chlorophyll is the photoreceptor, but what about other essential components of the photosynthetic apparatus? What are the working parts necessary to convert light energy, carbon dioxide, and water into complex organic compounds? Are these other essential parts located near chlorophyll molecules in the grana of the chloroplasts, or are some in the colorless stroma?

Although the electron microscope has been able to reveal the complexity and internal order of chloroplast structure, other methods and approaches were necessary to answer these questions and to ascertain the chemical composition and functioning of these organelles.

It has been known for some years that intact chloroplasts isolated from broken cells retain the capacity to carry out photosynthesis. More recently, isolated chloroplasts have been broken to determine how small a fragment still retains some photosynthetic capability. From this, it has been learned that the grana can be broken into subunits that can, when illuminated, evolve oxygen and store light energy as ATP (although they cannot utilize CO_2). These subunits have been named **quantasomes.**

Chemical analysis has shown these small, uniform pieces to have an overall molecular weight of approximately 2,000,000. They contain two kinds of chlorophylls, four kinds of carotenoid pigments, several plastoquinones (derivatives of benzoquinone, yellow compounds that can be reversibly oxidized and reduced), and a variety of lipids. About half of the weight of the quantasome is made up of protein. Some of the 12 iron atoms present

are associated with an iron-containing protein called ferredoxin, while others are part of cytochrome molecules. Copper atoms are found in an unusual green pigment called plastocyanin. It is obvious that the chemical machinery of photosynthesis is highly complex.

Quantasomes are the working parts of chloroplasts where light energy is converted to chemical energy (i.e. where ATP is generated), and they contain the machinery for oxygen evolution.

Because quantatomes do function in the presence of light energy but cannot utilize CO_2, i.e. are unable to "fix" it by the addition of hydrogen, it seems reasonable to suppose that the actual conversion of CO_2 into carbohydrate takes place in the colorless matrix and not in the chlorophyll-containing grana. Both classical and recent chemical investigations support this conclusion.

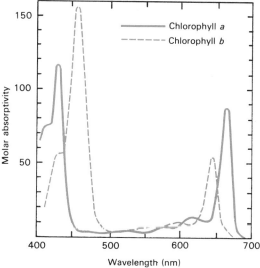

Fig. 10-4. Absorption spectra of chlorophyll *a* and chlorophyll *b* in ether. (From Moment, G. B., and H. M. Habermann. 1973. *Biology: A full Spectrum*. The Williams & Wilkins Co., Baltimore.)

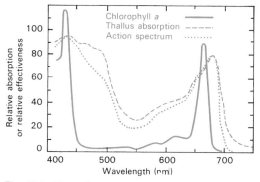

Fig. 10-5. Absorption spectrum of chlorophyll *a*, action spectrum of photosynthesis, and thallus absorption in the green alga, *Ulva taeniata*. Note the close correspondence between wavelengths promoting photosynthesis and those absorbed by chlorophyll. (From Moment, G. B., and H. M. Habermann. 1973. *Biology: A Full Spectrum*. The Williams & Wilkins Co., Baltimore.)

Effects of Environmental Factors on Rates of Photosynthesis: What They Reveal About the Process

Not only agricultural productivity on land (indeed even the feasibility of agriculture) but also the harvests of fish and other seafood from the oceans depend on a complex of environmental and climatic factors influencing the rates of photosynthesis.

The first systematic studies of the kinetics of photosynthesis (that is, the effects of factors such as temperature, CO_2 concentration, and light intensity on photosynthetic rates) were done early in the present century by Blackman and Mathgei. Their studies yielded important insights into the machinery of photosynthesis, including the realization that part of the photosynthetic process can take place in the dark and is therefore independent of light. That portion of the photosynthetic process not requiring light is sometimes called the Blackman reaction. The rationale behind all studies of the kinetics of a process is to vary a single factor at a time and to measure the effects of such variation with all other conditions held constant.

THE EFFECTS OF LIGHT INTENSITY

The dependence of photosynthesis on light as a source of energy would lead us to expect that photosynthetic rates should be zero in the dark and rates of oxygen evolution or CO_2 consumption should increase with increasing light intensity. Such is indeed the case at low light intensities. When rates are measured under physiologically appropriate conditions of constant temperature and CO_2 concentration, there is a range of light intensities where rates increase roughly linearly with increasing light (i.e. doubling intensity approximately doubles rate). There is also a light intensity that saturates the photosynthetic

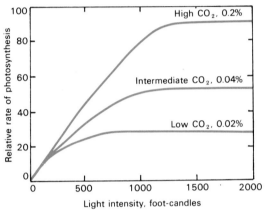

Fig. 10-6. Effect of light intensity on rate of photosynthesis. The intensity needed for and the rate at light saturation depend on the adequacy of other factors. In this graph the effects of light intensity on rates at three different concentrations of CO_2 are shown. (Redrawn from Johnson, W. H., and W. C. Streere, 1962. *This is Life, Essays in Modern Biology.* Holt, Rinehart and Winston, New York.)

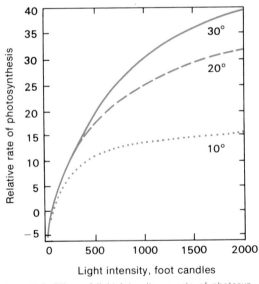

Fig. 10-7. Effect of light intensity on rate of photosynthesis at three different temperatures. In these curves the net rate of photosynthesis is plotted as a function of light intensity. At light intensities below the compensation point, rate of respiration exceeds that of photosynthesis and net rates of photosynthetic assimilation are negative. (Modified and redrawn from Noddack, W., and C. Kopp. 1940 Z. *Phys. Chem.* A187:79.)

process under these conditions. Beyond this saturating intensity further increase in light does not increase the photosynthetic rate. If we measure the intensities of light that saturate photosynthesis in most green plants, we find that they are in the range of 1,000–2,000 foot-candles, a figure that is much lower than the maximum light intensity attainable on a clear sunny day in most areas (which can be as high as 10,000 foot-candles). This raises challenging questions. What is the evolutionary explanation of this low saturation point? Did chlorophyll first evolve in deep or murky water? Do the laws of photochemistry limit a process based on chlorophyll to this level of efficiency? Would it be possible, perhaps in the distant future, for man to devise his own method of photosynthesis based on chlorophyll, or on some very different compound, that would have a far greater efficiency?

Another aspect of this problem is seen if we measure the intensity of light required for rate saturation at a higher but still physiologically tolerable temperature. We then find that the light intensity needed for saturation is higher. Thus, if rates are measured first at 10°C and then at 20°C, both the rate at light saturation and the light intensity needed for saturation are higher at 20°C than at 10°C (Fig. 10-7).

At low light intensities, still another phenomenon is evident. Below an intensity called the **compensation point,** the rate of photosynthesis is too low to be readily measurable in intact plant cells. The compensation point is the light intensity needed for the rate of photosynthetic oxygen production to equal the rate of oxygen consumption in respiration. The compensating intensity (roughly 200–300 foot-candles) is only a fraction of saturating light.

TEMPERATURE EFFECTS AND THE DISCOVERY OF DARK REACTIONS

The effects of temperature on photosynthetic rates were the evidence that led Blackman to suggest that photosynthesis is not a strictly photochemical process but rather one that involves both light and dark (or enzymatic) reactions. One characteristic of all strictly photochemical reactions is their temperature independence. However, an effect of temperature on rates of photosynthesis was already noted in our consideration of light intensity effects. This temperature dependence is unusual in a light-dependent process. In fact, it should be impossible.

Blackman found, at a constant *low* light intensity (above the compensation point but well below saturation), and at constant CO_2 concentration, that temperature has relatively

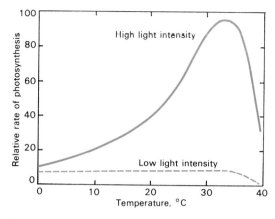

Fig. 10-8. Effect of temperature on rate of photosynthesis. At low light intensities temperature has little effect on rate of photosynthesis. At high light intensities photosynthesis responds to temperature much as enzymatic reactions do. Temperature studies provided the first clear indication that photosynthesis involves dark enzymatic as well as light reactions. (Redrawn from Johnson, W. H., and W. C. Steere, eds. 1962. *This is Life; Essays in Modern Biology.* Holt, Rinehart and Winston, New York.)

little effect on photosynthetic rates. Thus, at low light intensity, the photosynthetic process responds in a way expected for a light-dependent reaction (Fig. 10-8). However, at constant *high* light intensity (at or above saturation levels) with CO_2 concentration again held constant, he found that photosynthetic rates respond to temperature just as most enzymatic dark reactions do. That is, rates increase as the temperature is raised up to 30 or 35°C; at temperatures above 35°C rates drop due to the denaturation of enzymatic components of the photosynthetic apparatus. Clearly, there must be an important part of photosynthesis, or, more accurately, the production of carbohydrate from CO_2 and H_2O which is not directly dependent on light.

CO_2 CONCENTRATION

Carbon dioxide serves as a substrate, or raw material, for the process of photosynthesis and is the essential ingredient from which the carbon skeletons of all the complex organic components of plant and animal cells are manufactured. We would therefore expect that the carbon dioxide level would influence the rate of photosynthesis and that photosynthetic rates would be zero in the absence of this essential substrate.

Increasing CO_2 concentration does indeed increase the rates of CO_2 uptake and O_2 evolution; this effect depends on light intensity. Saturating levels of CO_2 are reached at about 0.05 per cent in low light intensity, whereas at saturating light intensities, CO_2 concentrations of 0.2 per cent or higher are needed for

maximum photosynthetic rates. When we consider that the normal level of CO_2 in our atmosphere is approximately 0.03 per cent, it is evident that this essential substrate is the factor most likely to be rate-limiting in photosynthetic processes on this planet. This knowledge has been used to increase the growth and productivity of plants in greenhouses, where CO_2 levels can be monitored and increased (up to several per cent) during the light period.

As the prevailing 0.03 per cent CO_2 is well below the optimum for green plants, it is tempting to speculate that photosynthetic organisms may have evolved, and that the photosynthetic apparatus must have developed to its present specifications, under conditions of considerably higher CO_2 concentrations. Perhaps this is also the evolutionary explanation for the low intensity of light needed to saturate the photosynthetic process: there simply was not enough CO_2 present to make a higher saturation level worthwhile, even though it certainly seems very likely that CO_2 levels were greater during past ages. During the Carboniferous era vast forests of primitive plants developed. From such vegetation deposits of plant material were laid down, subjected to heat and pressure, and later gradually transformed to coal and oil. Over millions of years, most of the earth's available carbon became bound in fossil fuels, in the form of carbonates in rocks (due in part to protozoa and mollusks secreting calcereous shells) or in the form of complex organic molecules in living organisms.

Today, this trend toward diminished atmospheric CO_2 levels may be reversing itself as the continued rapid depletion of coal and oil

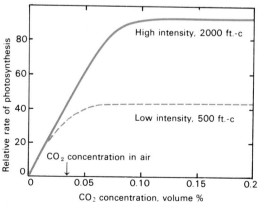

Fig. 10-9. Effect of CO_2 concentration on rate of photosynthesis. Note that the normal concentration of CO_2 in air (0.03 per cent) is well below that needed for saturation of photosynthesis even at low light intensities. (Redrawn from Johnson, W. H., and W. C. Steere, (eds. 1962.) *This is Life, Essays in Modern Biology.* Holt, Rinehart and Winston, New York.)

reserves is accompanied by the return of CO_2 to the atmosphere as a product of combustion. It is debatable whether average temperatures will change with these increasing CO_2 levels. However, we can predict that such changes should result in a general increase in photosynthetic rates and possibly in agricultural productivity.

Primary Reactions of Photosynthesis

As Blackman first demonstrated, modern studies confirm that photosynthesis consists of two kinds of reactions: those dependent on light and others that are dark (enzymatic) reactions. The light reactions utilize sunlight to produce energy-rich molecules of ATP, reduced molecules of the pyridine nucleotide NADP, and molecular oxygen. The dark reactions synthesize carbohydrate from CO_2, using ATP as the source of energy and reduced NADP as a source of hydrogens. We will first discuss the light-dependent, or **primary, reactions** of photosynthesis that take place within the quantasomes.

Current views of the nature of these primary reactions have come from a mass of experimental data that at times seemed very puzzling. It is the obligation of science to provide theories compatible with all experimental observations; in the case of photosynthesis, the observations that we must explain were downright mystifying. A close look at these strange facts will aid a student in understanding our contemporary ideas about photosynthesis as well as show something about the way science achieves knowledge of things once thought unknowable.

First of all, by the 1930's scientists knew that only a small fraction of the chlorophyll molecules present in a chloroplast were capable of doing anything. This became apparent when Robert Emerson and William Arnold at the Carnegie Institution in Stanford, California, studied the effects on photosynthetic rates of short flashes of light separated by relatively long dark periods. Essentially, their approach was a very ingenious extension of the old Blackman light-dark experiments. Their objective was to provide sufficient light to completely saturate the photosynthetic apparatus, then to follow this by a dark time long enough for all of the products of the light reaction to be used up. They were able to adjust the pattern of light and dark periods by placing a rotating disc between the light source and their experimental algae. The relative length of the light flash could be con-

Fig. 10-10. Robert Emerson (1903–1959). His experiments with Arnold on yields of photosynthesis in flashing light confirmed the concept of a photosynthetic unit. Later studies on quantum yield and his discovery of the phenomenon of enhancement provided a basis for present views of the two photoreactions in photosynthesis. (From *Plant Physiol.* 34(3):178, 1959.)

trolled by the size of the opening cut in the disc, and the number of flashes per unit time could be regulated by the speed of the disc's rotation.

Emerson and Arnold worked with their system until they found the conditions that resulted in the maximum rate of photosynthesis per flash. Then the amount of oxygen produced per flash of light and the amount of chlorophyll present in the experimental cells were measured for these conditions. If the photosynthetic apparatus was completely saturated with light and provided with sufficient time in the dark for all subsequent reactions to go to completion, they expected that one O_2 molecule would be produced and one molecule of CO_2 would be consumed for each molecule of chlorophyll. Instead, Emerson and Arnold consistently found that, under optimum conditions, the ratio of O_2 or CO_2 per chlorophyll molecule was in the range of one per several hundred to several thousand. The only possible conclusion they could draw from such observations was that only a small fraction of the chlorophyll present in a chloroplast can participate in a primary reaction. Stated another way, it appears that there are anywhere from several hundred to several thousand chlorophyll molecules per reactive site. This chlorophyll aggregate works as a **phososynthetic unit.** Quanta of light absorbed within the unit, which operates as an energy-gathering "antenna" (Fig. 10-11), can be transferred from chlorophyll to chlorophyll until one located next to the appropriate enzymatic machinery is reached.

A second peculiarity of the photosynthetic apparatus, that must be explained by any theories concerning its operation, is the observation that action and absorption spectra do not correspond perfectly at long wavelengths. Instead, photosynthetic rates and efficiency decrease in the far red region of the spectrum where there is still considerable ab-

Fig. 10-12. Otto Warburg, known for his classic studies of respiratory enzymes, was the first to demonstrate that multiple quanta must be absorbed for the incorporation of one CO_2 molecule into carbohydrate and the evolution of one molecule of O_2 in photosynthesis. (Courtesy of the Johns Hopkins University Institute of the History of Medicine.)

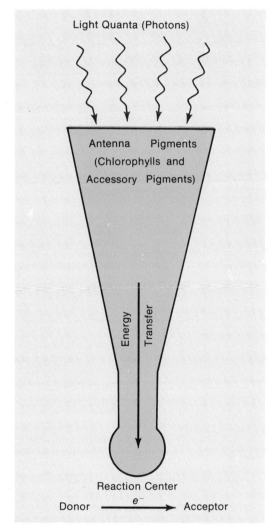

Fig. 10-11. Schematic diagram illustrating the concept of the photosynthetic unit. Photons absorbed anywhere within an aggregate of several hundred chlorophyll and accessory pigment molecules (antenna pigments) are transferred to a reaction center where the energy of light is utilized. At the reaction center a reduced electron acceptor is formed. (After Govindjee and R. Govindjee. 1974. The absorption of light in photosynthesis. *Sci. Am.* 231:73. Copyright © 1974 by Scientific American, Inc. All rights reserved.)

sorption of light by chlorophyll. This phenomenon of unexpected decrease in efficiency at wavelengths longer than approximately 680 nm is called the **red drop.**

The same Robert Emerson who, with Arnold, did the flashing light experiments already described, in later years became Professor of Botany at the University of Illinois. There he made painstaking measurements of the action spectra and the quantum requirements of photosynthesis in algae. During such studies, Emerson observed that small amounts of short wavelength (blue) light added to saturating amounts of long wavelength light increase rates of photosynthesis far in excess of what he expected from the sum of rates in low intensity blue light plus rates in high intensity far red light. This phenomenon, the stimulating effect of short wavelength light added to long wavelength light, is called the **Emerson enhancement effect.** This strange effect has now been explained and is known to result from the fact that there are two separate light-dependent steps in the photosynthetic process.

The evidence that two light-dependent events occur in photosynthesis came in part from studies of quantum requirements, experiments that point to a third peculiarity of the photosynthetic process. Many investigators, including the German Nobel laureate Otto Warburg, sought to answer the question of how many quanta (photons) of light are needed by the photosynthetic apparatus for the incorporation of one molecule of CO_2 or the evolution of one molecule of O_2. In other words, what is the **quantum requirement** of photosynthesis? We should recall at this point that the amount of energy per quantum of light varies inversely with wavelength. Thus, red light, with a wavelength of 660 nm, contains less energy per quantum than does shorter wavelength (450 nm) blue light (to review these energy relationships, see Chapter 7). We can extend our question about the quantum requirement of photosynthesis as follows: Does the quantum requirement remain the same for all wavelengths of light that are effective in promoting this process?

The quantum requirement for photosynthesis has been measured for all wavelengths of light from the ultraviolet to the infrared. One rather strange thing about these data is that when quantum requirements are plotted as a function of wavelength, the resulting curve looks very different from the action spectrum of photosynthesis: it is a flat line through most of the visible spectrum, increasing sharply in the far red region. Alternatively, these same data can be plotted as **quantum yield,** which for photosynthesis is the amount of O_2 produced or CO_2 consumed per quantum. This is nothing but the reciprocal of the quantum requirement so when plotted as a function of wavelength, it is a flat line dropping to zero in the far red (Fig. 10-13). The reason that these plots do not resemble action spectra is simple: in measurements of quantum requirements and yields, only the **absorbed** quanta count, but in measurements of action spectra the rates at constant intensities of **incident** quanta (whether absorbed or not) are important. You will recall from the absorption spectrum of chlorophyll (which can be viewed as a picture of the

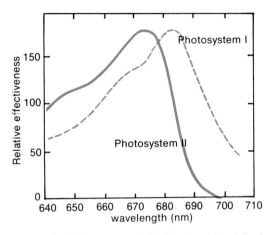

Fig. 10-14. Action spectra of Photosystems I and II of photosynthesis. Because the photoreceptor for Photosystem I absorbs at slightly longer wavelengths, it can be activated by quanta beyond approximately 680 nm that cannot be absorbed by Photosystem II. (From Moment, G. B., and H. M. Habermann. 1973. *Biology: A Full Spectrum.* The Williams & Wilkins Co., Baltimore.)

relative probability for light quanta of different wavelengths to be absorbed; see Fig. 10-4.) that blue and red quanta are readily absorbed by chlorophyll while green and far red light quanta are absorbed very poorly. The drop in quantum yield in the far red region (i.e. the need for additional quanta to accomplish a given amount of change) is obviously related to the decreased efficiency of light absorption by chlorophyll molecules in this long wavelength part of the spectrum. The discrepancy between the action spectrum of photosynthesis and the absorption spectrum of chlorophyll is due to the fact that there are two different light-requiring steps in the photosynthetic machinery. One of these involves a form of chlorophyll that can absorb light at longer wavelengths than the other, i.e. their **action spectra** are different (Fig. 10-14). These two primary reactions will be discussed in detail later.

The important point about quantum data, which we emphasize here, is that at least 8 quanta of light must be absorbed in order for one molecule of O_2 to be evolved. This **multiple quantum requirement** (most measurements range between 8 and 12 or more quanta per molecule of O_2) was perhaps the best indication that no simple, direct reaction between molecules of CO_2 and chlorophyll could explain the mechanism of photosynthesis.

In addition to the peculiarities of the photosynthetic apparatus already noted (the photosynthetic unit, the red drop, enhancement, and the multiple quantum requirement), there is a fourth and final fact coming from experi-

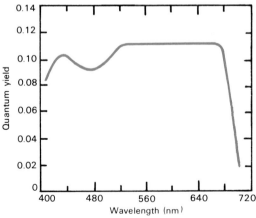

Fig. 10-13. The quantum yield of photosynthesis as a function of wavelength in *Navicula minima*, a diatom. (Redrawn from Tanada, T. 1951. *Am. J. Bot.* 38:276.)

ments with tracer oxygen that must be explained by any general theory. In the 1940's, Ruben, Randall, Kamen, and Hyde, working at Washington University in St. Louis, posed the question: Where does the O_2 evolved in photosynthesis originate? Both substrates, CO_2 and H_2O, are molecules containing oxygen. Using an isotope of oxygen (with an atomic weight of 18) to label either the CO_2 or H_2O, they ran experiments to determine the origin of the O_2 evolved. Their results are summarized by the two following equations:

$$C^{18}O_2 + 2H_2{}^{16}O \rightarrow |CH_2O| + {}^{16}O_2$$
$$C^{16}O_2 + 2H_2{}^{18}O \rightarrow |CH_2O| + {}^{18}O_2$$

The label appeared in the evolved oxygen only when it was introduced in the water in which photosynthetic cells were suspended. Therefore, the obvious conclusion from these experiments was that the oxygen produced in photosynthesis comes from water.

Van Niel's Unifying Concept: Light-dependent Splitting of Water

Until the last decade there was only one theory which could account for at least sev-

Fig. 10-15. C. B. Van Niel, whose concept of photolysis, or the light-dependent splitting of water, provided the first theoretical framework for understanding the mechanism of photosynthesis. (From *Ann. Rev. Plant Physiol.* 13: 1962.)

eral of the puzzling observations just discussed — that of C. B. Van Niel, a microbiologist working at the Hopkins Marine Station of Stanford University in Pacific Grove, California. His ideas provided the first theoretical framework for an understanding of the mechanism of photosynthesis. Van Niel examined an array of photosynthetic microorganisms, including algae and photosynthetic bacteria, and from such comparative studies he concluded that photosynthesis could be viewed as a sequence of reactions between a hydrogen donor and a hydrogen acceptor (CO_2). A hydrogen donor is a **reducing agent,** that is, a molecule that can contribute hydrogen or an electron to a receptor molecule (oxidant). In the green plants, H_2O serves as the hydrogen donor while alternate compounds serve this function in the bacteria (e.g. H_2S in the sulfur bacteria). Thus for any organism, photosynthesis could be summarized by the general equation:

$$CO_2 + 2H_2A \xrightarrow[\text{chlorophyll}]{\text{light}} (CH_2O) + 2A$$

where H_2A stands for the hydrogen donor and (CH_2O) represents carbohydrate. In green plants this equation becomes the familiar:

$$CO_2 + 2H_2O \xrightarrow[\text{chlorophyll}]{\text{light}} (CH_2O) + O_2$$

whereas in the sulfur bacteria it becomes:

$$CO_2 + 2H_2S \xrightarrow[\text{chlorophyll}]{\text{light}} (CH_2O) + 2S$$

A further generalization made by Van Niel was the suggestion that, in *all* photosynthetic organisms, the function of light is the same. He proposed that light quanta are utilized to split water (a step he termed **photolysis**) into a reduced product, [H], and an oxidized product, [OH]. The reduced product ultimately serves to reduce CO_2 to the level of carbohydrate (CH_2O). Oxygen is evolved in green plants from the oxidized product, and in the bacteria a variety of organic molecules are oxidized by the [OH]. Van Niel's scheme can be summarized as follows:

4 quanta ⤳ $4H_2O$

$$4H \vdots OH$$

CO_2 $4|H|$ $4|OH| \xrightarrow[\;+\;2H_2A\;]{\text{bacteria}} 4H_2O + 2A$

(CH_2O) + H_2O $\xrightarrow{\text{green plants}} 2H_2O + O_2$

This hypothesis can explain a number of the puzzling observations that were made during the past half century. However, it now is clear that light doesn't split water directly but rather is essential to the utilization of H^+ and OH^- ions in the way Van Niel proposed. His scheme is certainly consistent with the requirement for more than one quantum of light per O_2 evolved or CO_2 consumed. A minimum of 4 quanta would be required by Van Niel's scheme, since he postulated that four molecules of water are split, each by 1 quantum of light energy. Any discrepancy between this theoretical requirement of 4 quanta and actual measurements could be explained by a loss of quanta as heat, etc. Van Niel's theory is also completely consistent with the tracer studies indicating that the oxygen produced in photosynthesis comes from water. However, viewed from the critical vantage point of thermodynamics, Van Niel's original picture of the photolysis of water cannot be accepted, and for one very simple reason. There is just not enough energy in a single quantum of visible light to split a molecule of water. Photolysis, using one quantum of light, is thermodynamically impossible; using more than one quantum, it becomes a statistically improbable event because it would take too long for a single chlorophyll to be hit by two photons. Even without the insight of thermodynamics, Van Niel's theory has been open to re-examination for the further reason that it does not explain the enhancement effects already described. And, after all, a theory must be consistent with all of the facts. Today this generalization has not been abandoned, but, as so commonly happens with theories, refined with better knowledge of the mechanism by which light energy can separate the components of water.

THE TECHNOLOGY OF ISOLATED CHLOROPLASTS: THE HILL REACTION

There exists a long history of experimentation on photosynthesis based not on leaves or even on whole cells, but instead on isolated cellular organelles, the chloroplasts. Attempts to separate photosynthesis from the rest of cellular metabolism go back to the last century when Haberlandt in 1888 and Ewart in 1896 observed that chloroplasts isolated from leaves by grinding in water were capable of liberating small quantities of oxygen when illuminated. Later, in 1904 and again in 1925, Molisch reported the evolution of oxygen from preparations of dried leaf powders. Inman, in the 1930's, returned to the kinds of experiments done by Haberlandt and Ewart,

grinding fresh leaves to isolate chloroplasts. He demonstrated that the limited ability of his isolated chloroplasts to evolve oxygen in the light could be destroyed by factors such as high temperature or protein-digesting enzymes, and therefore suggested that oxygen evolution is enzymatic.

The experiments of all these pioneer investigators were disappointing, however, because the amounts of oxygen evolved by their preparations during illumination were never more than a small fraction of the photosynthetic rates of the living cells and tissues from which their chloroplasts were isolated. It was not until the late 1930's that the English investigator, Robert Hill, discovered how to sustain high rates of oxygen evolution in the light; a **hydrogen acceptor** (i.e. an **oxidant**) had to be added to suspensions of isolated chloroplasts. Using a mixture of hemoglobin and ferric oxalate as the hydrogen acceptor, Hill demonstrated that chloroplasts continued to evolve oxygen in the light until the oxidant (the ferric oxalate mixture) was completely reduced. Although the amount of oxygen evolved was limited by the amount of oxidant initially present, the *rate* of oxygen evolution depended on chloroplast concentration and light intensity. In terms of Van Niel's theory, Hill's chloroplast reaction can be summarized as follows:

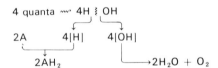

where A = the hydrogen acceptor or oxidant.

It is now clear that the mechanism of oxygen evolution by isolated chloroplasts is identical to that followed by photosynthesizing cells. Thus, Hill's discovery provided convincing evidence of the validity of Van Niel's view of the primary reactions of photosynthesis and was a landmark in our knowledge of this process.

In the years that followed Hill's discovery, many additional compounds were shown to be good oxidants for the family of reactions now referred to as **Hill reactions.** The list includes a variety of dyes, ferricyanide salts, quinones, and even molecular oxygen. The term Hill reaction is used for all chloroplast reactions in which a suitable oxidant reacts with a reductant generated by illuminated chloroplasts. Reduction of the hydrogen acceptor, or oxidant, is accompanied by evolution of oxygen. The oxidant essentially substitutes for the reactions involved in CO_2 reduction in intact photosynthesis, and for awhile it seemed that the CO_2-fixing system might even be located somewhere outside the chloroplasts. However, as methods were improved it eventually became possible to isolate unbroken chloroplasts and to demonstrate that they are capable of the entire set of reactions of photosynthesis including the synthesis of complex organic compounds. Once chloroplasts have been isolated, they can be broken down further with exciting results. A recent achievement of two Australian investigators, Boardman and Anderson, has been the separation of two distinctly different chlorophyll-containing fractions from broken chloroplasts. Each fraction contains the components of a unique part of the photochemical apparatus. According to current terminology, these two parts are called **Photosystems I** and **II**. They are at the crux of current theories about the action of light in photosynthesis, discussed at length in the next section.

With the perfection of a technology for isolating chloroplasts and individual parts of the photosynthetic apparatus from the rest of the cell, rapid strides have been made toward an understanding of the light-dependent reactions in photosynthesis. One major aspect of the process still poorly understood, however, is the mechanism of oxygen evolution. The only undisputed fact here is that manganese is essential. It has long been recognized that photosynthetic cells deprived of manganese lose their capacity to evolve oxygen, and that the enzymatic machinery for oxygen evolution can be rapidly restored by adding inorganic manganese salts to the medium in which deficient cells are suspended. Unlike the magnesium, which is known to be part of the chlorophyll molecule, or other elements associated with specific components of chloroplasts, the role of manganese is unknown. It is generally agreed, however, that the reactions involved in oxygen evolution are closely linked to that part of the photosynthetic apparatus called Photosystem II. Still, the solution of the mystery of the mechanism of oxygen evolution remains an accomplishment eagerly sought but yet to be achieved.

Present Views on the Photochemistry of Photosynthesis

Many disciplines have contributed to the current picture of what happens when light is absorbed within a chloroplast. Physicists tell us that light absorption results in the displacement of an electron from the photoreceptor molecule. This displacement is nothing more than movement of the electron to an orbit further from its atomic nucleus. Several options are possible thereafter. First, the electron can fall back into place (that is, back to its "ground state") with the result that its absorbed energy is dissipated as heat. A second possibility is that, as the electron falls back to its ground state, light can be re-emitted either as fluorescence or phosphorescence, and again lost. Finally, the electron can be captured by an appropriate acceptor molecule. Only under the last circumstance can the energy of the absorbed light quantum be conserved and used; that is, converted into potential chemical energy.

Perhaps the most significant progress in human understanding of photosynthesis during the past decade has been the realization that we can account for the strange characteristics noted earlier only by assuming the collaboration of two different primary reactions; that is, two distinct light-dependent steps, carried out by Photosystems I and II.

Energy, trapped anywhere within the aggregate of chlorophyll and accessory pigment molecules called the photosynthetic unit, is transferred to a reaction center (see Fig. 10-11). At such a center a chlorophyll molecule has associated with it an electron donor (**reductant**) and an electron acceptor (**oxidant**). The absorption of a quantum of light by the chlorophyll at a reaction center results in the displacement of an electron from the chlorophyll molecule; it is trapped by the acceptor molecule which thus becomes reduced. The chlorophyll is returned to electrical neutrality by receiving an electron from its donor molecule. In doing so, the donor becomes oxidized. Thus, the overall result of the absorption of a photon by the chlorophyll at a reaction center (i.e. a **primary reaction**) is the formation of a reduced electron acceptor and an oxidized electron donor. Donor and acceptor molecules that function similarly to those involved in the photosynthetic apparatus are also found in the electron transport system of the mitochondria (which contains cytochromes; see Chapters 3 and 11). Such molecules can be described by their oxidation-reduction potential, a quantity that is either positive or negative and is expressed in volts. Electrons, being negative, move spontaneously from a donor having a negative potential to an acceptor with a less negative potential. As electrons move along a gradient of carrier molecules toward the most electropositive acceptor, energy is released. In both the mitochondria and the chloroplasts, energy released by the movement of electrons along a chain of carriers is stored as ATP. The mitochondria and chloroplasts differ in one very important aspect, however. In the mitochondria, electrons removed during the breakdown of foodstuffs all flow "downhill" to the ultimate electron acceptor, oxygen. In the chloroplasts, two of the steps in electron transport are against the gradient of electrochemical potential, and for each of these two steps, a light quantum, i.e. a photon, is needed. Thus, there is an input of energy which accomplishes the feat of moving electrons "uphill," from water molecules (at a potential of +0.8 V, which serve as electron donors) to the pyridine nucleotide NADP, nicotinamide adenine dinucleotide phosphate (at a potential of −0.3 V), which serves as the terminal electron acceptor.

The two energy input steps occur at reaction centers having different action spectra (although a form of chlorophyll *a* is thought to be involved in both) and different electron donor and acceptor molecules. These two kinds of reaction centers are the Photosystem I and Photosystem II already mentioned (see Fig. 10-16). In Photosystem II, the primary reaction results in the generation of a strong oxidant and a weak reductant. In Photosystem I, a weak oxidant and a strong reductant are formed. The weak reductant formed by Photosystem II and the weak oxidant generated by Photosystem I are linked together by a chain of carriers along which electrons can move spontaneously.

An electron ejected from the chlorophyll of Photosystem II is transferred from its electron acceptor through intermediates including plastoquinone, plastocyanin, and cytochromes *f* and *b* to the positively charged chlorophyll remaining after removal of an electron from the chlorophyll of Photosystem I. The electron ejected from Photosystem I chlorophyll is trapped by its electron acceptor and transferred via ferredoxin to NADP.

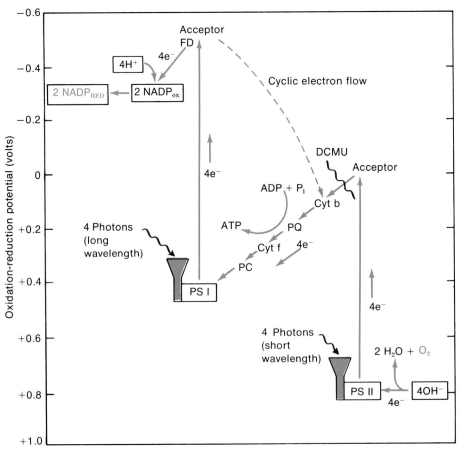

Fig. 10-16. The primary reactions of photosynthesis. The two photoreactions are linked by a chain of electron-carrying intermediates (see text for explanation). The following abbreviations are used: PQ, plastoquinone; PC, plastocyanin; FD, ferredoxin; DCMU, dichlorophenyldimethyl urea. (From Moment, G. B., and H. M. Habermann. 1973. *Biology: A Full Spectrum*. The Williams & Wilkins Co., Baltimore.)

But what about the remaining positively charged chlorophyll II? And how does **ferredoxin** (which accepts electrons from Photosystem I) reduce NADP? After all, protons as well as electrons are needed for the reduction of this pyridine nucleotide. Also, we have said nothing thus far about the involvement of water and CO_2, the recognized substrates of photosynthesis. Light quanta cannot be used to tear apart water molecules by any known mechanism. How, then, does water become involved in these events? Once again we must remind ourselves about a basic fact of chemistry—that a fraction of all water molecules, including those in the chloroplast, are ionized, or dissociated into negatively charged hydroxyl ions (OH^-) and positively charged hydrogen ions (H^+, or protons). These ions provide the means for restoring the photosynthetic apparatus to electrical neutrality following the absorption of light quanta. A negatively charged hydroxyl ion can donate an electron to restore the positively charged chlorophyll II molecule to its original state, while a proton (i.e. an H^+) joins the electron removed during the reoxidation of ferredoxin to reduce the pyridine nucleotide NADP. The hydrogens thus made available from reduced NADP are utilized in the sequence of reactions to be described later, where CO_2 is reduced to the level of carbohydrate. Two quanta of light are required for each electron moved through Photosystems I and II from water (i.e. OH^-) to the point where electrons and protons reduce NADP. It takes four hydrogens (i.e. four protons and four electrons) for the reduction of a single molecule of CO_2 to the level of carbohydrate and four [OH] radicals for the formation of one molecule of O_2 plus the two molecules of water needed to balance our chemical ledger. These aspects of the chemistry of photosynthesis are summarized in the following equations:

$$4H_2O \rightarrow 4H^+ + 4OH^-$$
$$4OH^- \rightarrow O_2 + 2H_2O + 4e^-$$
$$2NADP_{ox} + 4e^- + 2H^+ \rightarrow 2NADP_{red}$$
$$2H^+ + 2NADP_{red} + CO_2 \rightarrow 2NADP_{ox} + H_2O + (CH_2O)$$

Net Change: $H_2O + CO_2 \rightarrow (CH_2O) + O_2$

Thus, it appears that the experimentally measured requirement of at least 8 quanta of light for the evolution of one molecule of oxygen from, or the entry of one molecule of CO_2 into, the photochemical apparatus makes sense theoretically. Our present view of the primary reactions of photosynthesis is consistent with tracer experiments indicating that the oxygen evolved in photosynthesis comes from water. It must be emphasized, however, that water molecules are not broken apart directly through the action of light, but rather products of the ionization of water molecules can serve as electron acceptors and electron donors. The energy of light quanta is needed for only one purpose: to move electrons from one energy level to a higher energy level in a molecule of chlorophyll. Electrons thus displaced from their ground state can be removed from chlorophyll if caught by appropriate acceptor molecules.

There is a stepwise loss of energy as electrons move spontaneously along the chain of intermediates connecting the two photosystems (Fig. 10-16). As we have noted, the situation here is similar to that in the electron transport system of mitochondria. Coupled to electron transport in both chloroplasts and mitochondria are reactions in which ATP is synthesized from ADP and inorganic phosphate. In the chloroplast system, this formation of ATP is called **photosynthetic phosphorylation** or **photophosphorylation**. More will be said later about this light-dependent synthesis of ATP.

PRIMARY REACTIONS OF PHOTOSYNTHESIS AND THE COMPOSITION OF THE QUANTASOMES

One very comforting aspect of the present view of how photosynthesis works is that it indicates the usefulness of a number of the once strange-seeming components found in the quantasomes. In our discussion of the workings of Photosystems I and II we noted the involvement of chemicals listed earlier as constituents of the quantasomes, including copper in plastocyanin, a green pigment that is not a chlorophyll; quinones such as plastoquinone; iron in the cytochromes and ferredoxin; and pigments such as chlorophylls and carotenoids. An abundance of chlorophyll molecules is needed, of course, with a few of the more than 200 chlorophyll molecules per quantasome located near an appropriate electron acceptor. The carotenoids and other accessory pigments share with chlorophylls the job of transferring absorbed light quanta to reactive sites. The carotenoids also have a protective function in preventing the destruction of chlorophyll molecules by photo-oxidation. In most green cells there is a constant turnover of chlorophyll, with new synthesis more or less compensating for breakdown. In certain mutants of algae, photosynthetic bacteria, and higher plants that lack carotenoid pigments, the photodestruction of chlorophyll is so rapid that it cannot accumulate; such cells are albino in spite of their ability to synthesize chlorophyll.

The oxidants and reductants in the quantasomes include the various quinones. Of these, plastoquinone is located in the electron transport system between Photosystems I and II, close to the initial acceptor for Photosystem II. Plastoquinone becomes reduced very rapidly when chloroplasts are illuminated.

There is still uncertainty about the initial electron acceptor for Photosystem I. An early carrier of electrons is ferredoxin. There is evidence that a more electronegative compound is the initial acceptor and that this electron carrier in turn reduces ferredoxin. A rapid reduction of ferredoxin can be demonstrated spectrophotometrically in chloroplast preparations illuminated by light absorbed only by Photosystem I.

Present views about the sequence of components in the electron transport system of photosynthesis are based on very sensitive and complex optical experiments in which changes in absorption by individual compounds can be observed in times as short as milli- or even microseconds.

As a component of membranes, the various lipids of the quantasomes are involved in maintaining the structural integrity of the photosynthetic apparatus. It is within the lipid layers of the grana lamellae that the chlorophylls, carotenoids, and quinones are located. The proteins, which make up close to one-half of the total molecular weight of the quantasome, are involved not only in membrane structure, but also as enzymes and as proteins associated with components of the electron transport system such as the cytochromes and plastocyanin.

The components of our scheme that contain iron are ferredoxin and cytochromes *b* and *f*. The six atoms of copper per quantasome all appear to be associated with the green pigment plastocyanin, a component of the electron transport chain connecting Photosystems I and II. The two atoms of manganese are not associated with any known component of the quantasomes, but their recognized role in photosynthetic oxygen evolution leads us to assume that they are associated with an enzyme involved in conversion of the oxidized product of Photosystem II to molecular oxygen.

THE TWO PHOTOSYSTEM SCHEME AND THE PHENOMENA OF RED DROP AND ENHANCEMENT

Two peculiar characteristics of the photosynthetic apparatus noted earlier in this chapter must now be reexamined in light of the mechanisms we have been discussing:

red drop and enhancement. As we have already noted, the photoreceptors of Photosystems I and II are forms of chlorophyll with different absorption and action spectra. Through most of the visible spectrum, light quanta have approximately equal probability of being absorbed by either photosystem. It is at long wavelengths that the absorption and action spectra of the two photosystems differ significantly. The effectiveness of Photosystem I extends beyond that of Photosystem II into the long wavelength portion of the spectrum (i.e. in the far red; see Fig.10-14). Thus, at wavelengths beyond approximately 680 nm incident light quanta can be absorbed quite readily by Photosystem I but very poorly, if at all, by Photosystem II. The explanation of lower rates and the decrease in quantum yield (or increase in quantum requirement) seen at long wavelengths is a simple one: the absorption of light by system II chlorophyll drops rapidly at long wavelengths, while the absorption of light by system I chlorophyll continues. The differences in the absorption characteristics and in the action spectra of the two photosystems result in a deficiency of light quanta entering system II at long wavelengths. But for the entire process to continue to operate, light quanta must continually be absorbed by both photosystems. The stimulating effect of added short wavelength light is also readily explainable in that it provides the input of light energy needed for Photosystem II, which is rate-limiting under these conditions.

Photophosphorylation

The coupling of ATP synthesis to the electron transport chain connecting Photosystems I and II, i.e. **photophosphorylation** in chloroplasts, and the seeming parallel in mechanism of oxidative phosphorylation of mitochondria have already been noted. Both produce ATP, but ATP synthesis in chloroplasts has certain peculiarities indicating that this light-dependent synthesis is different from that in the mitochondria. To begin with, the entire photosynthetic apparatus need not be operational for photophosphorylation to take place. When an external electron donor is provided, **cyclic photophosphorylation** can proceed, utilizing light absorbed exclusively by Photosystem I. Cyclic photophosphorylation can also take place in the presence of an inhibitor such as dichlorophenyldimethyl urea (DCMU), which inhibits Photosystem II. The "short circuit" around Photosystem I involved in cyclic photophosphorylation can provide needed ATP under conditions where neither the reduction of CO_2 nor the evolution of oxygen can occur.

ATP synthesis also occurs when both photosystems are operational, and under these circumstances it is referred to as **noncyclic photophosphorylation.** In this case, ATP formation is accompanied by the production of reduced NADP and the evolution of oxygen.

For the complete photosynthetic process, the usual **stoichiometry** (or ratio of reactants and products) is that for each molecule of CO_2 entering the carbon cycle, one molecule of O_2 is evolved, two reduced NADP's are formed, and three ATP's are generated. It has been postulated that more than one photophosphorylation site must exist for such stoichiometry to be possible, but much controversy still exists concerning the exact location of these sites of ATP synthesis. Some light quanta may be utilized in cyclic photophosphorylation, while only the ATP's formed by noncyclic photophosphorylation are accompanied by NADP reduction and oxygen evolution. The measured quantum requirement for O_2 evolution (or CO_2 consumption) generally exceeds the theoretical minimum of 8 by 4 to 6 quanta. It is likely that some of these extra quanta are utilized for cyclic photophosphorylation.

Carbohydrate Synthesis: The Calvin Cycle

The enzymatic reactions involved in reducing CO_2 to the level of carbohydrates and other complex organic components of plants are thought to take place outside the grana, that is, in the colorless stroma of the chloroplasts. The reactions of CO_2 fixation are dark reactions dependent on reduced NADP and ATP generated in the light reactions which take place in the quantasomes.

Our understanding of the biochemical pathway of carbohydrate synthesis in photosynthesis comes largely from the work of Melvin Calvin and his associates at the University of California at Berkeley. For his achievement in elucidating this complex pathway, Calvin received the Nobel Prize in 1961. Calvin's work demonstrates the importance to scientific progress of the availability of appropriate materials and techniques. These experiments would have been impossible without the existence of a radioactive isotope of carbon, ^{14}C, which was used to label CO_2 entering the photosynthetic apparatus and to trace the in-

Fig. 10-17. Melvin Calvin, whose investigations of the path of carbon in photosynthesis resulted in elucidation of the complex reaction cycle bearing his name. (Courtesy of The Johns Hopkins University Institute of the History of Medicine.)

termediates and products formed. Paper chromatography was utilized to separate individual compounds from the complex mixtures isolated from cells and to identify compounds containing the radioactive tracer.

The technique of paper chromatography is based on the differences in solubility of different compounds in various solvents and in the degree to which they tend to be absorbed on paper. A small drop of sample is applied on one corner of a piece of filter paper and the edge of the paper is dipped into a solvent. As the solvent rises by capillary action, it carries the dissolved molecules with it. Molecules will move in accordance with their size, solubility, and other characteristics. Further separation can be attained by rotating the paper 90° and "developing" in a new solvent (Fig. 10-18).

Radioactively labeled components can be located by a technique called **radioautography,** which involves placing the dried chromatogram next to a sheet of photographic paper or film. After development, dark spots appear on areas of the film that were in contact with portions of the chromatogram where radioactively labeled compounds were located. Individual compounds can be identified from the location of labeled pure compounds subjected to the same conditions, or by chemical analysis of the spots themselves.

In the early experiments with ^{14}C-labeled CO_2, photosynthesizing cells were exposed to the tracer for relatively long periods of time. As a result, almost every compound in these cells became labeled. Therefore, it was apparent that exposure to the tracer had to be very short if only the early intermediates between CO_2 and carbohydrate were to be la-

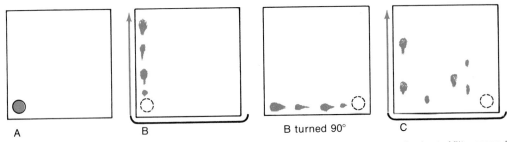

Fig. 10-18. The technique of paper chromatography. A, spot of sample is applied to one corner of a sheet of filter paper. B, separation of components in the sample after development in first solvent. Direction of solvent movement is indicated by arrow. C, after turning the chromatogram 90°, further separation of components with a second solvent (direction of solvent movement indicated by arrow).

beled. With sufficiently short exposure times, the label appeared in only the earliest intermediates; with longer exposure times more compounds became labeled. By varying the exposure times and by identifying all the labeled compounds, it was possible to piece together the sequence of reactions taking place.

Exposure of cells to tracer and light was done in a flat, round glass container appropriately called a "lollipop." After the desired reaction time, the contents of the lollipop (algal cells in a suspending medium containing ^{14}C-labeled CO_2) were drained rapidly into a beaker of hot alcohol. This treatment killed the cells and stopped all reactions instantly. The cells were then extracted, the extracts concentrated, and the components of the extract were separated and identified by paper chromatography.

Details of the intermediates and reactions discovered by these means are shown in Figure 10-19. There are certain aspects of the cycle that we should all understand and appreciate, although the detailed memorization of most intermediates is a task best left to the organic chemists. CO_2 molecules enter the cycle by attachment to a five-carbon, phosphorylated acceptor molecule, **ribulose diphosphate (RuDP)**. The resulting six-carbon intermediate is very unstable, and breaks into two molecules of **3-phosphoglyceric acid (PGA)** which is the first identifiable labeled compound. PGA is next reduced at the expense of reduced NADP generated in the light reactions to form two interconvertible triose phosphates, **dihydroxyacetone phosphate** and **3-phosphoglyceraldehyde.** At this point compounds have been formed that are familiar as intermediates in the respiratory breakdown of foodstuffs (see Chapters 3 and 11). In glycolysis these two triose phosphates are the products of the action of the enzyme aldolase on fructose-1,6-diphosphate. Their further breakdown to pyruvic acid provides the key substrate leading to the Krebs cycle.

Thus, it is evident that very early intermediates of the Calvin cycle can be diverted into the respiratory metabolism of the cell.

The remainder of the cycle is made up of a complex sequence of reactions in which a new acceptor molecule is generated. The regeneration of a new phosphorylated acceptor molecule and the formation of triose phosphates from PGA requires ATP, the other product of light-dependent reactions in the quantasomes that is essential for the continued operation of the Calvin cycle.

For every turn of the cycle, one CO_2 is incorporated into the organic matter of the cell, and one molecule of ribulose diphosphate (RuDP) is regenerated. For every molecule of CO_2 entering the cycle, two reduced NADP's and three ATP's are needed, a stoichiometry that agrees with the observation that, in the light reactions of the quantasomes, two reduced NADP's and three ATP's are generated for each O_2 evolved, and that an energy input of 8 or more quanta of light is required.

Finally, it is important to appreciate the intimate involvement of the photosynthetic carbon cycle with other aspects of cellular metabolism. The relationships between photosynthetic intermediates and the essential building blocks of plant cells are summarized in Figure 10-20. The key intermediate, 3-phosphoglyceric acid, the first stable product of CO_2 assimilation in photosynthesis, is a precursor common to pathways leading to carbohydrates, fats, and amino acids. All major plant constituents can be derived from PGA.

ALTERNATE ROUTES OF CARBON FIXATION: HIGH AND LOW EFFICIENCY PLANTS

Within the past decade, it has become apparent that, although the Calvin cycle is the major route of CO_2 into cellular intermediates, there are circumstances where CO_2 entering the photosynthetic machinery is diverted. One alternate route is utilized under conditions of low CO_2 concentration (not really such an unusual circumstance). Here the entry of CO_2 into the photosynthetic machinery appears to follow the Calvin cycle except that ribulose diphosphate is split into glycolic acid plus triose phosphate. The **glycolic acid pathway,** as its name implies, results in the accumulation of glycolic acid as an early product and is a more direct route to amino acid synthesis than is the Calvin cycle, in which carbohydrates tend to accumulate as excess products of photosynthesis. The glycolic acid pathway assumes a significant role in photosynthetic carbon metabolism primarily under what could be considered "CO_2 starvation" conditions for green cells.

When plants accumulate glycolic acid they have a tendency to oxidize this compound. The resulting **photorespiration** can be detected as an increased rate of oxygen uptake or CO_2 evolution in the light. Photorespiration occurs in cellular organelles called **leaf peroxisomes.** Those conventional plants in which CO_2 fixation is via the Calvin cycle and which exhibit photorespiration are referred to as **C_3 plants.**

There are in addition some plants, many of them tropical species, which have an alternate system for accumulating CO_2 from the atmosphere by incorporating it into four-carbon compounds, the dicarboxylic oxalacetic, malic, and aspartic acids. Such **C_4 plants** have an unusual pattern of leaf anatomy consisting of concentric cylinders: a central vascular bundle is surrounded by a **bundle sheath** (a dense single layer of dark green cells); a loosely packed layer of mesophyll cells lies outside the bundle sheath (Fig. 10-21). An all too familiar temperate plant exhibiting C_4 anatomy and metabolism is crabgrass.

In the so-called C_4 plants CO_2 is first incorporated into four-carbon dicarboxylic acids by the leaf mesophyll cells. These compounds later are decarboxylated and the CO_2 released is utilized to form the intermediates of the Calvin cycle by cells of the bundle sheath. Thus, the eventual accumulation of new carbon compounds is via a sequence of reactions identical to that in the C_3 plants (Fig. 10-22).

C_4 metabolism appears to be an evolutionary advance that provides a physiological adaptation to intense heat and water stress. In

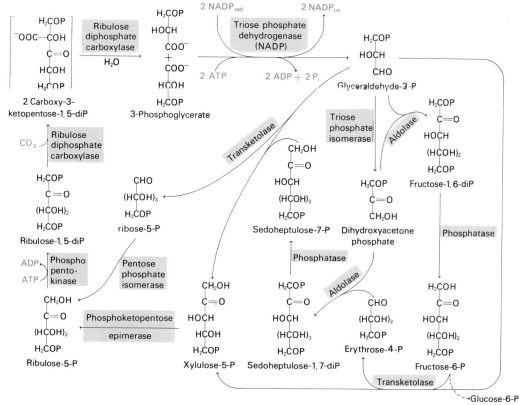

Fig. 10-19. The Calvin cycle. In each turn of this complex sequence of reactions, one molecule of CO_2 is incorporated into carbohydrate and one molecule of ribulose diphosphate acceptor is regenerated.

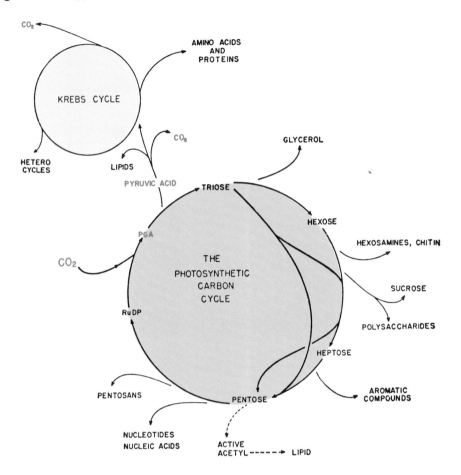

Fig. 10-20. Summary of pathways by which carbon compounds are synthesized from CO_2 by plants. (Redrawn from Bassham, J. A., and M. Calvin. 1957. *The Path of Carbon in Photosynthesis.* Prentice-Hall Inc., Englewood Cliffs, N.J.)

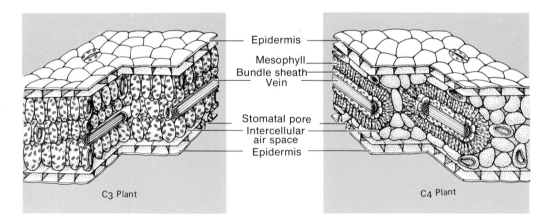

Fig. 10-21. Patterns of leaf structure in a "low efficiency" C_3 plant (left) and a "high efficiency" C_4 plant (right). In C_3 plants there is only one kind of photosynthetic cell in the mesophyll. C_4 plants have two kinds of chloroplast-containing cells: (1) those that have the enzymes of the Calvin cycle (bundle sheath cells) are located in a ring around the veins; (2) in the outer layer of cells around the vein and in other chloroplast-containing mesophyll cells CO_2 is trapped by reacting with phosphoenolpyruvic acid to form four-carbon acids. (Redrawn from Björkman, O., and J. Berry. 1973. High efficiency photosynthesis. *Sci. Am.* 229:80. Copyright © 1973 by Scientific American, Inc. All rights reserved.)

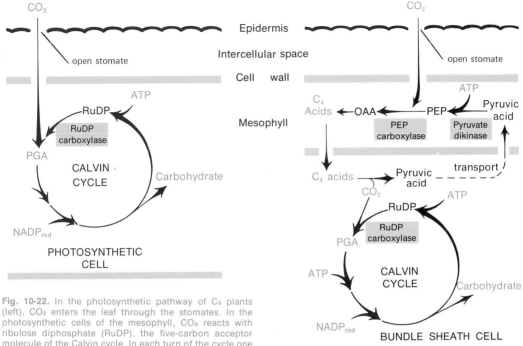

Fig. 10-22. In the photosynthetic pathway of C_3 plants (left), CO_2 enters the leaf through the stomates. In the photosynthetic cells of the mesophyll, CO_2 reacts with ribulose diphosphate (RuDP), the five-carbon acceptor molecule of the Calvin cycle. In each turn of the cycle one carbon enters the pathway to synthesis of carbohydrate or other end products of photosynthesis and a new RuDP acceptor molecule is regenerated. In C_4 plants (right), CO_2 entering the leaf via the stomates reacts with phosphoenolpyruvic acid (PEP) to form oxalacetic acid (OAA) in the mesophyll cells. From OAA other four-carbon organic acids (malic and aspartic) are formed which are transported to the bundle sheath cells. In the bundle sheath cells CO_2 is released from the four-carbon acids (CO_2 and pyruvic acid are formed). Pyruvic acid is transported back to the mesophyll cells, while CO_2 enters the Calvin cycle. An ATP is required for conversion of pyruvic acid to a new PEP acceptor molecule for CO_2. (Redrawn from Björkman, O., and J. Berry. 1973. High efficiency photosynthesis. *Sci. Am.* 229:80. Copyright © 1973 by Scientific American, Inc. All rights reserved.)

arid environments with intense sunlight and high day temperatures, C_4 plants exhibit higher photosynthetic efficiencies, higher temperature optima, and more efficient utilization of water. Stomatal resistance is high so that water loss from leaf surfaces is retarded. Although the import of CO_2 into the leaf is also slowed, the preliminary storage of CO_2 in dicarboxylic acids provides a mechanism for concentrating this essential substrate in photosynthetic tissues. Efficiencies are high because there is no diversion or loss of the intermediates of carbon fixation in the reactions of photorespiration. Many C_4 plants appear to lack the photorespiratory apparatus.

C_4 metabolism is a relatively recent discovery. Investigations of this new dimension of photosynthesis are being carried out in many parts of the world, especially in the United States and Australia.

The discoveries of photorespiration and the C_4 pathway of photosynthesis have led to speculation about ways to make photosynthesis more efficient. After all, photorespiration probably results in decreased efficiency in the utilization of light by plants, while in C_4 plants the two-stage mechanism for capturing and utilizing CO_2 permits very efficient conversion of light into chemical energy by providing a means to circumvent the shortages of CO_2 usually encountered by the photosynthetic apparatus.

Differences in productivity between C_3 and C_4 plants are by now well documented (see Table 10-1). The differences can be explained in part on the basis of high rates of photorespiration in low efficiency plants and in part on greater efficiency in CO_2 utilization by C_4 plants. Efforts are now under way to find selective inhibitors for photorespiration and to breed C_3 plants with lower than normal rates of photorespiration. An alternative approach to increased productivity has been to cross two related species, one of which is a C_3 plant and the other a C_4 plant. The goal would be to select progeny having the most desireable combinations of traits.

The need for more productive varieties of agricultural plants is all too obvious in a world with a finite amount of arable land and a rapidly increasing population. Eventually the recently developed techniques of plant breeding such as protoplast fusion and

TABLE 10-1
Rates of net photosynthesis in representative "high efficiency" and "low efficiency" plants. (Data of I. Zelitch published in Marx, J. L. 1973. Photorespiration: key to increasing plant productivity? *Science* **179:365)**

Species	Rate of Net Photosynthesis*
"High efficiency" (C_4) plants	
Maize	46–63
Sugarcane	42–49
Sorghum	55
"Low efficiency" (C_3) plants	
Spinach	16
Wheat	17–31
Rice	12–30
Bean	12–17

* Rate of net photosynthesis expressed as milligrams of CO_2 per square decimeter of leaf per hour. Rates were determined at high light intensity in air (0.03 per cent CO_2) at temperatures between 25 and 30°C.

growth of haploid plants (see Chapter 7) or transfer of specific genes by means of viruses may provide the means for developing "super plants." The increased yields from C_4 plants with the capability of fixing molecular nitrogen or from legumes with the C_4 mechanism for capturing CO_2 could provide at least a brief period of relief from the present threat of worldwide hunger.

Some Conclusions about Photosynthesis

In summary, the overall balance of raw materials consumed and products generated by the photosynthetic apparatus is as follows:

$$CO_2 + H_2O \xrightarrow[\text{chlorophyll}]{\text{8 quanta of light}} (CH_2O) + O_2$$

In order for this overall change to be accomplished, four [H]'s and three ATP's generated within the quantasomes are utilized in the reduction of CO_2 and the regeneration of a new CO_2 acceptor molecule.

In spite of the seemingly endless factual details that have been amassed by investigators of photosynthesis since the time of Priestley and Ingen-Housz, it is wise to remind ourselves that this is still an incompletely understood process. Neither the most skilled chemist nor the most complex chemical factory can duplicate what is now produced by green plants in their harvesting of the energy of sunlight. No man-made substitute for the process of photosynthesis is available at this time, nor will it be feasible for chemical factories to take over the work of photosynthesis in the foreseeable future. As long as test tube photosynthesis is still a dream, it is essential that we understand and appreciate the role of green plants (from the microscopic diatoms of the ocean to the giants of our redwood forests) in maintaining our atmosphere and food chains. The contribution of photosynthesis to the maintenance of life on this planet cannot be overemphasized. We could not exist without it.

USEFUL REFERENCES

Björkman, O., and J. Berry. 1973. High efficiency photosynthesis. *Sci. Am.* 229:80.

Fogg, G. E. 1972. *Photosynthesis,* 2nd ed. American Elsevier Publishing Co., Inc., New York. (Paperback.)

Govindjee, and R. Govindjee. 1974. The absorption of light in photosynthesis. *Sci. Am.* 231:68–82.

Hall, D. O., and K. K. Rao. 1972. *Photosynthesis.* Institute of Biology Studies in Biology No. 38. Edward Arnold (Publishers) Ltd., London. Distributed in U.S. by Crane, Russak and Co., Inc., New York. (Paperback.)

Levine, R. P. 1969. The mechanism of photosynthesis. *Sci. Am.* 221:58–70.

Rabinowitch, E., and Govindjee. 1969. *Photosynthesis.* John Wiley and Sons, Inc., New York.

Zelitch, I., 1971. *Photosynthesis, Photorespiration and Plant productivity.* Academic Press, New York.

Energy Release: Molecules and Energy Revisited

Modern knowledge of the chemistry of cells is both extensive and penetrating, reaching into every dimension of life. In Chapter 3 the essential components of life were considered and the broad outlines of cellular metabolism were described. In some ways the summary of cell chemistry presented there would be comparable to the understanding of the workings of a factory that can be gained from careful observation of the raw materials entering, the finished products leaving, and some shrewd guesses about what must occur within to accomplish the changes observed. We will now examine more carefully the machinery of life and the details of the changes that occur during the release of energy in respiration.

It is important to bear in mind that the chemistry of respiration is practically universal; it is the same in all eukaryotic cells, whether in the gills of a mushroom or a fish, in your little finger, or in the leaves of an oak tree. Because the details of energy release are so universally the same, they provide a major unifying theme for the life sciences.

We will begin with a look at enzymes, those marvelous cellular tools that make possible the chemistry of life by catalyzing each of the steps in metabolism. Then we will examine the individual chemical steps involved in the respiratory breakdown of foodstuffs. You will recall from Chapter 3 that three stages can be recognized. In the initial anaerobic stage, called **glycolysis,** breakdown of simple carbohydrates to pyruvic acid occurs. This three-carbon organic acid loses a carbon and the resulting two-carbon fragment enters the **Krebs (tricarboxylic acid) cycle** in which the final steps in disintegration of the carbon skeleton occur. The products of this cycle are CO_2 and electrons. The latter are fed into a chain of carriers in the **electron transport** system where finally they unite with protons (hydrogen ions) and molecular oxygen to form water.

The purpose in all this complexity is to release the energy stored in the molecules that serve as fuels for respiration at a controlled rate in accordance with the needs of the cell. An understanding of the capture and storage of energy in photosynthesis and of the release of energy in respiration will provide an overview of the major cycles of energy and matter on this planet.

Enzymes

Enzymes are of the greatest industrial, medical, and theoretical importance because they control the inner workings of cells. Most poisons are poisonous because they inhibit one or another enzyme; some antibiotics act on enzymes; and vitamins are essential because they form parts of enzyme molecules or cofactors for enzymes. Yet the 19th century understood as little about the inner workings of a cell as a child knows about the inner workings of a juke box. In one case you put in a coin and music comes out; in the other you put in oxygen and sugar, and carbon dioxide, water, and energy come out.

Enzymes as digestive agents have been known in a general way ever since Réné Réaumer (1683–1757) obtained gastric juice from his pet falcon by inducing it to swallow small sponges on strings. The gastric juice thus obtained could soften and dissolve meat. Modern knowledge about enzymes really began with the famous controversy between Louis Pasteur (1822–1895) in Paris and Justus Liebig (1803–1873) of the University of Heidelberg. Pasteur claimed that alcoholic fermentation was the result of the living activity of yeast cells. Without intact living yeast cells, there was no fermentation. Liebig, a brilliant chemist, claimed that fermentation was a purely chemical process analogous to the rusting of iron. Neither view is entirely wrong nor entirely true. It was Eduard Buchner (1860–1917) who found the answer. He was able to extract from yeast a cell-free juice that had the power of fermentation; that is, it could turn sugar into alcohol and carbon dioxide. He called that active ingredient in his yeast juice an enzyme which means, literally, *in yeast.* The alcohol-producing enzyme in yeast is zymase.

TERMINOLOGY OF ENZYMES

The terminology of enzymes is simple. The material on which an enzyme acts is called its **substrate.** In the case just mentioned, sugar is the substrate. The name of the enzyme is formed by adding *-ase* to the name of the substrate or the activity. Thus, an enzyme that acts on lipids is a lipase, an enzyme removing hydrogens, a dehydrogenase. Unfortunately various common enzymes were named before this system arose. Pepsin in gastric juice and ptyalin in saliva are examples.

ENZYME FUNCTION

Enzymes are involved in virtually all the chemical activities of cells, and thus with life itself. There are oxidizing and reducing enzymes, digestive (hydrolytic) enzymes that split carbohydrates, lipids, proteins, and nucleic acids, and synthesizing enzymes that build up these molecules. Enzymes are extremely specific in their activities and particular about the conditions under which they will work. Enzymes are specific in two ways. First, for a given substrate, many chemical changes might be thermodynamically feasible but only

one of these is catalyzed by a given enzyme. Second, enzymes are capable of acting on a limited number of substrates, sometimes only one. Specificity with respect to substrate may involve the linkage between units of a polymer, preference for one over another stereoisomer, or for a specific group within the substrate molecule. For enzymes to be active, proper conditions of temperature and pH are required as well. Since the enzyme molecule is not destroyed when it acts, only very few enzyme molecules are necessary to convert large amounts of substrate to product. A single enzyme molecule can unite or split many thousands of substrate molecules per second.

Most (but not all) enzyme molecules consist of two parts: a large protein portion, the **apoenzyme,** and a small nonprotein part, the **coenzyme** or **prosthetic group.** Coenzymes are generally not tightly bound to the apoenzyme and can be removed by dialysis. Prosthetic groups, on the other hand, tend to be tightly bound to the enzyme protein. Many vitamins serve as coenzymes or prosthetic groups of particular enzymes. The discovery that enzymes are proteins was the work of James B. Sumner of Cornell University. After 10 years of frustrating but single-minded labor he succeeded in crystallizing urease, an enzyme abundant in the jack bean. Crystalli-

Fig. 11-1. James B. Sumner, Cornell University biochemist who proved that enzymes are proteins. He purified the enzyme urease and demonstrated that the resulting protein is able to catalyze the breakdown of urea to CO_2 and NH_3. (From Moment, G. B. 1967. *General Zoology*, 2nd ed. Houghton-Mifflin, Inc., Boston.)

zation proved that he had a pure substance. It gave all the standard tests for a protein and, when dissolved, showed urease activity. At first no one believed that it could be true, but today his achievement stands as a landmark in biological science.

TOOLS FOR STUDYING ENZYMES

The classical method for investigating the action of respiratory enzymes on their various substrates has been to place sliced or finely minced tissue into a Warburg apparatus consisting of glass vessel attached to a manometer, which measures changes in gas pressure (Fig. 11-2). Changes in gas pressure within the flask can be converted to volumes of gas consumed or produced per unit time, i.e. respiration rates. In recent times more elaborate devices such as the oxygen electrode connected to an automatic recorder, or sensitive recording optical devices have been used in studies of specific reactions and intermediates of respiration.

Enzyme reactions can be controlled and the reaction stream guided into one metabolic pathway or another by several methods. Because most enzymes can function only within a very narrow pH range, their action can be blocked by relatively slight pH changes. Furthermore, many, if not all, enzymatic reactions can be made to slow down or even run backwards if the products of the reaction are allowed to accumulate or are added. Many enzymes are irreversibly blocked by combining with various poisons (inhibitors). Some poisons are highly specific but others, such as the heavy metals (lead, mercury, and arsenic), block many enzymes. The environmental consequences of careless release of heavy metals such as mercury and lead into waterways is now being realized.

THEORIES OF ENZYME ACTION

There is convincing evidence that enzymes produce their results at their surfaces and

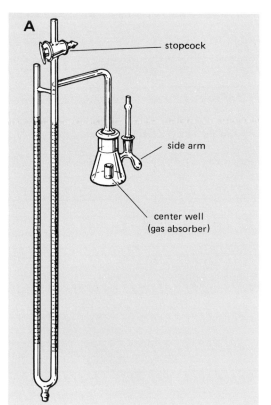
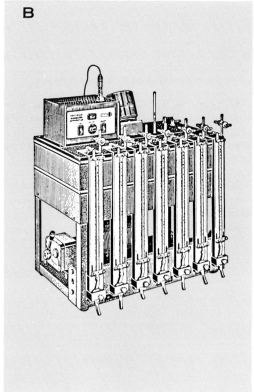

Fig. 11-2. A, flask and manometer for a Warburg respirometer. B, Warburg constant temperature water bath showing manometers with vessels attached. A shaking mechanism provides constant movement ensuring rapid equilibration between gas and liquid phases within the reaction vessel. (Courtesy of GCA/Precision Scientific Co.)

that there is what the first modern investigator of proteins, Emil Fischer, called a **lock-and-key relationship** between the shape of a particular substrate and the enzyme that acts on it. To be effective, the substrate must fit tightly against the enzyme. It now appears that enzyme molecules are not rigid but bend around their substrates. Hence, some prefer to use a **hand-in-glove** analogy. In any case, once molecular intimacy has been achieved, chemical forces cause a large substrate to split or two small ones to unite.

Part of the proof of this theory can be illustrated by succinic dehydrogenase, an enzyme important in respiration. A dehydrogenase is an enzyme that splits off hydrogen from its substrate (Fig. 11-3). In this case, as shown in the diagram, succinic acid fits neatly into the reactive site of the enzyme. Two hydrogen atoms are split off and fumaric acid is formed. The fumaric acid falls away from the enzyme, leaving it ready to repeat the act. If malonic acid, which is very similar to succinic acid, is substituted, the malonic acid is accepted by the enzyme, but since it is not a perfect fit, no reaction occurs. Because its reactive site is occupied, the enzyme is more or less permanently blocked, a situation known as **competitive inhibition.** Its discovery has led to an extensive hunt for such compounds, analogues of substrates which can be used to block unwanted reactions.

Another method of inhibiting enzymes is to give them a "phony" vitamin instead of a "phony" substrate. It will be recalled that many enzymes are really double structures, a large protein plus a nonprotein prosthetic group. If a given animal cannot synthesize the prosthetic group, it must have it in its diet. In such a case, the prosthetic group is known as a vitamin. If an animal or bacterium is fed a vitamin analogue, which the protein part of the enzyme does not distinguish from the genuine vitamin, a nonfunctional pseudoenzyme is produced. This is the way sulfa drugs work. Luckily, the enzymes of certain disease-causing bacteria are far more seriously damaged than are any essential human enzymes.

Certain metabolic poisons can inactivate enzymes either by blocking the prosthetic group or by changing the shape of the enzyme molecule so that the active site is modified more or less permanently. Such poisons are called **noncompetitive inhibitors.** A basic difference between competitive and noncompetitive inhibitors is that the effects of the former can be overcome by adding excess substrate. With a competitive inhibitor an abundance of substrate molecules can effectively compete for sites on the enzyme molecules, keeping at least some sites unblocked. Inhibitor molecules are not acted upon by the enzyme and therefore, once attached, they block further reaction by that site. Substrate molecules undergo change after attachment to a reactive site, and the products are released, freeing the site for attachment of a new substrate molecule (Fig. 11-4).

Details of the Energy Pathway

THE ANAEROBIC PHASE

In **glycolysis,** also known as the **Embden-Meyerhof pathway,** glucose is the usual starting point. (See Fig. 11-5; numbers used in text correspond to numbers of intermediates shown in the figure.) Glucose is converted into glucose-6-phosphate (1) by the enzyme hexokinase in the presence of magnesium ions and ATP as a source of energy. Glucose-6-phosphate is so called because the phosphate is attached to the sixth carbon in the glucose molecule. Hexokinase is the enzyme in the brain responsible for breaking down the glucose which provides the energy to read this chapter. Glycogen and starch are other common initial compounds. They are broken down to yield glucose-1-phosphate, which is then changed into glucose-6-phosphate. Glucose-6-phosphate is converted into fructose-6-phosphate (2) to which is added another phosphate, this time to the first carbon, making fructose-1,6-diphosphate (3). The first and third changes require an ex-

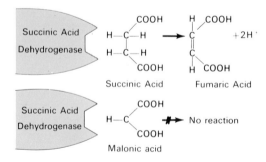

Fig. 11-3. Competitive inhibition of succinic acid dehydrogenase. Malonic acid is very similar to succinic acid and binds to the enzyme molecule. No reaction occurs, however, and the enzyme remains blocked by this competitive inhibitor. (From Moment, G. B., and H. M. Habermann. 1973. *Biology: A Full Spectrum.* The Williams & Wilkins Co., Baltimore.)

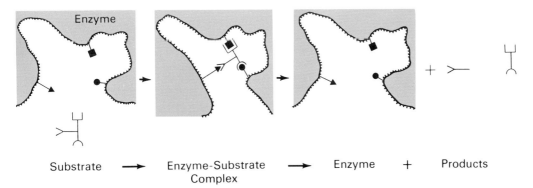

Fig. 11-4. Diagrammatic representation of enzyme-substrate interaction. (Modified from Mallette, M. P., C. O. Claggett, A. T. Phillips, and R. L. McCarl. 1971. *Introductory Biochemistry.* The Williams & Wilkins Co., Baltimore.)

penditure of energy which is provided by ATP.

These are the phosphorylating reactions discovered by Harden (see Chapter 3). Many authors use the term phosphorylation to mean any reaction in which a phosphate group is added to another compound. Others restrict the term to those cases where the phosphate group is connected by an energy-rich bond. The context will usually tell which is intended.

Fructose-1,6-diphosphate splits into two compounds, each with a single phosphate, namely, glyceraldehyde phosphate (also called phosphoglyceraldehyde or PGAL (4)) and dihydroxyacetone phosphate. These two compounds easily change into each other so there is a triangle here. Moreover, PGAL stands at one of the two chief metabolic crossroads in the entire pathway. By way of dihydroxyacetone phosphate it leads off into fats and alcohols or back from them into the energy-yielding pathway. PGAL can be formed from glucose via the route just outlined, from glycerol, in plants from early products of photosynthesis, and, of course, from dihydroxyacetone phosphate.

Continuing down the main pathway, PGAL is oxidized and gains an inorganic phosphate group to become diphosphoglyceric acid (5). This reaction requires the presence of the coenzyme nicotinamide adenine dinucleotide (NAD) which becomes reduced. Diphospho-glyceric acid is then converted into 3-phosphoglyceric acid (6) with the production of a molecule of ATP. By changing the position of the phosphate group, 2-phosphoglyceric acid (7) is formed. This compound is converted, by the loss of a molecule of water, into a simpler compound, enolphosphopyruvic acid (8), which becomes pyruvic acid (9), plus a second ATP. Each of the steps just reviewed is catalyzed by a specific enzyme.

THE AEROBIC PHASE

Aerobic respiration includes the **citric acid** or **Krebs cycle** and **terminal respiration** (also called **electron transport**). This portion of the metabolic pathway of energy production begins with the formation of CO_2 and acetic acid from pyruvic acid. The decarboxylation of pyruvic acid is a complex process in which NAD is reduced and the two-carbon fragment remaining after removal of CO_2 is bound to coenzyme A. The overall changes that occur are summarized in Fig. 11-6. The two-carbon acetate group formed from pyruvic acid is complexed to coenzyme A and later transferred to a four-carbon acceptor molecule (oxalacetic acid) in the Krebs cycle to produce a six-carbon citric acid molecule.

It will be recalled that acetyl-CoA is at the second major metabolic crossroad in the pathway, since not only the breakdown products of carbohydrates enter the system at this point but also products from metabolized amino acids and lipids.

Coenzyme A was discovered by Fritz Lipmann, who had studied under Meyerhof in Germany. He found that some phosphate bonds have only a low energy yield when broken while others such as those in ATP have a high energy yield. Coenzyme A, which controls the entrance of acetate into the citric acid cycle, turned out to be similar to ATP. It is a very large molecule, consisting of adenine, ribose, three phosphates, the vitamin panthothenic acid, and sulfur-containing thiolthylamine (also called cysteamine). The tiny two-carbon acetyl radical is linked via sulfur. Small wonder the complex is usually called acetyl-CoA.

What about the aerobic part of respiration? The clues and much of the evidence here were obtained by a remarkable Hungarian, Albert Szent-Gyorgyi (Fig. 11-7), who has be-

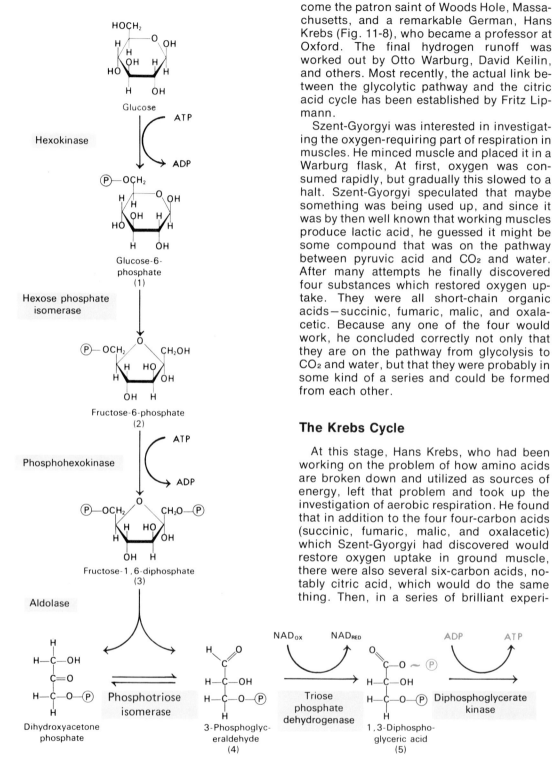

come the patron saint of Woods Hole, Massachusetts, and a remarkable German, Hans Krebs (Fig. 11-8), who became a professor at Oxford. The final hydrogen runoff was worked out by Otto Warburg, David Keilin, and others. Most recently, the actual link between the glycolytic pathway and the citric acid cycle has been established by Fritz Lipmann.

Szent-Gyorgyi was interested in investigating the oxygen-requiring part of respiration in muscles. He minced muscle and placed it in a Warburg flask, At first, oxygen was consumed rapidly, but gradually this slowed to a halt. Szent-Gyorgyi speculated that maybe something was being used up, and since it was by then well known that working muscles produce lactic acid, he guessed it might be some compound that was on the pathway between pyruvic acid and CO_2 and water. After many attempts he finally discovered four substances which restored oxygen uptake. They were all short-chain organic acids—succinic, fumaric, malic, and oxalacetic. Because any one of the four would work, he concluded correctly not only that they are on the pathway from glycolysis to CO_2 and water, but that they were probably in some kind of a series and could be formed from each other.

The Krebs Cycle

At this stage, Hans Krebs, who had been working on the problem of how amino acids are broken down and utilized as sources of energy, left that problem and took up the investigation of aerobic respiration. He found that in addition to the four four-carbon acids (succinic, fumaric, malic, and oxalacetic) which Szent-Gyorgyi had discovered would restore oxygen uptake in ground muscle, there were also several six-carbon acids, notably citric acid, which would do the same thing. Then, in a series of brilliant experi-

Fig. 11-5. Glycolysis, the stepwise breakdown of glucose to pyruvic acid by a sequence of reactions known as the Embden-Myerhof pathway. Numbers next to names of intermediates correspond to those used in the text. Names of enzymes are

ments, he showed how all these acids together in a cycle and that fresh fuel, i.e. acetate formed from pyruvic acid, enters the cycle by uniting with oxalacetic acid to form
citric acid. From citric acid all the other intermediates in the cycle are formed until finally
oxalacetic acid is reached and the cycle is
ready to begin again. The evidence for this is
obtained by several methods. Radioactive atoms can be used as labels and these markers
can be traced. By the use of appropriate inhibitors such as those described earlier in
connection with enzymes, the cycle can be
stopped at various points and the substances
which accumulate identified; or, various
compounds can be fed into the system to
determine which ones permit it to proceed.
Those that initiate further reactions are probably involved as intermediates

The citric acid cycle is made up of four six-
carbon acids (citric, *cis*-aconitic, isocitric,
and oxalosuccinic), one five-carbon acid (α-
ketoglutaric acid), and the four four-carbon
acids discovered by Szent-Gyorgyi (succinic,
fumaric, malic, and oxalacetic). Once again
the structures of the acids are rather simple.

Both α-ketoglutaric acid and succinic acid
are notable. α-Ketoglutaric acid is another
point at which amino acid breakdown products may enter the system; it is the starting
point in the formation of amino acids by transamination, that is, the transfer of an amino
group from one molecule to another. Suc-

cinic acid forms the backbone of porphyrin,
an important prosthetic group for hemoglobin, cytochrome, chlorophyll, and other compounds.

The intermediates, products and enzymes
involved in the Krebs cycle are summarized in
Fig. 11-9. When examined closely it is apparent that molecular oxygen is nowhere involved in this sequence of reactions. Why
then is the Krebs cycle considered to be part
of the aerobic phase of respiratory metabolism? In part this is because oxidations do
occur every time that hydrogens are removed
from intermediates and reduced NAD is
formed. Furthermore, aerobic conditions are
essential for the continued operation of the
electron transport system of respiration
where hydrogens removed in the Krebs cycle
finally join molecular oxygen to form water.

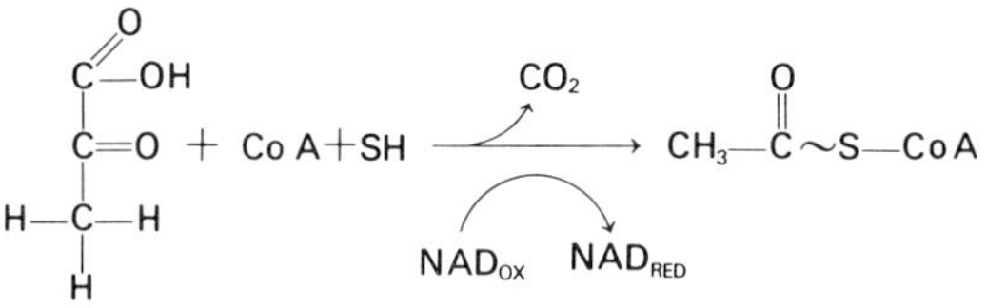

Fig. 11-7. Albert Szent-Gyorgyi, the Hungarian-American
biochemist now living in Woods Hole, whose studies of
oxygen uptake by minced muscle led to discovery of key
intermediates of the Krebs cycle. (Courtesy of The Bettmann Archive.)

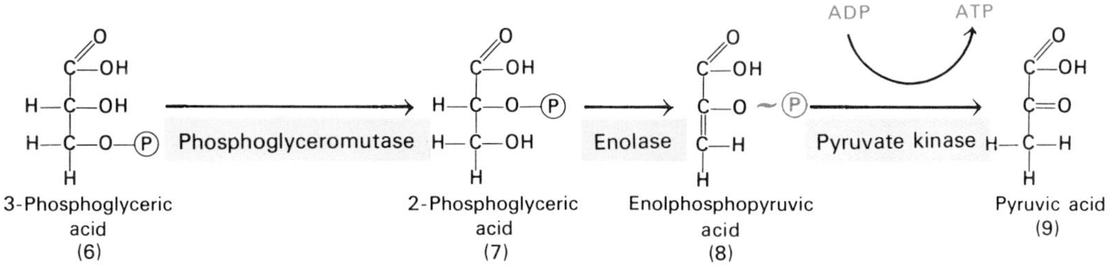

Fig. 11-6. Decarboxylation of pyruvic acid results in
release of a molecule of CO_2, reduction of an NAD, and
transfer of a two-carbon acetate fragment to coenzyme A.

designated by light color overlays. Dark color is used to show steps in which ATP is generated and high energy phosphate
bonds are formed.

Fig. 11-8. Hans Krebs in his laboratory at Oxford. His studies elucidated the cyclic reactions by which respiratory intermediates are broken down to CO_2. (Courtesy of The Bettmann Archive.)

Electron Transport—Terminal Respiration

What about terminal respiration, those steps at the very end of the series involving the dehydrogenases and cytochromes? This is the part which permits the citric acid cycle to keep running by taking care of the hydrogens and their electrons, and, incidentally, produces *three* ATP's per pair of electrons transported. The discovery of this part of the process was largely the work of Warburg and Keilin.

Warburg began to investigate the rate at which tissues fed carbohydrates and amino acids used oxygen. Working with some clever model systems he found that charcoal containing iron (which is true of charcoal made from dried blood) acts as a catalyst speeding up oxygen consumption. Charcoal which lacked iron (charcoal from sugar, for example) failed to do this. He concluded, therefore, that some iron-containing compound was an essential catalyst of oxygen uptake. He went on to show that his iron-containing charcoal plus carbohydrate made an amazingly useful model of cellular respiration. Cy-

anide in very low concentrations inhibits cellular respiration and also inhibited this model. Anesthetics which inhibit cellular respiration only in high concentrations had the same effect on his models. From such facts Warburg concluded that there is an enzyme containing iron which makes it possible for foodstuffs to be oxidized in the presence of free oxygen. He went on to isolate this enzyme which he called "the respiratory enzyme." It was the first enzyme known to be involved in aerobic respiration. All the others isolated up to that time were enzymes of fermentation, i.e. of glycolysis.

Meanwhile, David Keilin in Cambridge rediscovered a reddish pigment in the wing muscles of the horse botfly and in yeast cells. It is always a safe bet that any substance found in two such widely separated organisms will turn out to be important. Keilin named the substance **cytochrome** (*cyto,* cell, + *chrome,* color). When in the oxidized condition, produced by bubbling oxygen through a solution of the pigment, the cytochrome lost its color. However, if oxygen was excluded and the cytochrome kept in the reduced state, it appeared reddish and showed a characteristic absorption spectrum (see Fig. 11-10). If oxygen was readmitted, the bands of the spectrum faded away and the color disappeared. On analysis, cytochrome proved to be a compound similar to hemoglobin—an iron-containing porphyrin attached to a protein.

The really exciting facts came into view when cytochrome was compared with Warburg's respiratory enzyme. They both contain iron. They are both inhibited by cyanide and carbon monoxide. With these poisons cytochrome remains permanently reduced, i.e. reddish, no matter how much oxygen is bubbled through it. Hence, Warburg's enzyme was an enzyme which oxidized cytochrome. Such an enzyme is called a **cytochrome oxidase** because it oxidizes cytochrome in the presence of free oxygen.

In addition, it was found that anesthetics which blocked "respiration," at least oxygen uptake in Warburg's model, also blocked change in the cytochrome keeping it permanently oxidized. Anesthetics thus appear to block an enzyme which reduces cytochrome. Reduction, it will be recalled, is the addition of hydrogen or electrons. The enzyme, which can be blocked by anesthetics, evidently takes hydrogens or their electrons from something, presumably some breakdown product of a foodstuff, and transfers them to cytochrome. Such an enzyme which accepts hydrogens is commonly called a **dehydro-**

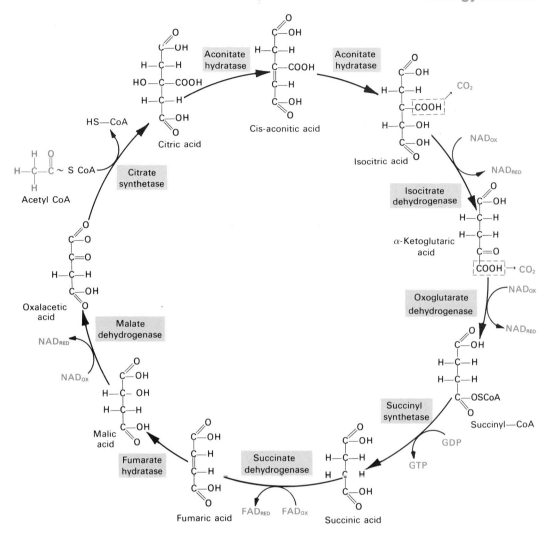

Fig. 11-9. The citric acid (Krebs) cycle. Names of enzymes are indicated by light color overlays. Dark color is used to designate substrates entering and products of the cycle.

genase. The series evidently runs via a dehydrogenase to cytochrome, and from cytochrome by means of an oxidase to oxygen. If hydrogens are somehow being moved along, what can the end product be but water? Subsequent work has shown that there are at least four slightly but significantly different cytochromes which act in series and that there are two dehydrogenases, which stand between them and the Krebs cycle, in this terminal segment of the respiratory pathway.

There is still disagreement about exactly how many carrier substances participate in the electron transport chain. Seven components are usually listed (Fig. 11-11). These are: FAD (flavin adenine dinucleotide), FP₂ (succinic acid dehydrogenase, another flavo-

protein), coenzyme Q (CoQ), cytochrome *b* (Cyt b), cytochrome *c* (Cyt c), cytochrome *a* (Cyt a), and cytochrome oxidase. The role of coenzyme Q (CoQ, also known as **ubiquinone**) has been controversial, but recent experiments support its proposed participation as a carrier of electrons from flavoproteins to cytochromes.

Each electron in the electron transport chain has been shown to have a characteristic "electron pressure" and each acceptor a characteristic "affinity" for electrons. These tendencies can be measured and expressed in terms of electromotive force or potential (usually under strictly standardized conditions). The resulting values for the **standard redox** (reduction-oxidation) **potential** of each electron carrier make it possible to arrange them in a series, starting with the carrier having the greatest electron "pressure" (the most negative redox potential) and ending with the one having the greatest electron "affinity" (most positive redox potential). Electrons move from more negative to less negative (more positive) carriers along this series. As a reduced carrier donates its electron to an acceptor, the donor becomes oxidized and the acceptor becomes reduced. The reduced acceptor in turn becomes an electron donor to the next intermediate along the chain. The ultimate electron acceptor (with a redox potential of +0.8) is oxygen.

For each pair of hydrogen atoms removed from each of four substrates in the Krebs cycle, two electrons are supplied to the respiratory chain. Each pair of electrons plus two protons ultimately reduce one atom of oxygen to form a molecule of water.

There are two points of entry for electrons into the electron transport chain. Reduced NAD formed by dehydrogenations of isocitric, α ketoglutaric and malic acids can donate electrons to flavoprotein. A fourth pair of electrons from the oxidation of succinic acid reduce the flavoprotein prosthetic group of the enzyme succinic acid dehydrogenase. Both the flavoprotein (FAD) accepting electrons from reduced NAD and the reduced flavin in succinate dehydrogenase (FP2) can transfer their electrons via coenzyme Q to the cytochromes.

As electrons move along the chain they lose free energy and the amounts of energy loss between intermediates have been measured (Fig. 11-12). For three of these steps the energy drop is relatively large and at these points high energy intermediates are generated which permit packets of energy to be conserved in the form of ATP. ATP is the "energy currency" that can be utilized by cells for the energy-requiring processes of synthesis, transport, or work.

ATP Synthesis by Oxidative Phosphorylation

One kind of evidence that energy released during the aerobic phase of respiration is conserved in the form of ATP comes from studies with isolated mitochondria. When carefully prepared so that they are intact, i.e. so that all enzymes of the Krebs cycle and electron transport are present and functional, it can be shown that phosphate and ADP are essential for continued high rates of electron transport. Not only are they necessary, but

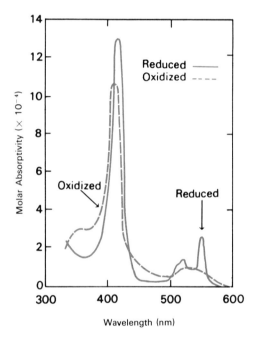

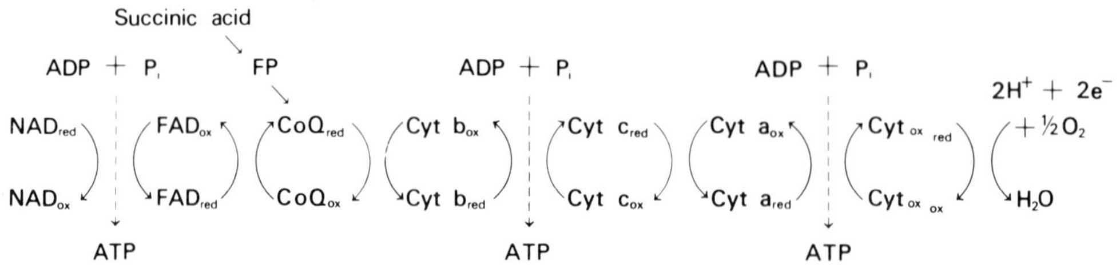

Fig. 11-10. Absorption spectra of oxidized and reduced forms of cytochrome c. The reddish color of the reduced form is a consequence of absorption bands in the 500 to 600-nm portion of the spectrum. (After E. Margoliash, quoted by Keilin, D., and E. C. Slater. 1953. *Br. Med. Bull.* 9:95.)

Succinic acid

ADP + P$_i$ FP   ADP + P$_i$ ADP + P$_i$

                                                                          2H$^+$ + 2e$^-$

NAD$_{red}$ ⟍ ⟋FAD$_{ox}$⟍ ⟋CoQ$_{red}$⟍ ⟋Cyt b$_{ox}$⟍ ⟋Cyt c$_{red}$⟍ ⟋Cyt a$_{ox}$⟍ ⟋Cyt ox $_{red}$⟍ ⟋ + ½O$_2$

NAD$_{ox}$ ⟋ ⟍FAD$_{red}$⟋ ⟍CoQ$_{ox}$⟋ ⟍Cyt b$_{red}$⟋ ⟍Cyt c$_{ox}$⟋ ⟍Cyt a$_{red}$⟋ ⟍Cyt ox $_{ox}$⟋ ⟍H$_2$O

ATP   ATP   ATP

Fig. 11-11. The electron transport chain. The following abbreviations are used: FAD (flavin adenine dinucleotide), FP2 (succinic acid dehydrogenase, another flavoprotein), Co Q (coenzyme Q, also known as ubiquinone), Cyt b (cytochrome *b*), Cyt c (cytochrome *c*), Cyt a (cytochrome *a*), and Cyt ox (cytochrome oxidase).

they are used up. Furthermore, for each pair of electrons moved from reduced NAD to oxygen, three molecules of ATP are formed. This production of ATP associated with transport of electrons along the respiratory chain in the mitochondria is called **oxidative phosphorylation.** The coupled processes of electron transport and ATP synthesis can be summarized by the following equation:

$$NAD_{red} + 3ADP + 3P_i + 2H^+ + \tfrac{1}{2}O_2$$
$$\rightarrow NAD_{ox} + 3ATP + H_2O$$

For each mole of ATP formed, 7,300 calories are conserved in the form of high energy phosphate bonds. Thus of the 52,000 calories of energy released during the transport of electrons from 1 mole of NAD_{red}, 3 times 7,300, or 21,900 calories (42 per cent), are stored in the form of ATP.

In the complete breakdown of glucose to CO_2 and water, the electrons from a total of 12 pairs of hydrogens enter the electron transport system. Thus, per mole of glucose, 12 times 52,000 calories are released during this stage of aerobic respiration. When a weighed sample of glucose is burned and the total energy released as heat is measured, the energy of combustion is found to be equal to 686,000 calories per mole. Thus of the total available energy in the molecule, over 90 per cent is released during electron transport, and as we have noted a significant fraction of

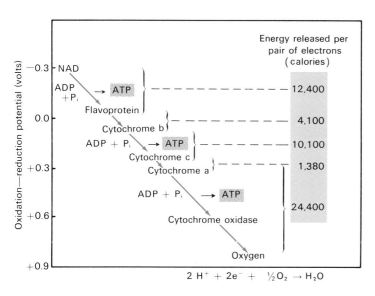

Fig. 11-12. Pathway of electron transport. Intermediates in this pathway can be arranged along a scale of oxidation-reduction potential. Electrons move in a stepwise fashion toward the most electropositive acceptor, oxygen. Energy released in each step per pair of electrons is indicated. In three steps, energy is conserved in the form of ATP.

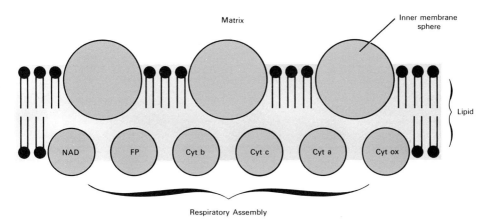

Fig. 11-13. Hypothetical arrangement of electron transport intermediates (respiratory assembly) on the mitochondrial inner membrane. Knoblike protein spheres protruding into the mitochondrial matrix are thought to be involved in ATP synthesis by oxidative phosphorylation.

the energy released is conserved in the form of ATP. How is such a remarkably high efficiency achieved? To a large extent the efficiency of the respiratory energy release is a consequence of mitochondrial organization at the molecular level. It is now apparent that the enzymes of the Krebs cycle are located in the mitochondrial inner matrix, while components of the electron transport chain are arranged in very specific arrays on the inner membrane. These arrays or clusters of dehydrogenases and cytochromes are complete functional units. On the inside of the inner membrane, knoblike spherical proteins protrude (Fig. 11-13). These **inner membrane spheres** are thought to be involved in the phosphorylation of ADP because when they are removed from isolated inner membrane fragments the capacity for ATP synthesis is lost.

Although the electron transport system is complex and still incompletely understood, what it accomplishes for the cell is clearly of great importance. By breaking down cellular fuels in relatively small, manageable steps and conserving some of the energy released, cells in a sense "deposit" a portion of this energy currency for later use. The resulting stored and convertible energy in the form of ATP can be expended to do work or to accomplish syntheses requiring an input of energy.

Respiration in Animals: Adaptations for Breathing

The necessity to breathe is a unique requirement of animals and an evolutionary consequence of the development of large and complex forms which are motile. Actions such as running and jumping require bursts of energy. That enormous quantities of food are required to fuel active animals can readily be appreciated by comparing the daily food intake of atheletes in training and the normal requirements of more sedentary adults. Regardless of the requirements, the chemistry of the breakdown of cellular fuels is universally the same. But to keep the process going, animals have had to evolve a mechanism for delivering adequate supplies of oxygen to all body cells. This has been accomplished by developing lungs and a circulatory system to move oxygen-carrying cells. The

specialized cells of the blood not only deliver oxygen required for aerobic respiration but also remove waste CO_2 released in the breakdown of foodstuffs.

From the earliest times respiration has seemed close to the essence of life itself. In fact, the word "spirit" is derived from the Latin *spirare* (to breathe). Hence there has always been an insistent philosophical interest in the meaning of respiration. In modern times urgent practical problems involving respiration have arisen in mining, in submarine and diving operations, in aviation, in the use of anesthetics, and in the diagnosis and treatment of metabolic diseases.

THE CLASSIC LAWS OF RESPIRATION

The overall laws of respiration were established in the 18th century largely by the work of Lavoisier, one of the discoverers of oxygen, and Simon Laplace, mathematician and astronomer. These men placed animals in ingenious **calorimeters** (heat meters) designed to measure the amount of heat produced. They also made extensive chemical tests of air before and after it had been inhaled by men and animals. This and subsequent work proved that:

1. In respiration, animals obey the **law of conservation of energy.** Both a breathing guinea pig and burning charcoal give off the same amount of heat energy when the same amount of oxygen is used up.

2. In respiration, animals obey the **law of conservation of matter.** During respiration the amount of O_2 consumed is equivalent to the amount of CO_2 produced. In a closed system there is no change in total weight.

The 19th century refined these fundamental observations and thereby added the third and fourth laws of respiration.

3. The **respiratory quotient,** that is, the volume of CO_2 given off divided by the O_2 consumed, is an indication of the type of food being burned. As can be seen from inspection of the equation:

$$C_6H_{12}O_6 \text{ (sugar)} + 6O_2 \rightarrow 6H_2O + 6CO_2$$

when sugar is burned the respiratory quotient is one. Fat, however, has far less oxygen in proportion to carbon and hydrogen than sugar does; hence, fat requires proportionally more oxygen to oxidize it completely into water and carbon dioxide. For example, the formula for beef fat is $C_{57}H_{110}O_6$. Simple arithmetic will show that 81.5 volumes of oxygen will be required for every 57 volumes of carbon dioxide produced. Hence the respiratory quotient (RQ) will be less than one, in this case 0.7. The RQ of proteins is intermediate between those of fats and carbohydrates. These laws of respiration are equally as valid for plants as they are for animals.

4. In mammals the **basal metabolic rate** (BMR) depends on size. The basal rate is the rate at which oxygen is used when the animal is at complete rest. The smaller the animal, the higher the BMR. The reason for this inverse relationship is clear. Mammals maintain a constant body temperature above their surroundings and hence lose heat to it. Any solid object can lose heat only from its surface. By the facts of solid geometry, the smaller an object is the more surface it has *in proportion* to its volume or mass. Therefore it follows that to maintain a given temperature a small animal must burn more glucose per pound of flesh than a large one.

Inspection of Figure 11-14, where the weight of several homoiotherms (warm-blooded animals) is plotted on semilog paper against their respiratory rate, will reveal that as size goes down, metabolic rate skyrockets. Because of their high rate of oxidation, very small mammals, like shrews, which weigh only 3 or 4 gm, must eat almost continuously and have voracious appetites. They consume approximately their own weight every 24 hours or starve to death. It is the equivalent of a 150-lb men eating three 50-lb meals a day! The graph also makes it clear that about 2.5 gm is the lower limit for size of a warm-blooded animal. Below that, food intake presents an impossible problem.

The 19th century also turned its attention to the actual mechanics and to the important physiological controls of respiration as well as the transport of O_2 and CO_2 in the blood.

MECHANICS OF BREATHING

There are three general mechanisms for getting oxygen into cells and carbon dioxide away. Feathery **gills,** which circulate blood in close proximity to water, are found both in vertebrates and invertebrates. **Tracheas** are found in insects and a few other invertebrates. They almost completely permeate the body with a treelike network of interconnected tubules. Vertebrates, except for most fish, possess internal and highly vascular sacs, the **lungs.** In the higher reptiles, birds, and mammals air is drawn into the lungs when a partial vacuum is created by the enlargement of the chest cavity. This is accomplished by the contraction of the muscles of ribs and diaphragm. The intercostal muscles between the ribs raise them and thereby enlarge the chest. When the **diaphragm,** a dome-shaped sheet of muscle separating chest from abdominal cavity, contracts it flattens the dome and thus enlarges the cavity of the thorax and the volume of the lungs. During expiration the size of the chest cavity is decreased.

Air travels from the pharynx down the trachea and via the right or left **bronchus** into the corresponding **lung.** Each lung is attached only in the region of its bronchus. The

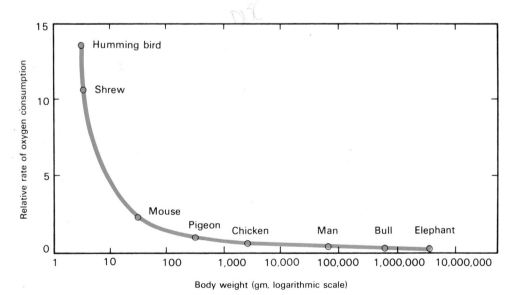

Fig. 11-14. The relationship between respiration and body weight. Rates of oxygen consumption are in milliliters of O_2 consumed per gram of body weight per unit of time. (From Moment, G. B., and H. M. Habermann. 1973. *Biology: A Full Spectrum.* The Williams & Wilkins Co., Baltimore.)

rest of the lung hangs freely in its own pleural cavity. If a stab wound is made into a pleural cavity, air will be sucked in, the vacuum will be destroyed, and the lung will collapse, making breathing impossible with that lung.

REGULATION OF BREATHING

Breathing is under both chemical and nervous control and constitutes a beautiful self-regulating or **feedback mechanism.** Increasing concentrations of carbon dioxide in the blood stimulate a breathing center in the medulla, part of the brainstem. This center is actually paired, one on the right and one on the left of the midline. Each of these centers is double with an expiratory and an inspiratory sub-center. There are important chemoreceptors in the aorta and the **carotid bodies** on the carotid arteries in the neck which sense the pH and probably O_2 and CO_2 concentration in the blood and inform the parts of the brain where breathing is controlled.

Impulses pass down the **phrenic nerves** from the brain to the diaphragm, cause it to contract, and thus produce inspiration. If the two phrenic nerves are cut, the diaphragm no longer contracts. Expansion of the lungs in turn stimulates sensory nerve endings, which send inhibitory impulses back to the brain via the two **vagus nerves.** Inhibition of the respiratory center permits the diaphragm to relax, producing expiration. Impulses leading to exhalation also arise from the carotid body, and from the aortic body. Exercise, involving movements of the joints of arms, hands, legs, and feet, produces stimuli that accelerate the rate of breathing.

The role of carbon dioxide in stimulating breathing is of paramount importance to both anesthetists and high-altitude flyers. If a man breathes into a closed system in which his expired carbon dioxide is allowed to accumulate, his rate of breathing will increase markedly. If the expired carbon dioxide is absorbed, for instance by KOH, his breathing will increase only slightly. Note that in a closed system the decline in oxygen concentration will be the same in both cases.

The effects of lack of oxygen depend on the rapidity with which it occurs. If an aviator suddenly loses his oxygen supply or a miner walks into a pocket of methane or other gas, he is likely to "black out" very suddenly and completely. If the loss of oxygen is gradual, the results are quite different, but more or less the same whether the loss is due to carbon monoxide poisoning, or ascent into high altitudes. At first there is commonly a sense of well-being and competence. As the oxygen lack persists, there comes a period of loss of judgment and unstable emotions, commonly accompanied by muscular incoordination, faulty vision, and poor memory. Fixed and irrelevant ideas are frequent. Finally, a feeling of sublime indifference and extreme weakness may end the series. In the case of continued deprivation of oxygen, these symptoms are followed by extreme nausea, convulsions, and finally, death.

Too much oxygen is also dangerous. Various symptoms of toxicity begin to appear when the oxygen pressure (concentration) in the gas breathed by most mammals begins to exceed 0.8 atm (i.e. about 4 times the normal level of O_2 in air). Excess oxygen is used in medicine to alleviate the effects of pneumonia, heart, and other diseases. Patients in oxygen tents are usually given air containing slightly over 50 per cent oxygen. An atmosphere of pure oxygen can be tolerated by man and in some conditions may be beneficial, but it may lead to extreme hypoventilation, resulting in coma because breathing stops due to lack of sufficient stimulation by CO_2.

The greatest hazard to human lungs, even in a smog-filled city, is tobacco smoke from cigarettes which, in some poorly understood way, leads to lung cancer. The evidence amassed in this country and in Europe is extensive and detailed. It would easily convince everyone if it were not for the understandable fact that no one likes to think that one of his established habits is potentially dangerous. The evidence is statistical, but it must be remembered that all evidence, in that famous final analysis, is statistical. Statistical reasons are the only ones for believing Koch's classic postulates about the germ theory of disease.

THE GRAND SYNTHESIS

The present century has built with amazing success upon the discoveries about respiration begun in the 17th century by Boyle and Lavoisièr. Consequently, for the first time in human history, we know how the food we eat and the air we breathe are united to provide energy—at the end of an electron transport system in our mitochondria. This modern insight into the secrets of what our predecessors called "the flame of life," not only represents a tremendous achievement, it is also a great simplification because on the biochemical level it sweeps into a single conceptual scheme all animals and plants from the lowest to the highest.

USEFUL REFERENCES

Bryant, C. 1971/73. *The Biology of Respiration*. Institute of Biology Studies in Biology No. 28. Edward Arnold (Publishers), Ltd., London. Distributed in the U.S. by Crane, Russak and Co., Inc., New York. (Paperback.)

Lehninger, A. L. 1975. *Biochemistry,* 2nd ed. Worth Publishers, New York.

Lehninger, A. L. 1971. *Bioenergetics*, 2nd ed. W. A. Benjamin, Inc., Menlo Park, Calif. (Paperback.)

Staub, N. C. 1969. Respiration. *In* Hall, V. C. et al., eds *Annual Review of Physiology,* Vol 31. Annual Reviews, Inc., Palo Alto.

Wynn, C. H. 1974. *The Structure and Function of Enzymes*. Institute of Biology Studies in Biology No. 42. Edward Arnold (Publishers), Ltd., London. Distributed in the U.S. by Crane, Russak and Co., Inc., New York. (Paperback.)

12

Soil, Water, and Air: The Modest Requirements of Plants

[Courtesy of UNICEF. Photograph by Bill Campbell]

Compared to animals, plants are amazingly modest in their nutritional requirements. Using water from the soil, CO_2 from the air, and energy from the sun, photosynthesis provides the carbon skeletons of all compounds needed as substrates for respiration or as structural components for plant and, indirectly, for animal cells. In addition, only a dozen or so elements absorbed from the soil as soluble inorganic ions are needed to sustain healthy plant growth and development. Thus, only about 16 out of the 100 or more elements found in the earth's crust are needed to make a plant.

Elements Required by Plants

Carbon and **oxygen** are obtained only from the atmosphere. **Hydrogen,** combined with oxygen in water, is obtained from the soil solution. From these three elements a plant can synthesize carbohydrates and lipids. With just one more element, **nitrogen,** most amino acids can be manufactured. Nitrogen is also a component of nucleic acids and porphyrins. Although nitrogen is absorbed from the soil in the form of NO_3^- (nitrate) or NH_4^+ (ammonium) ions, the ultimate source of this element is the gaseous nitrogen making up about 80 per cent of our atmosphere.

In recent years the great importance of a fifth element (**phosphorus**) has become clear. For the release of energy from reserve nutrients, plants need water from the soil and O_2 from the atmosphere. In order for carbohydrates to be metabolized, however, they must be phosphorylated. The element phosphorus is also a component of all nucleic acids, the important energy carriers adenosine triphosphate (ATP) and adenosine diphosphate (ADP), the pyridine nucleotides nicotinamide adenine dinucleotide (NAD) and nicotinamide adenine dinucleotide phosphate (NADP), and of the phospholipids essential in the structure of membranes.

If we think further about the details of cell chemistry discussed in Chapters 3, 10, and 11, we realize the need for other essential elements. **Calcium** is found in the calcium pectates of the middle lamella, the cement that holds together plant cells. **Sulfur** is found in the amino acids cystine, cysteine, and methionine. The sulfhydryl bonds formed between sulfur-containing amino acids have an important role in maintaining the special configuration of protein molecules. We have also encountered sulfur in the bond that holds acetate groups to the coenzyme A molecule in acetyl-CoA, the initial substrate for the Krebs cycle.

Magnesium is essential to plants because of its presence at the center of the tetrapyrrole nucleus of the chlorophyll molecule. Magnesium ions are also needed for the proper functioning of a number of enzymes. Mg^{2+} ions are involved in the aggregation of structural subunits of the ribosomes and thereby play a role in protein synthesis. If we add only one more element, **potassium,** that is essential for the activity of many enzymes and the maintenance of membrane permeability, we find that we have already listed the six essential mineral nutrients needed in largest amounts by plants, the so-called **major elements: nitrogen, phosphorus, calcium, sulfur, magnesium,** and **potassium.**

Another half-dozen or so elements are also essential but are needed in only trace amounts and therefore are usually referred to as the **minor elements.** The minor elements required by all plants are **iron, manganese, zinc, copper, molybdenum, boron,** and **chlorine.** Once again we recognize from considerations of the chemistry of cells or of special processes such as photosynthesis that the trace elements are components of molecules essential to the maintenance of cell metabolism. **Iron** is a component of the cytochromes and of ferredoxin and is present in the prosthetic groups of a variety of enzymes including catalase and cytochrome oxidase. **Manganese** is necessary for the activity of a number of enzymes including one involved in photosynthetic oxygen evolution. **Zinc** is

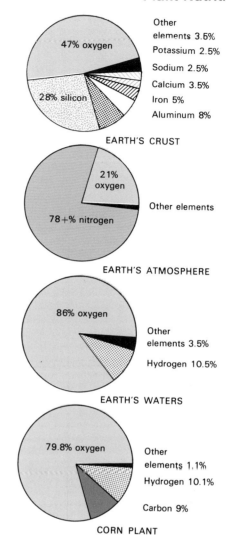

Fig. 12-1. Diagrammatic summary of the distribution of elements (per cent by weight) in corn plant and in the earth's waters, atmosphere, and crust. (Redrawn after Greulach, V. A., and J. E. Adams. 1967. *Plants: An Introduction to Modern Botany.* John Wiley & Sons, Inc., New York.)

needed for an enzyme in the biochemical pathway from the amino acid tryptophan to the plant hormone indoleacetic acid (auxin). **Copper** is found not only in plastocyanin in the photosynthetic apparatus but also in a variety of oxidizing enzymes including ascorbic acid oxidase, polyphenol oxidase, and cytochrome oxidase (an enzyme containing both iron and copper). The roles of molybdenum, boron, and chlorine are somewhat more obscure. **Molybdenum** is involved in nitrogen metabolism; **boron** in the translocation and metabolism of carbohydrates; and **chloride** ions are essential for activity of chloroplasts.

UNUSUAL REQUIREMENTS

In addition to the rather short list of elements universally needed, a few inorganic elements or ions are needed by plant cells having rather special requirements. For example, **silicon** is essential for the cell walls of diatoms and also accumulates in the stems of horsetails. The symbiotic nitrogen-fixing bacteria all require **cobalt,** a peculiarity of the enzymes associated with the steps in nitrogen fixation. It should be noted that except for the symbiotic and free-living nitrogen-fixing bacteria, plants are different from animals in not requiring cobalt. For animals, this element is an essential component of cyanocobalamin, or vitamin B_{12}. Although vitamin B_{12} is not synthesized by higher plants, it is manufactured by many soil microorganisms (bacteria and fungi) and by intestinal bacteria. In animals, vitamin B_{12} is bound to polypeptides. Animal proteins are a good source of this vitamin for human diets while plant proteins are not. **Sodium** is another element essential to animals that is ordinarily nonessential for plants. Only plants adapted to soils with a high salt content (the **halophytes)** absorb significant amounts of sodium. The nonessential Na^+ ions help to raise the internal osmotic pressure of plant cells to counteract that of their salty surroundings.

It is not unusual to find a variety of nonessential elements in plant tissues. Some 60 different elements have been detected in plants including gold, silver, lead, mercury, arsenic, and selenium. Some of these nonessential elements have no effect on plants but can be extremely toxic to herbivorous animals grazing on them. Certain plants, notably about two dozen species in the genus *Astragalus,* a legume somewhat resembling alfalfa, have the unique ability to extract selenium from the soil. These accumulator species have been shown to contain from 100 to 10,000 times higher levels of selenium than other plants native to the western plains of the United States. Because of the effects of the high selenium levels of accumulator species on animals grazing on them, they have been called "loco" weeds (from the Spanish *loco,* crazy). Cows and horses eating such selenium-containing plants develop alkali poisoning or "blind staggers" that can be fatal. Why the selenium analogues of sulfur compounds formed by accumulator plants are harmful only to the animals grazing on them is one of the mysterious differences between plants and animals. There is some evidence that selenium is a micronutrient essential for the accumulator species. They certainly grow more rapidly in the presence than

Fig. 12-2. A "loco" weed (*Oxytropus lambertii*). This species is closely related and very similar to *Astragalus*. After feeding on "loco" weed, range animals become nervous, jerky in their movements, and generally wild. (Courtesy of the U.S. Department of Agriculture.)

in the absence of selenium. Perhaps plants can tolerate the selenium analogues because of their ability to make more of the essential-sulfur compounds such as sulfur-containing amino acids while animals must survive on what they consume.

HOW MINERAL REQUIREMENTS WERE DISCOVERED

An interest in the conditions needed for healthy plant growth is as old as the science of agriculture. Controlled experimentation on the effects of fertilizers and a search for the elements essential for plants go back to the beginnings of our knowledge of chemistry. One of the earliest attempts to discover what plants need in order to grow was described by van Helmont in the early 1600's. He placed a 5-lb willow sapling in a container of dried soil weighing 200 lb. Nothing except rainwater was added to the container. After 5 years, van Helmont carefully removed the willow tree and dried and reweighed the soil. By this time the tree weighed 169 lb. After 5 years the soil weighed 199 lb, 14 oz, and van Helmont attributed this apparent weight loss to experimental error. In spite of the simplicity and logical nature of van Helmont's experiment, he arrived at a wrong conclusion because, on the basis of his experimental data, he believed that plants need only water and air.

HYDROPONICS: CULTURE OF PLANTS IN SOLUTIONS OF INORGANIC SALTS

By the 1800's, studies of the effects of fertilizers on agricultural yields had provided some evidence that the elements calcium, potassium, and phosphorus are taken up from the soil by the roots of plants. However, it was not until the late 1800's that a straight-forward methodology was developed for determining experimentally what elements are essential for plant growth. The German plant physiologists Sachs and Knop, using techniques similar to present-day water culture of plants (**hydroponics),** grew plants with roots immersed in nutrient solutions consisting of dissolved inorganic salts. Their objective was to determine the minimum number of elements needed to support plant growth. Knop found that a rather simple mixture of salts could support plant growth indefinitely. Knop's solution (which still is used with only minor modifications) contained the following salts: $Ca(NO_3)_2$, KNO_3, KH_2PO_4, $MgSO_4$, and $FePO_4$.

It is obvious to us that Knop did not add to his nutrient solution any of the trace elements except for iron. The relatively impure salts available in his day assured ample contamination with all of the other essential minor elements. This is a fine example of the necessity for long-term suspended judgment and a proper lack of dogmatism in interpreting scientific results, no matter how carefully obtained.

Table 12-1 shows the amounts of various component salts needed to prepare a liter of balanced nutrient solution in which a variety of common plants will grow well.

One of the practical advantages of nutrient solutions is that they enable us to delete just one element at a time to determine how a deficiency in one essential nutrient affects plants. The other important fact about the nutrition of plants that can be demonstrated by hydroponics is that for healthy growth, plants need only water, CO_2, sunlight, and a rather modest assortment of inorganic salts. Does this conflict with the present enthusiasm for organic gardening? Not really, because when organic materials, whether decomposing leaves, coffee grounds, etc., are

TABLE 12-1
Components of a balanced nutrient solution for plants
(From Moment, G. B., and H. M. Habermann. 1973.
Biology: A Full Spectrum. The Williams & Wilkins Co., Baltimore)

Salt	Grams per Liter
$Ca(NO_3)_2$	0.8205
KNO_3	0.50055
$MgSO_4$	0.24076
KH_2PO_4	0.13609
$FeCl_3$	0.0145
H_3BO_3	0.00286
$MnCl_2 \cdot 4H_2O$	0.00181
$ZnCl_2$	0.00011
$CuCl_2 \cdot 2H_2O$	0.00005
$Na_2MoO_4 \cdot 2H_2O$	0.000025

used as fertilizer, the gradual decomposition of these substances by the microorganisms of decay results in a release of inorganic nutrients and a final breakdown of organic compounds to CO_2 and H_2O. The decomposing organic materials contribute inorganic chemicals to the nutrition of plants and at the same time modify the consistency of soils, making clay soils lighter and more porous, sandy soils better able to absorb and retain water.

Symptoms of Mineral Deficiency

When grown in nutrient solutions or soils that are deficient in one or more essential elements, plants exhibit deficiency symptoms. Their growth is stunted and abnormal just as surely as the growth of humans is stunted and abnormal when their diets lack vitamins or contain the wrong kinds or insufficient amounts of protein.

The symptoms of specific mineral deficiencies are most readily demonstrated when plants are grown in nutrient solutions lacking one essential element but containing balanced and adequate amounts of all others (Fig. 12-3). Some symptoms, such as the **chlorosis** (yellowing due to the absence of chlorophyll) that develops when supplies of iron are inadequate, are so striking that they can be recognized even under the far more complex circumstances of home gardens or in agricultural crops. Most deficiencies are difficult to analyze under field conditions because soils are rarely poor in only one element. The deficiency symptoms characteristic of each of the essential mineral nutrients are listed in Table 12-2 along with information about the forms in which each is absorbed and its functions in plant metabolism.

Fig. 12-3. Symptoms of mineral deficiency in tobacco plants. All plants are the same age. The control plant (Ck.) received a balanced nutrient solution containing all the essential elements. Other plants received nutrient solutions lacking the element indicated on the label. (Photo by W. Rei Robbins, Rutgers University; from Greulach, V. A., and J. E. Adams. 1967. *Plants An Introduction to Modern Botany.* John Wiley & Sons, Inc., New York.)

INORGANIC NUTRIENTS AND CROP YIELDS

Soil deficiencies can be alleviated, and elements removed from soils in the form of harvested crops can be replaced by the application of appropriate amounts and kinds of fertilizers. One almost universal problem is how to determine what fertilizers to add. Because of its complexity, an analysis of the soil rarely provides an adequate answer to the question of what or how much should be added. Additionally, deficiency symptoms in the field are often difficult to interpret. The pH of the soil and the relative amounts of other elements can influence the availability and uptake of specific essential nutrients. For instance, calcium ions, because of their role in maintaining the integrity and functioning of cell membranes, can influence the uptake of other elements. Calcium deficiency sometimes results in the uptake of excess amounts of other nutrients and it is not uncommon when calcium is limiting to observe symptoms of magnesium toxicity rather than of calcium defi-

ciency. It turns out that the best indicator of what should be added to the soil to increase plant growth is the plant itself. Analyses of leaves have shown that for any crop there is a predictable relationship between mineral content and yield. Once this relationship has been established, the best way to determine whether more of a specific element should be added is to analyze a random sample of leaves for the element in question.

Commercially available fertilizers come in a variety of forms with a terminology mysterious to the uninitiated. The numbers used to describe them indicate nothing more complicated than the relative amounts of the three major elements present. For instance, in a fertilizer labeled 5-10-5, the first figure stands for the percentage of nitrogen in the total mixture, the second for available phosphate, and the third for potassium.

Generally, the yields of a field of corn or tomatoes in the home garden are dependent on a host of conditions including climatic factors such as temperature and rainfall, the presence or absence of predators and para-

TABLE 12-2
Elements essential for plant growth (From Moment, G. B., and H. M. Habermann. 1973. *Biology: A Full Spectrum*. The Williams & Wilkins Co., Baltimore)

Elements	Form in Which Absorbed	Biochemical Function	Deficiency Symptoms
Major			
Nitrogen	NO_3^- or NH_4^+	Present in all amino acids, nucleic acids, and porphyrins	Chlorosis, especially in lower leaves, with loss as deficiency becomes more severe; some species develop purple or red veins in leaves due to increased anthocyanin synthesis; excess nitrogen produces lush, dark green foliage with weak stems and abundant vegetative growth
Potassium	K^+	Maintenance of membrane permeability	Chlorosis, sometimes with mottling of leaves and development of dead areas in their tips and margins; stems often weak
Calcium	Ca^{2+}	Synthesis of middle lamella; enzyme activity	Rapid deterioration of terminal growing regions of shoots and roots; malformation of youngest leaves
Phosphorus	$H_2PO_4^-$	Present in phospholipids, nucleic acids, high-energy phosphate compounds, and coenzymes	Leaves dark green in color, tendency for red or purple anthocyanin accumulation; development of areas of dead tissues on leaves, petioles, or fruits, often resulting in leaf or fruit drop
Magnesium	Mg^{2+}	Present in chlorophyll; needed for enzyme activity	Chlorosis of leaves developing upward from the base of the plant, often accompanied by death of portions or entire leaves
Sulfur	SO_4^{2-}	Present in some amino acids and vitamins; involved in secondary and tertiary structure of proteins	Chlorosis, especially of younger leaves at tops of plants; even older leaves become pale green if deficiency is severe; root system may be larger than normal
Minor			
Iron	Fe^{2+}	Present in cytochromes, catalase, and other porphyrins; needed for chlorophyll synthesis	Chlorosis developing first in areas between veins of youngest leaves; chlorotic leaves persist and often remain on the plant for long periods; except for color, leaves appear healthy
Copper	Cu^{2+}	Present in plastocyanin, ascorbic acid oxidase, cytochrome oxidase, and polyphenol oxidase	Withering of tips of young leaves; plants often appear wilted even when ample supplies of water are available
Manganese	Mn^{2+}	Enzyme activity; needed for photosynthetic oxygen evolution	Young leaves become progressively paler and develop brown or gray dead spots; chlorosis and dead areas develop first in areas between veins of leaves and leaves are soon lost
Zinc	Zn^{2+}	Needed for auxin synthesis	Chlorosis of lower leaves at tips and margins; leaves remain clustered on short branches; "little leaf" disease of citrus crops
Molybdenum	Mo^{3+} or Mo^{6+}	Nitrate reductase	Chlorosis of leaves; early symptoms resemble those of nitrogen deficiency; with prolonged deficiency leaves become mottled, chlorotic areas are puffed in appearance, leaves become twisted and distorted
Boron	BO_3^3 or B_4O^{2-}	Involved in translocation and metabolism of carbohydrates	Death of growing regions, no new leaves formed; deficiency often called "top sickness;" fleshy organs such as fruits disintegrate with browning of internal tissues
Chlorine	Cl^-	Needed for photosynthesis	Reduced shoot and root growth; early foliar symptoms resemble those of manganese deficiency; later, depressions form in areas between veins
Essential to some plants but not to all			
Cobalt	CO^{2+}	Needed by nitrogen-fixing organisms	
Silicon	$H_2SiO_4^{2-}$	Present in cell walls of diatoms and horsetails	
Sodium	Na^+	Taken up by halophytes	
Vanadium	VO_3^-	Needed by certain algae and fungi	

sites, and the adequacy of minerals in the soil. Yield is influenced by the relative sufficiencies of each factor essential for growth, including each of the essential mineral elements. An inadequate supply of any one, even a trace element, can severely reduce yields. This generalization, also called the **law of the minimum,** is true for all living organisms. It was first enunciated by Liebig in 1846 at the time when fertilizers were being introduced into agricultural practice.

If agricultural yields can be increased drastically by applying the elements essential for plant growth, to what extent can we utilize our knowledge of mineral nutrition to increase world food supplies? Unfortunately, the application of fertilizers will not result in miracles because of a second universal law, the **law of diminishing returns** stated by Mitscherlich at the beginning of this century. Once adequate amounts of a given element are available, further increase will have no effect on growth or yield. In fact, overabundance of certain elements can even have undesirable effects on yield. Excess nitrogen leads to lush vegetative growth, a condition that is desirable in lawns or pastures, but which suppresses flowering and thus is not desirable when fruits or seeds rather than leaves are to be harvested. Many a home gardener has learned this lesson when his lush tomato plants failed to set fruit.

Although the application of excess inorganic fertilizers can increase crop yields up to a point, they can also lead to undesirable side effects. Surplus nitrates, phosphates, and the like unfortunately do not stay put after they are applied. Instead they are washed into lakes and streams. There they can stimulate algal "blooms" (explosive overgrowths that can cause fish kills). Where nitrates have accumulated in water supplies, they have approached toxic levels.

ACCUMULATION OF MINERALS FROM THE SOIL

To some extent, dissolved minerals are carried along in the transpirational stream (which will be discussed in Chapter 15). A look at the relative concentrations of elements within plant cells and in the environment of roots indicates that for many essential elements there is an active uptake from the environment and accumulation against concentration gradients. The mechanism of uptake and the ways in which ions are able to pass through selectively permeable membranes are poorly understood in spite of extensive research. Obviously if ions are accu-

mulated inside cells, work has to be done in order to move them against concentration gradients. It has been shown that factors inhibiting aerobic respiration (lack of oxygen or specific poisons) also inhibit the rates of ion accumulation. Therefore, the energy expended in the process of active uptake must come from the respiratory cycle. There is some evidence that there are specific absorption sites on cell membranes for particular ions. Enzymes at the absorption sites may participate in the transfer of ions across the membrane.

All minerals are taken up as positively or negatively charged ions. Sometimes the root cells secrete ions in exchange for ions absorbed. H^+ ions are exchanged for positively charged ions such as K^+, while the uptake of negatively charged ions such as NO_3^- is neutralized by secreting bicarbonate or the anions of organic acids such as malate ions. Secretion of large numbers of hydrogen ions will make the surrounding medium more acid. In this way the metabolism of plant cells can have yet another kind of effect on their environment.

Nitrogen Fixation

A great pool of nitrogen exists in the earth's atmosphere (which is almost 80 per cent nitrogen). In spite of the apparently inexhaustible supply of this essential element, plants often exhibit symptoms of nitrogen deficiency because they cannot use gaseous nitrogen. Most plants can only absorb this element through their roots in the form of nitrate or ammonium ions. A relatively small number of organisms have the ability to convert atmospheric nitrogen into a form that plants can absorb. The very few kinds of organisms, all prokaryotic, capable of doing this very essential conversion include: certain of the blue-green algae, notably members of the genus *Nostoc,* some of the photosynthetic bacteria such as *Rhodospirillum,* a few free-living bacteria existing on organic matter in the soil, including the aerobic *Azotobacter* and the anaerobic *Clostridium,* and finally species of such genera as *Rhizobium* living symbiotically in the root nodules of certain higher plants, especially legumes such as peas and beans. Although there is great diversity in the mode of nutrition among the nitrogen-fixing organisms, from saprophytic to photoautotrophic to symbiotic, they all share the distinct advantage of being able to utilize atmospheric nitrogen and being independent of the soil for supplies of this essential element. Because the symbiotic nitrogen-fixing bacteria add soluble nitrogen compounds to the soil, it is common agricultural practice to rotate crops, including a leguminous crop in every cycle of rotation.

In the nitrogen fixation process, atmospheric nitrogen is transformed to ammonia, which is either utilized directly or oxidized to nitrate by other microorganisms. Within plant cells, nitrates must be reduced back to ammonia through the action of the enzyme nitrate reductase. This enzyme contains molybdenum and is the only known component of plant cells for which this trace element is required. It has been shown that plants can grow perfectly normally in the absence of molybdenum if nitrogen is provided in the form of ammonia rather than nitrate. Plant cells utilize ammonia to make amino acids (which are then incorporated into proteins) and a host of other nitrogenous compounds including the nucleic acids.

The **nodules** formed in plant roots invaded by nitrogen-fixing bacteria (Fig. 12-4) are in-

Fig. 12-4. Root nodules formed by a strain of *Rhizobium leguminosarum* on the roots of a young pea plant (*Pisum sativum*). (From Dubos, R. 1962. *The Unseen World*. The Rockefeller Institute Press, New York.)

teresting both structurally and biochemically. The initial invasion probably takes place through the root hairs. An infection "thread" spreads to other parts of the root, and causes a curious proliferation of root cells. The interior of the resulting nodule contains large host cells, many of them filled with bacteria. The nodules of legumes usually are pink because of the presence of **leghemoglobin,** a red pigment chemically similar to animal hemoglobin, that appears to be necessary for nitrogen-fixing activity.

THE NITROGEN CYCLE

Through the work of the nitrogen-fixing microorganisms, vast amounts of atmospheric nitrogen are continually being transformed into soluble ammonia and nitrates in the soil. These fixed forms of nitrogen are absorbed from the soil solution by plant roots and are utilized by plants to form amino acids and ultimately proteins. The leaves, fruits, and seeds of plants are ingested by animals and transformed into animal proteins and nitrogenous wastes such as urea. The primary consumers of plants (herbivores) are not infrequently consumed in turn by other animals (carnivores) with the result that a given atom of nitrogen may pass through a complex and lengthy food chain. Ultimately all plants and animals die; their tissues are decomposed into simple breakdown products by microorganisms in the soil. Nitrogenous wastes may be absorbed by plants and recycled again through living systems, or free nitrogen may be released into the atmosphere through the action of soil denitrifying bacteria on nitrates.

Recycling — The Way of Nature

The nitrogen cycle illustrates a general strategy of nature. Nothing is ever really lost, including materials incorporated into plants or animals. At atom of an element is only temporarily lodged in a molecule within the cell of a leaf or a human big toe. The particular sort of molecule may have a rapid turnover (if it is a respiratory intermediate, for example) or it may be a more or less permanent part of a long-lived structure (perhaps cell wall or bone). Because death and decay are the ultimate fate of every living thing, the subsequent release and recycling of the chemical components of life are insured. It is interesting to ponder whether some of our own molecules might contain atoms that once resided in a dinosaur or a bird. Certainly

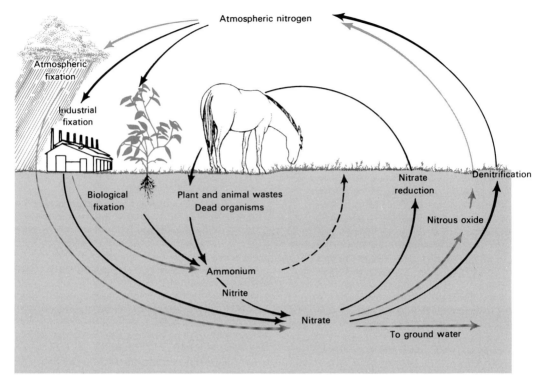

Fig. 12-5. The nitrogen cycle. The supply of atmospheric nitrogen is virtually inexhaustible, but it must be converted to NH_4^+ or NO_3^- by nitrogen-fixing microorganisms before it can be absorbed by plants. (From Moment, G. B., and H. M. Habermann. 1973. *Biology: A Full Spectrum.* The Williams & Wilkins Co., Baltimore.)

the carbon atoms (indeed the entire molecules) of the sugar on our breakfast cereal were synthesized rather recently in sugar cane or sugar beet leaves.

USEFUL REFERENCES

Epstein, E. 1972. *Mineral Nutrition of Plants: Principles and Perspectives.* John Wiley and Sons, Inc., New York.

Gauch, H. G. 1972. *Inorganic Plant Nutrition.* Dowden, Hutchinson and Ross, Inc., Stroudsburg, Pa.

Sprague, H. B., ed. 1964. *Hunger Signs in Crops,* 3d ed. David McKay Co., New York. (Contains excellent colored photographs of deficiency symptoms in crops.)

Steward, F. C., ed. 1963. *Plant Physiology:* A Treatise. Vol. III: Inorganic nutrition of plants. Academic Press, New York.

Sutcliffe, J. F., and D. A. Baker. 1974. *Plants and Mineral Salts.* Institute of Biology, Studies in Biology No. 48. Edward Arnold (Publishers), Ltd., London. Distributed in the U.S. by Crane Russak & Co., Inc., New York. (Paperback.)

13

Human Nutrition and World Food Problems

Nutrition is simultaneously a personal and an international problem of major importance. Not enough of the right kinds of proteins very early in life may result in permanent mental retardation. Too little of the right vitamins at any time of life will bring serious trouble. Too much fat, especially cholesterol, in middle age greatly increases the chances of heart disease. Little understood nutritional factors absorbed from intestinal bacteria seem to play an important role in health. Even intelligent people have been led to believe that calories in the form of carbohydrates are worthless because they are "empty," by which it seems to be meant that such things as polished rice or white bread are without nutritional value because they are not a complete and adequate diet in themselves. There are many proteins that by themselves cannot support human life because they lack certain essential amino acids. Nor can man live by vitamins alone.

There is also a philosophical interest in nutrition. If one stops to think, it is an amazing thing that what you eat turns into you, while the same food eaten by someone else may turn into a beautiful young girl, a crabby old man, or even a parrot.

Everywhere the problems of adequate nutrition are more than strictly scientific questions because political and economic factors are deeply involved. This is especially true in parts of the world where the population is larger than the resources of the country can support.

The problems of nutrition and respiration are basically inseparable because the purpose of eating is twofold, to build and maintain the structure of the body and also to furnish the energy that makes it function.

Food Processing

ANATOMY OF NUTRITION

For most vertebrates nutrition begins with the use of teeth. So clearly do teeth reflect the whole life of their possessors that a great comparative anatomist, Sir Richard Owen, paraphrased Archimedes' famous remark on discovering the principle of the lever, "Give me where to stand, and I will move the earth," by asserting, "Give me a tooth, and I will reconstruct the animal." This is an exaggeration, of course, but one that contains much truth. Carnivores, like the cat and seal, have large canines and pointed molars. Herbivores, like the deer and horse, have reduced canines or none, with molars that are grinders with flat, corrugated tops. Omnivores, like man and other primates, have generalized teeth.

The **pharynx** or throat follows immediately behind the buccal or mouth cavity which contains that versatile organ, the tongue. It is in the pharynx that the pathway of the food crosses the route of air to and from the lungs in reptiles, birds, and mammals.

The **esophagus** is merely a muscular tube conducting food from the pharynx past the heart and lungs into the stomach, which lies just below the diaphragm, a transverse muscular partition separating the coelom around the lungs from that surrounding the stomach, 227

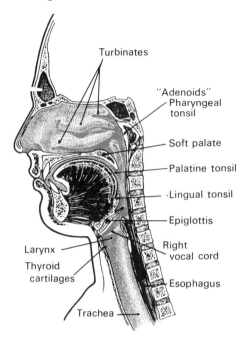

Fig. 13-1. Sagittal section through a human head to show the buccal and pharyngeal cavities where the pathways of air and food cross. (From Moment, G. B. 1967. *General Zoology*, 2nd ed. Houghton Mifflin Co., Boston.)

liver, and intestine. The structure of the wall of the esophagus is essentially the same as that of the stomach or intestine. The cavity or lumen of all three is lined by an epithelium composed of squamous, i.e. thin and flat, cells in the esophagus and tall, columnar ones in the stomach and intestine. This epithelium is supported by a thin connective tissue sheet, or lamina, and a very thin muscular sheet. These constitute the **mucosa,** although this word is sometimes used to refer solely to the epithelium. The submucosa encircling the mucosa consists of loose, fibrous connective tissue with numerous blood vessels and, in some parts of the digestive tract, glands. Outside of the submucosa is a layer of circular muscles, then a layer of longitudinal muscles, and finally, on the very outside, an epithelial covering. All these muscles are of the nonstriated type.

The **stomach** is an enlarged and specialized portion of the gut tube. The portion which the esophagus enters is known as the **cardiac region,** since it is nearer the heart. The opposite end is the **pyloric region.** The **pylorus** is the orifice between the stomach and the intestine, kept closed by a circular, i.e. sphincter, muscle.

The first region of the **small intestine** into which the stomach empties is the **duodenum.** The common bile duct, carrying the secretions of both pancreas and liver, enters the

duodenum not far from its origin at the pylorus. The remainder of the small intestine, i.e. the **ileum,** is much longer than the duodenum and leads into the shorter large intestine, or **colon.** The colon ends in a short straight section, the **rectum,** which leads to the exterior via the **anus.** Where the small intestine enters the large there is in most vertebrates a longer or shorter blind sac, i.e. a diverticulum, termed the **cecum,** or vermiform appendix in man.

CHEMISTRY OF DIGESTION

Mammals are equipped with an impressive battery of glands associated with the gut. Some are unicellular; others, like the pancreas, are large and complex. The function of all these glands is to secrete digestive enzymes that hydrolyze large molecules, i.e. split them with the addition of water, into smaller molecules that can be absorbed through the wall of the digestive tract. Carbohydrates are broken down into glucose and other **simple sugars,** lipids into **glycerol** and **fatty acids,** and proteins into **amino acids.**

Table 13-1 shows the main sources of mammalian digestive enzymes and their actions. Like other enzymes these are very sensitive to pH. **Ptyalin,** the amylase in saliva, works at a pH close to neutrality; the proteolytic gastric pepsin requires an acid pH; the pancreatic and intestinal enzymes require an alkaline environment for their activity.

Gastric juice is a clear fluid containing

TABLE 13-1
Principle digestive enzymes in mammals (From Moment, G. B., and H. M. Habermann. 1973. *Biology: A Full Spectrum.* The Williams & Wilkins Co., Baltimore)

Enzyme	Occurrence	Substrate	Chief Products*
Ptyalin	Saliva	Starch	Maltose
Pepsin	Gastric juice	Protein	Polypeptides
Amylases	Pancreatic juice	Starches	Maltose
Lipases		Lipids	*Glycerol and fatty acids*
Trypsin		Proteins	Peptides
Chymotrypsin		Polypeptides	Peptides
Peptidase		Peptides	*Amino acids*
Sucrase	Intestinal juice	Cane sugar	*Simple sugars*
Maltase		Maltose	*Simple sugars*
Lactase		Milk sugar	*Simple sugars*
Lipase		Lipids	*Glycerol and fatty acids*
Erepsin		Peptides	*Amino acids*
Peptidases		Peptides	*Amino acids*

* Substances shown in italics are absorbed through the walls of the intestinal villi.

about 0.4 per cent hydrochloric acid and two enzymes, **pepsin,** which splits proteins by hydrolysis into short chains of amino acids called peptides, and **rennin,** which clots milk. Recent evidence indicates that a lipase is also normally present in the stomach. Why doesn't the stomach digest itself? Such cannibalism is physiologically possible. The lining of the stomach is normally protected against pepsin by a coating of mucus. The enzyme-secreting cells form pepsinogen which becomes active pepsin only after it comes in contact with the acid gastric juice.

When the acid stomach contents pass through the pylorus into the duodenum, the pancreas pours out **pancreatic juice.** The duodenum secretes a hormone, **secretin,** which is carried by the bloodstream to the pancreas. Digestion in the intestine of mammals is due in part to the pancreatic juice which contains enzymes capable of hydrolyzing all three major classes of foodstuffs. Multicellular glands in the submucosa of the duodenum secrete mucus and probably more digestive enzymes.

The lining of the intestine is not only covered with ridges or folds, but the surface, including the surface of the folds, is covered with finger-like projections called **villi.** Each villus is covered with columnar epithelium and contains a core of fibrous connective tissue with capillaries and a lymph vessel. In between the villi are numerous glandular "post-holes," called **crypts of Lieberkühn.** The mucosal epithelium not only covers the villi and the general lining of the gut but extends down into the crypts where cell division is frequently seen. It has been estimated that the epithelial cells of the intestine are completely renewed every 1½–2 days.

Absorption

Most of the digested food is absorbed in the intestine by the villi just described. The glucose and amino acids are passed as such into the capillaries, whence they are carried to the liver in the hepatic portal vein. The absorption of fats is more complex and less understood. The glycerol is taken into the villi as such. The fatty acids unite with bile salts forming water-soluble compounds which enter the cells of the villi. Within the villi the fatty acids reunite with the glycerol, forming lipids again. Some lipids are absorbed directly as glycerides, i.e. neutral fats (esters) formed as usual by a union of glycerol and a fatty acid. Most of this fat passes as a milky emulsion in the lymphatics of the mesentery up to the thoracic duct and through it into the left subclavian vein to the heart.

The absorption of whole protein molecules is of negligible importance for nutrition but is significant in food allergies and in the absorption of bacterial toxins. By special staining for certain enzymes such as horseradish peroxidase, whole protein molecules can be detected passing into the interior of cells. Most absorption is due to **active transport,** requiring the expenditure of energy. This is evident because very low concentrations of glucose and amino acids can be absorbed from the intestine into the bloodstream where the concentration of these substances is much higher, and because metabolic poisons quickly stop absorption.

ROLES OF THE LIVER

The liver plays a key role in the fate of the digested food carried to it from the intestine by the hepatic portal vein. No organ, not even the brain, has been the object of such intense study by physicians and soothsayers from remote antiquity. Yet it was not until the latter

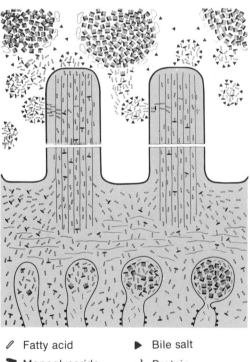

∥ Fatty acid	▶ Bile salt
⊤ Monoglyceride	∿ Protein
⬲ Diglyceride	ℭ Lipase
⩜ Triglyceride	

Fig. 13-2. Diagram of the process of lipid absorption into the epithelial cells lining the intestine. Note how lipase molecules break up oil droplets with the help of bile salts into fatty acids and (mono)glycerides which enter the cells and reassemble as intracellular lipid droplets. (From Bailey, F. R. 1971. *Textbook of Histology,* 16th ed. The Williams & Wilkins Co., Baltimore.)

years of the past century that something of its actual function was first learned.

Glucose is converted by the liver into glycogen under the influence of insulin. It is stored until the glucose in the blood falls to a threshold level when the glycogen is reconverted into glucose.

The fate of amino acids in the liver raises important questions and highly controversial views about how much protein is necessary in the diet. Only a small proportion of the absorbed amino acids reaching the liver via the hepatic portal vein from the intestine are passed through to be utilized in building body protein. Most amino acids have their amino groups taken from them, a process called **deamination,** a two-step process in which the amino acids are oxidized by removal of two H's and then hydrolyzed to form a keto acid and ammonia. The keto acids corresponding to the various amino acids are generally metabolized as energy sources in the Krebs cycle.

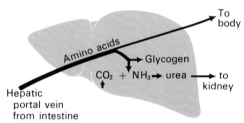

Fig. 13-3. Diagram of mammalian liver showing the course of amino acids from the intestine via the hepatic portal vein and their fate within the liver. Note some continue via the hepatic vein to the body in general. (From Moment, G. B., and H. M. Habermann. 1973. *Biology: A Full Spectrum*. The Williams & Wilkins Co., Baltimore.)

The synthesis of urea is a major function of the liver. The ammonia, whether from deamination or oxidation of amino acids, is poisonous except in very low concentrations. That "clever chemist," the liver, combines two molecules of ammonia with one of another waste product, CO_2, to form a colorless, odorless, and harmless compound, urea, which is carried by the bloodstream to the kidneys and there excreted. The synthesis of urea occurs when NH_3 and CO_2 unite with ATP to form a carbamyl phosphate which, in turn, unites with a long-chain amino acid called ornithine. The ornithine unites with citrulline and this compound with still another and then with arginine, after which the result-

$$2\,NH_3 + CO_2 \rightarrow H_2N\overset{\displaystyle O}{\overset{\displaystyle \|}{-C-}}NH_2 + H_2O$$

Fig. 13-4. The overall formula for the synthesis of urea from ammonia and CO_2.

ing compound plus water splits into urea, $(H_2N)_2CO$, plus ornithine. Hence urea is said to be formed in the **ornithine cycle.**

How could anyone even suspect that there is such a thing as the ornithine cycle? The discovery was made by the same Hans Krebs who later worked out the details of the citric acid cycle named after him. Krebs was studying the formation of urea by adding various amino acids to slices of liver fresh enough for the enzymes to be still active. He found that added ornithine or arginine greatly increased the production of urea, especially in the presence of added ammonia. This opened the door.

In addition to the synthesis of urea from the breakdown products of amino acids and the storage of glucose as glycogen, the liver secretes bile, destroys worn-out red blood cells, makes prothrombase, stores vitamins, detoxifies miscellaneous harmful substances, and plays a role in lipid metabolism, as well as synthesizing its own proteins. The liver is indeed a versatile as well as a clever chemist.

Human Nutritional Requirements

CARBOHYDRATES: THE PRIMARY FUEL

Carbohydrates are the primary fuel of life for man and most animals. Every major human culture has relied on some carbohydrate-rich food for energy—wheat, potatoes, rice, maize, bread, fruit. If both carbohydrates and fats, which are even richer in energy, are lacking, then body proteins will be utilized. It will be recalled that at the "lower" end of the glycolytic pathway, acetyl-coenzyme A serves as a central metabolic crossroads where many kinds of molecules as well as pyruvic acid from the anaerobic breakdown of glucose are fed into the Krebs cycle as acetyl-CoA. Lipids and amino acids from proteins are converted into acetyl groups, $CO\cdot CH_3$ which, linked to coenzyme A (a derivative of the vitamin pantothenic acid), enters that "meat grinder," the Krebs cycle. Lipids serve as essential constituents of cell membranes, as reservoirs of stored energy, and as carriers of the fat-soluble vitamins.

The **fuel** or **energy value** of any food is measured in heat units called **calories.** One large Calorie (kilocalorie), the unit usually employed, is defined as the amount of heat required to raise the temperature of 1,000 gm of water 1°C. The total caloric content of food can be determined by complete combustion in a heat-measuring device called a **bomb calorimeter.** The powdered food is ignited by an electric spark. The heat liberated is measured by the rise in temperature of the water in a jacket that surrounds the combustion chamber. A man who is doing hard physical labor requires from 3,500 to 5,000 calories per day. A person doing sedentary or semi-sedentary work burns about 2,500 calories daily. To maintain life, about 500 calories per day are necessary. One gram of carbohydrate equals approximately 4 calories, 1 gm of protein also equals 4 calories, while 1 gm of fat equals 9 calories. Although dietary proteins, unless present in minimal amounts, regularly serve as a source of energy, i.e. calories, the primary role of proteins is quite different, and to that we now turn.

PROTEINS: REALITIES AND MYTHS

No problem in nutrition is more important to every individual and to the world than the protein problem and none is more explosively controversial. Proteins form the framework of all cells and also the enzymatic machinery that puts together that frame and then carries on all the activities of life. Clearly, proteins are essential in the diet of any animal, or, to be more accurate, the amino acids needed to make proteins must be present. Remember, only plants can make amino acids de novo.

The most conspicuous and disastrous result of too little protein is seen in **kwashiorkor,** a disease of early childhood in which the victims fail to grow in weight or height and become enfeebled, usually with sparse reddish hair, abnormal skin coloration, swelling of the legs and feet, nausea, diarrhea, irritability, and other behavioral symptoms. Closely similar is **marasmus** which is due to a combination of protein and calorie deficiency, but is usually characterized by general apathy. Both are commonly lumped together as protein-calorie malnutrition. The afflicted are estimated to run into the hundreds of thousands spread over Central and South America, Africa, and Asia.

Far less dramatic than kwashiorkor, but probably much more prevalent, is subclinical marasmus which results in brain damage and permanent intellectual impairment in young children with inadequate protein in their diets. This conclusion has been reached in very careful studies of children in Central American villages by an international group of investigators including Rene Dubos of The Rockefeller University, Craviots, Ramos-Galvan, and others.

Heredity sets only the upper limits of intelligence in man as in any other primate. How

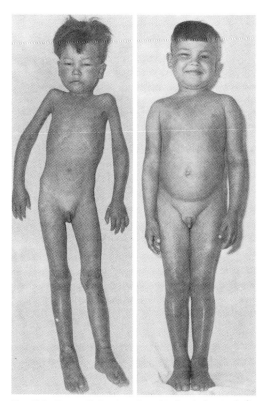

Fig. 13-5. Boy suffering from kwashiorkor before and after treatment with a diet of mixed plant proteins containing all the essential amino acids. (From Moment, G. B. 1967. *General Zoology*, 2nd ed. Houghton Mifflin Co., Boston.)

fully the hereditary potential is realized obviously depends on many factors. Too little thyroid hormone, for example, will turn an otherwise normally intelligent child into a retardate. In these protein deficiency studies, the effects of heredity were believed to have been ruled out by comparing these poverty-stricken, malnourished, and mentally retarded children with their more affluent and well-fed relatives who showed normal mental development. This conclusion has recently been challenged on the grounds that their poor, very simple environment may have damaged their intellectual growth. Rats raised from birth in empty cages are markedly inferior in intelligence as adults to rats raised on the same diet in cages enriched with ladders, wheels, and other toys. However, it must be remembered that an economically poor environment, however disagreeable, is not necessarily deprived from the behavioral point of view. For example, a jungle of fire escapes and street gangs is far from bland or oversimplified. Moreover, permanent, adverse chemical changes have been found in the brains of rats deprived of adequate protein from 7 to 17 days of age.

The causal relationship between excess dietary protein and kidney damage is well established in otherwise healthy children and in adults with liver or renal disease. Clearly it is not true that if some protein is not only necessary but highly beneficial, then a great deal must be better. What of carnivores which live on a diet of 90 per cent or more meat? It is thought that they possess kidney adaptations for this high protein diet just as they have teeth especially adapted for meat eating. Even so, cats, dogs, and other carnivores commonly die of kidney ailments in old age. Both rats and humans are omnivorous as shown by teeth, the length of intestine, and enzymes.

The useful questions are how much protein and what kind of protein make the best diet. The Food and Nutrition Board of the National Research Council of the National Academy of Sciences has proposed "recommended Daily Allowances," often called RDA's, for most nutritional components. The RDA for protein in the human diet is about 55 gm per person per day. Other authorities believe that 35 gm (about 1 oz) of protein per day is enough for the average man and 27 gm enough for the average woman. The U.S. Department of Agriculture reports that the average consumption of protein in the United States is about 99 gm per day, about 66 per cent of which is of animal origin. Clearly many Americans could safely cut their protein intake and probably with beneficial results. It will be recalled (see Fig. 13-3) that dietary protein is digested into its component amino acids and that most of these are deaminated by the liver and converted into the carbohydrate glycogen and the waste product urea, which contains the nitrogen. In healthy adults the amount of nitrogen in the protein ingested closely approximates that excreted. Such an individual is said to be in **nitrogen balance.** Only young children, pregnant women, and people recovering from severe illness characteristically show a positive nitrogen balance, excreting less than they ingest.

To eat proteins that will provide the amino acids you need is what is crucial. The amino acids which any animal must have in its diet for normal growth and health are called **essential amino acids,** and a protein which has them all is known as **complete.** This does not mean that no additional amino acids are needed by the animal but only that from the "essential" ones any necessary additional acids can be made by the animal itself. In man, 23 different amino acids have been identified. Ten are essential in the above sense. The remaining 13 can be made from the 10 by our metabolic machinery. In the body there are special enzymes, **transaminases,** which convert one amino acid into another.

In general, proteins from animal sources—meat, fish, eggs, cheese—are complete. Many plant proteins are not. This is the case with beans, peas, or grains from which the germ has been removed. Fortunately some plant proteins complement one another. For example, Indian corn, i.e. maize, is very poor in lysine but contains the other essential amino acids including methionine, which is deficient in beans. Beans, however, possess plenty of lysine. Obviously, when these two protein sources are mixed, their deficiences cancel out. Because there is no significant storage of amino acids in the body, these two proteins must be eaten at the same meal to be effective in supplementing each other. No wonder that the Algonkian Indians did so well on their succotash. A major breakthrough in very recent times is the development of a lysine-rich corn by a Purdue University team lead by E. T. Mertz. Unfortunately the high lysine mutant gene known as "opaque-2" is recessive so that breeding the corn for seed presents special problems. It also requires a somewhat different kind of milling and preparation for cooking. But when tested on piglets, those on opaque-2 grew over 300 per cent faster than those on "normal" corn. Spectacular results have now been obtained with malnourished children in Latin America.

Vegetarian diets which include eggs, milk and cheese present no problem. Those which lack these things require very careful planning.

FATS AND OTHER LIPIDS

Lipids vary greatly in molecular structure and composition from relatively simple triglycerides, the ordinary fats and oils, to complex phospholipids and glycolipids. None are more important than the sterols which include not only the sex hormones and other steroids but cholesterol and bile salts. The precursor of vitamin D is a sterol, ergosterol. Their functions are varied for they are essential components of all cell membranes, store energy, insulate against cold when in the form of subcutaneous layers of fat, serve as vehicles for fat-soluble vitamins, and even insulate nerve fibers.

Fortunately fats and lipids present few dietary problems because the body readily, too readily in some individuals, converts carbohydrates into triglycerides and other lipids. There is only one lipid known to be essential in the human diet, linoleic acid, a polyunsaturated lipid with 18 carbons. Its lack results in infant growth failure and in various skin disorders. Linoleic acid is so widespread that there is no danger of missing it unless your diet is highly unusual. Coconut, chocolate, and palm oils have almost none, but other vegetable oils are rich in this acid. Sunflower oil contains 70 per cent linoleic acid, olive oil 10 per cent, and most other vegetable oils fall somewhere in between. Poultry fat contains 23 per cent, most other animal fats 10 per cent or somewhat less.

The meaning of unsaturation of a lipid has been discussed in Chapter 3. Fats and oils, especially saturated oils and cholesterol, in the diet in large amounts appear to be very definitely related to heart attacks and vascular disease generally. Certainly cholesterol forms plaques on the inside of blood vessels in the heart and elsewhere which lead to blockages. The evidence is still controversial so the wise course would be to eat moderate amounts of egg yolks, liver, oysters, and other foods that are highly nutritious in other ways but high in cholesterol.

VITAMINS

While it has been commonly accepted that vitamins are essential for physical well-being, it is now evident that vitamins, specifically ascorbic acid and niacin, are also essential for mental health. How a knowledge of vitamins was won is an important part of our cultural heritage.

The discovery of vitamins grew out of the needs of explorers. For centuries, **scurvy** had been the nightmare of sailors and explorers as well as the curse of boarding schools and prisons. In his voyage around the Cape of Good Hope in 1498, Vasco da Gama lost almost two-thirds of his crew from scurvy. In the words of Dr. Logan Clendening, a noted medical historian:

"First a sailor's gums would begin to bleed. Then some of his teeth would fall out. The stench from his mouth would be horrible. Then great blotches would appear on the skin. The wrists and ankles would swell. A bloody diarrhea set in. Parts of the flesh would rot out. Finally, exhausted, delirious, loathsome, the poor wretch would pay his debt to nature."

The first recorded successful effort to control this scourge took place during an exploration of the St. Lawrence River in 1536 by Jacques Cartier. After he had lost 26 men, the rest were restored to health by a pine needle concoction prepared by the native Indians. In view of the dramatic nature of the need, it is difficult to understand why nothing came of this incident.

Eighty years later, John Woodhall, a surgeon in the employ of the East India Company which regularly sent ships around the Cape of Good Hope, mentioned in print the value of lemon juice in preventing scurvy, as did James Lind 137 years later. But such is the conservatism, stupidity, or perhaps mere inertia of men that it was not until 1768, over 230 years after Cartier's men were saved by pine needles, that the famous Captain Cook set out in *H.M.S. Endeavour* and proved the efficacy of lemons on a long voyage in the South Pacific. Even today British sailors are called "limeys" from the lemons and limes that formed a part of their diet. In comparatively recent times the agent in citrus fruits which is responsible for preventing scurvy was named vitamin C and has been identified as **ascorbic acid.**

The actual chemical identification of vitamins came about in a very interesting way. In the 19th century, a Russian investigator, Lunin, had found that when animals are fed on highly purified diets of carbohydrates, fats, and proteins, they sicken and die even though all three major food components are present. Many years later Hopkins, a pioneer biochemist working in a small basement laboratory at Cambridge, reinvestigated this old puzzle of Lunin and found in milk what he called an accessory food factor needed in minute amounts for health. This announcement stimulated work in several laboratories and by 1912 the first such substance. was isolated from rice polishings by Casimir Funk. It was the anti-beriberi factor, which turned out to be thiamine (B_1). Funk coined a new word by calling the factor a **vitamin.**

Modern Knowledge of Vitamins

Until rather recently, nothing was known about the biochemical function of vitamins. Then almost simultaneously, physiologists discovered a "yellow enzyme" essential for the oxidation of foodstuffs within the cells and nutritionists found a yellow vitamin, **riboflavin,** essential for normal growth in young

Fig. 13-6. Chicken showing symptoms of beriberi (polyneuritis) due to thiamine deficiency. (Courtesy of M. L. Sunde, the University of Wisconsin.)

birds and mammals and for the prevention of various eye and skin disorders in adults. The "yellow enzyme" and riboflavin were found to be intimately associated, and from this fact came the knowledge that vitamins can function as structural components of enzymes.

Most enzymes consist of two parts, a larger protein portion and a smaller nonprotein portion. The protein part of the enzyme molecule is called the **apoenzyme**; the smaller nonprotein part, the **coenzyme.** If the coenzyme is very firmly bound to the apoenzyme, it is called the **prosthetic group.** A coenzyme or prosthetic group is usually regarded as a vitamin for any particular animal only if the animal cannot manufacture it for itself.

The knowledge that vitamins form the nonprotein part of enzyme molecules is of much more than purely theoretical interest. It makes it possible to form medically useful drugs like sulfanilamide. Such substances are essentially fake vitamins, substances with which the protein part of the enzyme will unite but which will not make a functional enzyme. In such a way sulfanilamide blocks the enzymes of the bacteria.

Up to the present, somewhat more than a dozen vitamins have been discovered. Many are water-soluble, like vitamin C and the B vitamins. The rest are fat-soluble. Some are heat-stable; others are readily destroyed by heat. Table 13-2 gives the chemical name, the letter symbol (when one exists), dietary sources, and the function of most of the vitamins.

Niacin, the pellagra-preventing vitamin is of special interest because of its certain relation to a severe mental derangement. Pellagra is usually said to be characterized by three D's: dermatitis (a roughening of the skin), diarrhea, and dementia. It should be four, because death is the end result. Because niacin cures the severe psychosis of pellagra patients, it is now being prescribed for mental derangements not associated with pellagra. Some real success has been reported. Niacin forms part of two coenzymes essential for the utilization of carbohydrate in muscles, yeast, and presumably all cells. These are nicotinamide adenine dinucleotide (NAD) and nicotinamide adenine dinucleotide phosphate (NADP). In stability, niacin is at the opposite pole from ascorbic acid. It is stable to heat, light, acids, alkalis, and oxidizing agents. Yeast, lean pork or beef, and liver are rich in niacin. Corn (maize) is very deficient in it, and most cases of pellagra occur in regions where corn is the chief article of diet.

ROLE OF INTESTINAL FLORA

Most vitamins are obtained from food. However, evidence by Richard Barnes and his associates at Cornell and by other investiga-

TABLE 13-2
Important vitamins in human nutrition (From Moment, G. B., and H. M. Habermann. 1973. *Biology: A Full Spectrum*. The Williams & Wilkins Co., Baltimore)

Chemical Name	Letter Symbol	Solubility	Functions	Dietary Sources	Heat Stability
Ascorbic acid	C	Water	Anti-scurvy, mental health	Citrus fruits, tomatoes, leafy vegetables	Labile
Thiamine	B_1	Water	Anti-beriberi, coenzyme in pyruvate metabolism	Milk, meats, leafy vegetables	Fairly stable in acid
Cyanocobalamin	B_{12}	Water	Anti-pernicious anemia, synthesis of purines	Milk, whey, soybeans, cotton seed	
Niacin, nicotinic acid		Water	Pellagra-preventive, NAD and NADP nucleotides, mental health	Peanuts, liver, chicken, fish, whole wheat	Stable
Folic acid (folacin)		Water	Normal growth, blood formation	Meats, eggs, beans, yeast, leafy vegetables	Fairly stable
Aminobenzoic acid		Water	Normal growth	Many foods	
Riboflavin	B_2	Water	FAD respiratory coenzyme	Whey, most foods	Stable
Pantothenic acid		Water	Part of coenzyme A	Peanuts, lettuce, eggs	
Pyridoxin	B_6	Water	Anti-dermatitis, anti-acrodynia (in rat)	Egg yolk, wheat germ, yeast	
Inositol		Water	Anti-alopecia (in mouse), anti-fatty liver (in rat)	Rootlets, sprouts, fruit, lean meat, yeast, milk	
Biotin	H	Water	Anti-egg white injury	Egg yolk, kidney, liver, tomatoes, yeast	
Retinol	A	Lipid	Healthy skin and mucous membrane, night vision	Milk, yellow and green vegetables	Stable
Calciferol	D	Lipid	Anti-rickets	Fish liver oils, butter, egg yolk	Stable
Phytylmenaquinone	K	Lipid	Normal blood clotting	Most foods	Stable
Tocopherol	E	Lipid	Heme synthesis, anti-sterility, normal muscle development	Leafy vegetables, meat, yolks	Stable

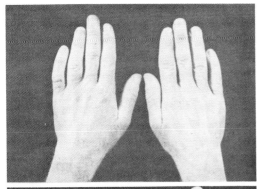
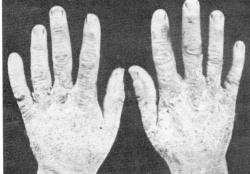

Fig. 13-7. The hands of a person with pellagra before and after treatment with niacin. (Courtesy of the Southern Medical Association.)

tors in this country and abroad indicate that the intestinal microflora contribute important amounts of vitamins to the host. **Gnotobiotic animals**, i.e. germ-free rats, chickens, and guinea pigs, show unmistakable symptoms of deficiencies of vitamins in the B group—thiamine, niacin, riboflavin, and folic acid—as well as ascorbic acid (vitamin C).

Antibiotics are regularly fed to many domestic animals to improve growth. The mechanism is poorly understood; the problem is complicated by evidence that some intestinal bacteria secrete toxic products. It is significant that antibiotics do not exert either a vitamin-sparing or a growth-promoting effect on germ-free animals.

In general, antibiotics like penicillin, which are easily absorbed and reach only the stomach and duodenum, increase vitamin production of the intestinal microflora. Antimetabolites like the sulfonamides, which are poorly absorbed by the small intestine and hence reach the large intestine and cecum, depress vitamin production.

There is still much to be learned about this whole problem. It was discovered in 1908 that germ-free ducks and chicks benefited from inoculation with intestinal bacteria, although this work was noticed by few. It has been known for many years that cows and other ruminants receive not only B vitamins but also proteins from their symbiotic stomach microorganisms.

World Food Problems and the Green Revolution

"Before 1985," says Paul Ehrlich, an environmentalist of Stanford University, "the world will undergo vast famines—hundreds of millions of people are going to starve to death." At the other extreme, Roger Revelle, a population and resource specialist of Harvard holds that "edible plant material (to feed) between 38 and 48 billion people," 10 times today's world population, could be produced if the earth's arable land were properly developed. It is a big if.

Some hold that the much talked of Green Revolution is withering. What are the facts and what is the possibility of more food production in the future? This agricultural revolution is not altogether new because hybrid corn (maize) has greatly increased corn yields per acre for many years (see Chapter 4), but recently there have been spectacular new breakthroughs. A new variety of wheat developed by Norman Borlaug, an achievement for which he was awarded a Nobel Prize, has increased harvests severalfold in parts of the world. A remarkable new type of potato has recently been developed at the International Potato Center in Lima, Peru which will grow to maturity in only 31 days and will flourish in hot, humid climates where the familiar varieties fail. According to Richard Sawyer, the developer, it will offer an alternative to the less nutritious cassava root because it has a higher protein content. Potatoes are already the world's fourth largest food crop and are expected now to move into third place.

These achievements of such great importance to the human race have been attained by the standard genetic methods of hybridization, the production and search for desirable mutations, and selective breeding. On the near horizon spectacular new methods are appearing. We have already talked about the potential benefits from working with haploid plants and with parasexual hybridization (see Chapter 7). A radical new answer to the urgent problem of adequate nitrogen for food crops seems possible when techniques, now under development, for introducing specific new genes into plants make it possible to "graft" the genes for nitrogen fixation into corn and other food plants. This is especially important because nitrogen fertilizer, if manufactured chemically, requires hydrogen which is obtained as a by-product of expensive petroleum.

An array of plant hormones and other chemical growth regulators is now available to improve yields. For example, TIBA (2,3,5-triiodobenzoic acid) sprayed on soybean or alfalfa leaves in very small amounts (300 gm per hectare; about 4 oz per acre!), will increase the yield by as much as 60 per cent! Important new biological methods of controlling plant-eating insects are coming into use. Pheromones, insect sex attractants, now lure males into death traps. Irradiated, and therefore sterile, male insects are raised by the millions and released. Vast numbers of females, which mate but once, then experience sterile matings and hence leave no offspring. For many decades there has been a running battle between researchers at agricultural stations and disease-causing fungi. In spite of success in breeding disease-resistant wheat, disease-producing fungi continue to mutate to forms able to penetrate plant defenses. Much the same has been happening with chemical pesticides. Who can doubt that the chemical industries will continue to produce new pesticides? But better yet, the biological methods just cited are very effective, as is the use of viruses, microbes, and crop rotation. When the highly destructive western corn rootworm developed almost total resistance to the major chemicals used against it, workers at the Illinois Agricultural Experiment Station found that the pest could be controlled by alternating corn with the also highly profitable soybean on which the corn rootworm could not live. New ways are being found to grow more than one crop a year on the same ground and new methods of farming which require little or no cultivation have been very successful under certain conditions. Such methods save energy and reduce erosion.

Probably future balance of population and food will lie somewhere between the predictions of Ehrlich and Revelle. Unfortunately the Green Revolution has done little, and in some places nothing, to reduce malnutrition and hunger in such densely populated countries as India, Pakistan, and the Phillipines. The statistics are frightening. On a *per capita* basis grain production in India has actually fallen below levels attained in the early 1960's, even though the total production had been increased from 68.5 to 92.8 million metric tons by the early 1970's. In 1960–1961, 68.5 metric tons equalled 0.1556 metric tons per person but, due to population increase, in 1970–1971 a 35 per cent larger harvest amounted to only 0.1473 metric tons per individual. For several years after the introduction of Green Revolution varieties, the Phillipines exported grains but, with an increase in population, is now again an importer. Clearly the consequences of uncontrolled population growth are inexorable and can never be permanently counteracted by miracle grains. Moreover, these new crop plants usually require modern agricultural methods, fertilizers, pesticides, herbicides, equipment, and training that are not readily available to the farmer on a small peasant holding.

Obviously enough, the problem of world food supply is not only a scientific problem but, to an even greater extent, a complex social, economic, and political problem.

USEFUL REFERENCES

Best, C. H., and N. G. Taylor. 1968. *The Living Body*, 5th ed. Holt, Rinehart, and Winston, New York. (None better.)

Brazier, M. A. B., ed. 1975. *Growth and Development of the Brain: Nutritional, Genetic, and Environmental Factors*. Raven Press, New York.

Brown, L. R. 1975. The world food prospect. *Science* 190:1053–1099.

Consumer & Food Economics Institute, Agricultural Research Service. 1974. *Family Fare: A Guide to Good Nutrition*. U.S. Department of Agriculture, Washington, D.C. (Good common sense.)

Cooper, J. P., ed. 1975. *Photosynthesis and Productivity in Different Environments*. Cambridge University Press, New York and London.

Guthrie, H. A. 1975. *Introductory Nutrition*, 3rd ed. C. V. Mosby Co., St. Louis. (Complete and authoritative.)

Hotzel, D., and R. H. Barnes. 1966. Contributions of the intestinal microflora to the nutrition of the host. *In* R. S. Harris, ed. *Vitamins and Hormones*, vol. 24. Academic Press, New York. (Some important and often overlooked facts.)

Martin, E. A. 1971. *Nutrition in Action*, 3rd ed. Holt, Rinehart, & Winston, New York. (Excellent and highly readable.)

Olson, R. E., ed. 1975. *Protein-Calorie Malnutrition*. Academic Press, New York. (Kwashiorkor, marasmus, and mental damage.)

Pirie, N. W., ed. 1975. *Food Protein Sources*. Cambridge University Press, New York and London. (Everything from algae to wild herbivores.)

Underwood, E. J. 1971. *Trace Elements in Human and Animal Nutrition*. Academic Press, New York. (From Australia with some surprises.)

Wittwer, S. H. 1974. Maximum production capacity of food crops. *BioScience* 24:216–224. (A mine of information.)

14

Circulation in Animals

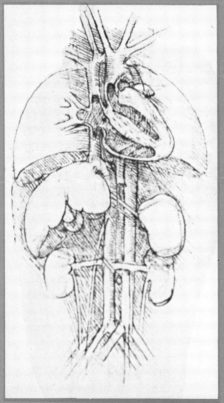

Heart and blood vessels with lungs, stomach, spleen and kidneys. [Reproduced by gracious permission of Her Majesty the Queen from Clark, Sir K. 1969. *The Drawings of Leonardo da Vinci in the Collection of Her Majesty the Queen at Windsor Castle.* Phaidon Press, Ltd., London. Copyright reserved.]

A vascular system to carry oxygen and food to all parts of the body and to take away carbon dioxide and other wastes is a clear necessity for any animal much larger than a flea. Furthermore, a sound knowledge of the vascular system is a prerequisite for understanding almost any other aspect of animal physiology. It was no accident that no one knew how the human body functions until Harvey demonstrated convincingly that the blood is pumped by the heart via the arteries to the body and returns to the heart via the veins. Before his time everyone thought that the blood merely surged back and forth in the veins and that the arteries contained air.

Harvey's Discovery of Circulation

After graduating from Cambridge University, William Harvey (age 19) left for northern Italy, then the world center of scientific learning. There he made one of the great landmark discoveries in the growth of human knowledge. Like his contemporary, Newton, and like Darwin he combined the work of his predecessors with original observations of his own to achieve a new synthesis. At Padua, Harvey became familiar with the work of Vesalius who had published, a generation earlier, the first accurate account of human anatomy based on actual dissection. He also learned that there are flaplike valves in the veins which had recently been discovered by his own teacher, the famous Hieronymus Fabricius.

Possibly the most crucial contribution to 237

Harvey's thinking was the argument of Michael Servetus that there is a pulmonary circulation in which blood passes from the heart to the lungs via the pulmonary arteries and returns to the heart via the pulmonary veins. Servetus had been burned at the stake before Harvey arrived in Padua for his political and religious beliefs, and all his books consigned to the flames. But it is difficult to burn all copies of a book. Matheus Columbus (a professor of anatomy and no relative of Christopher) stole the ideas of Servetus word for word. Reprehensible as this was, it nevertheless made the concept of pulmonary circulation available to Harvey.

Harvey marshaled his argument along three main lines. The first was anatomical. He noted the correspondence in size between the arteries and veins of any organ. That was not proof, but did constitute a necessary condition for circulation. He showed that the valves in the veins prevented blood from flowing in any direction except toward the heart. He also showed that the valves, at the place where the arteries leave the heart, permit blood to leave but not to enter the heart. As

seen in Harvey's own figure (Fig. 14-1), if the veins in the forearm are made to stand out by applying a tourniquet about the elbow, it can be demonstrated by squeezing blood along the vein with a finger that blood will flow in a direction toward the heart but not in the opposite direction.

Harvey's second line of reasoning was quantitative. He counted the number of times the heart of a dog beat per minute; then he killed the animal and measured both the amount of blood the heart forces into the arteries with each beat (by measuring ventricular capacity) and the total volume of blood in the animal. By simple arithmetic he then showed that in a dog or a sheep the heart pumps out many times this total volume every hour. In man, the heart pumps out 3–5 liters per minute. In other words, the volume of blood passing through the heart every minute is as great as the total volume of blood in the body. Where can all this blood pumped by the heart come from or go to, if it does not travel within a closed circuit?

Last, Harvey backed up these arguments by a wide array of experiments on dogs, snakes, snails, and even insects and shrimp. If the veins entering the heart are tied off, the heart becomes empty of blood. If the arteries leaving it are tied, the heart remains permanently gorged.

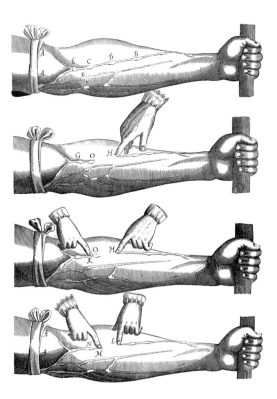

Fig. 14-1. Harvey's experiment to show that blood can be made to flow only toward the body in the veins. (From Leake, C. D. 1928. *Harvey's Exercitatio Anatomica: De Motu Cordis et Sanguinis in Animalibus.* courtesy of Charles C Thomas, Publisher, Springfield, Ill.)

Vascular Systems

Animals have evolved three major patterns of blood vessel arrangement. There is the annelid-arthropod pattern, seen in earthworms and lobsters, which Harvey studied in shrimp to supplement his observations on snakes and dogs; a molluscan system with some very remarkable features; and, third, the vertebrate system.

ANNELID-ARTHROPOD SYSTEM

In the annelid-arthropod system (Fig. 14-2), blood is pumped by a pulsating dorsal blood vessel located above the gut. In an earthworm the dorsal vessel squeezes blood toward the hearts in a series of waves of contraction. In the anterior part of the body are five pairs of sausage-shaped hearts which pump blood downward (ventrally) into a ventral blood vessel beneath the gut. From this, smaller arteries carry blood to fine capillaries ramifying throughout the body. These capillaries empty into veins which lead back to the dorsal blood vessel. This is a **closed circulatory system** such as is found in man. In a newly hatched

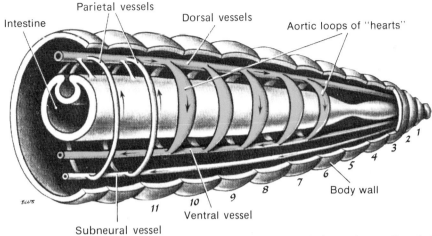

Intestine
Parietal vessels
Dorsal vessels
Aortic loops of "hearts"

2 1
3
4
5
6
7
8
9
10
11

Body wall
Ventral vessel
Subneural vessel

Fig. 14-2. The basic annelid-arthropod vascular system as seen in the anterior end of an earthworm. The arterial portion is colored. (From Moment, G. B. 1967. *General Zoology,* 2nd ed. Houghton Mifflin Co., Boston.)

earthworm the scarlet blood can easily be seen under a dissecting microscope as it follows the course just described. In crustaceans and insects the system is similar to the earthworm's, except for the oblong-shaped heart in place of the long pusating dorsal vessel, and the lack of capillaries. Arteries empty blood into tissue spaces, **sinuses,** from which it is carried back to the heart via veins. This is an **open system.**

VERTEBRATE SYSTEM

All vertebrate vascular systems are modifications of a basic closed system pattern clearly recognizable in fish and salamanders and in the embryos of birds and men. In discussing vertebrate anatomy it is very useful to understand the terms ventral and dorsal. **Ventral** means toward the belly side. **Dorsal** means toward the back side. The backbone is always dorsal to the gut and heart. The gut is always ventral to the backbone, whether the animal is standing on four feet, when the gut is below the backbone, or lying upside down on a dissecting table, when the gut is above the backbone. In human anatomy ventral is often called anterior and dorsal called posterior for the obvious reason that we stand on our hind legs. The heart is ventral to the pharynx or throat in fish and the embryos of higher vertebrates. In adult reptiles, birds, and mammals it has moved posteriorly during development to occupy the familiar position in the chest. This is why the heart is innervated by the **vagus nerve** which comes directly from the brain, enters the heart while it is under the pharynx during embryonic development and then is pulled down into the chest with the heart.

From the vertebrate heart a ventral artery runs anteriorly below the pharynx. In fish and in embryos of the higher forms, six pairs of **aortic arches** curve up on either side of the pharynx between the gill slits and join to form the **dorsal aorta** above the pharynx (Fig. 14-3). The dorsal aorta is the main blood vessel of the body and runs to the tail just ventral to the backbone and dorsal to the gut. Branching off are a pair of **carotid arteries** to the head, **subclavian arteries** to the front legs or arms, **renal arteries** to the kidneys, a **coeliac** and other unpaired arteries to the intestine and other viscera, a pair of **iliacs** to the hind legs.

The arteries divide into smaller and smaller vessels and finally into thin-walled **capillaries.** It is only through the capillary walls that exchanges between the blood and the tissue cells take place. Food is absorbed from the intestine and distributed to other parts of the body, oxygen is obtained in the lungs, and carbon dioxide given off. As soon as microscopes were invented people began looking for the capillaries. Leeuwenhoek observed the comb of a young rooster, a rabbit's ear, and even a bat's wing. When he finally used a tadpole's tail he wrote:

A sight presented itself more delightful than any mine eyes had ever beheld; for here I discovered more than 50 circulations in different places. . . . For I saw not only that in many places the blood was conveyed through exceedingly minute vessels, from the middle of the tail toward the edges, but that each of the vessels had a curve or turning, and carried the blood back toward the middle of the tail, in order to be again conveyed to the heart.

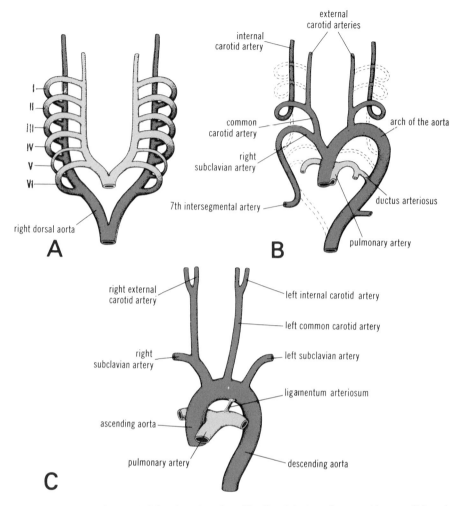

Fig. 14-3. Diagram of the vertebrate arterial system. A, as found in all vertebrate embryos and in some fish and amphibian adults. B, as seen in some amphibians and in mammalian embryos. C, as seen in an adult mammal. By international convention in anatomy, these figures are drawn as though the animal were being dissected lying on its back. Therefore, the dorsal aorta is represented as beneath the ventral. Light color represents oxygen-poor blood. (From Langman, J. 1975. *Medical Embryology*, 3rd ed. The Williams & Wilkins Co., Baltimore.)

Blood is returned to the heart in vertebrates by a system of veins, the **postcaval vein** from the abdomen being the largest. In the lower vertebrates there are two portal systems: one, the **renal portal veins** which take blood from the hind legs or fins to the kidney, and another, the **hepatic portal veins** which carry blood from the intestine to the liver. In mammals like ourselves only the hepatic portal system is present. A portal vein is one which carries blood to any structure other than the heart.

Vertebrate Heart

The vertebrate heart always begins in the embryo as a straight, pulsating tube squeez-ing blood anteriorly. In fish it remains a straight tube consisting of a linear series of three single chambers, viz. a thin-walled **sinus venosus** which receives blood from all of the body, an **atrium** which receives the blood from the sinus, and a muscular **ventricle** which takes the blood from the atrium and pumps it out into the **ventral aorta** and up through the pairs of aortic arches into the dorsal aorta.

In mammals the embryonic tubular heart bends in the course of development into a U-shaped structure so that the arteries and veins all leave or enter the heart at the same end. At the same time the interior of the heart becomes divided into right and left halves, making two atriums and two ventricles. The sinus venosus becomes reduced to a small area called the **pacemaker** on the wall of the right atrium, so called because it initiates the heartbeat. Oxygen-poor blood from the body enters the right atrium, passes into the right ventricle, and thence to the lungs via the pulmonary arteries. Blood returning from the lungs, as doomed Michael Servetus noted so long ago, enters the left atrium, goes into the left ventricle, and thence out into the dorsal aorta. The blood passes through the heart twice to complete one circuit, but the time required is very short. In a rabbit it takes only 7 or 8 seconds; in a human only a bit over ¹/₄ minute.

Because of its great medical importance and intrinsic interest, the study of the heart (**cardiology**) has produced an enormous body of knowledge. We will mention only the most significant aspects. Heart muscle is striated and composed of single cells like skeletal muscle but the fibers form a network

rather than all being parallel. Physiologically, cardiac muscle exhibits four outstanding properties.

First, it is inherently **rhythmic.** Isolated heart cells from an embryo beat when grown in vitro and even bits of adult heart muscle isolated from nerves will beat. Second, after each contraction, heart muscle has an extremely long refractory period during which it cannot be stimulated to contract. This ensures that **diastole** (dilatation) and filling of the heart will occur. It is impossible to throw the heart into tetanus, or cramp, which happens easily to skeletal muscle.

A third characteristic is the way the heart automatically adjusts the strength of its contraction to the amount of work it has to do. If it is filled with a large volume of blood, as in violent exercise, its walls are stretched and the contractions become correspondingly

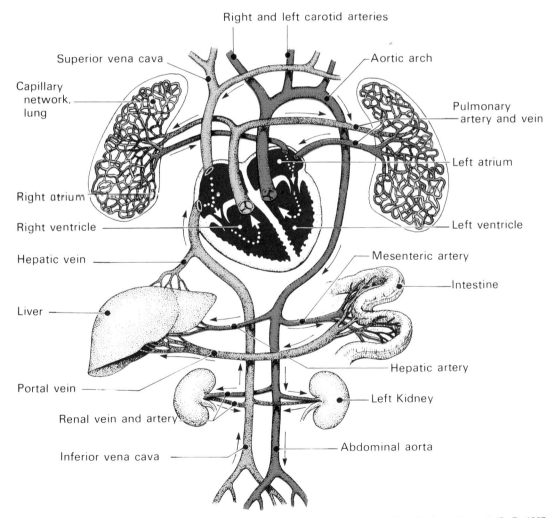

Fig. 14-4. The mammalian vascular system. Lighter color represents oxygen-poor blood. (From Moment, G. B. 1967. *General Zoology,* 2nd ed. Houghton Mifflin Co., Boston.)

more powerful. This is Starling's "law of the heart."

Last, heart muscle requires a proper salt balance to beat rhythmically. Sodium, potassium, and calcium must all be present in the fluid surrounding the organ. The ratio of these necessary ions is closely similar to that in sea water. The molecular mechanisms of the effects of salts are still unclear, but in general, high concentrations of calcium or sodium stop the heart in **systole** (contraction); high concentrations of potassium stop the vertebrate heart in diastole. These discoveries were originally made on the frog heart by Sidney Ringer and led directly to all the physiological salt solutions now in use for intravenous administration and many other procedures.

The rate of heartbeat in adults is under nervous control. The heart can be completely stopped by stimulation of the vagus nerve. The heart also receives accelerator nerves from the sympathetic system. One of the most dramatic indications of neural control of heart rate is the slowing i.e., **brachycardia,** which occurs when a seal dives or even when its nose is held under water. The onset is prompt, so fatigue cannot be involved. A seal's heartbeat falls from 80 to 7 or 8 beats per minute. Similar brachycardia has been found in every diving animal studied, including ducks, crocodiles, porpoises, hippopotami, and men. Try holding your face under water while a friend takes your pulse. The adaptive value of diving brachycardia is obscure, but a brachycardia also occurs during hibernation. In the light of these facts, it is not surprising the heart rate can be controlled by conditioned reflexes in dogs and humans.

Lymphatics

The fluid directly bathing the cells is not whole blood, which is confined within capillaries, but **lymph.** The composition of lymph is essentially that of blood minus the red and white cells. The lymph vessels resemble capillaries and veins, possessing valves which permit lymph to flow only toward the heart. The flow is one-way and empties via the great veins in the chest into the right side of the heart. Lymph is continually replenished by seepage from the capillaries.

Where the legs join the body, in the armpits and neck, and at several other sites, the lymph passes through glandular **lymph nodes.** The lymph nodes are the site of pro-

duction, during adult life, of **lymphocytes,** the white blood cells which produce antibodies against foreign proteins, invading bacteria, etc. Lymph nodes also "filter" out foreign material.

Blood

Blood consists of cells and a fluid, the **plasma.** In mammals, red cells, or **erythrocytes,** lack nuclei, except during their developmental stages which occur chiefly within the marrow of the long bones and sternum.

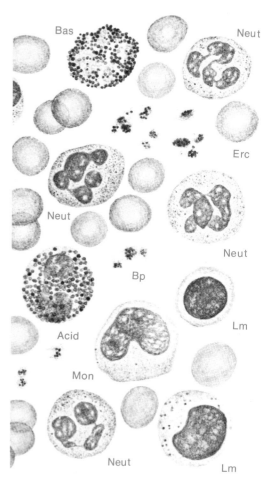

Fig. 14-5. Cells from normal human blood lightly centrifuged to concentrate the leukocytes and stained with the commonly used Wright's stain. Acid, acidophil; Bas, basophil; Bp, blood platelet; Erc, erythrocyte; Lm, lymphocyte; Mon, monocyte; Neut, neutrophil. (From Copenhaver, W. M., R. P. Burge, and M. B. Burge. 1971. *Bailey's Textbook of Histology,* 16th ed. The Williams & Wilkins Co., Baltimore.)

The white cells are called **leukocytes.** About 75 per cent have a granular cytoplasm and a very irregular, or polymorphic, nucleus. These are the **polymorphonuclear leukocytes,** popularly called polymorphs. About 70 per cent of all leukocytes are polymorph **neutrophils,** so called because their cytoplasm stains in most types of dyes. Polymorph **acidophils** (or eosinophils), which stain with acid dyes, constitute about 4 per cent, and the polymorph **basophils** about 1 per cent or less. The polymorphs ingest bacteria and cell debris. The remaining approximately 25 per cent of the leukocytes are mostly either small **lymphocytes** or large **monocytes.** Monocytes easily become phagocytic, engulfing foreign particles. They probably are the same cells, known since the time of Metchnikoff about a century ago as macrophages, which play an important role in the first steps of immune reactions. The lymphocytes (plasma cells) manufacture the antibodies. The absolute and relative abundance of various leukocytes is a useful diagnostic tool in medicine. In addition, there are small particles called blood **platelets** which are essential for the formation of blood clots.

Oxygen-Carbon Dioxide Transport

Water will carry some dissolved oxygen. When exposed to ordinary air, which is approximately 21 per cent oxygen, water holds only about 0.6 ml of oxygen per 100 ml of water at 30°C. The red hemoglobin of some annelids will make it possible for 100 ml of blood to hold about 10 ml of oxygen, the hemoglobins of fish about 15 ml, and of birds and mammals up to 20 ml, or even 30 ml of oxygen in the case of porpoises, which are adapted for lengthy dives. The significance of hemoglobin is that it combines easily with oxygen when there is a high oxygen pressure, becoming scarlet oxyhemoglobin, and gives it up as readily when the oxygen pressure (concentration) falls as it does in the tissues of the body some distance from gills, lungs, or moist skin. Hemoglobin is thus a very special kind of compound.

Deep within the body, carbon dioxide (more accurately carbonic acid) produces acid conditions which facilitate the breakdown of oxyhemoglobin into oxygen and hemoglobin. The H_2CO_3, which is more acidic than hemoglobin, takes a potassium ion away from the hemoglobin and becomes potas-

sium bicarbonate, $KHCO_3$. Once back in the lungs, the hemoglobin gains oxygen, becoming more acidic and therefore able to recapture the potassium. This means that potassium bicarbonate is reconverted to carbonic acid which, under the influence of the enzyme carbonic anhydrase in the red cells, breaks down into H_2O and CO_2 which is lost to the surrounding atmosphere or water.

$$CO_2 + H_2O \; \underset{}{\overset{\textit{anhydrase carbonic}}{\rightleftharpoons}} \; H_2CO_3$$

Oxygen-carrying pigments present many intriguing puzzles. **Hemoglobin** is found not only in vertebrates but in earthworms, many marine annelids, water fleas among crustaceans, some pond snails, *Ascaris* (a common intestinal parasitic worm), and even certain ciliates. Chemically hemoglobin is a double molecule consisting of heme united with a protein, globin. Heme is a close chemical relative of chlorophyll, of cytochrome, and of vitamin B_{12} (cyanocobalamin). It is a typical porphyrin having iron in the center where chlorophyll has a magnesium atom.

The other most widespread respiratory pigment is **hemocyanin,** which is a copper-containing compound lacking porphyrin. It is a light French blue when oxygenated, colorless when reduced. This is the pigment in the blood of lobsters, *Limulus,* the horseshoe "crab," some spiders, the octopus, and various other animals.

Coagulation

The coagulation of the blood appears simple, yet careful investigation has shown it to be one of the most complex of all known biological processes. At least 30 factors may play some role. From the point of view of the organism, the problem is that unless the blood clots promptly, death by hemorrhage might follow even a minor scratch. On the other hand, if the clot blocking a cut should extend into the arteries or veins and cause even a small fraction of the blood within the body to clot, death would also result.

From the welter of observations and theories the following facts stand out. The jellylike clot or **thrombus** is composed of a fibrous protein, **fibrin,** which is present in normal plasma in an unpolymerized form called **fibrinogen.** In order for fibrinogen to polymerize into fibrin, calcium ions must be present. This is why citric acid, which combines with calcium to form calcium citrate, can prevent clotting. What produces the conversion of fibrinogen to fibrin? This is done by another blood protein, the enzyme **thrombase** sometimes called simply **thrombin.**

USEFUL REFERENCES

Adolph, E. F. 1967. The heart's pacemaker. *Sci. Am.*, 216 (March): 32–37

Guyton, A. C. 1963. *Circulatory Physiology: Cardiac Output and Its Regulation*. W. B. Saunders Co., Philadelphia.

Ingraham, V. M. 1963. *The Hemoglobins in Genetics and Evolution*. Columbia University Press, New York.

Spain, D. M. 1966. Atherosclerosis. *Sci. Am.*, 215 (August): 48–56.

Vander, A. J., J. H. Sherman, and D. S. Luciano. 1970. *Human Physiology*. McGraw-Hill Book Co., New York.

Vroman, L. 1968. *Blood*. The Natural History Press, Garden City, N.Y. (Easy reading and highly informative paperback.)

15

Transport of Water and Solutes in Plants

Life arose in the sea so it is not surprising that all living things, whether plants or animals are dependent on water for survival. Plants share with animals many of the following reasons for needing water: (1) Many constitutents of protoplasm, including carbohydrates, proteins, and nucleic acids, are hydrated. Their physical and chemical properties change if water is removed. Ordinarily protoplasm contains up to 95 per cent water. Dehydration usually results in the death of cells. (2) Water molecules are essential components of many of the chemical reactions that occur in protoplasm. Cleavage of polymers such as carbohydrates and proteins by addition of water (**hydrolysis**), adding together molecular subunits by splitting out water (**condensation**), utilization of water as a substrate for photosynthesis, and the formation of water from molecular oxygen, electrons, and protons in the terminal oxidation step of respiration are but a few examples of the involvement of water in cellular metabolism. This unique liquid also serves as the medium within which many chemical reactions occur. (3) Water is a solvent, carrying in dissolved form all of the inorganic salts needed to sustain plants. (4) Water also has a universal role in maintaining the shape and rigidity of plants, particularly structures such as leaves, flowers, and fruits that lack woody supporting tissues. Most plant cells have large central vacuoles filled with water plus some dissolved chemicals. These relatively massive bags of fluid within the thin layers of protoplasm are responsible for maintaining the turgor of cells, tissues, and whole organs. The phenomenon of **wilting** is no more than a loss of turgor due to excessive loss of water from the vacuoles of plant cells. (5) Water is essential not only to the proper internal functioning of plant cells, but also must be present in thin films covering cell surfaces. Water molecules even fill the spaces between cellulose microfibrils in the cell walls. Extracellular water in plants provides a medium for transferring dissolved substances from the environment into cells (gases from the atmosphere as well as minerals from the soil) and from cell to cell.

The Cellular Basis for Water Movement

Plant cells are enclosed in cellulose walls, very porous boxes that offer practically no resistance to the movement of water molecules or solutes. The cellulose walls of plant cells provide a physical limit to the expansion of the protoplasm and vacuole. Control over what substances enter and leave plant cells rests in the membranes on the outer and inner surfaces of the protoplasm, the **plasma membrane (plasmalemma)** and **vacuolar membrane (tonoplast)** (Fig. 15-1). These membranes are **selectively permeable;** that is, they permit practically unhindered passage of some molecules such as water, are impassable to others such as proteins, and permit intermediate degrees of movement of other molecules such as dissolved sugars and salts. The advantages of selective permeability are obvious. If a cell must import essential substrates from its environment, it

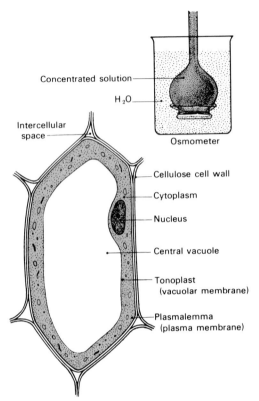

Concentrated solution

H_2O

Intercellular
space

Osmometer

Cellulose cell wall

Cytoplasm

Nucleus

Central vacuole

Tonoplast
(vacuolar membrane)

Plasmalemma
(plasma membrane)

Fig. 15-1. The semipermeable membranes of plant cells, the plasma membrane and the vacuolar membrane, control the movement of substances into and out of cells. These membranes obey the same physical laws as a simple thistle tube osmommeter. (From Moment, G. B., and H. M. Habermann. 1973. *Biology: A Full Spectrum.* The Williams & Wilkins Co., Baltimore.)

joined end to end, provide a means for movement of water and solutes from the roots to the top of the shoot. Xylem elements are structurally and functionally analogous to the plumbing system in our homes. Within both systems there is a mass flow of liquid under hydrostatic pressure gradients (Fig. 15-2).

Another phenomenon involved in the transfer of water and solutes is **osmosis.** When a concentrated solution is separated from pure solvent by a semipermeable membrane, i.e. one that is permeable to the solvent but not to the solute molecules, there is a net tendency for solvent molecules to diffuse into the concentrated solution. The extent to which solvent molecules move into the solution is proportional to the concentration of solute and to the absolute temperature. The **osmotic pressure** of the internal solution can be expressed either in terms of concentration (molarity of solute) or as pressure (atmospheres). It turns out that the potential osmotic pressure of a solution inside a semipermeable membrane follows the same physical laws as do gases. In other words, the osmotic potential of a solution is dependent on the total concentration of dissolved molecules and ions. A 1 M solution of sugar (sucrose) at 0°C has an osmotic pressure of 22.4 atm. A 1 M solution of salt (NaCl) under the same conditions would have twice this osmotic pressure value because of its complete dissociation into Na^+ and Cl^- ions.

Each plant cell is, in effect, a tiny osmometer. When placed in pure water, cells take up water molecules due to the osmotic potential of substances dissolved in their vacuoles and protoplasm. Uptake is accompanied by a swelling of cell contents until they press against the cell walls, i.e. become **turgid.** In a fully turgid cell the counterpressure exerted by the cell wall on its contents just balances the tendency for water molecules to enter the cell in response to the osmotic pressure of the vacuole and cytoplasm. If the external environment of the cell has a higher osmotic potential than its contents, water molecules diffuse outward at a rate faster than they enter. Consequently the contents of the cell decrease in volume. If these circumstances persist, the vacuole and protoplasm contract away from the cell wall. The cell becomes **flaccid** and finally **plasmolyzed** (Fig. 15-3).

The phenomenon of **plasmolysis** provides one of several possible means of determining the osmotic concentration of the contents of cells. Individual cells or thin strips of tissue placed in a series of solutions of differing molarities can be observed under the microscope. That concentration in which cells are

must provide for the entry of these materials. On the other hand, a cell must prevent the loss of its essential structural components and nutrients.

Several kinds of mechanisms are involved in the entry, movement, and loss of water and solutes such as sugar and minerals in plants. Both the uptake of water by the root and the loss of water vapor by evaporation from the leaves follow the principles of **diffusion.** Diffusion is nothing more than a random movement of molecules from a region of higher to one of lower concentration. This is probably the only physical mechanism involved in the movement of water molecules from the moist intercellular spaces of stems and leaves outward through openings in the surfaces of these structures, the lenticels and the stomates.

The structural components of a plant's vascular, i.e. conducting, tissues, the **xylem** and **phloem**, are long and tubular. The xylem elements, with their cylindrical vessel cells

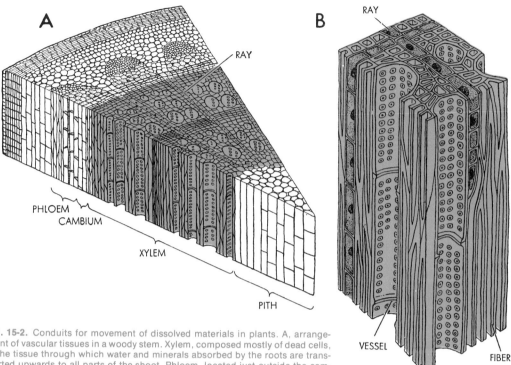

Fig. 15-2. Conduits for movement of dissolved materials in plants. A, arrangement of vascular tissues in a woody stem. Xylem, composed mostly of dead cells, is the tissue through which water and minerals absorbed by the roots are transported upwards to all parts of the shoot. Phloem, located just outside the cambium, is the route of transport for a variety of dissolved substances (including carbohydrates and minerals) to all parts of a plant. B, enlarged section of xylem showing arrangement of vessels, fibers, and ray parenchyma cells. (From Biddulph, S., and O. Biddulph. 1959. The circulatory system of plants. *Sci. Am.* 200:44–49. Copyright © 1959. Scientific American, Inc. All rights reserved.)

at the state of incipient plasmolysis is the one having an osmotic concentration equivalent to that of the cell contents. An alternate method involves the use of uniform pieces of tissue (potato tubers are favored materials for student laboratories because they contain a single type of thin-walled parenchyma cell). Weighed slices of tissue are placed in a series of solutions of different concentrations and are left to equilibrate for several hours. They are then blotted dry and reweighed. The solution causing neither gain nor loss of water (and weight) is equivalent in osmotic concentration to that of the cell contents in the experimental tissue.

Transpiration: The Mass Flow of Water through Plants

The discovery of blood circulation in animals by William Harvey in 1628 provided an impetus in the search for a comparable circulatory system in plants. The first major publication dealing with the possible circulation of sap and related aspects of plant metabolism was the book *Vegetable Staticks* published in 1726 by Stephen Hales, who concluded that there is no circulation of sap in plants comparable to the circulation of blood in animals. Instead the movement of fluids in plants has been found to be like the action of a wick: massive amounts of water are drawn from the soil solution into the roots, pulled upward through the stems, and finally move out through the stomates of leaves and the lenticels of the stems as water vapor. This unidirectional mass flow of water in plants with a continual loss from aerial parts is called **transpiration.**

Transpirational water loss can be measured very simply by using methods first introduced by Stephen Hales. All one has to do is follow the loss of weight of an entire potted plant or of a shoot or even a single leaf by weighing at regular intervals. The only necessary precaution in the case of a potted plant is to cover the pot so that only the water loss from the shoot is measured. When working with shoots or single leaves, the cut end of a shoot or the petiole of a leaf is immersed in a reservoir of water so that the excised piece of plant

A. 4% solution (turgid)

B. 6% solution

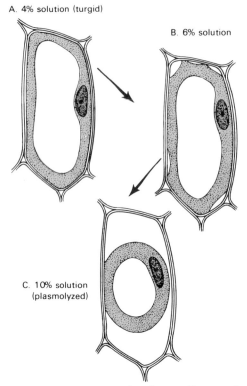

C. 10% solution
(plasmolyzed)

Fig. 15-3. Plant cells respond to the osmotic concentration of their environment. A, in water or a hypotonic solution of sugar a plant cell is turgid. B, in a solution of osmotic concentration equal to that of its contents a cell is in the state of incipient plasmolysis. C, in a hypertonic solution there is a net loss of water from the cell, the contents contract away from the cell wall, and the cell becomes plasmolyzed. (From Moment, G. B., and H. M. Habermann. 1973. *Biology: A Full Spectrum.* The Williams & Wilkins Co., Baltimore.)

does not dry out. In the latter case it is the water loss from the closed reservoir that is measured as water is drawn from it to replace that lost in transpiration.

As soon as scientists began to measure transpiration rates, they discovered that rates vary not only with the species of plant used in the experiment but also with a number of environmental conditions and even the time of day. Obviously, there must be some mechanism by which the rate of water loss from the leaves may be regulated. If we look at the surfaces of leaves, we find that most of the epidermal cells look much like pieces of a jigsaw puzzle, forming a tightly interlocking covering. Dispersed in the layer of epidermal cells are pores called **stomates**, openings between two kidney-shaped **guard cells**. The entire complex of opening plus surrounding cells is called the **stomatal apparatus**. When the guard cells are turgid, their inner surfaces pull apart and the pore or **stoma** is open.

Conversely, when the guard cells become flaccid, the surfaces forming the edge of the stoma collapse and the stoma is closed (Fig. 15-4). Stomates regulate the movement of gases between the intercellular spaces of the leaf and the atmosphere. Closed, they drastically reduce the exchange of gases and loss of water vapor from leaves. With intermediate degrees of opening, the rate of gas exchange is proportional to the extent of opening. There is a diurnal pattern of stomatal opening and closing that corresponds closely to diurnal changes in rates of transpiration, strongly

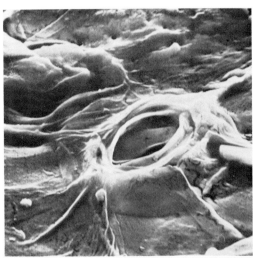

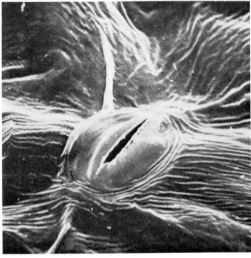

Fig. 15-4. Stomatal apparatus of *Vicia faba* (broad bean) as seen in the scanning electron microscope, × 1,800. Top, after a period of illumination guard cells are turgid and the stoma is open. Bottom, guard cells are flaccid and the stoma is closed. This occurs in darkness or in this case after incubation in $Al_2(SO_4)_3$ solution. (From Schnabl, H., and H. Ziegler. 1974. Über die Wirkung von Aluminiumionen auf die Stomatabewegung von *Vicia faba*-Epidermen. *Z. Pflanzenphysiol.* 74:394–403.)

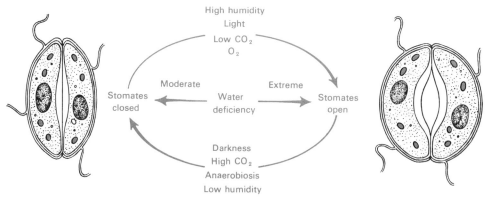

Fig. 15-5. Summary of stomatal responses to environmental factors. (From Moment, G. B., and H. M. Habermann. 1973. *Biology: A Full Spectrum*. The Williams & Wilkins Co., Baltimore.)

suggesting a cause and effect relationship. Assuming that the opening and closing of stomates is the means for regulating rates of transpiration, it appears that the crucial question is: What mechanism regulates the opening and closing of the stomates?

FACTORS CONTROLLING STOMATAL OPENING AND CLOSING

Stomatal opening and closing provides a mechanism not only for the regulation of water loss from the leaves but also the regulation of gas exchange between the internal tissues of the leaf and the atmosphere so important in photosynthesis.

Environmental factors such as high humidity, light, and low CO_2 tension promote opening while dry air, darkness, and high CO_2 tension promote stomatal closing (Fig. 15-5). Some studies have indicated that the action spectrum for light-induced stomatal opening corresponds closely to that of photosynthesis with maxima in the blue and the red portions of the spectrum. It was therefore concluded that photosynthesis is somehow involved, although the fact that this light-dependent response saturates at very low light intensities was quite mystifying. A further observation, that light-dependent stomatal opening requires oxygen, added to the mystery.

Recently some new observations have reopened the question of what mechanism regulates the light-dependent changes in permeability of guard cells. It has been shown in several laboratories that the increased turgor of guard cells in the light is accompanied by a rapid uptake of potassium ions. Other studies with an albino mutant sunflower have revealed that light-dependent stomatal opening can occur even in nonphotosynthetic albino leaves. Far red light promotes, while red light inhibits, opening of the mutant sunflower sto-

mates. Wild type sunflower stomates respond in the same way to red and far red light, but in addition exhibit a high intensity opening in blue light. It seems likely that the low intensity opening in far red light, which, appears to be phytochrome-mediated, may be responsible for permeability changes resulting in the flux of potassium ions into the guard cells.

There is a striking parallel between light-dependent stomatal opening and the sleep movements of leaves in leguminous plants such as *Mimosa pudica* (the sensitive plant) and *Albizzia julibrissen* (the mimosa tree). Leaflet orientation is controlled by the turgor of motor cells located in the upper and lower sides of the swelling (**pulvinus**) at the base of the petiole. Dark closing (or folding together) of leaflets is accompanied by a movement of potassium ions out of the ventral motor cells and into the dorsal motor cells. Pre-irradiation with red light is necessary for this movement of potassium. Opening of the leaflets is promoted by blue light (and also by far red light) and is accompanied by a reverse flow of potassium. Ruth Satter and Arthur Galston of Yale University, the scientists who have studied these leaflet movements, have suggested that "potassium flux is a general prerequisite for turgor changes in all species with moving leaves or leaflets." The same condition may also apply to turgor changes in the guard cells.

ADAPTATIONS THAT AFFECT TRANSPIRATION RATES

Plants exhibit a variety of adaptations that influence transpiration rates. Many such adaptations have to do with the spacing and placement of stomates or with the shape and the surface to volume ratio of leaves. Land plants are able to survive in a wide array of habitats ranging from very wet to very dry.

Often the amount of available water varies with the season. One way that plants manage to survive during parts of the year when water supplies are limited is to shed their leaves and then grow a new set of these appendages when conditions are again favorable. Winter months are usually a time of water deficit for terrestrial plants, not because of any lack of precipitation but rather because water in its solid state cannot be absorbed. Even in areas having very cold winters, there are plant species that retain their leaves, including most of the conifers, the rhododendrons, and hollies. If we look at the structure of the leaves of these plants we can understand why they can be retained. In the conifers, needle-like leaves are much reduced in surface area and the stomates are located not on the outer surface but rather in cavities below the surface (Fig. 15-6). Leaves of the conifers and of the evergreen rhododendrons and hollies have very thick waxy cuticles covering the epidermal cells. Also, there are fewer stomates per unit surface area than in species that lose their leaves at the end of the warm growing season. Thus, there are three aspects of leaf anatomy that affect transpiration rates: surface area, the number and location of stomates, and thickness of the cuticle.

The water problems of many desert species have been resolved by drastic reduction of leaves and development of fleshy, water-storing stems. Plants adapted to very moist or aquatic habitats do not have the problems of water conservation faced by species growing in deserts or even in areas of moderate supplies of soil moisture. Although the stomates of aquatic species are not particularly important for the regulation of water loss, they still serve an important function in facilitating the

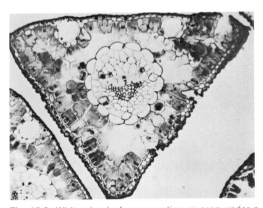

Fig. 15-6. White pine leaf cross section as seen under a microscope. Note reduced surface relative to volume, sunken stomates, and cutinized epidermal cells. (From Fisk, E. J., and W. F. Millington. 1959. *Atlas of Plant Morphology: Portfolio I—Photomicrographs of Root, Stem, and Leaf.* Burgess Publishing Co., Minneapolis, Minn.)

exchange of gaseous substrates and products of photosynthesis and respiration. Therefore, it is not surprising to find that the floating leaves of water lilies have stomates only on their upper surfaces. Many aquatic plants have internal tissues made up mostly of air spaces (the so-called **aerenchyma** tissues) that provide gaseous reserves and often help to assure buoyancy so that leaves will float.

Water Movement from the Roots to the Shoots

Vascular tissues, both xylem and phloem, provide a continuous conducting system from the root, through the stem and into the leaves where branching strands, the **veins**, extend to all parts of the leaf blade. There are several kinds of evidence that the xylem (the woody center of tree trunks) is involved in the transport of water. First, of the two kinds of vascular tissue in plants, xylem and phloem, only the xylem has sufficient volume to account for the massive amounts of water that must be moved. Second, it is possible to remove the bark of a tree (and thereby the phloem, a procedure usually referred to as **girdling**) without interfering appreciably with the upward transport of water. On the other hand, destruction of the xylem results in immediate cessation of upward movement of water accompanied by obvious wilting of the leaves. Third, water can be labeled, either by means of hydrogen or oxygen isotopes or by means of soluble dyes. The label can be seen to rise in the xylem but not in the phloem. (This can be neatly demonstrated in a stalk of celery.) Finally, the contents of xylem cells are dilute and watery compared to phloem cells. Thus, the major function of xylem is the transport of the massive amount of water needed to replace that lost in transpiration.

Phloem, on the other hand, is involved in the transport of solutes; its major function is the distribution of the excess products of photosynthesis to all parts of the plant. Evidence of phloem transport will be presented later when the translocation of sugars and other metabolites is discussed.

Water enters plants through the **root hairs** (Fig. 15-7). These extensions of individual epidermal cells of the root provide a large surface in close contact with the films of water that surround soil particles. Water molecules entering the root through the root hairs and other surface cells are transferred through the cortex to the endodermis. Walls of most

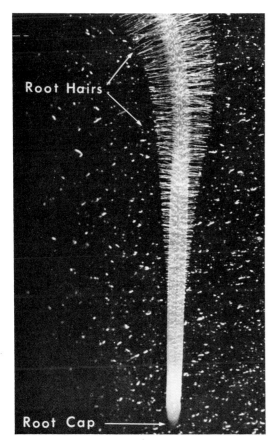

Fig. 15-7. Roots hairs are the site of entry of water and dissolved minerals from the soil into a plant. (From O'Brien, T. P., and M. E. McCully. 1969. *Plant Structure and Development*. The Macmillan Co., Collier-Macmillan, Ltd., London.)

cells in the endodermis are **suberized**, i.e. they contain a waterproof corky material and present a barrier to the movement of water. Some cells of the endodermis, called passage cells, remain thin-walled and are thought to provide a pathway for free movement of water and solutes to the central core of xylem tissue.

Once in the xylem, water is carried upward through nonliving cells, **vessels** and **tracheids**. These tubular cells are arranged one on top of another. The end walls of vessel elements usually disintegrate at maturity so that there is no barrier to water movement. Although tracheid end walls remain intact, they have tapered ends with many thinner areas called **pits**. Pits, also found on the side walls of tracheids and vessels, facilitate the passage of water from cell to cell.

ROOT PRESSURE AND GUTTATION

The crucial question which we must now attempt to answer is: What forces propel water upward through the rather remarkable "plumbing" system just described? We could sum up the state of our ignorance by simply saying that movement of water is a consequence of **root pressure** and **transpirational pull** exerted by the leaves.

After their shoots are cut off, the stumps of plants often exude liquid from their cut stem surfaces. The amount of such exudates can be measured and is often surprisingly large: in 1936 Crafts collected 551.7 ml (well over $\frac{1}{2}$ quart) of exudate from a 450-gm (about 1 lb) squash root in 24 hours. If a tube is connected to the cut stem, liquid will rise in the tube to amazing heights. In 1726, Stephen Hales reported in *Vegetable Staticks* that liquid rose to a height of 21 ft in a tube connected to the stem of a cut vine. Furthermore, Hales observed that the liquid rose more rapidly during the day than during the night.

Root pressure is related to the metabolic activity of roots and is correlated with an accumulation of dissolved substances in the xylem cells. Root pressure decreases when temperatures are lowered, when any of the essential inorganic nutrients are lacking, or if certain metabolic poisons are present. This would not be true of a purely physical process. It is probably correct to conclude that root pressure is a consequence of the active accumulation of nutrients from the soil solution at the expense of energy made available through respiration. A positive pressure on contents of the xylem can occur not only in severed stems but also in intact plants; it is responsible for the phenomenon of **guttation**, or the exudation of liquid water from leaves (Fig. 15-8). Water is expelled either through the stomates or from special structures called **hydathodes** at the edges of leaves. Guttation usually occurs at night when water loss through transpiration is reduced. It can readily be seen in the laboratory or classroom

Fig. 15-8. Strawberry leaf with drops of water exuded by guttation. (Photograph by J. A. Herrick; from Greulach, V. A. and J. E. Adams. 1967. *Plants: An Introduction to Modern Botany*. John Wiley & Sons, Inc., New York.)

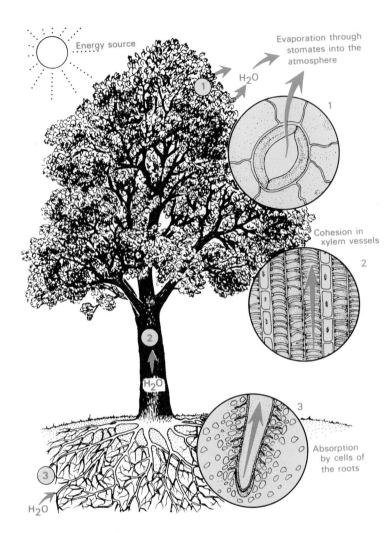

Fig. 15-9. The cohesion-tension theory. Evaporation of water from leaves (1) creates a water deficit. To replace water lost in transpiration, water is withdrawn from the conducting tissues. This results in a tension which is transmitted downward through the xylem (2). The "pull" on water in the xylem is a result of cohesive forces holding water molecules together and draws water into the roots (3). A mass flow of water thus moves up the plant to replace water lost in transpiration. (From Moment, G. B., and H. M. Habermann. 1973. *Biology: A Full Spectrum*. The Williams & Wilkins Co., Baltimore.)

when seedlings of corn or oats are placed under a bell jar.

Although root pressure is probably a contributing factor in the ascent of sap in plants and certainly is involved in the uptake and translocation of dissolved mineral nutrients, it cannot explain how water gets to the tops of trees. At best, roots can exert a hydrostatic pressure of only a few atmospheres. One atmosphere can push a column of water to a height of 10.4 meters. Pressures of 10 atm or more would be required to move water to the heights of over 100 meters attained by the tallest trees.

COHESION-TENSION THEORY OF SAP ASCENT

It is reassuring in a time when new scientific knowledge is accumulating at an almost explosive rate to find that some theories are durable. The best available explanation for the ascent of sap, the **cohesion-tension theory** was proposed in 1894 by a plant physiologist, H. H. Dixon, and a physicist, J. Joly. According to this theory, the evaporation of water from leaf cells creates a water deficit in them. To replace water lost in transpiration from the leaves, water is withdrawn from the ends of nearby conducting tissues. This results in a tension which is transmitted downward on the continuous columns of water in the vessels and tracheids of the xylem. This tension, or "pull" on the water within the xylem, is a result of the cohesive forces holding the water molecules together and draws water molecules into the roots. The cohesiveness of water molecules is a consequence of their tendency to form hydrogen bonds with other water molecules. A mass flow of water thus moves up the plant to replace water lost in transpiration (Fig. 15-9). This theory predicts that tensions on the xylem ought to cause a decrease in stem diameter during times of maximum transpiration. It is relatively easy to demonstrate that tree trunks contract slightly during the day and are of measurably greater diameter at night when rates of transpiration are low.

Further support for the cohension-tension theory comes from a physical model constructed by Askenasy in 1895. A porous clay cup filled with water is connected by a long glass tube to a reservoir of mercury. Cohesive forces exist between water molecules in the glass tube and there are adhesive forces between water and mercury at their interface. As water molecules evaporate from the surface of the clay cup, capillary forces draw water into the pores of the cup. In such a model system, the column of mercury can be drawn upward far in excess of the height it can be raised by atmospheric pressure. Comparable results are obtained if the porous clay pot is replaced by a transpiring shoot with leaves.

Translocation: Transport of the Products of Plant Metabolism

One further aspect of the transport of materials in plants remains to be considered. Thus far we have concentrated on the movement of raw materials, water and dissolved minerals, from their point of entry, the roots, to the tops of plants. In the shoot, these raw materials are utilized. The leaves of the shoot have in their chloroplasts the factories where raw materials from the soil and the atmosphere are converted to usable organic compounds. The shoots of plants import simple inorganic materials which are converted at the expense of absorbed solar energy into the more complex substrates for respiration, protein synthesis, and the complex structural components of all plant cells. The products of the leaves are "exported" to the rest of the plant. In the absence of a closed circulatory system and with a "pumping" system that appears to be unidirectional, how is this accomplished?

The oldest evidence of a movement of materials from the leaves was obtained in the late 1600's by Malpighi, the Italian who also discovered blood capillaries and many other items of microscopic anatomy. He removed rings of bark from the stems of various kinds of trees. Such **girdling** often resulted in a swelling of the tissues above the ring because the movement of dissolved substances through the phloem was blocked. Malpighi further noted that such swelling above a girdle occurred only during those seasons of the

year when the trees had leaves. It was not until 1837 that Hartig discovered the sieve cells of the phloem and only in 1860 did he demonstrate that sap flowed from the cut bark of trees. Hartig's evidence that phloem cells in the inner layers of the bark are involved in the downward conduction of substances exported from the leaves was refuted or ignored until the 1920's—mostly because it was contrary to the diffusion theories of translocation in vogue in his day.

Phloem, a complex tissue made up of a number of different cell types including sieve elements, companion cells, phloem parenchyma, sclereids, and albuminous cells (Fig. 15-10), is located in the inner portion of the bark region. It is a difficult tissue to study because as stems increase in diameter during secondary growth, the layers of phloem outside the cambium get pushed outward. Consequently, older parts of the phloem are crushed and distorted. **Sieve elements**, also

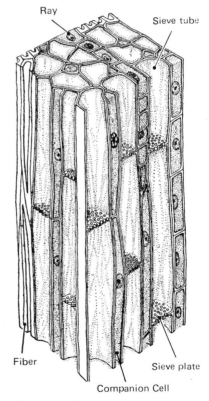

Fig. 15-10. Conduits involved in transport of products of plant metabolism. Enlarged section of phloem showing arrangement of sieve tubes, companion cells, ray parenchyma, and phloem fiber cells. See Figure 16-2 for location of phloem in woody stem. (From Biddulph, S., and O. Biddulph. 1959. The circulatory system of plants. *Sci. Am.* 200:44–49. Copyright © 1959. Scientific American, Inc., All rights reserved.)

Fig. 15-11. A, an aphid feeding on a willow stem. (Photograph by J. S. Redhead, University of Durham, England.) B, photomicrograph showing stylet of an aphid inserted into a phloem sieve cell. (Photograph by R. Kollmann, Botanisches Institut, Universität Bonn.) (From Richardson, M. 1968. *Translocation in Plants*. St. Martin's Press, New York.)

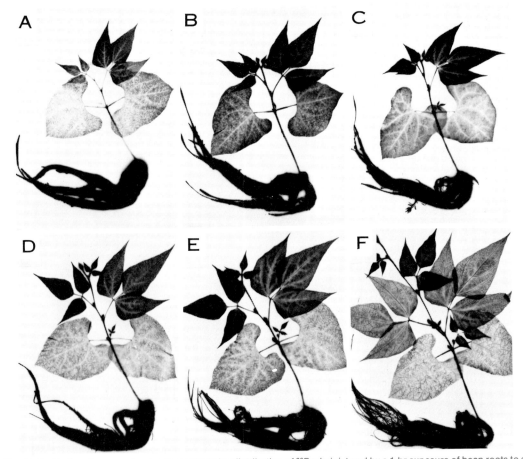

Fig. 15-12. Radioautograms showing the progressive distribution of ^{32}P administered by a 1-hr exposure of bean roots to a nutrient solution containing 88 μCi of ^{32}P per liter. Bean plants were harvested at the following times after exposure to ^{32}P: A, 0 hr; B, 6 hr; C, 12 hr; D, 24 hr; E, 48 hr; F, 96 hr. Harvested plants were dried and then left in contact with X ray film for 17 days prior to development of the film. Note continuing high concentration of labeled phosphorus in young leaves and stem apices indicating great mobility of this element. (Courtesy of O. Biddulph, from Biddulph, O., S. Biddulph, R. Cory, and H. Koontz. 1958. Circulation patterns for phosphorus, sulfur and calcium in the bean plant. *Plant Physiol.* 33:293–300.)

called **sieve tubes,** are the conducting cells of the phloem. At maturity, the nucleus of the sieve tube disintegrates; pores develop in the end plates of these cells and their cytoplasm becomes disorganized. Recently there have been reports describing strands of protoplasm, present in mature sieve elements. These **transcellular strands** which appear in some materials to be tubular and bound by a membrane while in others to consist of slime or fibrils, connect to strands in adjacent cells through pores in the end plates. It remains to be seen whether the transcellular strands of sieve elements are indeed functional structures or only artifacts resulting from the methods used to prepare phloem for observation in the light or electron microscope.

SAMPLING THE CONTENTS OF PHLOEM CELLS

The contents of phloem cells are rich in carbohydrates. The concentrations of sugars in phloem exudates are highest during the day, when photosynthetic rates are greatest, and are lowest at night. It has recently become possible to sample and analyze the contents of single sieve tubes by taking advantage of the feeding habits of aphids and other insect parasites (Fig. 15-11). After they have inserted the tips of their minute tubular stylets into the phloem, the aphids are anesthesized with a stream of CO_2. The insects are cut away with a fine splinter of glass, leaving the severed stumps of stylets still inserted. Sap continues to exude from the severed stylets for several hours and can be collected with a capillary tube.

Another approach to studies of what is translocated in plants and by what route involves labeling the products of photosynthesis by introducing radioactive substrates. Not only the identity of the components translocated but also their location and rate of movement can be determined by tracer techniques.

The first application of radioisotopes to a biological problem was in a study of the uptake of lead isotopes by bean seedlings done by the Danish investigator Hevesey in the early 1900's. All tracer studies are based on the assumption that living cells do not discriminate between radioactive and nonradioactive isotopes of an element in their environment. Thus, the uptake of any isotope is in proportion to its relative abundance. Using radioactive forms of several essential elements including phosphorus, Orin and Susan Biddulph and their associates at Washington State University followed the uptake and distribution of these elements in intact bean plants. By harvesting plants at intervals after

exposure to the tracer, they were able to follow the movement of elements absorbed through the roots to all parts of the plant and to follow the redistribution of tracer following its incorporation into the products of metabolism. (Fig. 15-12).

Plants grown in an atmosphere containing ^{14}C-labeled CO_2 form radioactively labeled products of photosynthesis. In many plants it has been shown that radioactively labeled compounds are translocated both upward and downward from leaves. The sugar present in greatest amount is the simple disaccharide sucrose. If a piece of stem is killed by steam or hot wax, radioactivity no longer moves past the dead tissue. If thin sections cut from a stem translocating labeled sugars are placed next to pieces of photographic film, the film becomes exposed only in the area of the phloem cells.

Tracer techniques have also been used to estimate the amounts of sugars and inorganic

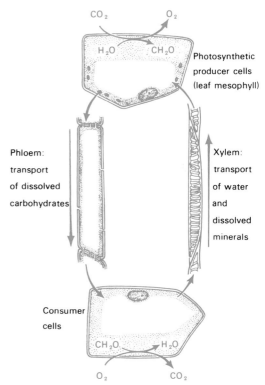

Fig. 15-13. The mass flow hypothesis. Water is drawn up through the xylem and into leaf cells by the transpirational pull of leaves. Hydrostatic or turgor pressure of leaf cells forces a flow of solutes (mostly carbohydrates) through the phloem to the roots. Photosynthetic cells of the leaf are producers of carbohydrates that are translocated to the consumer cells of the roots which utilize the products of photosynthesis for respiration, growth, or storage. (From Moment, G. B., and H. M. Habermann. 1973. *Biology: A Full Spectrum.* The Williams & Wilkins Co., Baltimore.)

nutrients translocated per unit time and their velocity of travel. The movements of amino acids, plant hormones, insecticides, herbicides, and viruses have been studied.

THEORIES OF PHLOEM TRANSPORT

The facts of translocation are firmly established. The redistribution of essential nutrients within plants is well documented. A demonstration of the uptake of radioactive phosphorus by the roots, its movement first to mature leaves, and somewhat later accumulation in the apical meristems, has become a routine exercise for beginning students of plant physiology. But we have not yet answered the question of how translocation, frequently in a direction counter to the transpirational stream, actually works.

Of the several available theories of phloem transport, none is completely satisfactory. It has been suggested that sugars move through sieve tubes by a process of simple diffusion. Unfortunately, diffusion is much too slow to account for measured rates of solute movement. A second theory suggests that substances moving in the phloem are carried by actively streaming protoplasm. Even though an end-to-end surging of the contents of young sieve tubes has been observed, it is difficult to imagine how such a mechanism can account for the rapid move-

ment of dissolved substances in the phloem sap.

Mass Flow Hypothesis

The best available hypothesis of phloem transport was proposed in 1930 by Münch. According to the mechanism suggested by Münch (which differed very little from Hartig's view in 1860), photosynthesizing leaves act as a "source" of translocated material and the roots as a "sink." Water drawn upward through the xylem by the transpirational pull of the leaves is drawn into the leaf cells which contain high concentrations of carbohydrates. This uptake of water due to osmotic forces causes increased hydrostatic or turgor pressure in the leaf cells. At the same time the roots are utilizing the products of photosynthesis for respiration, growth, or storage, and because the osmotic pressure thus tends to be lowered in the root cells, these "sinks" for the products of photosynthesis tend to have a reduced hydrostatic or turgor pressure. According to this picture, there is a mass or pressure flow of solutes from their source to sites of utilization.

The **mass flow** concept can be illustrated most easily in terms of a physical model: Two cells, A and B, permeable only to water, are connected by tube C. Cell A, representing the photosynthetic leaf cells, has a high osmotic

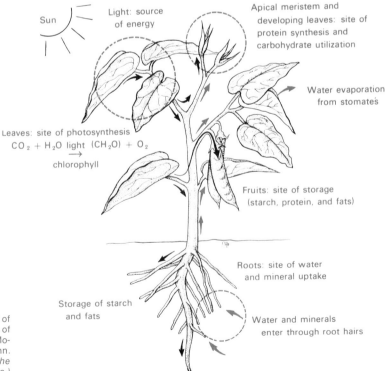

Sun

Light: source of energy

Apical meristem and developing leaves: site of protein synthesis and carbohydrate utilization

Water evaporation from stomates

Leaves: site of photosynthesis
$CO_2 + H_2O$ light $(CH_2O) + O_2$
$\xrightarrow{}$
chlorophyll

Fruits: site of storage (starch, protein, and fats)

Roots: site of water and mineral uptake

Storage of starch and fats

Water and minerals enter through root hairs

Fig. 15-14. Patterns of movement of water, minerals, and the products of photosynthesis in plants. (From Moment, G. B., and H. M. Habermann. 1973. *Biology: A Full Spectrum.* The Williams & Wilkins Co., Baltimore.)

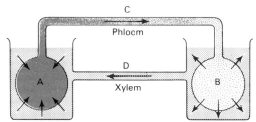

(From Moment, G. B., and H. M. Habermann. 1973. *Biology: A Full Spectrum.* The Williams & Wilkins Co., Baltimore.)

concentration while B, representing the root cells, has a low osmotic pressure. If such cells are placed in containers of distilled water, water will enter cell A, creating a hydrostatic pressure in A that forces fluid through connecting tube C into cell B. Water will be forced out of cell B and the flow just described will continue until the osmotic concentrations of cells A and B are equalized. If we add to our model the circumstances existing in plant cells, of production of soluble carbohydrates in A and their continual utilization in B, then we see how this process can persist so long as the sources and sinks of translocated material remain operative (Fig. 15-13). In the veins of the leaf are thin strands of vascular tissue made up of xylem and phloem. Through these strands there are simultaneous import of water and dissolved minerals through the xylem and export of the products of photosynthesis through the phloem.

One of the requirements of this hypothesis for phloem transport is that a gradient of turgor pressure must exist between shoot and root cells. Although it is difficult to demonstrate gradients in turgor pressure, it is relatively simple to show gradients in concentration of dissolved substances. When leaves are present and photosynthesis is taking place, concentrations of sugar in the contents of phloem cells are highest near the top of a stem and progressively decrease toward the base. When leaves are removed experimentally, or in trees during the late fall and winter (after leaves have been dropped), such gradients no longer exist. In early spring there is usually a massive transport of reserve carbohydrates upward. Advantage is taken of this upward movement when sap is collected from maple trees and evaporated to make maple syrup and sugar.

Differences between Transport in Plants and Animals

In considering the means that plants have evolved for solving the problems of transporting water and nutrients, we have noted the absence of a pump and a system of circulation that is open rather than closed as it is in animals. Viewed at the level of individual cells, we find practically no difference between plants and animals: the same laws of diffusion and osmosis plus active transport prevail and control the movement of molecules through selectively permeable membranes. The problems of transport become increasingly complex for multicellular forms primarily because greater distances separate the site of entry or synthesis of a given ion or molecule and the site of utilization. In plants, translocation takes place through the specialized cells of the vascular tissues, the xylem and phloem. Transfer of molecules from one part of a plant to another is an intracellular as well as intercellular phenomenon.

Although crude outlines of the mechanisms of transport in vascular plants have emerged, they are still incompletely understood. The hypotheses discussed in this chapter are not completely adequate to account for what can be observed in nature, and many problems remain to be solved even though the circulation of water, minerals, and metabolites was one of the first aspects of plant physiology to be attacked experimentally. The new technologies of radioactive tracers and electron microscopy have stimulated renewed efforts to answer questions first posed centuries ago.

USEFUL REFERENCES

Biddulph, S., and O. Biddulph. 1959. The circulatory system of plants. *Sci. Am.* 200:44–49.

Crafts, A. S., and C. E. Crisp. 1971. *Phloem Transport in Plants.* W. H. Freeman and Co., San Francisco.

Meidner, H., and T. A. Mansfield. 1968. *Physiology of Stomata.* McGraw-Hill Book Co., New York.

Richardson, M. 1974. *Translocation in Plants,* 2nd ed. Institute of Biology Studies in Biology No. 10. Edward Arnold (Publishers), Ltd., London. Distributed in the U.S. by Crane, Russak & Co., Inc., New York. (Paperback.)

Steward, F. C., ed. 1959. *Plant Physiology: A Treatise.* Vol II: Plants in relation to water and solutes. Academic Press, New York.

Sutcliffe, J. 1968/73. *Plants and Water.* Institute of Biology Studies in Biology No. 14. Edward Arnold (Publishers), Ltd., London. Distributed in the U.S. by Crane, Russak & Co. Inc., New York. (Paperback.)

STRATEGIES: Part D

INTERNAL CONTROLS

The porpoise: a marvel of internal and behavioral control. [Reproduced from Moment, G. B. 1967. *General Zoology*, 2nd ed. Houghton Mifflin Co., Boston]

16

Maintaining Internal Stability in Animals: Temperature, Immunity, and Excretion

Homeostasis is the tendency of living organisms to maintain relatively constant internal conditions, an essential for survival. The maintenance of homeostasis is the purpose of eating and breathing, and hence of all the innumerable adaptations in teeth and claws and in muscles and enzymes wonderfully fitted for carrying on the business of life. All the escape and protective devices of animals and plants, from the thick blubber beneath the skin which protects a whale against the frigid polar seas to the waxy cuticle which protects a desert plant from water loss, are ultimately involved in maintaining homeostasis.

Within animals there is a wide diversity of control mechanisms upon which their lives depend. One of the most important and best known mechanisms is the excretory system which, by removing waste and water from the blood, maintains a constant internal environment. In birds and mammals, a high and constant body temperature is of great importance, giving them a freedom from the environment and conferring a superiority over cold-blooded or poikilothermous animals like insects and reptiles, which slow down when the temperature falls. Other homeostatic mechanisms are concerned with eating and drinking.

Many constant conditions in vertebrates are maintained by beautifully balanced pairs of centers within the hypothalamus in the lower part of the brain. The endocrine system is in large part a homeostatic system of many feedback circuits which regulate the concentration of hormones in the bloodstream. The ability of animals to produce antibodies maintains the integrity of the organism against bacteria and other foreign invaders. This chapter will be devoted to three of the major regulating systems: those concerned with excretion, with the control of body temperature, and with the production of immunity.

Excretion and a Constant Internal Environment

Life arose in the sea. Consequently, protoplasm has an osmotic concentration like that of the ocean from which it arose. As a result, any organism living in fresh water, or descended from one which did, has a problem. Because the osmotic concentration of fresh water is very low, cells tend to take up water through the cytoplasmic membrane and will burst unless this excess water is excreted. From amoeba to man, the primary function of excretory organs is the elimination of excess water. In marine invertebrates there is no such problem because the osmotic concentration of their salt water environment is much higher. Thus excretory organs, although present, are much less important.

The distinction between secretion and excretion is a fuzzy one. **Excretion** is usually defined as the concentration and elimination of metabolic waste materials, including water, which have been within cells. **Secretion** is defined as production by a cell, gland, or tissue of a substance synthesized or at least concentrated within cells and released through their membranes. Thus urine can properly be called an excretion, although wastes are secreted into it.

VERTEBRATE EXCRETION

In higher vertebrates the three chief excretory organs are the lungs, the liver, and the kidneys. The **lungs** eliminate waste carbon dioxide and other volatile substances. Ether can be smelled on the breath of postoperative patients for many hours after anesthesia. The breath of long untreated diabetics smells of acetone, a volatile product of acetoacetic acid, which in this disease accumulates in the blood faster than it is converted into acetyl-CoA and fed into the Krebs cycle.

The **liver** synthesizes urea from CO_2 and NH_3, which is later excreted by the kidneys. It

is also responsible for the chemical processing of a multitude of substances in the blood and the changing of many into forms which are then eliminated by the **kidneys.**

The first function of the kidneys is to hold constant the **osmotic pressure of the blood** at a physiologically appropriate level. Were the blood to become too concentrated, the cells would shrivel; if too dilute, they would swell up and burst. Happily, you cannot dilute your blood no matter how much water you drink. With great precision, the kidneys will excrete exactly the correct amount of additional water. When water intake is restricted or water loss by evaporation increased (a man walking on a desert at 110°F loses a quart of water per hour from lungs and skin), then the kidneys excrete a smaller volume of more concentrated urine.

A second important activity of the kidneys is to regulate the **pH of the blood plasma**. If the plasma becomes acidic, i.e. if the pH falls, the kidneys excrete more hydrogen ions. If the blood becomes more alkaline, the kidneys excrete more bicarbonate. Since the concentration of sodium, calcium, and potassium bicarbonates in the plasma influences the osmotic pressure of the blood, it is obvious that the regulation of blood pH is closely related to the regulation of its osmotic pressure.

The elimination of **waste nitrogen** is a third major function of the kidneys. Most of the protein in the human diet is deaminated (has its nitrogen removed) as its constituent amino acids pass through the liver. This excess nitrogen is in the form of ammonia, which is combined with CO_2 and converted into **urea** by the liver. Mammals also excrete nitrogen in the form of **creatinine** in a small but constant daily amount which is independent of the amount of protein ingested but depends on the total muscle mass of the individual. This is not too surprising since creatinine is a constituent of muscles.

Most mammals also secrete small amounts of nitrogen as **uric acid**. This compound may

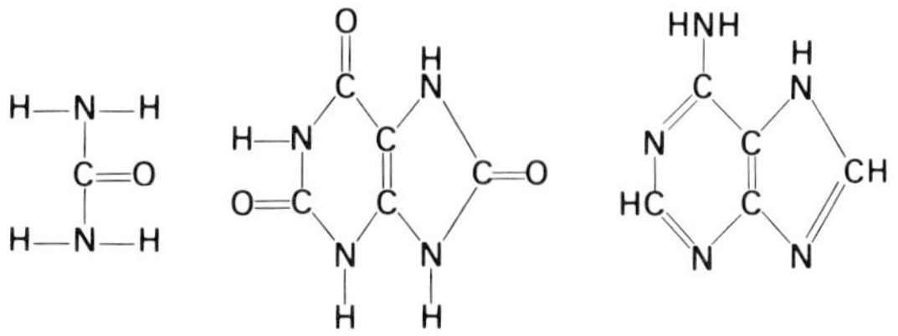

Fig. 16-1. Structure of urea, uric acid, and the purine adenine. Note the resemblance of uric acid and adenine familiar in nucleic acids.

be built up from ammonia or other simple nitrogen compounds or it may be derived from the breakdown of the purines in nucleic acids. Uric acid is itself a purine not too different from adenine and guanine.

The kidney excretes a very wide variety of additional substances, mostly **waste products** of liver metabolism or injurious substances processed by the liver. Included is **urochrome**, a yellow pigment which gives urine its characteristic color. Urochrome is a degradation product of bile pigments ultimately derived from worn out red blood cells. Among the virtually endless number of substances removed from the blood by the kidneys are glucose (if its concentration exceeds a threshold level), ketones, and other products of deranged lipid and carbohydrate metabolism, breakdown products of coffee and many other substances, including vitamins, hormones, and even a few amino acids.

The Anatomy of Vertebrate Excretion

The gross anatomy of the excretory system is simple and essentially the same from fish to man. A pair of kidneys lie on either side of the backbone, in mammals a short distance posterior to (below, in man) the diaphragm. Each kidney is served by a short **renal artery**, coming directly from the dorsal aorta, and by a renal vein. From each kidney a **ureter** carries urine to the **urinary bladder** in the pelvis (Fig. 16-2). From the bladder the urine is conducted to the exterior by the **urethra**.

The microscopic anatomy of the kidney was first explored by Malpighi. He noticed small round structures near the edges of the kidney. They can be seen with a good hand lens without dissection close to the thin edges of a frog's kidney. These little bodies are termed Malpighian or **renal corpuscles**. Each renal corpuscle is at the beginning of a long **tubule**. Together they form the functional unit of the kidney, the **nephron** (Fig. 16-3).

Each renal corpuscle consists of a tuft of capillaries called the **glomerulus** and its surrounding double-layered capsule called **Bowman's capsule**. The cavity between the capsule layers opens into the beginning of the uriniferous tubule which produces urine. Immediately after its formation at Bowman's capsule the tubule becomes convoluted, and then forms a long straight loop about 3 cm long known as **Henle's loop**. When a loop gets back to the renal corpuscle it forms a second or distal convolution. The distal convolutions empty into collecting tubes which in turn empty into the pelvis of the kidney where the ureter originates.

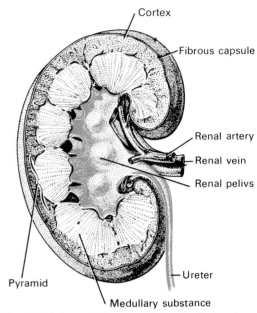

Fig. 16-2. A human kidney seen in longitudinal section. The functional units, called nephrons, lie primarily in the cortex. (From Moment, G. B., and H. M. Habermann. 1973. *Biology: A Full Spectrum*. The Williams & Wilkins Co., Baltimore.)

How the Kidney Works

How does the kidney do all of these remarkable things? This question is not only of evolutionary interest but also of the greatest medical importance. Over a century ago, William Bowman discovered the capsule that bears his name and immediately proposed a theory of kidney action: **glomerular filtration** followed by **tubular secretion** of wastes. Carl Ludwig, the leader of the great Leipzig group of physiologists, immediately proposed a different theory, glomerular filtration followed by **reabsorption** of useful molecules. Thus arose a classic controversy which stimulated research the world over for decades and finally led to our present knowledge.

Filtration takes place from the glomerulus as long as the hydrostatic pressure of the blood exceeds its osmotic pressure. A hydrostatic pressure of about 75 mm of mercury is imparted to the blood by the beating of the heart and tends to force its fluid constituents through the thin membranes of the capillaries of the glomerulus into Bowman's capsule. On the other hand, blood proteins, urea, and other osmotically active blood components all tend to pull liquid back into the blood. This osmotic pressure is about 30 mm of mercury, roughly equivalent to a 0.1 per cent solution of sucrose.

Filtration can be proved in several ways.

If the hydrostatic pressure of the blood falls to a point where it is no greater than its osmotic pressure, due to a heart weakness, bleeding, or drugs, the formation of urine will stop. If back pressure is applied up the ureter into the kidney, the formation of urine will stop when the applied pressure plus the osmotic pressure equals the hydrostatic pressure. Finally, direct proof can be obtained by collecting fluid from Bowman's capsule. This fluid on analysis shows the same concentrations of salts, glucose, amino acids, urea, and other substances of small molecular size as does the blood plasma. It lacks the blood cells and the blood proteins which are filtered out.

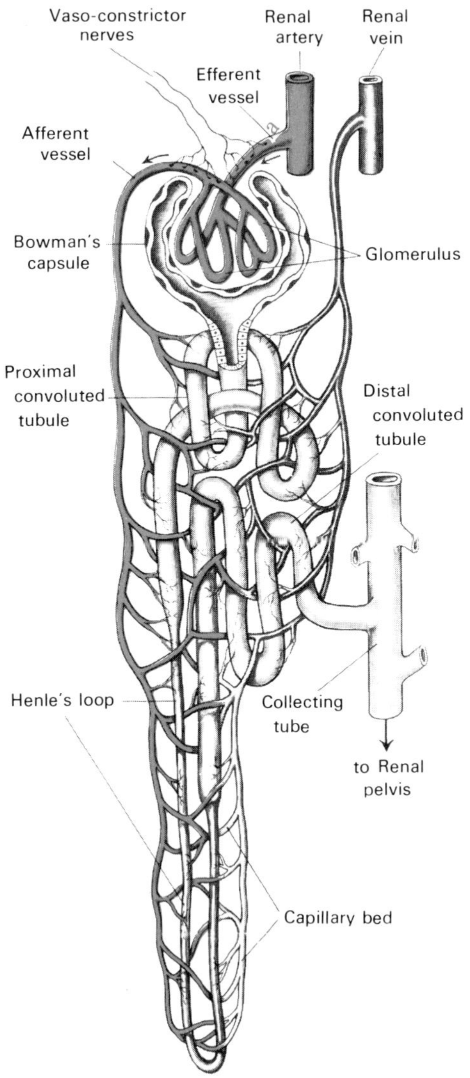

What about **reabsorption**? This obviously occurs because if the glucose concentration in the blood does not exceed the threshold value of about 160 mg per ml, none of it appears in the urine. Water, also, is reabsorbed, for the urine may be more concentrated than the fluid in Bowman's capsule. Water absorption is under the control of the antidiuretic hormone (ADH) of the posterior lobe of the pituitary gland.

What about **secretion** of materials into the tubules? The first convincing evidence that tubular secretion as well as reabsorption can occur came with the discovery that salt water teleost fish have either no glomeruli whatever or have reduced ones. The ugly little toadfish, *Opsanus tau*, of north Atlantic coastal waters must form its urine by secretion, for it completely lacks renal corpuscles, i.e. glomeruli and Bowman's capsules.

The first clear evidence for **active transport**, i.e. active secretion, in the mammalian kidney was gained by the use of an obvious method (obvious, once you think of it). The dye phenol red is readily excreted in the urine after injection into the bloodstream. The dye could be in the urine entirely due to filtration and then concentration in the tubules by its failure to be reabsorbed with the watery part of the filtrate. The dye could be in the urine because of active secretion into the tubules. Or its presence might be the result of both processes. To test whether or not active secretion was taking place, the blood pressure in an experimental animal was lowered until filtration stopped. Phenol red was injected and later the tubules were observed to see if they were concentrating the dye. They were, especially in the proximal convolutions. In mammals, at least, there appears to be an active secretion of urea (in addition to that present due to filtration) and other substances into the tubule.

Filtration into Bowman's capsule from the glomerulus is a passive process as far as the cells there are concerned. The necessary energy is provided by the heartbeat. In contrast, reabsorption and secretion against a concentration gradient are both active processes which require energy at the site, i.e. in the tubule cells. Numerous mitochondria and abundant ATP would, therefore, be expected in tubular cells engaged in these activities. This expectation is fulfilled by electron microscope studies. Indeed the kidney has been found to require more energy per hour per gram of tissue than does an active muscle like the heart.

Fig. 16-3. Diagram of a single nephron. The glomerulus plus Bowman's capsule together constitute a renal corpuscle. Arteries are shown in color. (From Moment, G. B. 1967. *General Zoology*, 2nd ed. Houghton Mifflin Co., Boston.)

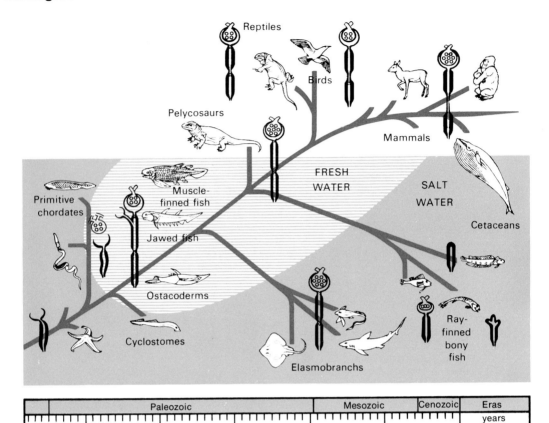

Fig. 16-4. Evolution of the vertebrate kidney. Note that in bony fish which return to salt water (shown on the right) glomeruli which filter water from blood are reduced or even lacking. This is an adaptation to the high osmotic pressure of sea water which prevents fish blood from absorbing large amounts of water. Only the sharks retain well developed glomeruli but they counter the osmotic pressure of the sea by retaining urea in their blood. (Redrawn after Homer Smith.)

Oxygen Supply Equilibrium

One of the most important and subtle of human feedback regulations is the control of the number of red blood cells which adjusts so that an adequate supply of oxygen is delivered to the tissues over a wide range of atmospheric oxygen concentrations. If you have been living approximately at sea level and move to a high altitude in the Rockies, your red cell count will rise from about 5,400,000 per mm³ if you are a man and about 4,800,000 if you are a woman to 6,500,000 or more. This is the most important thing that occurs during the usual period of adjustment to living at high altitudes. If your red cell count should fall because of severe hemorrhage, the number will return to normal levels within a week or so. In both cases, low oxygen levels in the tissues triggers this homeostatic response. Surprisingly, lack of oxygen does not stimulate the formation of new red cells at a higher rate directly but stimulates

the kidney to secrete a hormone-like substance, **erythropoietin,** which in turn stimulates the stem cells in the bone marrow to proliferate and produce a larger number of red cells, erythrocytes. Why the regulator for this continual balance should be located in the kidney is a question students of evolution have not answered.

Body Temperature

Animals are easily divided into those which have little or no control over their body temperature, the so-called cold-blooded animals, or **poikilotherms**, and those which have an internal control which maintains their body temperature at a constant level, the warm-blooded birds and mammals or **homoiotherms**.

BEHAVIORAL, ANATOMICAL, AND PHYSIOLOGICAL ADAPTATIONS TO ENVIRONMENTAL TEMPERATURES

Both poikilotherms and homoiotherms have behavioral ways of controlling their body temperature. The animals of a sand dune are active mostly at night and burrow in the sand to avoid the searing heat of midday. By night the white ghost crabs search the beach for scraps of food but only their tracks are visible by day. Desert insects, toads, snakes, and rats keep out of the sun.

Most homoiothermous animals possess anatomical adaptations to preserve heat. Fur, wool, and feathers are all excellent insulators, as is a thick layer of subcutaneous fat common to seals, walruses, porpoises, and whales.

From the medical standpoint and for the life of any bird or mammal, it is the physiological thermostat in the brain that is of paramount importance. The temperature control center, like so many other centers regulating bodily activities, is located in the **hypothalamus**. It has been known since the beginning of this century that a brain center controls temperature because brain injuries sometimes result in high fevers with few if any other symptoms. The precise location of the control site had to wait for modern instrumentation. The use of delicate probes like those used to locate the eating, drinking, and "pleasure" centers showed that body temperature was indeed regulated by cells in the hypothalamus, located at the base of the brain above the optic chiasma where the nerves from the eyes cross.

But we all know that temperature changes are keenly felt by sensory nerve endings in the skin. Stimulation of these endings leads to shivering, ruffling of feathers or panting, and other cold- or heat-combatting actions. What kind of thermoregulating center is in the hypothalamus? Does it receive and coordinate reports coming to it from the skin or is it a thermostat which has its own way of knowing what the temperature is before sending out the appropriate neural messages?

The best available evidence indicates the brain center takes its own temperature as well as sending out appropriate messages. If the blood supply to this part of the brain is somewhat raised, skin capillaries dilate and the temperature of the body as a whole will fall. If cooled, skin capillaries contract, shivering begins, and the body temperature rises. What does this leave for the cold and heat sensors in the skin to do? Apparently impulses from the skin lead animals into overt behavior which helps control temperature. For example, a dog too near a hot stove moves away.

Immunity

The phenomena of immunity include not only our ability to resist diseases but also the problems of skin and organ transplants, of blood groups, and of a host of allergies. If cancers are due to some action of viruses, as now seems likely, then it may be possible to develop artificial immunity to them as it has been for poliomyelitis.

Human knowledge about immunity grew out of attempts to overcome a loathsome and often fatal disease, smallpox. In the 18th century Lady Mary Montagu brought back to England from Turkey the practice of acquiring immunity against smallpox by inoculation with material from a mild case. If the inoculated individual came down with a mild case, he would be permanently immune to this dread disease. However, much too frequently severe disfigurement and even death was the result, so that this sort of inoculation was a kind of Russian roulette. About 50 years later a teen-age medical apprentice, Edward Jenner, wondered about combining the Turkish method of inoculation using actual smallpox material and the folklore that anyone was immune to smallpox who had ever had cowpox (a very mild bovine ailment resembling smallpox). In May of 1796 Jenner, by then a doctor, began a historic test. In so doing he faced the same kind of acute ethical problems that heart transplants and genetic counseling involve. He inoculated a healthy boy with cowpox material. After the boy had well recovered from the cowpox, Jenner injected him with material from a case of smallpox. The boy failed to contract smallpox and Jenner became a hero throughout Europe. But what if the theory had been wrong and the boy had developed the disease and died?

KINDS OF IMMUNITY

Today, as a result of the work of Louis Pasteur, Paul Ehrlich, and a host of others, a vast body of knowledge about immunity has been gained although the gaps in this knowledge are enormous. Immunity is of two sorts. **Natural immunity** is that which is characteristic of a species. Humans do not get bird malaria or distemper which is highly contagious and often fatal for dogs. **Acquired immunity** involves some kind of response by an individual to a foreign invasion. Three different mechanisms of immune response are known. **Cellular immunity** is due to leukocytes which phagocytize i.e. engulf, foreign bacteria. **Humeral immunity** is due to proteins, specifically gamma globulins, circulating in the blood and lymph. These globulins recognize foreign proteins and somehow unite with them, causing agglutination and precipitation.

A third type of immunity is due to interferons. The **interferons**, a newly discovered class of proteins produced by animals in response to viral (and also some nonviral) substances, hold much promise as a new approach to the control of virus diseases. Interferons are produced naturally in all vertebrates in response to virus infection. Their production is very different from the production of antibodies which follows vaccination. It appears that practically all body cells can produce interferon. Although synthesis is very rapid (within hours of stimulation), induced interferon disappears within hours or days. Reserve interferon (stored in an inactive, higher molecular weight form) seems to be manufactured by the reticuloendothelial system. The latter form is released in response to certain nonviral inducer substances. Instead of reacting directly with viruses to destroy infectivity as do antibodies, interferon protects cells by inhibiting the replication of viruses. As the yield of new virus particles is thus drastically reduced, the spread of the virus is slowed.

We cannot soon expect viral diseases to be treated with interferons because these antiviral agents are species-specific; thus, to be effective in treating human diseases interferon has to be produced by human cells. Human cell or organ cultures will not help the supply problem because of the very small quantities that could be synthesized in this way. Possibly interferon proteins may some day be artificially synthesized. The alternative approach is to find chemical agents that will stimulate interferon production—either to provide immunity or to help cure a disease after infection has occurred. None of the inducer substances now known such as the synthetic RNA, poly I (a polymer of inosinic-cytidylic acid), or bacterial endotoxins (lipopolysaccharides from the cell walls of certain gram-negative bacteria) can be used because of their undesirable side effects. However, the search for nontoxic inducers continues. Perhaps someday we may be able to avoid common colds or influenza simply by swallowing an effective interferon inducer.

BASIC TERMINOLOGY

Any substance which elicits an immune response is called an **antigen**. Most antigens, although probably not all, are proteins or closely associated with proteins. The proteins which are produced in response to antigens are known as **antibodies**. These are small protein molecules called **gamma globulins** or immunoglobulins. They consist of four polypeptide chains linked by disulfide bonds (see Fig. 16-5). Two of the chains are longer and hence heavy, two are short and light. There are five classes of these gamma globulins, termed IgG, which is the most abundant in the blood, IgA and IgM plus two which occur in very low concentrations, IgD and IgE. The latter is involved with allergies such as hay fever and other sensitivities. The amino acid sequence in these globulins has been determined in the hope of being able to synthesize them.

THE PROBLEM OF ORGAN GRAFTS

Rejection of grafts may be a very special case of antigen-antibody reactions. **Autografts** of skin or other structures like the kidney from one part of an individual to another part of the same individual present no problem. But **homografts** between two individuals of the same species or **heterografts** between individuals of different species are rejected after a period of weeks. It is clear that this rejection is an immunological phenomenon because if, after the rejection of the first graft, two new grafts are made, one from the same donor and the other from a new donor, the graft from the original donor is rejected much more quickly than it was the first time, while the graft from the new donor takes as long as the first did to be rejected the first time. This experiment shows that there is an **immunological memory** and that the antibodies re-

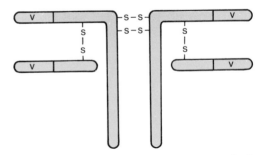

Fig. 16-5. A highly schematized diagram of an antibody molecule, a gamma globulin or immunoglobulin. Note that the separate polypeptide chains are held together by disulfide bonds. The ends of the polypeptide chains labeled V vary from one antibody to another.

sponsible for these rejections are highly specific. Because identical twins can accept each other's skin and other grafts permanently it seems certain that genetic factors control tissue compatibility. Studies with inbred strains of mice and analysis of human data show that mammals carry at least a dozen of these **histocompatibility genes**. They seem similar to the various sets of genes which produce the different blood groups.

There are several ways of **suppressing immunity**, which is important in organ and skin transplant operations. If an animal is injected with cells of another at or around the time of birth, when adult it will not be able to recognize tissues of the donor as foreign and will consequently accept grafts from that individual. X irradiation or chemicals which destroy the lymphoid cells which produce antibodies will result is a loss of ability to synthesize antibodies. Various antimetabolites which interfere with protein synthesis also inhibit antibody production. Finally, the removal of the thymus gland very early in life results in a lack of the various sorts of lymphocytes and, hence, no antibodies. Obviously all these methods, except the first, have serious side effects, not the least of which is a high susceptibility to infections.

Theories of the Origin of Immunocompetent Cells

There are thousands, and perhaps millions, of different proteins against which an individual can make specific antibodies. How is this feat possible? How, indeed, do the cells know not to make antibodies against the other cells composing the body of which they themselves are a part? Many unanswered questions remain, but if the thymus gland is removed at birth, the lymph glands remain very small—apparently because they have not received via the bloodstream the seed cells which produce lymphocytes. At the same time, ability to produce antibody is reduced or absent. By various labeling techniques is it possible to show that globulins are produced by small lymphocytes called **plasma cells** and that large white blood cells called **macrophages** play some role.

There are two theories to account for how all the thousands of kinds of antibodies can be formed. According to the **instructional theory** of Pauling and Haurowitz, foreign protein somehow acts as a template or model which instructs the lymphocytes to form the specific antibody globulin to match the foreign protein. The discovery that foreign protein complexes with host RNA supports such a view. Possibly, the normal protein-synthesizing machinery can be adapted so that a protein can make another protein which is complementary to it, as RNA is complementary to DNA.

The other theory, which is now in general favor, is the **clonal selection theory** proposed by the Australian investigator, Burnet. According to this view, during very early development approximately 10,000 different kinds of lymphocyte stem cells are produced by mutation. (Highly mutable genes are well known in corn and other organisms.) Each stem cell and all its descendants are able to produce some specific antibody. Presumably, any cells which have mutated in such a way that they form antibody against their own organism die or are killed. This may help explain how permanent tolerance can be obtained by early exposure to a foreign protein. In any case, according to the selection theory, when a foreign protein enters the bloodstream it comes into contact with macrophages and small lymphocytes. The lymphocytes which are of the right sort are somehow stimulated to proliferate, which may account for the delay in antibody response, and then to synthesize the appropriate gamma globulins.

It has recently been discovered, in what is a beautiful example of one of the ways biological science advances, that there are two types of lymphocytes involved. Graduate student Timothy Chang, in Ohio, attempted to demonstrate for a class the development of immunity in chickens after they are inoculated with salmonella bacteria. He obtained his chickens from fellow student Bruce Glick but all failed to develop immunity! It turned out that Glick had previously removed the bursa of Fabricius, a finger-shaped gland attached to the chicken hindgut, in an effort to discover its function. Soon after, Professor Robert Good in Minnesota, learning of these facts, organized a team to remove the thymus from some newly hatched chicks and the bursa from others.

As adults the chickens developed some immunity but those without a thymus could not reject foreign skin or organ grafts and those without a bursa did not make gamma globulin. Both types of lymphocytes are present in humans, the T or thymus-dependent cells and the B or bone marrow-derived cells. (A bursa of Fabricius is not present in mammals.) The whole field abounds in important and challenging problems.

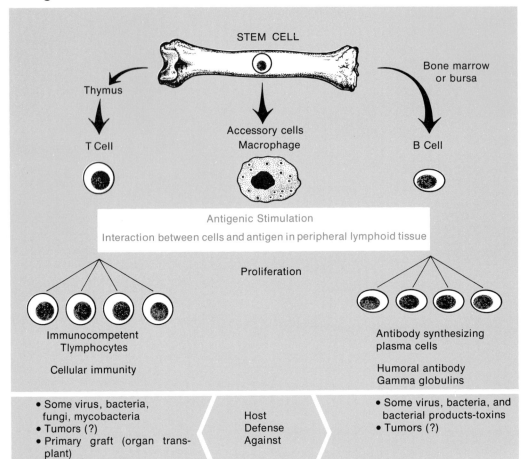

Fig. 16-6. Diagram to illustrate the origin and development of cells of the immune system. (After William Adler.)

USEFUL REFERENCES

Cannon, W. B. 1970 (reprint). *Bodily Changes in Pain, Hunger, Fear and Rage,* 2nd ed. McGrath Publishing Co., College Park, Md. (The great classic in the field and still valid.)

Hoar, W. S. 1966. *General and Comparative Physiology* Prentice-Hall Inc., Englewood Cliffs, N.J.

Ingram, V. M. 1963. *The Hemoglobins in Genetics and Evolution.* Columbia University Press, New York.

Langley, L. L., ed. 1973. *Homeostasis.* Dowden, Hutchinson, & Ross, Stroudsburg, Pa. (Readable and important papers from Claude Bernard to the present.)

Smith, H. W. 1951. *The Kidney: Structure and Function in Health and Disease.* Oxford University Press, New York.

Snively, W. D. 1960. *Sea Within Us: The Story of Our Body Fluids.* J. B. Lippincott Co., Philadelphia.

Gutyon, A. C. 1971. *Basic Human Physiology.* W. B. Saunders Co., Philadelphia.

Vroman, L. 1968. *Blood.* The Natural History Press, Garden City, N. Y.

17

Hormonal Control in Animals (Endocrines)

The most important fact about hormones is that knowledge of them can confer enormous power over the lives of men, animals, and plants. Too little thyroid hormone and you will become a virtual vegetable; too much and you will become so hyperexcitable that life is almost unbearable. Without the pituitary growth hormone, we would all be Tom Thumb midgets; with too much, either circus giants or victims of acromegaly with enormous bony hands and feet, heavy elongated faces, and short lives. The Pill, one of the most widespread agents of birth control, is a hormone product.

Insect pests that annually cause damage to crops worth billions of dollars will probably be controlled in the future not by DDT, a long-lasting and indiscriminate killer, but by insect hormones which either force the bugs to molt too often or prevent their attainment of maturity. Such hormones are both **selective** and **biodegradable**.

Behavior in animals including man is profoundly influenced by hormones. A female canary will never sing unless a pellet of male sex hormone is implanted under her skin. She then sings so well she can be sold as a male but will continue to sing only as long as the pellet lasts, 6–8 weeks. Hormone deprivation in humans can result in marked psychological and personality changes.

Newer knowledge about hormones is advancing in three principal areas. First, it is becoming clear that in both vertebrates and invertebrates endocrine glands are under the control of the nervous system at some point. Second, the mechanisms by which hormones produce their results are being discovered and, third knowledge of the effects of hormones on behavior is being greatly expanded.

Animal glands fall into two main classes: exocrine and endocrine. The digestive glands, for example, are known as **exocrine** glands because they possess ducts which carry the chemical agents from the gland of origin and empty it into some body cavity. The glands which produce hormones are called **endocrine** because they lack such ducts. Their products are taken up by the bloodstream and are carried to one or more "target organs," perhaps the comb and feathers of a rooster or the lining of the uterus of some mammal. It is on the target organ that hormones have their effects.

Among vertebrates, most endocrine glands secrete under the stimulus of specific hormones produced by the pituitary gland (itself an endocrine gland) on the underside of the brain. If the pituitary is removed, the thyroid, gonads, and adrenal glands stop secreting and may even degenerate. On the other hand, these glands exert an inhibitory influence on the pituitary by way of the hypothalamic brain centers which control the pituitary. In other words, the pituitary turns these glands on and they reach back, so to speak, and turn the pituitary off with their own hormones, although indirectly. The result is a self-regulatory **feedback mechanism** which may result in a periodic rise and fall in hormone level. This is the explanation of the familar female reproductive cycles in mammals. The same feedback mechanism may also maintain a steady-state level of hormones. Over these

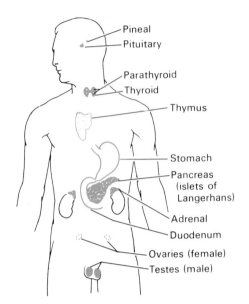

Fig. 17-1. Position of the human endocrine glands. (From Moment, G. B., and H. M. Habermann. 1973. *Biology: A Full Spectrum*. The Williams & Wilkins Co., Baltimore.)

various cycles and responses the nervous system imposes a measure of control, sometimes slight, sometimes complete.

Foundations of Endocrinology

The effects of castration on the growth and behavior of man and other vertebrates have been known from remote antiquity. Eunuchs were commonplace in ancient civilizations and often held important posts in government and the military services. Castrati were used as professional singers in Europe until about a century ago. The practice of removing the gonads from horses, bulls, and chickens is also very ancient.

It has always been known that there is a great deal of variation in the results of castration. Some of this variation may be due to the activities of the adrenal glands, which are known to produce masculinizing hormones under certain conditions. The bearded lady of the circus is usually a victim of misbehaving adrenal glands. Psychological factors also play an important role.

The first clear and convincing proof that a substance secreted by the cells of one organ could be carried by the bloodstream and produce a specific effect on a distant target organ was provided by the work of Bayliss and Starling in 1902. They discovered that as food from the stomach enters the first portion of the intestine, i.e. the duodenum, it liberates a chemical, **secretin**, which signals the pancreas to secrete digestive enzymes. With the help of a professor of Greek, they coined the word hormone and the science of endocrinology was truly launched.

METHODS OF STUDY

A more or less standard procedure for the identification and study of endocrine glands has been developed. The first step is usually examination of the suspected cells with an ordinary light microscope. The cells of endocrine glands—pituitary, gonad, adrenal, and the rest—suggest their secretory function by their structure. The cells tend to be cuboidal and filled with granules. The gland is well vascularized. With an electron microscope an extensive endoplasmic reticulum and Golgi complex are visible.

The orthodox test for actual endocrine activity is a double one. The gland is removed surgically. A sharp lookout is kept for accessory glands in unexpected places. Then the results are observed. The gland is implanted in another animal, or better yet, an extract is made and injected. Both aqueous and fat-soluble extracts must be tried. Often the results have been spectacular. Sometimes, however, the effects of glands have been missed completely because no one knew what effects to look for or else removed the glands at the wrong time in the life of the animal. This happened over and over again in the case of the adrenal, the thymus, and the pineal glands.

The third phase in the analysis of a hormone is chemical identification. To accumulate enough hormone sometimes requires almost superhuman efforts. To get enough of the insect growth hormone, for example, the glands of several barrels of insects have to be dissected out, homogenized, and the extract concentrated. Once the active material has been identified, the fourth step is the laboratory synthesis of the hormone molecule. With the identification of any hormone goes a study of its role in the life of the animal and of how its secretion is controlled.

CHEMICAL NATURE OF HORMONES

Chemically, hormones fall into two major groups. Some are steroids; others are small proteins, polypeptides or simple amino acids.

The sex hormones and the hormones of the adrenal cortex are **steroids**; so also is ecdysone, the molting hormone of insects. This means they are fat-soluble compounds closely related to vitamin D, cholesterol, and

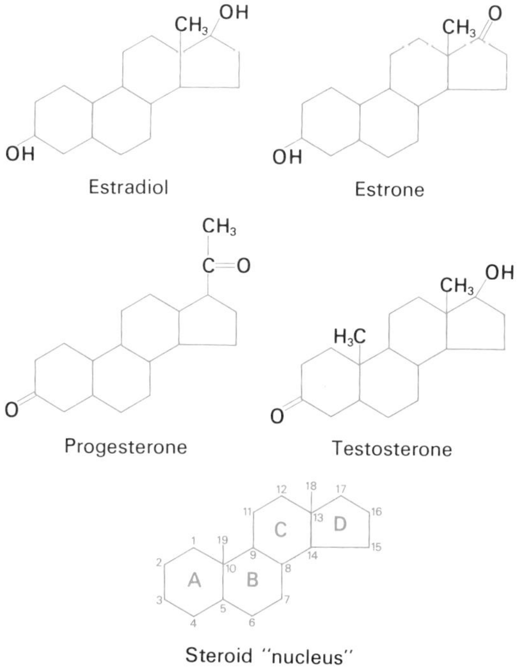

Fig. 17-2. Structural formulas of several steroid hormones and of the steroid nucleus. Estradiol, estrogen, and progesterone are female vertebrate hormones; testosterone is a male hormone.

bile salts. Other steroids include potent cancer-producing agents, embryonic inductors, and the very useful heart stimulant, digitalis.

All steroids are built around the same chemical nucleus consisting of four joined carbon rings. Three are six-carbon rings and one is a five-carbon ring. These rings are given conventional letters and their carbon atoms are given numbers for identification (Fig. 17-2). A keto group (which is merely a carbon atom with two bonds to an atom of oxygen and two single bonds to other atoms),

$$\begin{array}{c} O \\ \parallel \\ -C- \end{array}$$

in position 17, for example, makes a 17-ketosteroid, a substance excreted by men. All 19 carbons in the basic steroid configuration are derived via acetyl-CoA, mostly from acetic acid. Once again the pivotal position of acetyl-CoA becomes apparent. The various hormones are produced by adding keto, hydroxy, methyl, or other groups to one of the 19 carbons.

The hormones of the pituitary, thyroid, and pancreatic glands are proteins, or, in the case of the thyroid, simply an amino acid. These hormones are water-soluble. The hormone insulin from the pancreas, for example, consists of two chains, one of 21 and one of 30

amino acids, held together at two fixed points by disulfide bonds, i.e. the $-S-S-$, between two cystines. It will be recalled that cystine is an amino acid containing sulfur. The whole insulin molecule has a molecular weight of about 5,700. Glucagon, the other pancreatic hormone, which has an effect on blood sugar opposite to that of insulin, consists of a single chain of 29 amino acids and has a molecular weight of 3,485.

HOW DO HORMONES WORK?

The way hormones produce their results, conferring on a male a beautiful beard or a handsome set of antlers and on a female a lovely smooth skin or perhaps characteristic feathers, is still incompletely understood. However, there is good evidence that cells possess "receptor sites" for protein or polypeptide hormones on their surfaces and for steroid hormones within the cytoplasm. This fact can be determined by radioactively labeling hormones and feeding or injecting them into animals. Such tests show, for example, that TSH, the thyroid-stimulating hormone of the pituitary, is picked up by the surface of the thyroid gland cells but not to any measurable extent by other cells.

A second and highly important group of facts was uncovered by E. W. Sutherland and his colleagues for which they were awarded a Nobel Prize in 1971. In investigating the mechanism by which epinephrine (adrenaline) stimulates the liver to release glucose into the bloodstream, they found that before any glucose is released there is a marked increase in **cyclic AMP,** cyclic adenosine monophosphate, in the cytoplasm of the liver cells. As soon as cyclic AMP is available, the appropriate enzymes begin to convert glycogen, the form in which carbohydrates are stored in the liver, into glucose. Cyclic AMP is made from the familiar energy-rich ATP by an enzyme called adenyl cyclase. An exciting part of this work is that the enzyme adenyl cyclase is a part of the cell membrane. It is therefore easy to believe that it is part of the specific receptor site for epinephrine. The hormone apparently fits into the surface of the adenyl cyclase and activates it, probably by causing a slight change in its shape.

A third important group of facts concern some, and perhaps all, of the steroid hormones. This discovery began with work in Germany on the insect molting hormone, ecdysone. Several investigators showed that ecdysone, when injected, activated specific genetic loci on the chromosomes of the salivary glands of certain insect larvae and that this activation revealed itself by puffing of

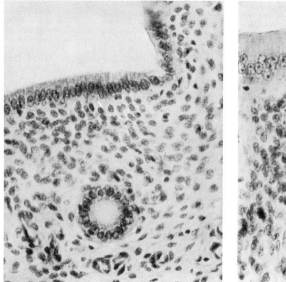
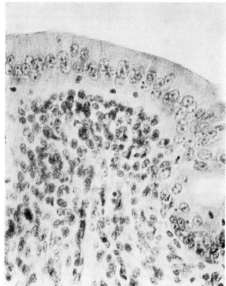

Fig. 17-3. Rat uterus. A, appearance after castration, more correctly, spaying, i.e. removal of the ovaries. Note the thin line of epithelial cells (along top edge). B, the appearance after administration of estrogen. Note much thicker endometrial epithelium while the underlying mesodermal cells remain much the same. (Courtesy of Sheldon Segal.)

specific regions (Fig. 6-15). Other workers investigating bacteria demonstrated that the antibiotic actinomycin blocks the formation of new RNA. Here, then, was the idea, i.e. hormones turn genes on, and the tools to test it.

In several laboratories in the United States and abroad investigators working with different hormones on different animals have obtained convincing evidence that hormones initiate DNA-dependent RNA synthesis, presumably by combining with genetic repressors. New mRNA, of course, means new enzymes. New enzymes mean changes in metabolism and a wide range of possible new products.

Clear evidence comes from studies of female sex hormone. If the ovaries of a rat are removed, the endometrium lining the uterus remains permanently in the reduced, anestrous condition. If estrogen, female sex hormone, is then injected into the rat, an increase in uterine mRNA can be detected within half an hour. After 3 or 4 hours an increase in protein can be measured and the uterus begins to grow into the estrous condition. These facts in themselves support the theory. After hormone injection, new mRNA appears first, followed later by new protein. Confirming evidence has been obtained by administration of actinomycin D, an inhibitor of DNA-dependent RNA synthesis. If actinomycin is injected before the estrogen, no new RNA appears and there is no subsequent in-

crease in protein. The uterus remains anestrous (Fig. 17-3).

Thus it appears that certain polypeptide hormones such as epinephrine and the pituitary hormones bind to membrane receptors on the cell surface and activate the adenyl cyclase-cyclic AMP system, while steroids enter the cell and bind to cytoplasmic receptors and then move into the nucleus where they interact with chromatin.

Endocrine Glands of Mammals

THE THYROID

A mammal's thyroid gland forms two lobes pressed against either side of the trachea and connected with each other by a narrow band of thyroid tissue, so that the entire gland resembles a pair of saddle bags. The thyroid glands of all vertebrates are found in the same general region of the throat.

Microscopic examination reveals that the gland is composed of thousands of more or less spherical follicles made up of secretory cells enclosing a colloidal material. The gland is so highly vascularized that more blood flows through it, in proportion to its size, than through any other organ except perhaps the adrenal glands.

Marked underfunction, **hypothyroidism,** in small children produces **cretins.** These pitiful

individuals are greatly stunted, woefully fee-
ble-minded, and have characteristically
bloated faces, bodies with loose, wrinkled
skin, and coarse, sparse hair. Basal metabolism, body temperature, and heart rate all are
abnormally low. Usually the thyroid gland is
abnormally small but if thyroid deficiency occurs in adult life, the thyroid may enlarge
greatly and form a swelling in the throat
called a **goiter.** Sometimes goiter is accompanied by many symptoms of cretinism; in
other cases it is not, perhaps depending on
whether or not the enlarged gland supplies
minimal needs.

Cretins and goiterous persons used to be
common in localities far from the sea, such as
isolated valleys in the Alps and Pyrenees and
in the interior of continents, where there is a
marked deficiency of iodine in the diet. Dramatic recoveries can be accomplished by
feeding thyroid gland, iodine, or the syntheti-
cally made hormone thyroxin to hypothyroid
patients. The blank face gains expression,
the bloated body assumes a normal shape,
and the mind brightens.

Hyperthyroidism, oversecretion of the thyroid, sends the basal metabolism to abnormal
heights and produces a hyperactive and un-
pleasantly irritable animal or person with protruding eyeballs (exophthalmia), a rapid
heartbeat, high metabolic rate, and some thyroid enlargement, i.e. goiter.

The hormone of the thyroid gland is an
amino acid, **thyroxin.** It is an iodine-carrying
modification of another amino acid, tyrosine.
That iodine is concentrated by the thyroid
gland can be readily shown by injecting radioactive iodine into a mammal. Within an
hour the iodine shows up in the cells constituting the walls of the thyroid follicles. Later
the radioactive material appears within the
follicles.

Thyroxin causes mitochondria to swell and
increase in number, but whether this is the
cause or a result of the increase in rate of
metabolism is uncertain.

The activity of the thyroid gland is dependent on a thyrotropic hormone from the anterior lobe of the pituitary. Thyroxin itself inhibits the formation of thyrotropin either by
acting directly on the anterior pituitary or
through inhibiting cells in the hypothalamus,
the action of which is required for the formation and release of thyrotropin by the pituitary.

THE PARATHYROIDS

The parathyroids are four small glands
either on or embedded in the thyroid, two on
each side (Figs. 17-5 and 17-6). In some way

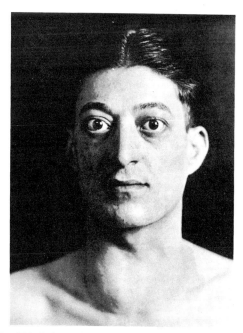

Fig. 17-4. Patient with exophthalmic goiter. Note the protruding eyeballs and swollen neck. (Courtesy of Massachusetts General Hospital; from Moment, G. B. 1967. *General Zoology,* 2nd ed. Houghton Mifflin Co., Boston.)

they control calcium and potassium metabolism. The target organs are the bones and the
kidneys. In dogs and carnivores generally,
parathyroidectomy leads to a serious lower-
ing of blood calcium levels, muscular spasmo
increasing in severity into total tetany, and
ultimate death. The symptoms are usually
much less severe in omnivorous and herbivorous animals.

CALCITONIN

It has recently been discovered that both
the thyroid and the parathyroids secrete a
hormone essential for health, named **calcitonin.** The effect of calcitonin is just the opposite of that of the parathyroid hormone. It
depresses the concentration of calcium ions
in the blood, causing them to be deposited in
the bones.

THE PANCREAS

The pancreas is an elongated gland located
beside the intestine close to the stomach. It is
sometimes sold, along with the thymus
gland, as sweetbreads. It secretes a digestive
pancreatic juice via the pancreatic duct into
the intestine, and also two hormones, insulin
and glucagon, synthesized in nests of cells
called the **islets of Langerhans.** These islets

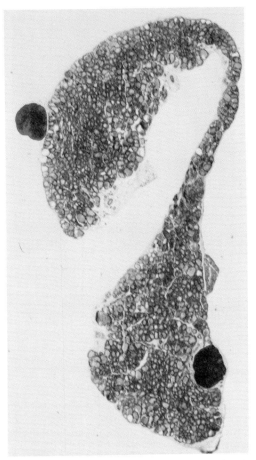

Fig. 17-5. Section through a rat thyroid gland showing the two lobes connected by the isthmus. Parathyroid glands (color) can be seen attached superficially to one lobe and almost completely inbedded in the other. × 20. (From Moment, G. B., and H. M. Habermann. 1973. *Biology: A Full Spectrum.* The Williams & Wilkins Co., Baltimore.)

usually lie scattered throughout the body of the gland.

Lack of **insulin** causes diabetes mellitus, a disease described by the ancient Greeks as the ailment in which "the flesh melts away into urine." Nearly 2,000 years later a 17th century English anatomist and physician noticed that the urine of diabetics is sweet, so to the classic symptoms of excessive thirst, excessive urination, and great emaciation, a fourth was added, excretion of sugar.

About 200 years later still, in typical 19th century experiments, von Mering and Minkowski made a surprising discovery. Removal of the pancreas of a dog causes diabetes! It could not be from lack of the digestive juice, because if the pancreatic duct was brought to the surface of the body so that all of the pancreatic juice merely dropped to the

ground, the dogs did not develop diabetes. In 1900 Eugene Opie, a member of the first class of The Johns Hopkins Medical School, discovered that in diabetic patients the islet cells were atrophied.

The two kinds of cells, arbitrarily called alpha and beta cells, make up most of the islets of Langerhans. The **alpha cells** are somewhat larger than the beta cells, lie near the periphery of the islets, and have a granular cytoplasm which stains differently from the granules of the smaller, more centrally placed beta cells. It is the **beta cells** that atrophy in diabetes, and they are the ones that secrete insulin. There are two types of diabetes mellitus. The juvenile type which appears in childhood and usually requires insulin for effective treatment, and the adult type, which appears in middle age, and can often be controlled by diet. At least in the adult type, there is an important genetic factor.

The alpha cells secrete a hormone, **glucagon,** which has the opposite effect on blood glucose from insulin. Instead of decreasing the level of glucose, glucagon increases it. The existence of such a hormone came to light after is was found that patients from whom the entire pancreas had been removed (because of cancer) required less insulin to keep the blood sugar down to normal levels than patients with naturally occurring diabetes. This indicates that the pancreas produces something which tends to increase the concentration of glucose.

The Adrenals

The adrenal glands of mammals are a pair of more or less rounded, highly vascularized structures situated either close to or against the kidneys—hence the name **adrenal.** The central part of each gland is called the **medulla** and secretes two similar hormones which are modifications of the amino acid tyrosine and are often grouped together under the term **adrenaline.** The outer part of the gland, the **cortex,** secretes several steroids of which the best known is **cortisone.**

The hormones synthesized by the medulla are **epinephrine** and **norepinephrine.** They differ only in that epinephrine has a methyl group, i.e. a CH_3, which norepinephrine lacks. Both are catecholamines, important compounds in biochemical psychology and will be discussed in connection with the nervous system. Adrenaline is the hormone of strong emotion, of "fight or flight" which gives added strength both to the fox to chase and the rabbit to flee or to a man to meet an emergency with feats of strength he did not know was possible for him.

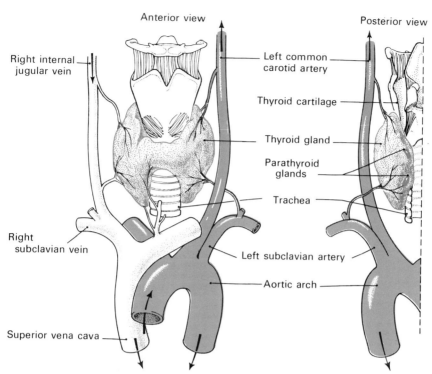

Anterior view

Posterior view

Right internal jugular vein

Left common carotid artery

Thyroid cartilage

Thyroid gland

Parathyroid glands

Trachea

Right subclavian vein

Left subclavian artery

Aortic arch

Superior vena cava

Fig. 17-6. Human thyroid gland and associated structures. Note the arteries and veins serving the gland and also the pair of parathyroid glands on the posterior surface of the left thyroid lobe. (From Moment, G. B., and H. M. Habermann. 1973. *Biology: A Full Spectrum.* The Williams & Wilkins Co., Baltimore.)

The effects of lack of cortical hormones have been known since Thomas Addison described the strange and fatal disease that results. Oversecretion or deranged secretion of the adrenal cortex is also a serious affliction. Some of the cortical hormones have a masculinizing effect. If oversecretion begins in childhood, the voice deepens, and facial, axillary, and pubic hair develop in a typical masculine way even in genetic girls. Muscular development may produce an adult-like dwarf of herculean conformation. In genetic boys, although the penis may attain adult size, the testes remain infantile or even abnormally underdeveloped. Under normal conditions, cortisone promotes healing, plays an important role in muscle contraction, promotes fat deposition (often in the face), and affects excitation thresholds in the nervous system and the proportions of the different types of leucocytes. It has various other poorly understood functions such as counteracting shock.

In populations of mammals such as woodchucks and mice, enlargement of the adrenal glands and decrease in reproductive competence accompany great crowding, i.e. stress. This response may act as a brake on population growth.

The secretion of the adrenal cortex is under the usual feedback control by the anterior pituitary, which secretes a polypeptide ACTH, **adrenocorticotropic hormone.** Without it the cortex fails to secrete.

THE GONADS

It seems highly probable that gonadal hormones, which are steroids (Fig. 17-2), produce their effects on the growth and pigmentation of their various target organs by initiating the transcription of new mRNA. This certainly seems to be the case with female sex hormone and the lining of the uterus. The effects which sex hormones have on the behavior of vertebrates appear to be due to direct action of the hormone on specific cells in the brain, specifically in the hypothalamus.

It should be noted that although the principal **estrogen** (female sex hormone) produced by the Graafian follicles of the ovary is **estradiol**, this hormone is often found in a slightly modified form called **estrone** in many tissues including the ovary itself, the placenta, the adrenal cortex, and even the testis. Paradoxical as it seems, the testes and urine of stallions are among the richest sources of estrogen. Of course they carry even more male sex hormone, **androgen.** There has never been

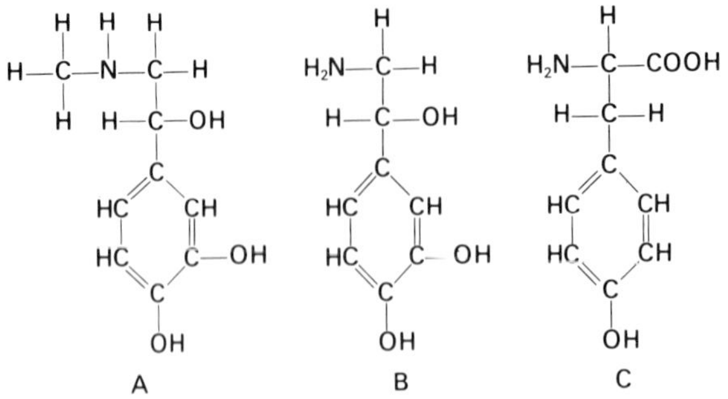

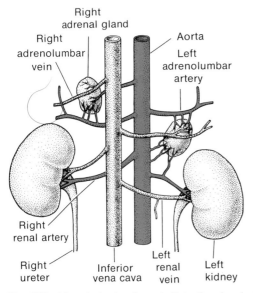

Fig. 17-7. The hormones of powerful emotion. A, epinephrine (adrenaline); B, norepinephrine (noradrenaline); C, tyrosine, the amino acid from which they are formed.

Fig. 17-8. Adrenal glands of a cat. Note the abundant blood supply. (From Moment, G. B. 1967. *General Zoology,* 2nd ed. Houghton Mifflin Co., Boston.)

any satisfactory explanation for this fact. One of the commercial sources of estrogen is the Mexican yam or barbasco root. Why this plant should produce such a hormone is also a mystery.

Progesterone is secreted by the corpora lutea of the ovary after fertilization; it is necessary for the continuance of the pregnancy and has a molecular structure similar to estradiol. So also do the androgens, **testosterone,** the form in which male sex hormone is secreted by the interstitial cells in the testis, and **androsterone,** the slightly modified form in which it is excreted in the urine.

All of these gonadal hormones are under the regulation of the anterior lobe of the pituitary, which secretes gonadotropic hormones essential for the formation of sex hormones. They in turn inhibit the secretion of their respective gonadotropins, stimulating hormones from the pituitary.

THE PITUITARY

The pituitary gland, or **hypophysis,** is commonly called the "master gland," for it exerts a regulatory control over most of the other glands of internal secretion. It is beginning to appear, however, that the pituitary is not so much a master as an executive officer carrying out the instructions of the **hypothalamus.** In adults, the pituitary is located on the floor of the skull just behind the optic chiasma where the optic nerves cross as they enter the underside of the brain—a very awkward place to reach surgically. This is an inconvenience we apparently owe to our remote, marine, tunicate-like ancestors.

All the hormones of the pituitary are polypeptides. Indeed some of them are molecules large enough to rank as proteins. The **anterior lobe** of the pituitary secretes somatotropin (the growth hormone), gonadotropins, thyrotropin, adenocorticotropin, a corpus luteum-stimulating, or luteinizing, hormone (LH), and a lactogenic hormone. In general these **anabolic hormones,** especially somatotropin, promote protein synthesis and a general buildup of body structures. There is occasional talk that athletes are injected with these hormones to increase their strength. How effective such measures are depends in part on how close an individual is to his maximum size. ACTH stimulates the synthesis of cortical hormones and inhibits the synthesis of fat in adipose tissue.

The **posterior lobe** of the pituitary stores and discharges two hormones, the antidiuretic hormone (ADH) and oxytocin. These are produced in neurosecretory cells located in the hypothalamus above the optic chiasma. The secretion moves down into the posterior lobe via the axons of these neurosecretory cells. The **antidiuretic hormone** controls the formation of urine by increasing water reabsorption in the kidney tubules. Failure of the posterior pituitary results in diabetes insipidus, which is unrelated to a

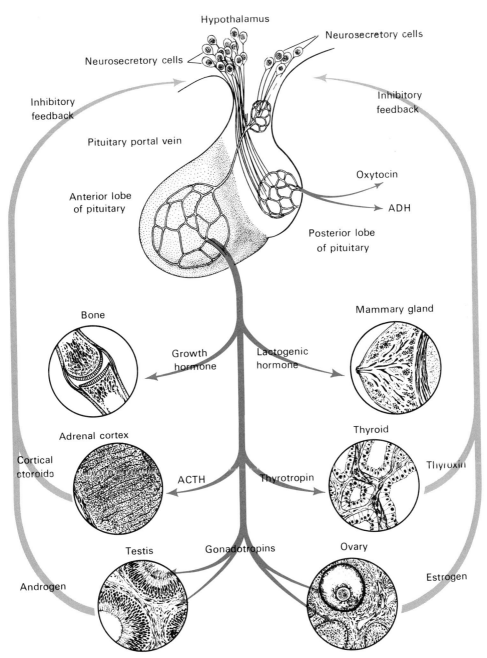

Fig. 17-9. Feedback control between the anterior pituitary and other endocrine glands. (From Moment, G. B., and H. M. Habermann. 1973. *Biology: A Full Spectrum.* The Williams & Wilkins Co., Baltimore.)

lack of insulin. Large amounts of sugar-free urine are produced. **Oxytocin,** also known as **vasopressin** because in large doses it causes arteries to constrict, is a hormone which stimulates smooth muscles. It is important in the release of milk on suckling and almost certainly in stimulating uterine contractions during childbirth.

The **intermediate lobe** of the pituitary pro-

duces a melanophore-dispersing hormone, **intermedin,** in lower vertebrates and probably in mammals as well. It enables fish and frogs to become darker.

The cells of the **anterior lobe** are of three major types according to staining and other characteristics. There are **acidophils,** so called because their cytoplasmic granules stain with acid dyes like eosin or orange G.

TABLE 17-1
Animal hormones (From Moment, G. B., and H. M. Habermann. 1973. *Biology: A Full Spectrum*. The Williams & Wilkins Co., Baltimore.)

Hormone	Source	Chemical Nature	Action
Growth and other hormones, GH, FSH, LH, TSH	Anterior pituitary (adenohypophysis)	Protein	Stimulates thyroid, gonads, and mammary glands; bone growth; protein synthesis and anabolic metabolism generally
ACTH		Protein	Stimulates cortisone synthesis and fat breakdown
Antidiuretic hormone, ADH, and oxytocin	Posterior pituitary (neurohypophysis)	Protein	Stimulates renal reabsorption of water; contraction of uterine and mammary musculature
Intermedin	Intermediate lobe of pituitary	Polypeptide	Darkens skin by causing melanophore expansion
Thyroxin	Thyroid	Tyrosine derivative	Stimulates metabolism; promotes protein synthesis; metamorphosis in amphibia
Calcitonin			Lowers blood calcium
Parathyrone	Parathyroid		Maintains blood calcium
Insulin	Islets of pancreas	Protein	Controls carbohydrate utilization; depresses blood sugar level; promotes protein synthesis; anabolism
Glucagon			Increases blood sugar level
Gastrin	Stomach	Protein?	Stimulates secretion of gastric juice
Melatonin	Pineal gland	Tryptophan derivative	Slows ovarian development and estrous cycle, etc.
Epinephrine (adrenaline)	Adrenal medulla	Tyrosine derivative	Raises blood pressure; increases heart rate; liberates liver glycogen as glucose
Norepinephrine	Adrenal medulla	Tyrosine derivative	Similar to epinephrine
Acetylcholine	Nerve cell endings		Transmits information across synapses
Serotonin	Brain, octopus salivary gland, plant nettles	Cyclic organic compound	Similar to adrenaline but more violent; neural functions uncertain
Prostaglandins	Prostate, etc.	Fatty acid	Affects blood pressure, gastric secretion, birth, etc.
Cortisone	Adrenal cortex	Steroid	Promotes healing, fat deposition, muscular development; influences neural excitation thresholds, white blood cells
Aldosterone	Adrenal cortex	Steroid	Activates production of enzymes increasing Na reabsorption in kidney and elsewhere
Estrogen (estradiol)	Ovary	Steroid	Female secondary sexual traits; in adult estrous cycle; growth of uterus and mammary glands
Progesterone	Corpora lutea of ovary	Steroid	Stimulates growth of uterine lining and milk-secreting cells
Androgen (testosterone)	Testis	Steroid	Masculinizes the brainstem reproductive rhythm center; embryonic development of male reproductive structures; male secondary sexual traits in adult
Ecdysone	Insect prothoracic glands	Steroid	Induces molting hormone
Juvenile hormone	Insect corpora allata	Steroid(?)	Prevents attainment of adulthood
Ecdysone or similar hormone	Crustacean Y organ		Induces molting
Juvenile or similar hormone	Crustacean X organ and brain	Steroid(?)	Prevents molting

These cells are often called eosinophils or alpha cells. There are **basophils,** often called beta cells, in which the cytoplasm stains with basic dyes like methylene blue. Third, there are cells which resist staining and are therefore called **chromophobes.** They constitute roughly half of the cells present.

Many attempts have been made to identify which cells secrete which hormone. Some, if not all, of the acidophils secrete **growth hormone.** Several genetic strains of dwarf mice show a hereditary lack of alpha cells and no pituitary growth hormone. Human pituitary (Tom Thumb) midgets are also deficient in alpha cells. Some of the beta cells or basophils evidently secrete a hormone related to the gonads. After removal of either testes or ovaries, many of the basophils become enlarged as though engorged with secretion.

The importance of the anatomical relationships of the pituitary gland can scarcely be exaggerated. Not only does it lie adjacent to the underside of the hypothalamus but each lobe is functionally tied to the hypothalamus in a very special way. The posterior lobe is directly innervated by nerve fibers coming

Fig. 17-10. Five brothers and sisters, three hypopituitary dwarfs and two normal adults. Their ages in years from left to right are: 31, 27, 25, 24, and 20. The failure of the pituitary to secrete sufficient growth hormone is due either to an autosomal or an X-linked recessive gene. Such dwarfs, often called midgets, are sexually immature and contrast markedly with achondroplastic dwarfs (see Fig. 19-4). (From Rimoin, D. L. 1971. Genetic forms of pituitary dwarfism. *In* Bergsma, D., ed. *The Endocrine System*. Birth Defects: Orig. Art. Ser., Vol. VII, No. 6. Published by The Williams & Wilkins Co., Baltimore, for The National Foundation-March of Dimes.)

down the stalk of the gland from the hypothalamus. These are called **neurosecretory cells** because secretion granules can be observed to form in the cell bodies which lie in two groups (each called a "nucleus") in the brain. These granules then move down the axons into the posterior lobe of the pituitary.

The anterior lobe of the pituitary is also connected to its own set of neurosecretory cells in the hypothalamus, but via the bloodstream. These nerve cells send their axons down into the stalk of the pituitary where they end among the meshes of a small capillary network. The capillaries pick up the secretion from this group of neurosecretory cells and carry it in a **pituitary portal system** which almost immediately breaks up into a second set of capillaries within the anterior lobe. The blood supply of the posterior lobe is quite separate from the system of the anterior lobe.

The evidence for a regulatory **feedback control** between the pituitary and the hormones produced by ovaries, testes, thyroid, and adrenal cortex is unassailable. If the anterior pituitary is removed, the thyroid gland atrophies, the gonads fail to produce sex cells or sex hormones, reproductive cycles stop, and the adrenal cortex fails. If one of these glands, a target organ for the pituitary, is removed, the level of the corresponding pituitary trophic hormone increases greatly. In the case of removal of either male or fe-

male gonads, characteristic "castration cells" appear in the anterior lobe. Injection of sex hormones or thyroxin, for example, depresses the amount of gonadotropin or thyrotropin secreted by the pituitary.

It is not yet clear whether the inhibitory effect of the various hormones is directly on the cells of the pituitary or on the neurosecretory cells in the hypothalamus, or on both. Here is some of the evidence. When part of the thyroid of a rat is removed, the pituitary increases its release of thyrotropin. The remaining piece of the thyroid gland then increases in size and ultimately the level of thyroxine in the blood is restored to normal levels. At the same time the level of thyrotropin returns to normal. So far there is no hint of a role for the hypothalamus and its neurosecretory cells. However, this sequence of events can be completely blocked by electrically destroying the cells in a very small region of the hypothalamus. If the thyroid is then removed, the pituitary behaves as though it knew nothing about it.

THE PINEAL BODY

The pineal body is a small rounded structure on the dorsal side of the thalamic portion of the brain opposite the pituitary on the underside of the thalamus. In the adult frog the pineal body underlies the skull, but in mammals it is overgrown by the enormous cerebral hemispheres and the cerebellum. In many lizards the pineal body is a third eye; in frogs it contains **cone cells** closely similar under an electron microscope to the cones of the visual retina.

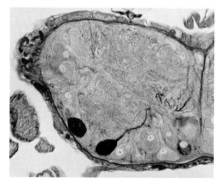

Fig. 17-11. Neurosecretory cells in the central nervous system of an earthworm stained dark purple by the Scharrer-Gomori aldehyde-fuchsin method. Such neuroendocrine cells appear identical with those seen in vertebrate brains. Note the axon carrying the darkly stained secretion. × 50. (From Moment, G. B., and H. M. Habermann. 1973. *Biology: A Full Spectrum*. The Williams & Wilkins Co., Baltimore.)

Over the years many investigators have attempted to discover what the pineal gland does. It has long been known that tumors of this structure in children are commonly associated with precocious sexual maturity, or paradoxically, sometimes with delayed sexual development. The answer was found only recently. Precocious sexual maturity is associated with tumors of the supporting tissue around the pineal gland which apparently choke the structure into inactivity while overgrowth of the pineal itself is associated with delayed puberty. These facts suggests that the pineal exerts an inhibitory effect on gonadal development.

A real breakthrough came with the discovery by Virginia Fiske at Wellesley that continuous illumination results in a decrease in the weight of the rat pineal body and an increase in the size of the ovaries. Other investigators have shown that exposure of male hamsters to cycles of 1 hour of light and 23 hours of darkness will cause atrophy of the testes but that this effect can be prevented by removal of the pineal gland. It was then discovered that there is a circadian (roughly 24-hours) cycle in the formation of a pineal enzyme which produces **melatonin.** Melatonin is a hormone derived from the amino acid tryptophan. It is worth noting as a typical example of how laborious the identification of a hormone can be that extracts of over 200,000 cattle pineals were prepared and purified in this work by A. B. Lerner and his co-workers. This hormone slows down the estrous cycle and inhibits ovarian development; it is produced only in the pineal body.

THE THYMUS

The thymus, like the pineal body, has been an enigmatic gland, often thought to play some role in retarding sexual development because it undergoes marked regression with the onset of sexual maturity. Many experimenters have removed the gland, which lies in the upper chest beneath the breastbone. Extracts have also been injected, but no significant results have followed either procedure. It has now been found that if the thymus is removed promptly at birth, antibody-forming lymphocytes (plasma cells) will be lacking. Apparently the lymphocyte stem cells in the various lymph glands, which produce these blood cells in later life, come from cells which originate in the thymus. There is also good evidence that the thymus secretes some hormone-like material essential for proper lymphocyte development.

PROSTAGLANDINS AND ANABOLIC HORMONES

The recently discovered prostaglandins have been widely hailed in the popular press as the wonder hormones of the future and perhaps correctly. Only the future can tell. They were first discovered in seminal fluid where their concentration is about 100 times what it is in any of the many other tissues where they are produced. Because most of the ejaculate is formed by the prostate gland, the name prostaglandin seemed appropriate although it has turned out to be somewhat misleading.

The effects of prostaglandins cover a broad spectrum, although many of them seem to be due to an initial action on smooth muscle. All the prostaglandins are 20-carbon fatty acids. Some raise and some lower blood pressure. Others, or perhaps the same, facilitate birth, reduce the secretion of progesterone by the corpora lutea of the ovary, and thus prevent implantation of the egg. Still others alleviate severe asthma, gastric ulcers, and may affect neural transmission.

Occasionally there are accounts in the press of the use of anabolic hormones by athletes and others to increase their prowess. It has already been noted that insulin is a hormone which promotes many anabolic, i.e. building up, processes. Androgens and, to a lesser extent, estrogens promote muscle growth. The pituitary growth hormone is also anabolic in action. Two things are important to remember about the use of anabolic hormones. Most individuals probably produce as high a concentration of their own as their muscles are capable of responding to and, secondly, all hormone treatments involve some danger. The endocrine system is full of complex feedbacks and side effects. Consequently hormones should only be used under the supervision of a physician who understands endocrinology.

USEFUL REFERENCES

Barrington, E. J. W. 1964. *Hormones and Evolution*. D. Van Nostrand Co., Princeton.
Guyton, A. C. 1969. *Function of the Human Body*. W. B. Saunders Co., Philadelphia.
Hunt, C. 1972. *Males and Females*. Penguin Books, Baltimore. (Paperback. A broad spectrum from chromosomal basis to the question of innate differences in creativity.)
Kruskemper, H. L. 1968. *Anabolic Steroids*. Academic Press, New York.
Lissak, K., ed. 1973. *Hormones and Brain Function*. Plenum Press, New York. (Hormonal effects on brain function, masculinization of mammalian brain by male sex hormone, control of pituitary hormones, etc.)
Turner, C. D., and J. T. Bagnara. 1971. *General Endocrinology*, 5th ed. W. B. Saunders Co., Philadelphia.

18

Hormonal Regulation in Plants

Before recorded history some farmer probably wondered how seedlings achieve the proper orientation with shoots growing up and roots down, regardless of whether the seeds are planted right side up or upside down. About a century ago Charles Darwin asked the same question and began a long investigation into the effects of light on the direction in which plants grow. This work on plant **tropisms**, as such movements are called, led directly to a series of investigations which are continuing today, for it led to the discovery of plant hormones. The concept of specific chemical communication between different parts of plants was proposed by the founder of experimental plant physiology, Julius Sachs, in the 1880's; however, the idea of plant hormones was generally rejected until the isolation of auxin in 1928 by Frits Went, a Dutch army draftee who worked at night in his father's laboratory in Utrecht.

It is now clear that growth and development in both plants and animals are controlled by chemical regulators, the hormones. The synthesis of a hormone usually occurs at a site removed from the target organ on which it has its effect. Characteristically, very small amounts of these specific chemicals produce amazingly large effects. Consequently, it is not surprising that natural and synthetic growth-regulating hormones have been used for many purposes, ranging from the control of shape and appearance of ornamental plants and the chemical weeding of crops and lawns to the defoliation of jungles as a tactic of warfare. Whether used to promote rooting and thus speed the propagation of horticulturally useful species or in the eradi-

cation of unwanted weeds, the chemical control of plant growth provides a powerful tool for agriculture. In many ways the era of "made-to-order" plants is at hand.

Naturally Occurring Plant Hormones

At the present time there are three major classes of hormones known to be naturally occurring and essential for normal development of plants. These are the **auxins**, the **cytokinins**, and the **gibberellins**. Recent investigations indicate that two additional classes of compounds fall into the category of naturally occurring regulators of plant growth: **abscisic acid** and **ethylene**. Still other kinds of plant hormones have long been postulated, such as the flowering hormone, **florigen**, and **wound hormones**. It is probably safe to predict that our list will expand in the years ahead. In addition to the naturally occurring chemical regulators of plant growth, a number of synthetic herbicides and defoliants have been developed. These products of the chemical industry have enabled us to grow weed-free lawns or crops and have made the harvesting of certain plant products much easier. The widespread use of plant poisons and defoliants by military forces has been the subject of much controversy and concern over long-term ecological effects. The subject of plant hormones is of general human concern and has raised serious ethical questions extending beyond the realm of pure science.

279

Fig. 18-1. Julius Sachs, founder of experimental plant physiology who, in the 1880's, proposed the idea of plant hormones. (Courtesy of the Hunt Botanical Library.)

EXPERIMENTAL APPROACHES TO THE NATURE OF HORMONE ACTION

One of the first steps in studying any hormone is to prove that a particular tissue extract or chemical can function as a plant (or animal) hormone. This is done by applying criteria similar to those proposed by Robert Koch in establishing the relationship between a disease and its causative agent: (1) removing the source of the substance must change the pattern of growth or metabolism; (2) applying the substance exogenously must restore normalcy; and (3) effects of the substance must be specific.

To answer questions about the effects of a hormone and how these results are achieved requires information about what happens to the gross structure of a plant and also what happens on the cellular and biochemical levels. Effects on the whole plant can readily be seen in most cases. In recent years evidence has been obtained that at the molecular level many (if not all) hormones act as derepressors of genetic information. Thus, in the dogma of molecular biology, a hormone is an effector molecule that turns on genes. Other mechanisms of hormonal action have been proposed in addition to the currently fashionable hypothesis that they are involved in the transcription and translation of genetic information. Hormones may be involved in changing the configuration of proteins or in the activation of enzymes. There is considerable evidence that they affect cell membranes and thereby regulate the entry of essential substrates into cells.

Auxins: The Molecular Basis For Tropisms

Tropisms are growth movements of plants in response to environmental factors such as light or gravity. **Phototropism** is a growth response to unilateral light. Ordinarily this is a positive response in the shoots of plants, although certain fungi exhibit a negative phototropism. **Geotropism** may be either positive or negative: shoots grow upward (negative response), while roots grow downward (positive response).

Tropisms are of vital importance during the early stages in plant development. Nutrient reserves available to the embryo in a germinating seed are limited. Before these reserves are exhausted, the roots must be oriented so that they can absorb water and inorganic nutrients from the soil; the shoots must emerge from the soil so that developing leaves will be exposed to the sun. Unless the appropriate orientation is achieved within a few days, the seedling will die. Older plants also respond to light and gravity, although their responses may be slower. Because of their sensitivity and the rapidity of their growth responses,

Fig. 18-2. Frits Went, the first plant physiologist to isolate auxin and to devise an assay for this plant hormone. (Courtesy of the Hunt Botanical Library.)

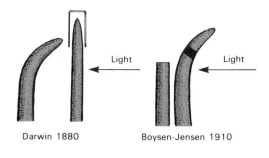

Darwin 1880   Boysen-Jensen 1910

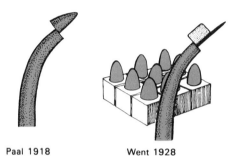

Paal 1918   Went 1928

Fig. 18-3. Early studies of plant hormones. (Redrawn from Thimann, K. V. 1957. Growth and growth hormones in plants. *Am. J. Bot.* 44:49–55.)

most studies of the effects of light and gravity have been performed with seedlings.

The first studies of tropisms were reported in 1881 by Charles Darwin in a book called *The Power of Movement in Plants.* Experimenting with young seedlings of canary grass, Darwin observed that "when seedlings are freely exposed to a lateral light, some influence is transmitted to the lower part causing it to bend." Such a bending response could be prevented by either cutting off the tip of the seedling or by shading it. The bending response actually occurs some distance below the tip. Clearly there must be some means for transmitting the signal from the tip, where light is perceived to the lower portion of the stem that responds. Additional insights into such bending responses accumulated very slowly. In 1913, Boysen-Jensen demonstrated that cut tips of seedlings could be pasted back in place with a layer of gelatin without interfering with transmission of the stimulus from the tip. Although the stimulus passed readily through gelatin, Paal in 1919 found that mica or foil would prevent its transmission. Paal also showed that severed tips replaced to one side caused curvatures of seedlings resembling those in response to unilateral light. Then in 1928, Frits Went, in his father's laboratory at the University of Utrecht, succeeded in isolating the chemical

signal first postulated by Darwin (Fig. 18-3). The technique of isolating the active substance turned out to be amazingly simple: it diffused out of plant tissues into a small block of agar. This technique provided not only a method for separating auxin from plants but also the basis for a sensitive assay technique for auxins used to this day.

The assay developed by Went is based on the bending response of decapitated oat (*Avena sativa*) seedlings in response to auxin in a tiny block of agar applied to one side of the cut tip. The bending of seedlings in response to known amounts of auxin is measured to obtain a standard curve for this bioassay. By comparing the response to extracts of plant tissues with the response in the standard curve, an investigator can determine the concentrations of extractable auxin.

The availability of an assay for auxin led to further insights into the nature of auxin effects on plant growth. Attempts to isolate and chemically identify auxin obtained from plant materials led to the realization that this thermostable substance that is soluble in water, ether, and alcohol is a rather simple molecule, indole-3-acetic acid:

The relationship between auxin and tropisms was clarified in the 1930's. Blaauw had demonstrated that phototropism in *Avena* seedlings is due to a difference in rate of growth between the illuminated and dark sides of the seedling. Went cut off the tips of illuminated oat seedlings and placed them on agar blocks separated by a razor blade so that the auxin from the illuminated and shaded sides diffused into two separate blocks. When the auxin activity of these blocks was tested in the *Avena* assay, blocks from the shaded sides caused greater curvature. Thus the curvature of intact seedlings appeared to be the result of higher auxin concentration in the shaded side (Fig. 18-4).

At about the same time, Dolk, another Dutch investigator, reported parallel experiments demonstrating that an asymmetric distribution of auxin is also involved in geotropism. Of the total auxin diffusing from the tip of an oat seedling placed on its side, 62.5 per cent was in the lower half. When similar assays were performed on roots, a comparable redistribution of auxin was found. But roots

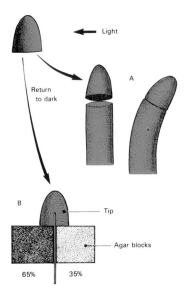

Fig. 18-4. Experiments of Went demonstrating asymmetric distribution of auxin caused by light. After unilateral illumination, an oat coleoptile tip placed on a decapitated seedling in the dark can cause a bending response (A). When placed on agar blocks separated by a sheet of mica (B), more auxin is collected from the shaded than from the illuminated side of the coleoptile tip. (From Moment, G. B., and H. M. Habermann. 1973. *Biology: A Full Spectrum.* The Williams & Wilkins Co., Baltimore.)

turn downward while shoots grow up. It was soon realized that increased auxin concentrations may promote growth under some circumstances but inhibit under others.

OTHER EFFECTS OF AUXIN ON PLANT GROWTH AND DEVELOPMENT

In 1933, Thimann and Skoog proposed that auxin produced in the plant apex can inhibit the development of lateral buds. Thus began the accumulation of a long list of ways in which auxin is involved in the morphology of plants. **Apical dominance**, a growth pattern in which lateral bud development is inhibited, can easily be shown to be the consequence of high concentrations of auxins exported from the apical regions. If the terminal portions of a plant are removed, lateral buds develop; if an external supply of auxin is applied to the cut ends, development of lateral buds will remain suppressed. The extent of lateral bud development establishes the branching pattern of a plant. The upright conical form of trees such as the spruces results from suppression of lateral growth near the apex of the plant. In other trees, terminal buds often form flowers; further shoot growth is due to development of lateral buds and a much branched pattern results. The relationship between the source of auxin production

and growth form is the basis of an old horticultural practice, **pruning**, or cutting back the terminal branches of plants to promote a fuller, bushier growth.

It was also learned in the 1930's that externally supplied auxin stimulates the formation of **adventitious roots**. This knowledge has been used to promote the rooting of cuttings, the method used to propagate many horticultural species. For example, hollies are notoriously hard to propagate, yet they form roots more readily when auxin-containing dusts or pastes are applied to the ends of cuttings.

In 1937, Thimann proposed a generalized scheme for the effects of auxin on the growth of various plant parts. Auxin is effective not only at amazingly low concentrations but also over a very wide range of concentrations, encompassing 10 orders of magnitude from 10^{-11} to 10^{-1} molar. Distinctly different concentrations of auxin are needed to stimulate the growth of stems, buds, and roots (Fig. 18-5). For example, a concentration that promotes stem elongation can inhibit the development of buds or roots. From this generalized picture we can more readily understand why the increased concentration of auxin on the lower side of a seedling has opposite effects on the shoot and the root. The higher auxin content stimulates shoot elongation but inhibits root growth.

RECENT STUDIES ON THE NATURE OF AUXIN ACTION

What seemed like a very neat and comprehensive picture of auxin action by the end of the 1930's left unanswered some crucial questions that could not be approached experimentally until the development of modern tracer techniques. Using ^{14}C-labeled indoleacetic acid, a number of modern workers

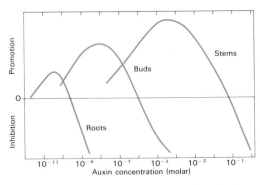

Fig. 18-5. Summary of auxin concentration effects on growth of roots, buds, and stems. (From Moment, G. B., and H. M. Habermann. 1973. *Biology: A Full Spectrum.* The Williams & Wilkins Co., Baltimore.)

Fig. 18-6. K. V. Thimann, whose investigations have contributed much to our understanding of auxin action. (From *Ann. Rev. Plant Physiol.,* Vol. 14, 1963.)

tried to repeat the classic experiments on tropisms of Went and Dolk. Radioactively labeled auxin was applied to young oat seedlings that were then illuminated from one side or placed horizontally. To everyone's dismay and confusion, no differences in radioactivity could be detected between the upper and lower or lighted and shaded sides of the seedlings. Were the earlier experiments that so clearly showed asymmetric distributions of auxin in error and the classic theory of auxin involvement in tropisms invalid? Even if the old theory were still true, it did not explain how more auxin could be extracted from the shaded or lower side of a plant.

The key to ending the confusion over experiments with ^{14}C-labeled auxin was the realization by Thimann and his associates that the method of labeling is critical. Indoleacetic acid (IAA) is oxidized by an enzyme in plant cells. Uniformly labeled IAA, i.e. having ^{14}C in both the acetate and indole portions of the molecule, when oxidized by this oxidase-peroxidase enzyme, loses it carboxyl group and is converted to a derivative of oxindole that has no auxin activity. Yet most of the initial radioactivity remains in the inactive breakdown product. Under these conditions there is no relationship between the presence of radioactivity and the location of active hormone. If, however, the ^{14}C label is introduced only in the carboxyl group of indoleacetic acid, then the radioactivity and biological activity of the auxin remain together. Recent tracer experiments indicate that there is a lateral movement of auxin in both geotropism and phototropism. In the response to light,

neither photodestruction nor a light influence on longitudinal transport is detected.

RECEPTORS FOR TROPIC RESPONSES

It has now been firmly established that the effector molecule in both geotropism and phototropism is indoleacetic acid and the asymmetric distribution of this effector molecule is a consequence of its lateral transport. Two further basic questions remain: What are the receptors for these responses to the external environment and what organelles are responsible for detecting gravity and light? For gravity to be detected, there must be something within plant cells that falls in response to this force. To detect light, there must be a colored photoreceptor that absorbs light energy.

At various times it has been proposed that starch grains or plastids, structures which can become oriented along the bottoms of cells, might be involved in the geotropic responses of plants. However, the response to gravity occurs just as well in the absence as in the presence of these structures. The best available hypothesis at this time is that gravity is detected by the movement of cytoplasm and vacuoles. Pressures against membranes could affect the rates of auxin transport.

Although there is still some uncertainty about the receptor for phototropism, its action spectrum has been measured with great precision by Thimann, Curry, and others. The most effective wavelengths are in the blue region of the spectrum with peaks at 445 and 475 nm and a shoulder at 425 nm. There is an additional region of effective wavelengths in the ultraviolet between 360 and 380 nm (Fig. 18-7). While the visible portion of the action spectrum for phototropism closely resembles the absorption spectra of certain of the carotenoid pigments (Fig. 18-8), the additional

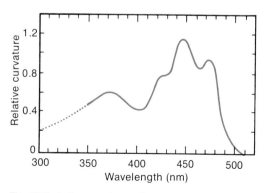

Fig. 18-7. Action spectrum of phototropism. (From Moment, G. B., and H. M. Habermann. 1973. *Biology: A Full Spectrum.* The Williams & Wilkins Co., Baltimore.)

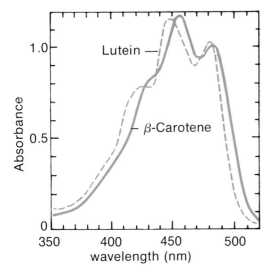

Fig. 18-8. Absorption spectra of two common carotenoids, β carotene and lutein. (From Moment, G. B., and H. M. Habermann. 1973. *Biology: A Full Spectrum*. The Williams & Wilkins Co., Baltimore.)

peak in the ultraviolet suggests that a flavin type compound may be involved. Unfortunately, no single known compound has an absorption spectrum identical to the measured action spectrum. Not only is there uncertainty about the chemical nature of the photoreceptor, but also about where it is located. Recently Thimann reported observing rectangular or rhomboidal bodies, smaller in size than mitochondria, in *Avena* and also in the photosensitive parts of certain fungi. These structures are bounded by a single membrane and are filled with crystalline material made up at least in part of protein. They are found between the cell wall and the protoplasm. In time it may be shown that these particles contain a carotenoid-protein complex that serves as the photoreceptor for phototropism. The question will then arise: How does the absorption of light energy by these particles control the lateral transport of auxin?

CHEMICAL SPECIFICITY OF AUXINS

Since the identification of auxin as indoleacetic acid, a large number of related chemicals have been shown to possess hormonal activity. For example, adding two chlorines to phenoxyacetic acid vastly increases its growth-promoting activity to the point of making 2,4-dichlorophenoxyacetic acid (2,4-D) one of our most potent weed killers (Fig. 18-9). The nature of the side chain is also important: an even number of carbons must be present because side chains with an odd number of carbons are readily broken down. A knowledge of how molecular structure affects auxin activity has made it possible to manufacture made-to-order herbicides and growth regulators. The 2,4-D mentioned above is one of the oldest herbicides and still the most widely used. By 1966, 2,4-D production in the United States exceeded 100 million pounds per year. This weed killer appears to work by promoting growth to such an extent that the plant outgrows its resources. An interesting aspect of the way 2,4-D affects plants is that it is a **selective herbicide**, toxic to dicots (broad leaved plants) at concentrations that have no noticeable effects on monocots (parallel veined grasses and their relatives).

The fact that dicots are generally far more susceptible to the auxin herbicides than are monocots is usually explained by differences in the morphology of these two groups of seed plants. In the dicots, meristems (or the actively growing regions) of the shoot are relatively more exposed, located terminally rather than near the base of the plant as in grasses. Furthermore, their broad leaves provide more surface for the absorption of herbicides applied as sprays. In addition, factors such as differences in the wettability of leaf surfaces (which would affect the rates of absorption) and in rates of translocation following absorption would influence susceptibility. Because of the great differences between species in the effects of particular compounds, the auxin herbicides are highly selective. Their use as weed killers has another decided advantage. The herbicidal action of these compounds is rapidly lost in the soil; that is, they are biodegradable and leave no toxic residues. They are effective in extremely low concentrations and are therefore inexpensive. Most important of all, they are relatively nontoxic to animals and humans.

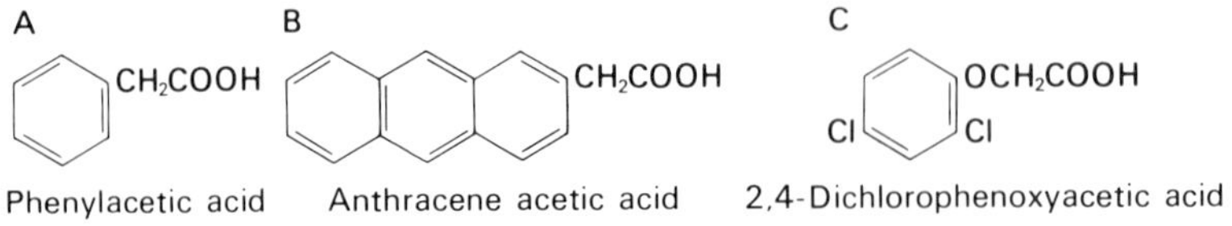

A
CH₂COOH
Phenylacetic acid

B
CH₂COOH
Anthracene acetic acid

C
OCH₂COOH
Cl   Cl
2,4-Dichlorophenoxyacetic acid

Fig. 18-9. A, phenylacetic acid; B, anthracene acetic acid; C, 2,4-dichlorophenoxyacetic acid, a widely used herbicide that is more toxic to dicots than to monocots.

Although 2,4-D is relatively "safe," it is important to reserve judgement about whether a new and untested but chemically similar substance is or is not toxic. There is increasing evidence that some auxin-related herbicides may be **teratogenic** (i.e. may affect development and thereby cause birth defects). Thus, what is harmless for most of the population could cause serious abnormalities in the offspring of exposed pregnant women. It is generally advisable to avoid exposure to any chemical that has not been thoroughly tested.

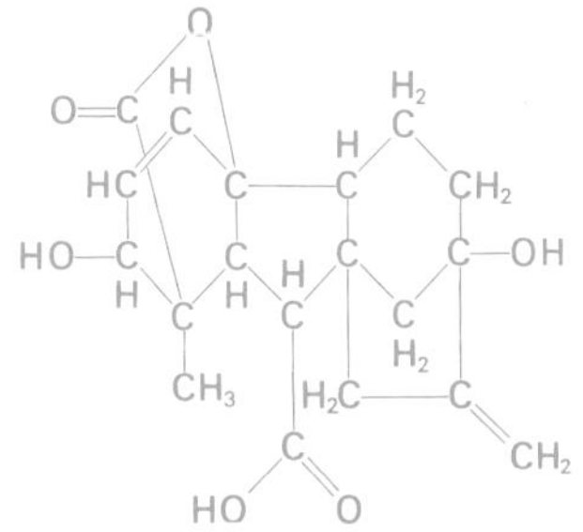

Fig. 18-10. Structure of gibberellin.

The Gibberellins: Promoters of Elongation

Although the **gibberellins** were first discovered at about the same time as auxin, an awareness of the existence of this kind of plant hormone and research on its role in the growth of plants lagged for a quarter century. Scientists in Europe and the United States remained preoccupied with auxin research while ignorant of the existence of the gibberellins because the latter were discovered by Japanese scientists who published their findings in Japanese journals. These investigators were trying to solve a problem of great economic importance. Early in this century Japan suffered serious agricultural losses from a fungus disease of rice, the *bakanae* ("foolish seedling") disease. Infected plants were unusually tall, weak stemmed, pale green, and at times failed to set fruit. With crop losses as high as 40 per cent, the Japanese had good reason for wanting to learn the cause of this disease and how to control it. It was soon discovered that the cause is a Fusarium type fungus called *Gibberella fujikuroi*. In 1926, Kurosawa demonstrated that a cell-free extract of the fungus applied to rice plants could produce the symptoms of excessive growth characteristic of the *bakanae* disease. It was not until 1935 that another Japanese investigator, Yabuta, crystallized the active ingredient in the fungus extract and called it **gibberellin**. Gibberellin research continued in Japan, unnoticed by Western scientists until after the end of World War II. In the early 1950's, the gibberellins were suddenly "discovered" by the rest of the world, and intensive studies began in the United States, England, and elsewhere. A number of gibberellins have been characterized chemically and they have been extracted not only from fungi but from a number of higher plants as well. Like the auxins, the gibberellins are considered to be naturally occurring plant hormones.

Chemically, the gibberellins are far more complex than auxin. They are diterpenoids having five rings (Fig. 18-10). Biological assays for gibberellin activity are based on the fact that these compounds are able to reverse genetic dwarfism. Thus the promotion of stem growth in dwarf peas and the elongation of dwarf corn have been used to determine the amounts of gibberellin in extracts of plant materials. The changes in growth patterns resulting from application of gibberellins provide a ready explanation for dwarf and rosette patterns that are seen in plants like lettuce and cabbage. A head of cabbage is essentially a tall plant with many leaves at intervals (nodes) along the stem, but in which the growth of the internodes has failed. The result is that successive new leaves are produced immediately above the older ones. Application of gibberellins to such species results in extensive stem growth (internode elongation) and 4 to 5 meters tall cabbages have been produced (Fig. 18-11). The normal growth patterns of cabbage, lettuce, and dwarf corn or peas are undoubtedly a result of the low levels of gibberellins normally produced in these species.

In addition to their obvious role in promoting stem elongation, gibberellins have far-reaching effects on seeds. Their application can break dormancy, an indication that the natural termination of dormancy is accompanied by an increase in the gibberellin content of seeds. In certain seeds such as lettuce, where light is needed for germination, gibberellins can substitute for light in promoting germination. The synthesis of enzymes needed to digest food reserves during the germination of seeds is triggered by gibberellin produced in the embryo. This last example of how gibberellins affect seeds provides an amazing story of how a plant hormone can act on the molecular level.

GIBBERELLINS AND THE DEREPRESSION OF GENES

Although by 1940 the Japanese had established that gibberellin promotes the germination of barley and rice seeds, it was not until 1960 that a Japanese (Yomo) and an Australian (Paleg) simultaneously reached the conclusion that gibberellin is the chemical which activates genes for production of digestive enzymes. The seeds of cereals consist mainly of an embryo and the endosperm, or food storage tissue. The latter is surrounded by cells of the **aleurone** layer (Fig. 18-12). During germination, the starch stored in the endosperm is hydrolyzed. As long ago as 1890, Haberlandt had recognized that in order for digestion of the endosperm to occur, the cells of the aleurone layer had to secrete an enzyme (α-amylase). This enzyme is formed only if the embryo is present. If barley seeds are cut in half, starch in the embryoless halves does not liquify. But, as Yomo and Paleg showed, if gibberellin in concentrations as low as 2×10^{-11} molar is applied to the embryoless halves, the digestion of starch can be observed within 48 hours (Fig. 18-13). More recently, Varner has proved that the α-amylase secreted by the aleurone layer is formed by de novo synthesis of the enzyme. Here we have a clear-cut case of a hormone, gibberellin, acting as an effector molecule to derepress the genetic information needed for the synthesis of a specific enzyme protein. Without the aleurone layer, gibberellins have no effect on the endosperm tissue. Gibberellin synthesized in the embryo triggers the production of many other enzymes during germination. In addition, it activates enzymes involved in the breakdown of cell walls, an effect that not only speeds the digestion of food reserves but also weakens the seed coats and facilitates the emergence of the root of the developing embryo.

OTHER EFFECTS OF GIBBERELLINS ON DEVELOPMENT

Gibberellins have been shown to affect many other aspects of plant development. They can induce flowering in certain species requiring specific photoperiods or temperature regimes. The bypassing of photoperiodic requirements for flowering or the light requirements for germination by gibberellin is not particularly mysterious. It has been shown that the induction of flowering as well as the start of germination is accompanied by increases in gibberellin content. Thus, light may promote germination by stimulating the synthesis of gibberellin in the embryo. Developing fruits and seeds are good sources of

Fig. 18-11. Cabbage plants treated with gibberellic acid grow to heights of several meters. Untreated plants are shown at bottom left. Cabbage is a biennial that "bolts" and forms flowers only during its 2nd yr of growth. Gibberellins cause flowering during the 1st yr. (Courtesy of S. H. Wittwer, Michigan Agricultural Experiment Station, Michigan State University, East Lansing, Mich.)

gibberellins, and it appears likely that the gibberellins are involved in the development of these structures. One poorly understood effect of gibberellins is their influence on the

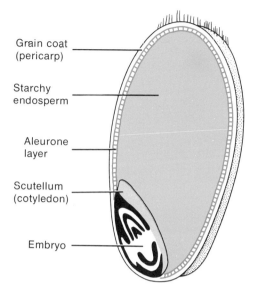

Grain coat (pericarp)

Starchy endosperm

Aleurone layer

Scutellum (cotyledon)

Embryo

Fig. 18-12. Structure of barley seed. Longitudinal section showing position of embryo, endosperm, aleurone layer, and seed coat.

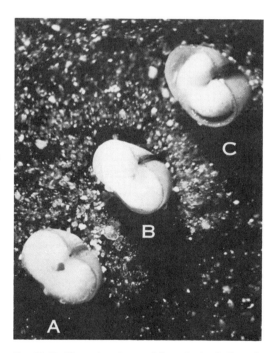

Fig. 18-13. The cut surfaces of three barley half seeds (embryo-containing halves removed) have been treated as follows: A, water control; B, gibberellin solution, 1 part per billion; C, gibberellin solution, 100 parts per billion. Forty-eight hours after treatment, digestion of endosperm tissue is very evident in C, less so in B, while there is no evidence of digestion in the control (A). (Photograph by J. E. Varner. From van Overbeck, J. 1968. The control of plant grower. *Sci. Am.* 219:75–81. Copyright © 1968. Scientific American, Inc., All rights reserved.)

sex of flowers. Perhaps this effect is in some ways parallel to the hormonal control of secondary sex characteristics in animals. Gibberellins promote the formation of **staminate** (i.e. male) flowers in cucumbers and squashes. In these species auxin has the opposite effect: promoting the formation of **pistillate** (i.e. female) flowers.

SYNTHETIC ANTIGIBBERELLINS

In recent years the manufacturers of agricultural chemicals have produced compounds which have effects on the growth patterns of plants just opposite to those produced by gibberellins. **AMO-1618** (an ammonium salt of pyridine carboxylate), **CCC** (a chlorinated choline) and **Phosphon** (a chlorinated phosphonium salt) act as inhibitors of stem elongation. Application of these compounds to plants grown in greenhouses during the winter results in shorter, stockier plants with larger, darker green foliage. The effects of these plant growth regulators can be counteracted by gibberellins. Recently it has been shown that growth retardants such as AMO-1618 and CCC act by suppressing gibberellin synthesis. They are now being used as selective inhibitors to investigate the physiological and molecular action of gibberellins.

The Cytokinins: Cell Division Factors that Act Synergistically with Auxin

Although the cytokinins are a class of naturally occurring plant hormones that have been recognized rather recently, the idea of chemical control of cell division is an old one. Early in this century Haberlandt discovered that phloem exudates could induce cell division. In potato tuber cells, crushed cells promote cell division near a wound, while wound healing does not occur as readily if the contents of injured cells are rinsed away. After plant tissue culture techniques had been developed in the 1940's and 1950's, investigators, including Steward at Cornell and Skoog at the University of Wisconsin, attempted to devise completely defined media capable of sustaining indefinite growth of plant tissues in vitro. Both investigators found that a factor in coconut milk was necessary for sustained growth of excised plant tissues and the existence of yet unknown plant hormones was suspected.

In 1955, C. O. Miller separated a compound from yeast DNA that stimulated cell division.

This compound was found to be a purine derivative, 6-furfuryl amino purine. It was called kinetin, and the generic name kinin was proposed because it promoted **cytokinesis** (cell division). The name kinin was not retained, however, because at about the same time, zoologists had proposed that this term be used for a different class of biological products—polypeptides that act on smooth muscles and nerve endings as do insect and snake venoms. The plant hormones having the same effects as the initially isolated purine derivative are now referred to as the **cytokinins**.

The cell division-promoting compounds isolated originally from such strange sources as yeast and herring sperm DNA must now be viewed as artifacts. However, the biological activity of these breakdown products of nucleic acids was very real. By 1964 Letham had isolated and characterized an active factor from young kernels of sweet corn. This substance was the first naturally occurring cytokinin identified and has been called **zeatin** (Fig. 18-14). Chemically, it is 6-(4-hydroxy-3-methyl-trans-2-butenylamino) purine, not surprising because DNA contains much purine. Zeatin has since been obtained from peas and spinach. Furthermore, zeatin analogues and closely related chemicals have been extracted from a variety of sources. One such close relative, 6-(γ,γ-dimethylallylamino)purine, has been obtained from hydrolysates of the transfer RNA's for serine and tyrosine from a number of animal and plant sources.

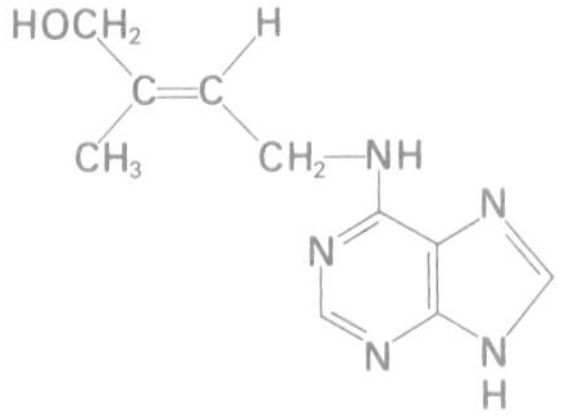

At first, the only known effect of the cytokinins was their ability to promote cell division in cultures of plant tissues such as tobacco **callus** (tissue of thin-walled cells developed on wound surfaces) where growth is dependent on added cytokinins. A very strange aspect of this promotion of cell division is that cytokinins alone have relatively little effect, yet they have a remarkable capacity to act synergistically with other hormones. This is illustrated by the absence of growth in tobacco callus in the presence of kinetin alone and the great promoting effects of kinetin in the presence of auxin. The range of effective kinetin concentrations is dependent on the level of auxin present. In addition to promoting cell division, cytokinins were soon found to have profound effects on the differentiation of cultured plant tissues (Fig. 18-15). The ratio of auxin to cytokinin in the medium is a controlling factor in determining the kind of differentiation that will take place: high auxin to cytokinin ratios favor root formation while low auxin to cytokinin ratios lead to shoot

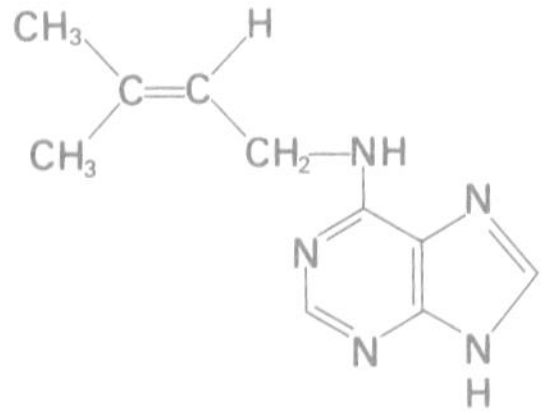

Zeatin

6-(γ,γ-Dimethylallylamino)-purine

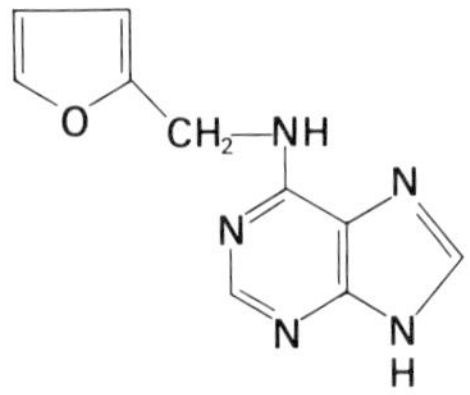

Kinetin

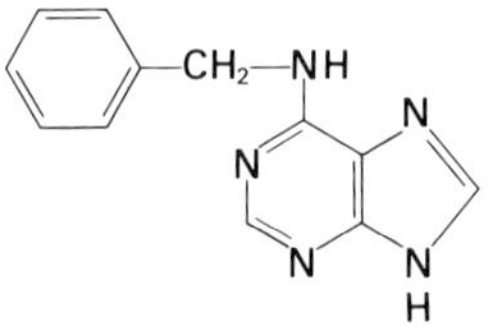

6-Benzylaminopurine

Fig. 18-14. Structures of four cytokinins. Zeatin and 6-(γ,γ-dimethylallylamino)purine occur naturally in plants. Kinetin and 6-benzyladenine are synthetic cytokinins.

initiation. This kind of relationship was not too surprising once the effect of **auxin to cytokinin ratio** had been established because in nature new shoots form on the stumps of trees and adventitious roots are produced on the base of a cutting or in the region just above a girdle on a stem. A plant's main source of auxin is its shoot system with major sites of synthesis in the apical meristems and leaves. A girdle interferes with the transport of auxin down the stem and auxin accumulates above the girdle. The high auxin cytokinin ratio above the girdle promotes formation of adventitious roots. Cutting off the supply of auxin by girdling or by chopping down a tree results in a low auxin to cytokinin ratio below the girdle or at the top of the stump. The low ratio promotes the formation of buds and development of new shoots.

It is also evident that auxin to cytokinin ratios are important in the control of **apical dominance,** a phenomenon we have already discussed in terms of auxin concentration. As you will recall, lateral buds develop when apical meristems (and the source of auxin) are removed. The development of lateral buds can be inhibited by applying auxin to replace that formerly provided by the apical meristem. The action of applied auxin can be overcome by simultaneously applying cytokinins.

CYTOKININS AND DELAY OF SENESCENCE

An additional and perhaps very different effect of the cytokinins that is of obvious commercial application is their generally inhibiting action on changes associated with senescence. The synthetic cytokinins, kinetin and 6-benzylaminopurine, have been used in the postharvest preservation of vegetable crops such as asparagus, broccoli, and celery. Not only is the fresh appearance of such crops maintained, but also their storage life is extended and their nutritional value is preserved. In part, these **antisenescence effects** of the cytokinins are the result of reduced respiratory rates. The breakdown of chlorophylls as well as hydrolysis of proteins is retarded. When cytokinins are applied to the leaves of intact plants, a transport of nutrients toward the treated areas can be observed.

Initially, the only available bioassay method for the cytokinins was their promoting effect on cultures of plant tissues in vitro, a slow and laborious technique. More recently, rapid assays based on the expansion of young leaves employing discs cut from radish or etiolated bean leaves have been developed. It has also been demonstrated that assays can be based on the delay in chlorophyll breakdown in detached leaves.

Interactions of Auxins, Gibberellins, and Cytokinins

Some specific interactions of the naturally occurring plant hormones have already been noted. Especially intriguing is the fact that in regulating differentiation, the cytokinins and auxins interact. It is the ratio of their concentrations rather than the absolute concentration of either that determines what pattern of growth will occur. Studies with plant tissue cultures have clearly shown that not only auxin and cytokinin but also gibberellins must be present for maximum growth. With completely defined media it is now possible to obtain at will practically any desired pattern or rate of growth in isolated plant tissues.

In the development of plants from seeds, the plant hormones tend to act in sequence. In the earliest phases of germination, the production of enzymes needed for hydrolysis of food reserves is triggered by the gibberellins. As cell division begins in the developing embryo, the cytokinins are essential for the promotion of nucleic acid and protein synthesis. Later, when the seedling has emerged and the proper orientation is essential, the auxins play a major role in tropic responses to light and gravity. The levels of hormones in developing plants are continually changing. Be-

Fig. 18-15. By controlling the relative amounts of cytokinin and auxin in culture media, Skoog and his collaborators at the University of Wisconsin were able to grow roots and shoots from cultures of undifferentiated tobacco stem cells. Flasks (left to right) contained the following concentrations of 6-(γ-γ-dimethylallylamino)purine: 0, 0.04, 0.2, 1.0, 5.0, and 25.0 μmol per liter plus standard amounts of minerals, sucrose, vitamins, and auxin. Cultures had grown for 6 wk. (From van Overbeck, J. 1968. The control of plant growth. *Sci. Am*. 219:75–81. Copyright © 1968. Scientific American, Inc. All rights reserved.)

cause of the ways in which the auxins, gibberellins, and cytokinins interact, a change in the concentration of one can affect the action of constant levels of the others.

Ethylene: The Gaseous Hormone Involved in Ripening Fruits and Leaf Abscission

Ethylene, $CH_2 = CH_2$, is a simple hydrocarbon occurring in natural gas. Between 1860 and 1870 there were reports of defoliation of trees and many varieties of herbaceous plants after accidental exposure to illuminating gas. It was not until the early 1900's, however, that ethylene was identified as the biologically active component of this fuel. For many years the striking effects of ethylene on plants, including the promotion of **leaf abscission,** induction of **epinasty** (the downward bending of petioles, Fig. 18-16), and a hastening of the ripening of fruit were regarded as interesting curiosities.

With the introduction of the technique of gas chromatography, a sensitive means for measuring the concentration of ethylene in mixtures of gases became available. It is now a proven fact that ethylene is a product of plant metabolism. Its production is prevented by temperature extremes (i.e. near 0°C or above 40°C) and low oxygen tensions. In the natural ripening of fruits there is a sharp increase in the rate of ethylene formation. This so-called **climacteric,** when respiration rates increase sharply, can be induced by adding ethylene to the atmosphere where harvested fruits are stored, or it can be delayed by lower temperatures and rapid air circulation which prevents the synthesis and accumulation of naturally produced ethylene.

Fig. 18-16. Epinasty of tomato petioles caused by ethylene. Plant at right was exposed to ethylene; plant at left is a control. (Courtesy of F. B. Abeles.)

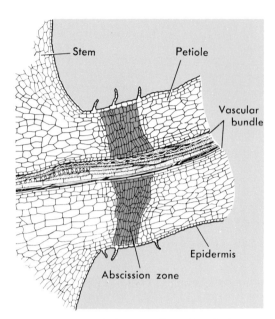

Fig. 18-17. The abscission zone differentiates in tissues at the base of the petiole. As a leaf becomes senescent, the cells in this zone separate. Ethylene promotes these changes by inducing the synthesis of cellulases. (From Torrey, J. G. 1967. *Development in Flowering Plants.* The Macmillan Co., New York.)

The production of ethylene generally occurs in those parts of plants where auxin content is high. In fact, the production of ethylene can be increased very rapidly (within 15 to 30 min) by applying auxin. The capacity to produce ethylene in response to exogenous auxin is related to the age of the tissue and falls off rapidly during senescence. In promoting leaf abscission, ethylene has its greatest effects on old leaves and leaf drop can be prevented by applying auxin.

LEAF ABSCISSION

A century ago von Mohl had shown that two kinds of changes are involved in leaf fall: (1) There is a differentiation of cells at the base of the petiole with the formation of an **abscission zone** (Fig. 18-17), and (2) the cells in this zone separate from each other with their walls intact. The work of Abeles and his associates has contributed much to the understanding of how ethylene is involved at the molecular level in the process of leaf abscission. According to the aging-ethylene hypothesis proposed by Abeles, the essential role of ethylene is to serve as an effector molecule triggering the synthesis of enzymes responsible for cell separation in the abscission zone. As would be predicted from this hypothesis, the stimulation of mRNA and protein synthesis (including increases in cellulase) by ethylene have been detected in the abscission zone. Inhibitors of DNA-dependent RNA synthesis (actinomycin D) and protein synthesis (cycloheximide) retard abscission and block the usual ethylene-triggered changes in the separation layer.

The realization that ethylene should be classified as a naturally occurring plant hormone has greatly simplified the interpretation of many diverse observations concerning the effects of specific chemicals on leaf abscission. It has now been established that the many seemingly unrelated chemicals that are effective as defoliants all have the common effect of stimulating ethylene production and therefore have a single physiological basis for their action. Although the impetus for research on the chemical control of leaf abscission initially came from military needs, peaceful applications of the results of such chemical warfare research are numerous. The machine harvesting of some crops such as cotton is much simpler if leaves are absent. If some easy means were available for controlling the time when leaves are shed from trees in the fall, homeowners would be everlastingly grateful for the simplification of their autumn chore of raking leaves.

Abscisic Acid: A Naturally Occurring Inhibitor of Development

Abscisic acid (also called **abscisin II** or **dormin** in the literature of plant growth-regulating chemicals) is a carboxylic acid with a structure related to that of vitamin A (Fig. 18-18). It was originally found in the leaves of woody plants grown under conditions of short daily photoperiod and in mature cotton fruit. Present evidence indicates that the occurrence of this hormone is practically universal in higher plants. Abscisic acid induces dormancy in buds, inhibits germination, retards plant growth (especially elongation) and hastens senescence. As might be expected, it interacts in complex ways with other plant hormones. There is a direct an-

tagonism between abscisic acid and the gibberellins: abscisic acid inhibits and the gibberellins promote elongation of stems, germination of seeds, and the sprouting of buds. The action of one can be counteracted by applying the other. Abscisic acid and the cytokinins have antagonistic effects in lettuce seed germination, leaf senescence, and the over-all growth of plants.

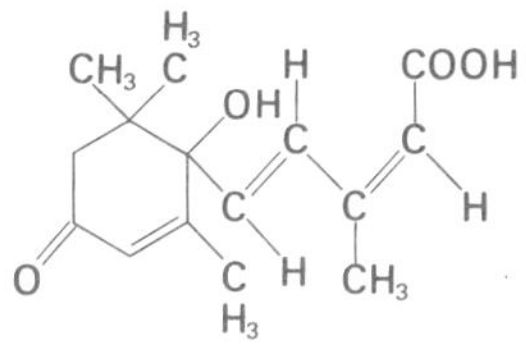

The mode of action of abscisic acid has not been established with certainty, but its antagonistic action toward other hormones, particularly the gibberellins, strongly suggests that it acts as a repressor that prevents the synthesis of messenger RNA specific for enzymes essential in the developmental process. Its effectiveness in causing dormancy of buds and seeds is of obvious advantage to plants during the winter months. The absence of dormancy in the buds and seeds of temperate plants would mean sure death for both the individual plant and its progeny.

Morphactins: Recently Synthesized Chemicals Affecting Plant Growth and Development

During the past several decades, chemical companies have begun a systematic testing of newly synthesized products for activity as plant growth regulators. Among the most interesting of these are fluorene-9-carboxylic

acid, and its derivatives, collectively referred to as the **morphactins** (from the term *mor-pho*genetically *acti*ve substances). The over-all effect of these compounds is a general retardation of growth. When dilute solutions are sprayed on vegetation, they have little

Fig. 18-18. Structure of abscisic acid. This naturally occurring growth retardant is also called abscisin II and dormin.

Fig. 18-19. Effect of morphactin (1T 3233, 6 × 10⁻⁵ M) on the phototropic response of oat seedlings. Treated seedlings on left, controls on right. (Courtesy of A. A. Kahn; from Kahn, A. A. 1967. Physiology of morphactins: effect on gravi- and photo-response. *Physiol. Plantarum* 20:306–313.)

Fig. 18-20. Effect of morphactin (IT 3233, 6×10^{-5} M) on geotropism in seedlings of timothy (*Phleum praetense*) after 96 hr of growth. Left, treated; right, control. (Courtesy of A. A. Kahn; from Kahn, A. A. 1967. Physiology of morphactins: effect on gravi- and photo-response. *Physiol. Plantarum* 20:306–313.)

effect on the parts of plants already formed, while further growth is inhibited. The advantages of such compounds for those of us who appreciate green lawns but are too lazy to mow them are obvious. The most intriguing effects of the morphactins for those who are interested in how plants grow are their effects on phototropism and geotropism. Seeds germinated in the presence of these compounds show absolutely no response to unilateral illumination or gravity (Figs. 18-19 and 18-20).

The Future: Made-to-Order Plants?

With currently available herbicides, natural and synthetic hormones, and plant growth regulators and inhibitors, it is possible to eliminate all but a single wanted species from a farmer's fields, to grow plants of practically any desired shape and dimensions, and to induce abscission of plant parts such as leaves that interfere with the mechanical harvesting of other structures; in short, to grow plants to order. Although this approach to agriculture is still in its early trial and error stages, the future application of our expanding knowledge of the chemical control of plant growth and development has tremendous potential. Compared to some insecticides that tend to persist and accumulate in plants and later in herbivores and carnivores in the food chains, many of the herbicides and chemical regulators of plant growth tend to be degraded rapidly in the soil and are relatively nontoxic to animals. Some exceptions to this generalization have been found, however, and the indiscriminate use of weed killers is to be avoided. The destruction of undesirable plants often has profound effects on host-parasite and predator-prey relationships. Furthermore, as herbicides are more thoroughly tested for possible harmful effects on animals, we are beginning to learn that some can influence animal development, particularly early stages in the growth of the embryo. With growing concern about the environment, there is a healthy increase in caution about the introduction of any new chemical into the biosphere. No matter how desirable the short-term effects might appear to be, we must thoroughly test all new agricultural chemicals for their possible long-term consequences.

USEFUL REFERENCES

Galston, A. W., and P. J. Davies. 1970. *Control Mechanisms in Plant Development.* Prentice-Hall Inc., Englewood Cliffs, N. J. (Paperback.)

Galston, A. W., and P. J. Davies. 1969. Hormonal regulation in higher plants. *Science* 163:1288–1297.

Hegelson, J. P. 1968. The cytokinins. *Science* 161:974–981.

Hill, T. A. 1973. *Endogenous Plant Growth Substances.* Institute of Biology Studies in Biology No. 40. Edward Arnold (Publishers), Ltd., London. Distributed in the U.S. by Crane, Russak and Co., Inc., New York. (Paperback.)

Letham, D. L. 1967. Chemistry and physiology of kinetin-like compounds. *Ann. Rev. Plant Physiol.* 18:349–364.

Phillips, I. D. J. 1971. *The Biochemistry and Physiology of Plant Growth Hormones.* McGraw-Hill Book Co., New York. (Paperback.)

Steward, F. C., ed. 1972. *Plant Physiology: A Treatise.* Vol. VIB: Physiology of Development: The Hormones. Academic Press, New York.

Thimann, K. V. 1967. Tropisms in plants. *Embryologia* 10:89–113.

van Overbeek, J. 1968. The control of plant growth. *Sci. Am.* 212:75–81.

STRATEGIES: Part E

COMPO- NENTS OF BEHAVIOR

The octopus: its doughnut-shaped brain has proven exceptionally useful in analyzing the relation of brain to behavior. [Courtesy of M. Woodbridge Williams. Reproduced from Moment, G. B. 1967. *General Zoology*, 2nd ed. Houghton Mifflin Co., Boston]

19

Skeletal and Muscular Systems

For the life of man and the other vertebrates, the importance of a skeleton can hardly be exaggerated. The same holds true for the insects and other arthropods. On a planet having the gravitational pull of the earth, we would all be as hopelessly doomed without our skeletons as astronauts projected into orbit without spacesuits, although for different reasons. Without skeletal support, terrestrial animals of any appreciable size would rapidly suffocate, for there would be no rib cage to support the lungs of vertebrates or framework for the trachea of arthropods. In fact, under the pull of gravity, we would all slump down into more or less formless blobs.

Some knowledge of the skeleton is a prerequisite for any complete understanding of how most animals function. The skeleton comes first in the book which marks the beginning of the modern study of human anatomy, Andreas Vesalius' *De Humani Corporis Fabrica* (1543), and it remains first in many studies of anatomy.

Major Types and Functions of Skeletons

Two major types of support have appeared in the course of evolution: **exoskeletons** and **endoskeletons.** Exoskeletons, covering the entire surface of an animal, have been enormously successful. The vast majority of species and of individual animals are enclosed in an exoskeleton, since this is a characteristic of all the insects, crustaceans, and other arthropods. Except for the sponges there are only two important groups of animals with endoskeletons—the starfish and other echinoderms, and the vertebrates. Both possess calcareous, i.e. bony, support which is always covered by living tissue.

THE SKELETON'S FUNCTION

In addition to providing necessary **support,** a major function of the animal skeleton is to make possible **locomotion** of a more rapid and powerful sort than is found in any of the worms and all but a few of the mollusks. The notochord, running along the length of all chordates, and the backbone which develops around it in the vertebrates, is what gave our primitive ancestors a decisive advantage over the worms. The paired fins of fish and the paired limbs, whether legs or wings, of other vertebrates (though all are built on the same pattern), show wide diversity of adaptations for different modes of locomotion. The same is true of the legs of insects and of crustaceans. In both groups parts of the skeleton act as levers for the pull of muscles; in vertebrates the muscles are outside the skeleton, whereas in arthropods they are within.

In vertebrates the formation of red blood cells (and some white ones also) takes place in the marrow of the long bones of the limbs and in the sternum between the tips of the ribs.

In arthropods a major function of the skeleton is **protection.** The entire body is encased in armor, chitin, which is strengthened, in the case of lobsters and other crustaceans, by calcium. In vertebrates, this protective function is largely limited to the skull, which protects the brain, and the neural arch of the backbone, which encases the nerve cord down the back.

The advantages of an exoskeleton are clearly very great, as is shown by the enormous success of arthropods. An exoskeleton, however, has one great disadvantage. The obvious difficulty with living inside an exoskeleton is that growth in size is impossible without molting. This process, termed **ecdysis,** is a highly dangerous and physically difficult feat. A soft crab, for example, is at the mercy of its enemies in the brief period immediately after shedding its old skeleton but before its new exoskeleton has been secreted and hardened. Why are there no really big terrestrial arthropods? For any animal that must shed its skeleton to grow, there is an ecological reason imposed by the relationship of size (weight) to the pull of gravity on our planet. To be without any skeletal support is bad enough even temporarily for an aquatic organism, but for a terrestrial one it would be disastrous, because lungs or tracheal sacs

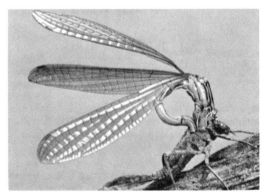

Fig. 19-1. An arthropod emerging from its old exoskeleton. The damsel fly in the middle figure has experienced one of the sometimes fatal accidents of life with an exoskeleton: its head became caught during ecdysis. (From von Frisch, K. 1964. *Biology.* Harper & Row, Publishers, New York.)

and cavities in which oxygen and carbon dioxide exchanges occur would collapse.

STRUCTURE AND COMPOSITION OF EXOSKELETONS

The structural material of exoskeletons is a nitrogen-containing polysaccharide called **chitin.** It clearly rivals the best plastics in many respects and surpasses most because fortunately it is biodegradable. Were it not, the world would be a vast trash pile of discarded insect and arthropod shells. Chitin is not too difficult to analyze. On hydrolysis in acid, it breaks down into glucosamine, i.e. the sugar glucose with an $-NH_2$ group added close to the keto end of the molecule. Glucosamine is found in numerous components of the living world, including the antibiotic streptomycin and the blood anticoagulant, heparin. Chitin itself is found in a wide variety of places: the bristles of annelids, the teeth and beaks of the octopus, and the cell walls of many fungi.

The glucosamines which make the nitrogen-containing polysaccharide, chitin, are linked in a long chain much as in cellulose. Furthermore, chitin is often combined with other components—sometimes pigments that confer the familiar browns, blacks, reds, and yellows to the bodies of insects and their relatives, and sometimes with calcium or some other strengthening material.

The properties of chitin vary enormously, depending on the other materials incorporated with it. It may be glass-clear as in the cornea of the eyes of insects and crustaceans, more or less permeable to water and other solvents. It may be thin and very flexible or else quite rigid. It is indigestible by most animals.

The **exoskeletons** of insects and crustaceans are very similar in structure. There is a noncellular basement membrane separating the exoskeleton from the rest of the body. Over this is a single layer of secretory cells, the epidermis. Outside the epidermis lies a relatively thick chitinous cuticle covered in turn by a thin epicuticle. The latter is nonchitinous and contains waxy substances, fatty acids, and cholesterol. No wonder insects are so resistant to poisonous sprays.

THE VERTEBRATE ENDOSKELETON

The endoskeleton of vertebrates and, more particularly, of mammals consists of two basic divisions. The **axial skeleton** (color, Fig. 19-2) consists of the skull and the backbone plus the ribs, sternum, and the remnants of the gill arches seen in the throat. The **appendicular skeleton** is composed of the pectoral and pelvic limb girdles and the two pairs of limbs.

The **skull**—especially the human skull—has been an object of study for many centuries. Classical Greeks, medieval Arabians, the men of the Renaissance, the poet Goethe, and many others have contributed to our knowledge, but it was T. H. Huxley, Charles Darwin's champion, who established the modern view of the nature of the mammalian skull. The skulls of all mammals, including man, are

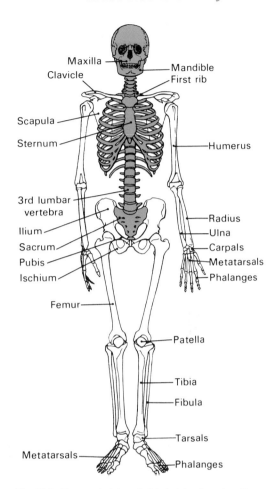

Fig. 19-2. Human skeleton. Axial skeleton in color. (From Moment, G. B., and H. M. Habermann. 1973. *Biology: A Full Spectrum.* The Williams & Wilkins Co., Baltimore.)

built on the same plan, have the same number of bones, and bone for bone are homologous throughout. Only the relative sizes and shapes of the individual skull bones are different.

A comparison of the skull of a man and a cat will demonstrate this fact. Notice (Fig. 19-3) that the cartilages and bones of the throat, the hyoid bone at the base of the tongue, the thyroid cartilage (Adam's apple), and the other derivatives of the seven embryonic gill arches are basically the same. The chief difference between the two skulls lies in the much greater growth of the bones which protect the human brain, viz. the frontal, temporal, parietal, and occipital bones.

In mammals, the **backbone** is divided into five regions—cervical, thoracic, lumbar, sacral, and caudal. With the exception of only two or three species, there are precisely seven **cervical vertebrae** in the necks of all

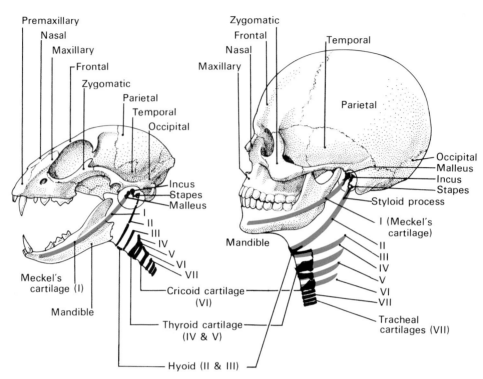

Fig. 19-3. Human and cat skulls showing their basic homology bone for bone. The Roman numerals indicate the gill arches, which are represented with solid black where ossified. (From Moment, G. B. 1967. *General Zoology,* 2nd ed. Houghton Mifflin Co., Boston.)

mammals—men, giraffes, whales, or mice. The differences in length in these animals are due entirely to differences in lengths of the seven cervical vertebrae. Why this is so is unknown and is all the more puzzling because birds with long necks, like swans and flamingos, possess many more cervical vertebrae than do birds with short necks, like owls. The **thoracic vertebrae** bear the **ribs,** one pair per vertebra. There are usually 12–14 ribs, depending on the species. There are 12 in both sexes of *Homo sapiens.*

The **appendicular skeleton,** consisting of **limb girdles** and **limbs,** has been highly modified in many of the mammals, but the basic pattern is unchanged from the ancestral reptiles. Each limb is supported by a tripod of three bones. The bases of the tripods press against the axial skeleton, i.e. against the backbone. At the apex of the tripod, the three bones meet the first—i.e. the most proximal—bone of the leg. This is the humerus in the front leg or arm, and the femur in the hind leg. The pair of tripods that support the front legs or the arms constitute the **pectoral girdle;** the pair that support the hind legs, the **pelvic girdle.**

It is often asked why the pelvic girdle of mammals, through which the young must be born, should be fused in such a rigid manner. The pectoral girdle, in marked contrast, is very loosely connected with the axial skeleton via shoulder muscles and ligaments. It seems probable that the very different construction of the two limb girdles and their contrasting connection with the backbone is the result of adaptation for running and jumping. The rigid pelvis transmits the main push of the hind legs to the backbone and thus to the whole skeletal frame simultaneously. However, after making a leap, mammals from horses to rabbits land on their front legs. The flexible muscles of the pectoral girdle can help absorb the shock of landing.

The skeletons of front legs (or wings or arms) and hind legs are homologous bone for bone throughout the entire range of mammalian adaptations. In the forelimb are a single **humerus,** then an **ulna** and **radius** side by side, with the ulna making a hinge joint at the elbow, a set of **carpals** in the wrist, five **metacarpals,** at least in the embryo, and five **digits** also, at least in the embryo. In the hind limb are a single **femur,** a **tibia,** and slim **fibula** side by side in the shank, a set of **tarsals** in the ankle, and the five **metatarsals** and five **digits,** at least in the embryo. Compare the arm of a man, the flipper of a seal, the wing of a bat, and the column or foreleg of an elephant (see Fig. 22-5).

Units of the skeleton are held together by **ligaments,** bands of tough connective tissue composed of a protein, largely collagen. Similar bands that attach muscles to bones or to each other are called **tendons.**

STRUCTURE AND GROWTH OF BONES

The **growth of bones** is controlled by many factors going back ultimately to the genetic make-up of the individual. There is a regular order in which the epiphyses of the various human bones appear and then later fuse with the shaft. X ray photographs of the bones of the hand and wrist give an excellent indication of physiological age as contrasted with chronological age. Bone formation is profoundly influenced by vitamin D, by the pituitary, thyroid, and parathyroid glands, and by mechanical forces. Vitamin D, a sterol related structurally to the sex hormones, appears to facilitate the action of an enzyme concerned with the absorption of calcium and phosphorus from the gut and their reabsorption in the kidney. This enzyme is also active in bone, where it breaks down organic phosphorus

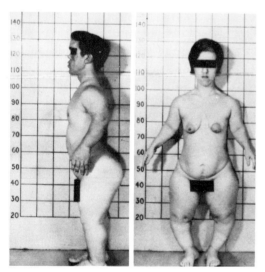

Fig. 19-4. Typical achondroplastic dwarfs. Note the short arms, normal trunk and generally powerful build. The Nibelung dwarfs of Norse mythology are usually described as achondroplastic. The condition is due to a dominant gene but how it acts to prevent normal development of the skeleton is obscure. (From Gamstrop, I. 1971. Characteristic clinical findings in some neurogenic myopathies and in some myogenic myopathies causing muscular weakness, hypotonia and atrophy in infancy and early childhood. *In* Bergsma, D., ed. *Birth Defects Original Article Series. Part VII, Muscle.* Vol. VII, No. 2. Published by The Williams & Wilkins Co., Baltimore, for The National Foundation-March of Dimes.)

compounds into the inorganic phosphates which help to form bone. Lack of vitamin D leads to rickets, i.e. weak, poorly developed bones which tend to become bent under the weight of the body. Estrogens favor bone formation and are sometimes used in the treatment of osteoporosis due to pathological bone absorption commonly seen in elderly women.

Principles of Muscle Action

A muscle can do only one thing: contract. It cannot push. The result of this basic fact can be clearly seen in the way muscles are organized in animals from hydra to man. Almost everywhere muscles are arranged in **antagonistic sets.** This is the meaning of all the groups of **flexors** which bend arms, wings, and legs and of the **extensors** which straighten them. It is why the iris of the eye is provided with both a circular **sphincter** muscle which closes the pupil and a radiating **dilator** muscle which enlarges the pupil. In a simple coelenterate like hydra or in an earthworm there is a set of longitudinal muscles running the length of the animal. Their contraction makes the body shorter and plumper. The antagonists are a set of circular muscles at right angles to the longitudinal ones. Their contraction makes the animal longer and thinner.

BASIC VERTEBRATE MUSCLE PATTERN

The muscles of a shark, the marine dogfish (*Squalus acanthias*), for example, show the basic vertebrate muscle arrangement with great clarity. The body muscles are divided into a series of more or less zig-zag segments or **myotomes** on each side of the body. The significance of such segmentation is that if there were only a single muscle along each side, the fish could only bend into the shape of the letter C. With a series of muscle segments, each of which is able to contract independently, waves of contraction can pass along the animal, pushing the body and tail against the water so that forward motion results. This muscular segmentation and the paired segmental series of nerves that go with it persist in all vertebrates.

In the course of evolution from fish to man, the segmented muscles of the trunk have undergone relatively little change, but the muscles which move the legs have fanned out over the muscles of the trunk. Muscles on what are merely simple paired fins in a shark have undergone a great increase in size and complexity over the legs, arms, or wings of the higher vertebrates.

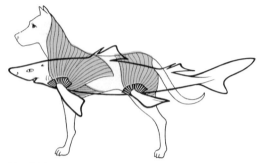

Fig. 19-5. Evolutionary development of the muscles moving the paired limbs from the dogfish (black) to the dog (color) and to man shown in Fig. 19-6. (From Moment, G. B. 1967. *General Zoology*, 2nd ed. Houghton Mifflin Co., Boston.)

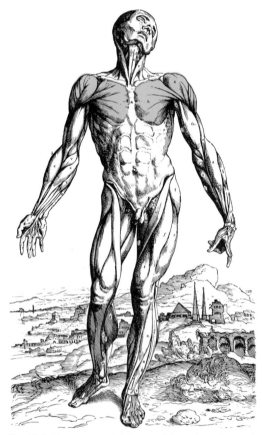

Fig. 19-6. Human superficial musculature, from the work of Andreas Vesalius, *De Humani Corporis Fabrica*, "Of the Structure of the Human Body." Published in 1543, this immense work marks the beginning of modern biology. Compare the colored shoulder and chest muscles which move the arms with the corresponding muscles in the dog and shark in Fig. 19-5.

Histologically, i.e. as seen under a light microscope, muscles are of two types, striated and smooth. The striations of striated muscle are very fine and incredibly regular cross markings at right angles to the direction of contraction. They can be seen in ground beef on a glass slide under a compound microscope. Until recently the meaning of the striations was completely unknown. **Striated muscles** are often called **skeletal muscles,** because they are usually attached to bones, or **voluntary muscles,** because for the most part they are under the control of the will. These muscles can contract rapidly, but they also fatigue rapidly.

Smooth muscles lack fine cross striations. Smooth muscles are also called **visceral muscles,** because they are found in the walls of the stomach, uterus, and other viscera, or **involuntary muscles,** because they cannot be directly controlled by the will. They contract and fatigue more slowly than striated muscle.

Heart or **cardiac muscle** has special properties. Although striated, cardiac muscle is physiologically involuntary. Its fibers appear to branch, although the evidence from electron microscope studies and from growing heart muscle cells in vitro clearly shows that this is not so. The cells are not lined up in parallel like ordinary skeletal muscle; neither do the individual cells fuse into a giant multinucleate network as seems to be the case under a light microscope.

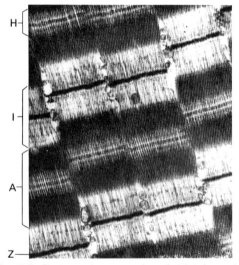

Fig. 19-7. Electron microscope view of striated muscle. A represents a dense anisotropic band transected by the lighter H (Hensen's) band, I an isotropic band transected by the very dense Z line (disk) bounding a sarcomere. (Courtesy of H. E. Huxley; from Copenhaver, W. M., R. P. Bunge, and M. B. Bunge. 1971. *Bailey's Textbook of History*, 16th ed. The Williams & Wilkins Co., Baltimore.)

The relationships of muscle types is shown in the following scheme:

Voluntary	Skeletal	}	Striated
Involuntary	Cardiac	}	
	Visceral		Smooth

MODERN KNOWLEDGE OF MUSCULAR CONTRACTION

For centuries there has been an impassable gulf between the anatomist who studied form and the physiologist who investigated function. The anatomist knew about the arrangement of extensors and flexors. The physiologist knew about such things as the speed of contraction and relaxation, the characteristics of fatigue, and even some of the chemical prerequisites and end products of muscle action. But no one knew what goes on inside the muscle itself. A muscle was what a physicist would call a "black box,"—impenetrable, perhaps unknowable. Modern biochemistry and the electron microscope have now combined to give us a look inside that "black box."

It has long been known that a muscle can contract in the absence of oxygen but that under such anaerobic conditions the **glycogen,** which represents stored energy in the muscle, disappears and **lactic acid** accumulates. After a relatively short time the muscle becomes fatigued, or at least will no longer respond to further stimulation. If oxygen is supplied, the lactic acid does not accumulate, even though the glycogen is broken down into glucose. As we now know, the glucose passes down the Embden-Meyerhof glycolytic pathway and is finally completely oxidized in the Krebs cycle and electron transport system of the mitochondria. It is hardly surprising that muscles are rich in mitochondria. They require energy in abundance.

When muscle contractions take place faster than the mitochondrial enzymes can provide energy as ATP packets, pyruvic acid accumulates at the end of the glycolytic pathway faster than the Krebs cycle can take it up. This excess pyruvic acid is promptly converted into lactic acid:

$$CH_3-\underset{O}{\overset{O}{C}}-COOH \underset{\substack{\text{lactate} \\ \text{dehydrogenase}}}{\overset{NAD_{Red} \quad NAD_{Ox}}{\rightleftharpoons}} CH_3-\underset{H}{\overset{OH}{C}}-COOH$$

Pyruvic acid   Lactic acid

The lactic acid is later reconverted into pyruvic acid or even glycogen in either the muscle itself or the liver, and oxidized in the mitochondrial Krebs cycle when oxygen becomes available. This is the chemical basis for being out of breath after violent exercise. One is "repaying" an "oxygen debt" as the Krebs cycle catches up with the material supplied to it by glycolysis. That ATP and the numerous mitochondria play an important role in contraction is certain.

A spectacular breakthrough into the "black box" of muscular contraction was achieved by the Hungarian-American Nobel prize winner, Albert Szent-Györgyi. He succeeded in extracting two kinds of protein from striated muscle, **actin** and **myosin.** When the two are mixed and then squirted out of a fine pipette into very dilute KCl solution, the combined **actomyosin** will contract when ATP is added as an energy source. Neither actin nor myosin alone will contract, no matter how much ATP is added; moreover, ATP is degraded to ADP in the process. These experiments strongly suggest that contraction is due to some kind of relationship between the two proteins. Actomyosin, or some closely similar protein, has now been extracted from the muscles of animals as far removed from man as sea anemones. Apparently the structural basis of muscle action is generally the same throughout the animal world.

A second breakthrough came from analysis with an electron microscope by H. E. Huxley. The electron photomicrographs of striated muscle confirm the presence of dark bands alternating with lighter ones as revealed by the light microscope. Each dark band is bisected by a narrow light one, and each lighter band by a very narrow dark line. Terminology has varied a bit but the wide dark bands are usually called **A bands,** A standing for anisotropic, meaning that the molecules here are so completely oriented that they transmit polarized light only in a particular plane. The A bands can be shown by chemical extraction techniques to be mostly myosin. The lighter stripe across the A band is called an **H (Hensen's) band.** The large light bands are usually known as **I bands,** I for isotropic. The very dark narrow lines bisecting the I bands are terminal or **Z lines.** The I bands turn out to be mostly actin. A single unit from Z line to Z line is called a **sarcomere.** Because muscle fibers are elongate cylinders, each line really represents a disk.

With the electron microscope H. E. Huxley was able to show—and many others have since confirmed and extended his work—that one end of each of the thinner actin fibers is attached to a Z line (disk) and the other end extends out into the sarcomere between the thicker myosin fibers. Thus, the actin and myosin fibers mesh much as the bristles of

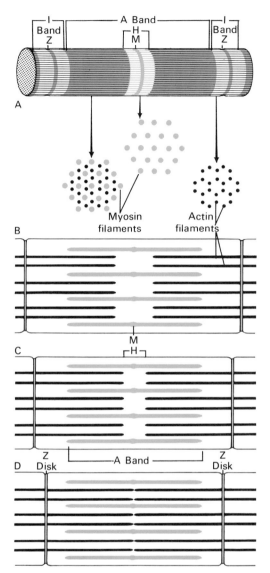

Fig. 19-8. Diagram showing the structure of a muscle fibril as revealed by the electon microscope in a relaxed condition, A and B, and during contraction, C and D. The thicker myosin filaments are shown in color, the thinner actin filaments in black. Note that a section through the H band shows only myosin filaments and through the I band close to the Z disks only actin. (From Moment, G. B. 1967. *General Zoology*, 2nd ed. Houghton Mifflin Co., Boston.)

two brushes would if pushed together. Cross sections viewed with an electron microscope show that when a muscle contracts, the actin fibers move further in between the myosin fibers and thus shorten the sarcomere much as an extension ladder shortens when the two parts slide together. Cross sections through the I bands near the Z lines at the ends of sarcomeres show only thin actin filaments. Sections through the H stripe in the center of

the A band show only heavier myosin filaments. Sections through the portion of the A bands adjacent to the I bands show both types of filaments. If this theory is correct, the A bands should not become narrower during contraction but the I bands should. They do.

This **sliding filament theory,** in contrast to any accordion-like folding theory, is also compatible with what the electron microscope has shown to be true of vertebrate smooth muscle, invertebrate striated and smooth muscle. Thus, once again an advance in biochemistry brings not only new knowledge but also a simplification and order on a vast scale.

What causes the actin and myosin filaments to slide along each other thus shortening the muscle? The whole story is still obscure but, under the higher powers of an electron microscope, myosin filaments show finger-like projections reaching out and attaching against **receptor sites** on the thinner actin filaments. It has been shown that myosin is not only a structural protein but also an enzyme, ATPase, and that actin carries ATP. When a nervous impulse stimulates a muscle, calcium ions are released into the intramuscular fluid initiating the ATPase activity of the myosin. Where the myosin projections are attached to the ATP-rich receptor sites, ATP is split, i.e. dephosphorylated, with the release of the energy that makes contraction occur. Apparently a structural change takes place in the molecular configuration of the myosin finger causing it to bend, and since all the myosin projections are structurally oriented in the same direction and therefore bend in the same direction, the myosin and actin filaments slide along each other. Once the bending has occurred, the myosin is no longer bound to the actin receptor site and springs back to its original straight position, ready to start another cycle.

MUSCULAR DYSTROPHY

One of the most tragic and baffling of human diseases is muscular dystrophy (*dystrophy*-faulty nutrition) in which the muscles gradually waste away. Ability to walk is commonly lost within 10 years after onset followed by muscular contractures leading to skeletal distortions and death from inanition, i.e., inability to assimilate food or respiratory failure, often complicated by infections. There are three types. Duchenne muscular dystrophy usually appears within the first 3 years of life. It is an X-linked recessive trait and like hemophilia, which is also X-linked, is usually seen in boys. There is genetic evi-

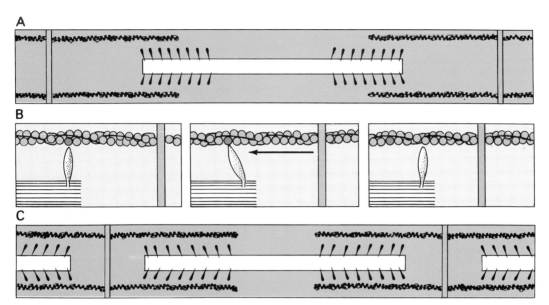

Fig. 19-9. A highly schematized diagram of present ideas about how the sliding filaments are pulled, or pushed, along each other during muscular contraction. A, myosin "fingers" extending toward the actin filaments. The vertical bands represent Z disks. B, under higher magnification, the initial attachment (leftmost) of a single finger to a specialized site on the actin filament. Energy is then released causing the finger to change its molecular configuration, i.e. bend, forcibly moving the filaments after which the finger is released. C, position of the filaments after pulling the Z disks closer together.

Fig. 19-10. A 7-yr-old girl with severe congenital dystrophy and bilateral facial weakness. Note the severe muscular atrophy and contractures. (From Bergsma, D., ed. 1971. *Birth Defects Original Articles Series.* Part VII: Muscle. Published by The Williams & Wilkins Co., Baltimore, for The National Foundation-March of Dimes.)

dence for linkage with red-green color blindness.

The face-shoulder type of muscular dystrophy may appear at any time of life but usually before middle age. Beginning with the face and shoulders it often spreads to the pelvis. Mild cases are common. It is inherited as an autosomal dominant. The limb girdle type appears due to an autosomal recessive gene and is intermediate in severity between the Duchenne and the face-shoulder forms.

The causes, i.e. the ways the mutant genes destroy the muscles, are unknown. There seems to be no deficiency in the anabolic hormones such as somatotropin from the pituitary which promotes protein synthesis. There is no evidence that muscular dystrophy is an autoimmune disease, at least as it occurs in mice. There is some evidence that it may be primarily a defect in the nervous system. For many years nerves have been known to be essential for the regeneration of legs in salamanders. Muscles which have been deprived of their nerve supply degenerate. Recently it has been shown that in chickens and mice with this disease there is a defect in acetylcholinesterase, an enzyme essential for the proper functioning of nerves.

USEFUL REFERENCES

Gray, H. 1959. *Anatomy of the Human Body*, 25th ed. Lea & Febiger, Philadelphia.

Wilkie, D. R. 1968. *Muscles*. St. Martin's Press, New York. (Paperback.)

Barzel, U. S. 1970. *Osteoporosis*. Grune & Stratton, New York. (A general presentation of bone secretion and absorption in relation to human problems.)

Alexander, R. M. 1968. *Animal Mechanics*. Univ. of Washington Press, Seattle. (How we all work, fish to athletes, plus the birds and the bees.)

Falls, H. B., ed. 1968. *Exercise Physiology*. Academic Press, New York. (Complete, from diet and exercise to drugs and athletic performance.)

20

The Nervous System

The nervous system presents problems of a depth and complexity not found elsewhere in the living world. Of all the objects known to science, none even remotely approaches the brain in intricacy of structure; and within it lies that ultimate mystery, consciousness. Long after the problem of how the genetic code in DNA transforms an egg cell into an organism is solved, the problems presented by the nervous system will remain to challenge scientist and philosopher alike. Anyone eager to explore areas beyond the fringe of knowledge—for example, "psychokinesis," the direct action of mind on matter—should remember that whether or not thought can directly influence the way dice fall, conscious purposes and physical actions are united within the central nervous system in some utterly unknown way. Answers to these deep-set questions are for future centuries. Meanwhile, we can look at the nervous systems found on this planet and learn something about the ways in which they work. Much actually has been learned already. A beginning has been made even on the biochemical basis of memory.

Neurons, the Basic Units

Although the nervous systems found in higher animals—an octopus, a man, or a honeybee—are very different in gross organization, the unit in every case is the same: the single nerve cell or **neuron.** Neurons occur in a great variety of forms but all are characterized by threadlike cytoplasmic extensions, often called **processes.** Some are elaborately branched, others are unbranched; some are

microscopic, others are over a meter long. These are the units which compose both the simplest and the most intricate wiring patterns of animal nervous systems from jellyfish to man. The telephone and computer analogies to the nervous system are useful and valid, as far as they go. What must be remembered is that neurons are more complex than the component parts of a computer and that the versatility and range of abilities of a complex nervous system not only exceed those of a computer but transcend them into new dimensions. The arguments in favor of the continued presence of astronauts in the space program are based on this greater range of capabilities and adaptability of the human nervous system compared with computerized packages of instruments.

Each neuron consists of a cell body, the nucleus with its surrounding cytoplasm a region often called the **perikaryon,** and from one to numerous filamentous extensions. Within the cytoplasm are numerous elongate granules called **Nissl bodies,** which are now known to contain RNA. The precise function of this RNA is uncertain. The only function ever suggested for RNA beyond protein synthesis is some role in memory, but this is far from proven and would seem unlikely, although not impossible, in the case of motor neurons.

A nerve fiber carrying impulses toward the cell body of a neuron is a **dendrite;** a fiber conducting impulses away from the cell body is an **axon.** Each neuron usually has several short dendrites, often extensively branched, and one axon which may or may not be branched.

Despite their almost bewildering variety of form, nerve cells fall into three great classes.

Motor or **efferent neurons** conduct impulses outward, away from the central nervous system to effector end organs, muscles or glands. (Note that both "efferent" and "exit" begin with e.) **Sensory** or **afferent neurons** conduct impulses toward the central nervous system from free sensory nerve ends or from nerve endings in complex sense organs. **Association, internuncial,** or **interneurons** (all three terms are in common use), conduct impulses from one neuron to others within the central nervous system or some of its ganglia.

The visible nerves, readily seen on dissection looking like silvery-white cords, are essentially cables composed of a few dozen to hundreds or even thousands of individual nerve fibers, i.e. axons or dendrites.

Most of the cranial nerves to and from the brain and the spinal nerves coming off the spinal cord are **myelinated.** That is, each nerve fiber is covered with an insulating **myelin sheath** made of a series of **Schwann cells** that have wrapped themselves around the nerve fiber. It is the myelin, composed of fatty material, which gives nerves their characteristic glistening appearance. Myelinated nerve fibers carry impulses faster than unmyelinated ones. Among the nerve cells, both around and between cell bodies, axons, and

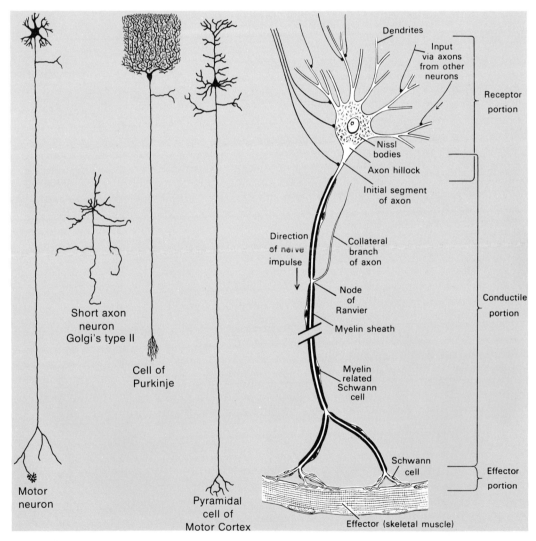

Fig. 20-1. Some common types of nerve cells. The large neuron at the right represents a typical motor, i.e. efferent, neuron having its nucleus and cell body within the spinal cord with its axon extending out to innervate a muscle perhaps in the foot. The other neurons are representative of those found in the brain. (From Copenhaver, W. M., R. P. Bunge, and M. B. Bunge, eds. *Bailey's Textbook of Histology,* 16th ed. The Williams & Wilkins Co., Baltimore. 1971.)

dendrites, are numerous supporting cells called **glia cells**. The interneurons or association cells occur in many forms. Compare, for example, a pyramidal cell from the cerebral cortex with a Purkinje cell from the cerebellum (see Fig. 20-1).

The Major Divisions

The vertebrate nervous system consists of two closely integrated parts. The **central nervous system, CNS,** is comprised of the brain and the spinal cord enclosed within the skull and backbone, respectively. The **periph-** **eral nervous system** is comprised of efferent or motor nerves carrying impulses from the CNS to all parts of the body and afferent or sensory nerves carrying impulses to the CNS. The famous **gray matter,** which forms the outer layers of the brain (Fig. 20-2) and the core of the spinal cord, is composed of the cell bodies of nerve cells, the **white matter** within the more central areas of the brain and which forms the outer layers of the spinal cord is made up of myelinated nerve fibers. These are commonly organized in bundles or tracts. The corpus callosum which connects the two sides of the brain is such a tract.

The peripheral nervous system of vertebrates is itself separable into two divisions:

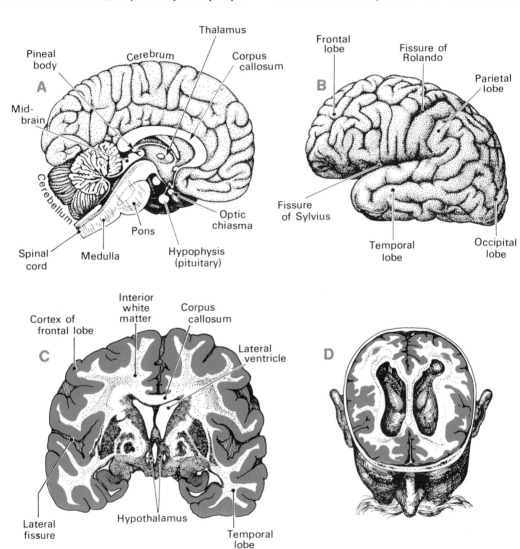

Fig. 20-2. The human brain. A, midsaggital section; B, view of left side; C, cross section; D, coronal section. The gray matter composed of the bodies of nerve cells is in color. (From Moment, G. B., and H. M. Habermann. 1973. *Biology: A Full Spectrum*. The Williams & Wilkins Co., Baltimore.)

the **somatic nervous system** which innervates the skeletal muscles, the skin, the major sense organs, etc., and is generally under the control of the will; and the **autonomic nervous system** which innervates the smooth muscles, especially of the viscera, the heart and blood vessels, etc., and is not under conscious control. (If it be objected that the concept of consciousness is not a scientific concept, the authors will not argue. One can avoid using the term, but at some cost in convenience and perhaps honesty.) The autonomic system in turn is composed of two divisions: the **sympathetic** and the **parasympathetic.**

THE SOMATIC PERIPHERAL NERVOUS SYSTEM

The peripheral nerves leave the central nervous system in pairs, one member of each pair going to the right and one to the left side. Nerves leaving the brain are known as **cranial nerves;** those leaving the spinal cord as **spinal nerves.**

In mammals there are 12 pairs of cranial nerves numbered consecutively starting from the most anterior and continuing back to the beginning of the spinal cord. They also bear names sometimes indicative of their function and sometimes of their anatomical relationships.

The pairs of spinal nerves vary in number: there are several dozen in some fish, 10 in a frog, and 31 in a man. In mammals, they are commonly grouped as **cervical, thoracic, lumbar, pelvic** or **sacral,** and **caudal** spinal nerves.

Regular segmentation of the nerves is a conspicuous feature of the human as well as of the shark's nervous system. Each nerve has a double root by which it connects to the spinal cord. There is a dorsal (posterior in an erect man) sensory root with a ganglion and a ventral (or anterior) unganglionated motor root. The nuclei and cell bodies of the sensory nerves constitute these ganglia. The nuclei and cell bodies of the motor nerves lie within the spinal cord, specifically within the lower wing of the "butterfly," often called in medicine the **anterior horn** (Fig. 20-3). The spinal nerves innervate the muscles of the trunk, limbs, and skin.

The separate functions of the dorsal and ventral roots were discovered in the early years of the last century by Charles Bell in England and Francois Magendie in France. Bell severed the ventral roots of spinal nerves in cats and observed paralysis, which demonstrated that the ventral roots are motor or

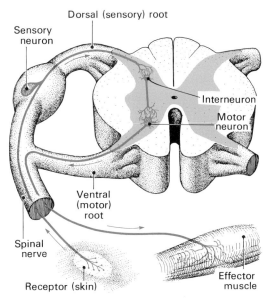

Fig. 20-3. Cross section of a spinal cord showing a pair of spinal nerves each with its dorsal and ventral roots. The gray matter where the cell nuclei are located is colored. (From Moment, G. B., and H. M. Habermann. 1973. *Biology: A Full Spectrum.* The Williams & Wilkins Co., Baltimore.)

efferent. A decade later Magendie showed that cutting the dorsal roots abolished sensitivity and so established what has been termed ever since the **Bell-Magendie law:** dorsal roots are sensory; ventral roots are motor.

THE AUTONOMIC PERIPHERAL NERVOUS SYSTEM

This system presides over organs and responses which most people have long believed are not under the direct control of the will — tear glands, heart, genitalia, gastrointestinal tract, blushing, sweating, and the like. In general this is true and not contradicted by the fact that most people know the effectiveness of the relation between the autonomic system and the highest centers of the brain. For example, involuntary blushing can be induced by sophisticated causes of embarrassment such as being seen out with a date by someone you had told you needed to spend the evening on a term paper. The real shocker came several years ago when it was found in several laboratories that you can learn to control several autonomic functions by an act of the will. Heart rate is one. To learn, you merely sit in front of a pair of lights, one red and one green. Your heart rate is monitored electronically. If you want to learn how to slow your heart rate, the apparatus is so arranged that as long as it is slowing or

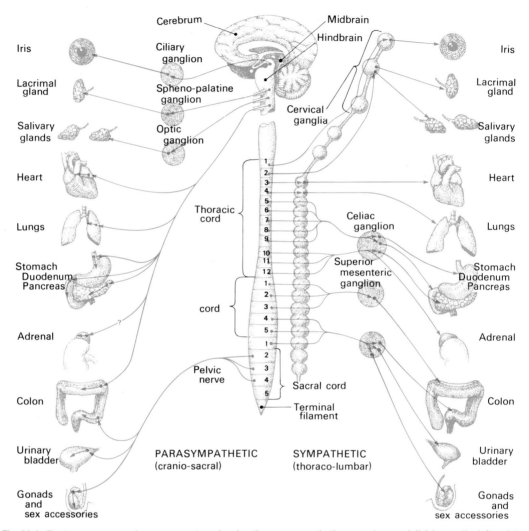

Fig. 20-4. The human autonomic nervous system showing the parasympathetic or cranio-sacral division on the left and the sympathetic or thoraeo-lumbar on the right. (From Moment, G. B., and H. M. Habermann. 1973. *Biology: A Full Spectrum.* The Williams & Wilkins Co., Baltimore.)

remains slow, the green light shows. Whenever the heart begins to beat faster, the red light comes on. With practice it is possible to become very skilled in such control. In a group of about 25 medical students tested in Baltimore, it was found that about one-third learned to regulate their heart rates, the others couldn't seem to learn. If you ask those who succeeded, they say they really don't know how. This is not really surprising. If you were to be asked how you move your arm, what could you say?

What is the two-parted autonomic nervous system like? Both the parasympathetic and sympathetic nerve fibers consist of two neurons in tandem, **preganglionic neurons** that have their cell bodies within the central

nervous system and **postganglionic neurons** that have their cell bodies in ganglia outside the central nervous system. The two parts are antagonists and each end organ, heart, stomach, sweat gland, etc., is innervated by both. Where the sympathetic excites, the parasympathetic inhibits although sometimes their roles are reversed.

The **sympathetic division** is composed of nerves having preganglionic neurons with cell bodies within the gray matter of the thoracic and lumbar regions of the spinal cord, and axons which pass out of the cord in the ventral motor roots of spinal nerves and then reach one of the chains of **sympathetic ganglia** by what is called a ramus or branch of a spinal nerve to a ganglion. The postganglionic fibers then extend for some distance, often considerable, to reach their end organ.

Preganglionic sympathetic fibers release a chemical, acetylcholine, and are hence called **cholinergic.** Acetylcholine stimulates the appropriate postganglionic neurons. Postganglionic sympathetic fibers do not release acetylcholine at their ends but release epinephrine (adrenaline) and therefore are called **adrenergic.** Their stimulation may produce dilation of the pupil, acceleration of the heartbeat, inhibition of gastrointestinal mobility, constriction of the pyloric, ileocolic, and anal sphincters, relaxation of the urinary bladder, secretion of sweat, and erection of hair. Acetylcholine is very widely distributed throughout the animal kingdom. So far epinephrine appears to be confined to vertebrates and earthworms.

The **parasympathetic division** consists of nerves having preganglionic neurons with cell bodies, some located within the brain and others within the pelvic region of the spinal cord. The preganglionic fibers travel all the way to the end organ—heart, stomach, colon, etc. Within their end organ they synapse with very short postganglionic fibers which terminate in the end organ. The parasympathetic nerves to the eye and salivary glands are exceptions, for they do enter ganglia and there synapse with postganglionic fibers which travel some distance to their end organs. Both pre- and postganglionic, parasympathetic nerves, like preganglionic sympathetic fibers, release acetylcholine and are termed **cholinergic.** The effects of the parasympathetic nerves are opposite to those of the sympathetic nerves.

The Nervous Impulse

A question of great human interest from ancient times has been: "What is a nervous impulse?" Only a little over a century ago many people believed that the nature of the impulse which carries messages along a nerve could never be known. Whenever it is asserted that a particular problem can never be solved, the case of the nervous impulse is worth remembering. The most eminent physiologist of the first half of the 19th century, Johannes Müller, maintained that even the speed of a nervous impulse was unknowable. Yet the print was hardly dry on Müller's assertion when one of his own students, Hermann Helmholtz, showed that this could indeed be discovered and measured.

Helmholtz's method was simplicity itself. He used the frog which Müller had reintroduced into physiology (Bell and Magendie had used mammals) and exposed the great sciatic nerve which innervates the gastrocnemius muscle in the calf of the leg. He stimulated the nerve at two points, one fairly close to the muscle, the second at a carefully measured distance along the nerve from the first. By very accurately measuring the time elapsing between application of the stimulus and contraction of the muscle, he discovered that the greater the distance of nerve over which the impulse had to travel, the longer the time between the stimulus and the contraction. Simple arithmetic gives the speed of the nervous impulse. It varies somewhat from species to species and from nerve to nerve. Nerves lacking the myelin sheath conduct impulses very slowly. In mammalian nerves at body temperature, the speed is about 100 meters (325 feet) per second, faster than a man can run but far slower than sound (about 300 meters per second).

New facts and new theories about the action of nerves have been accumulating ever since. These fall logically into three groups: what happens to a nerve cell at the time of stimulation; transmission of the nervous impulse along the nerve, and the means by which a nerve delivers its message either across a synapse or to an end organ such as a muscle.

Irritability is a common and presumably universal attribute of protoplasm. In nerves, this property has become specialized and heightened. A nervous impulse is a wave of **electronegativity** easily detected with a student-type galvanometer and a pair of electrodes. This wave of negativity is apparently not a mere accompaniment of the nervous impulse but is itself part of the driving mechanism. With the wave of negativity goes a sudden and profound **change in permeability** of the cell membrane. A nervous impulse is an **"all-or-none"** affair. Either a neuron fires off or it doesn't. The only requirement is that the exciting stimulus—whether mechanical, chemical, or electrical—must attain a certain intensity, usually called the **threshold.** The wave of negativity neither increases nor decreases as it moves. The action at every point along the fiber is a local action.

The old analogy of the powder train is a good one here. In fact a spectacular demonstration can be made by pouring a very narrow path of smokeless gunpowder along the top of a laboratory table and then carefully igniting one end. The gunpowder train analogy illustrates the fact that the impulse does not vary with the strength of the initial stimulus, provided only that the stimulus was sufficient to ignite the reaction. If one or more branches are made in the powder path, the initial spark may result in two, four, or many burning powder trains. This illustrates the **principle of amplification.** Because nerve action is local, total response depends on branching neural connections, not incident energy at a sense organ.

Modern ideas about the nature of this self-propagating event, the nervous impulse, are based on the theory of Julius Bernstein at the turn of the century. He pointed out that the wave of negativity could be explained on the following basis. If the outside of the nerve cell and its fiber are positive and the inside negative, and if the cell membrane is semipermeable, keeping the more negatively charged ions inside and the positively charged ones outside, then a nervous impulse could be a wave of breakdown in selective permeability permitting a relatively free passage of ions through the membrane so the membrane potential disappears.

To test the theory that the outside of a nerve fiber is positive and the inside negative, a fiber large enough to place one electrode actually inside the fiber and the other outside is required. The best place to find such giant nerve fibers is in the squid, where they're on the inner side of the mantle, a muscular body wall loosely surrounding the animal. No dissection is necessary, as the fibers lie on the surface.

The diameter of a single fiber is as large as an entire vertebrate nerve made up of hundreds of nerve fibers. Using these very large fibers, A. L. Hodgkin and A. Huxley at the Marine Biological Laboratory in Plymouth, England, were able to show, 40 years after Bernstein had proposed the electrical depolarization theory of the nervous impulse, that it was in fact correct. The electrical potential between inside and outside falls from about +70 mV outside, to zero and then to a reversed polarity, with the inside actually positive sometimes by as much as 50 mV before the normal resting condition is restored. The characteristics of the electrical change as an impulse passes along a nerve fiber can be handsomely observed if electrodes on a nerve are connected to an amplifier and then to an oscilloscope (an instrument similar to a television tube) (Fig. 20-5).

The explanation of how this wave of negativity is propagated has not been easy to obtain, and all the answers are not in yet. In a resting cell there is about 20 times as high a concentration of potassium ions on the inside as on the outside. The concentrations of sodium and chloride ions are far higher outside than inside. The wave of negativity induces a transient change in permeability which permits positive sodium ions to enter. Once the membrane is sufficiently depolarized, an electrical "hole" is produced into which the positive charges "fall," and the collapsing edge progresses like a row of falling dominos.

There is some evidence that acetylcholine is important in this process. Metabolic energy

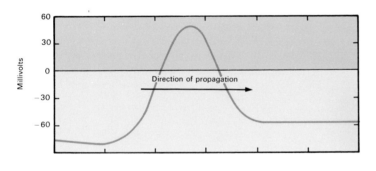

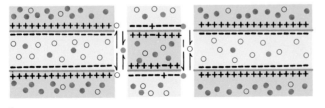

● Sodium ions
○ Potassium ions

Fig. 20-5. Diagram of the electrical charges which accompany and in large part constitute a nervous impulse. The electropositive regions are shown in darker color than the electronegative. (From Moment, G. B., and H. M. Habermann. 1973. *Biology: A Full Spectrum.* The Williams & Wilkins Co., Baltimore.)

is required to restore the resting potential and the marked differences in ion concentration. This can be shown by the use of metabolic poisons. The mechanism which maintains the higher external concentration of sodium by the expenditure of metabolic energy is referred to as the **sodium pump** (or sodium-potassium pump).

Synapses

A synapse is the place where one nerve cell speaks to another. They are of utmost importance because without them no nervous system could function. Furthermore, synapses are the sites of action of many drugs which can intensify, soften or wildly distort the neural message. Synapses may be located at the tips of the dendrites of the receiving nerve cell, directly on the cell body or at various intermediate points (Fig. 20-6). The transmitting ends of the efferent fibers swell into tiny **synaptic knobs** or **end bulbs** containing several small mitochondria and several dozen

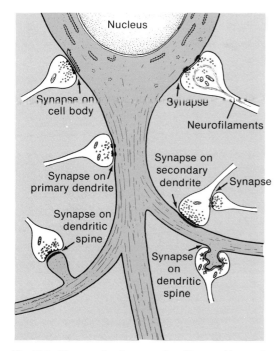

Fig. 20-6. Diagram showing several positions and types of synapses. Note that most are between an axon and a dendrite but that some of the end bulbs impinge directly on the receiving cell body. Both mitochondria and secretory granules are visible in the end bulbs. (From Copenhaver, W. M., R. P. Bunge, and M. B. Bunge, eds. 1971. *Bailey's Textbook of Histology*, 16th ed. The Williams & Wilkins Co., Baltimore.)

small synaptic vesicles which seem to contain the transmitter chemicals. These end bulbs do not actually come in contact with the receiving neuron, but are separated from it by a cleft visible with a electron microscope.

Transmission can pass only in one direction at a synapse, from the synaptic end bulb to the receiving cell. Thus, synapses act like valves without which the functioning of the nervous system would be thrown into chaos.

A very important fact about large motor nerve cells is that they receive impulses from hundreds of end bulbs of many different sensory and internuncial nerve fibers which form synapses directly against their cell bodies. This anatomical arrangement has two functional results. Impulses arriving from many different sources may cause the same motor nerve cell to fire off a message over what is clearly the **final common pathway** to a given muscle. A second result is that if several weak stimuli arrive from several different sources, none of sufficient strength by itself to cause the motor nerve to discharge, their **summation** may attain the necessary threshold for a motor nerve response.

Synaptic Transmission of Nervous Impulses

The properties of synapses, where an impulse passes from one neuron to another, and of myoneural junctions, where a nervous impulse passes from a nerve fiber to a muscle, are of the utmost importance. Some of the most powerful poisons known exert their deadly effects at these points. Among these poisons is curare, which South American Indians use on the tips of their arrowheads, the modern anticholinesterase type of nerve war gas, and some of the most dangerous insecticide sprays. Drugs of various sorts exert their effects here also. It is even thought by some that changes in synaptic connections are at the basis of learning. In the everyday functioning of the nervous system, synapses act as valves which transmit in one direction only, and of course the nervous system could not function without its thousands of billions of connecting points.

What happens when an impulse reaches a synapse at the end of its run? Axons terminate in one or a cluster of small **synaptic knobs** (or end bulbs) which press within about 20 nm (20 × 10^{-9} m, the kind of gap only an electron microscope can see) of a dendrite or nerve cell body. Within the knob, the vesicles evidently discharge a transmitter substance, either acetylcholine or epinephrine, when the impulse arrives. If the nerve fiber is a stimulating fiber, the transmitter substance lowers the membrane potential of the receiving or post-synaptic cell on the far side of the gap. The lowered potential of that cell is called by neurophysiologists and some psychologists the **excitatory postsynaptic potential** or **EPSP.** In some instances the discharge of a single end knob can lower the postsynaptic potential to the approximately 50 mV necessary to trigger a wave of depolarization, i.e. a wave of electronegativity commonly called a nervous impulse. If the discharge of a single knob is not sufficient to produce an EPSP as low as 50 mV, then there will be no impulse shot off unless stimuli from several other nerve fibers reach the cell membrane. If enough end knobs discharge transmitter substance within a short enough interval of time, the excitatory postsynaptic potential will be low enough to initiate a nervous impulse. When several nerve fibers act together, it is called **spatial summation.** If the EPSP is produced by repeated discharge of a single end knob, it is called **temporal summation.** These types may occur together.

After the discharge of acetylcholine at a synapse, a special enzyme, acetylcholinesterase, destroys the acetylcholine and permits the receiving nerve to restore its membrane potential and thus be ready to respond to another stimulus, i.e. another dose of acetylcholine.

If the nerve carrying the initial impulse is an inhibitory nerve, like the vagus branch to the heart, the acetylcholine does not lower the membrane potential of the receiving neuron to any appreciable extent. It may even raise it, causing the membrane to be hyperpolarized. In some way an inhibitory message over a synapse stabilizes the membrane potential and either prevents or makes more difficult the triggering of an impulse. This new potential is called the **inhibitory postsynaptic potential** or **IPSP.**

CHEMICAL CONTROL OF SYNAPSES

Much study has been devoted to effects of biochemically active agents on synaptic transmission. Presumably, any agent which interferes with acetylcholine, epinephrine, norepinephrine or any other naturally occurring transmitter substance would interfere with the functioning of the nervous system. Such agents might facilitate or block transmission by combining with a natural transmit-

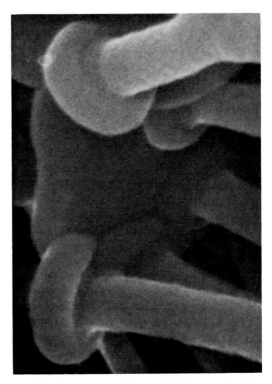

Fig. 20-7. Synaptic end bulbs, × 60,000, as seen with a scanning electron microscope. These are from a California marine snail but are closely similar to those in vertebrates. (Courtesy of E. R. Lewis.)

ter or acting on enzymes necessary for its removal or by affecting the receiving membrane. Lysergic acid diethylamide (LSD), for example, mimics serotonin (5-hydroxytryptamine), which is a derivative of the amino acid tryptophan. Serotonin in very small amounts is found in neural tissue of vertebrates, mollusks, and arthropods. It is the poison in the deadly saliva of an octopus, in the sting of a hornet, and in the venom of some spiders. Mental diseases, especially schizophrenia, have been attributed to disturbances of serotonin metabolism. Anticholinesterases are so effective that a few drops in the eye can kill a man. As with the hormones, there are specific receptor sites on the surfaces of nerve cells for many and perhaps all such drugs. Presumably if there were none, the drug would be ineffective. Whether or not this knowledge can ever be useful in the treatment of drug addicts is an open question. Knowledge of such potent and often psycho-pharmacologically active chemicals can furnish a two-edged sword able to wreak havoc or bring healing into human lives.

TRANSMITTERS   HALLUCINOGENS   TRANQUILIZERS

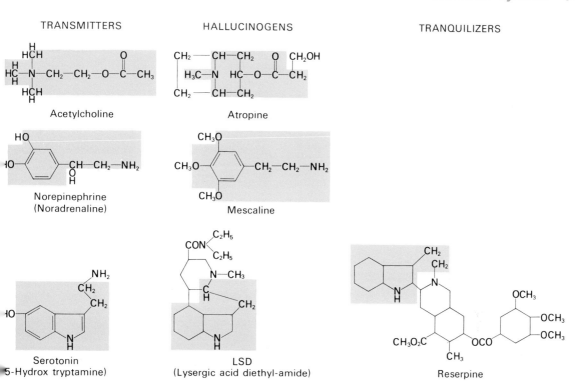

Fig. 20-8. Part of the evidence that hallucinogens and tranquilizers act at the synapses. Similarities of molecular structure are shown in color. (Read from left to right.)

Reflexes

Modern ideas about the action of the nervous system began with the French mathematician and philosopher René Descartes in the 17th century. He proposed that the unit of action of the nervous system is the **nervous reflex.** By this he meant that a stimulus, such as light striking the eyes, would set off an impulse to the brain which would be reflected back from the brain through nerves to muscles in the arm or some other organ.

One of the first men to find experimental confirmation for such a view was a versatile 18th century clergyman and amateur biologist, Stephen Hales, who also discovered blood pressure. Briefly, Hales found that one can destroy a frog's brain, by decapitation if necessary, and still not destroy responses to stimuli. A frog so treated is known as a **spinal animal.** If a small square of paper soaked in vinegar is placed on the back of such a frog, one of the hind legs will flick it off with great accuracy. Of course there is little flexibility in the reaction, for all "originality" is lost with the brain. If the spinal cord is destroyed by

pushing a wire down inside the backbone, these spinal reflexes cease permanently. Here then is clear proof that such responses can be mediated solely by the spinal cord, without any assistance from the brain.

Spinal men have occasionally been produced by war or other accidents. In the case of one army lieutenant, a piece of shrapnel completely severed his spinal cord between the 5th and 6th ribs. He retained control of his arms, but his legs were completely without sensation, nor was he able to move them at will. However, if a toe was pinched, his leg would bend at the knee and pull the foot back.

Since Hales' day much has been learned about reflexes. A number of reflexes concerned with breathing and with the heartbeat has been discussed in previous chapters. Salivation at the sight or odor of food is a famous reflex in dogs studied by Ivan Pavlov, but common to man as well. An easily observed reflex is the pupillary light reflex, although the neural pathways involved have turned out to be more complex than in a spinal reflex. Stand before a window and cover one eye completely for about 2 minutes. Then look in a mirror as your readmit light to the eye. You will observe that the pupil becomes smaller. This is due to a contraction of the sphincter muscle in the iris.

Although reflex action is simple to demonstrate in the pupil of your eye or with a spinal frog, the neural mechanisms underlying reflexes are complex. Think for a moment of the intricate interconnections and delicate adjustments necessary to place the toes of a frog's foot on a precise spot anywhere on the frog's back! Moreover, in an intact frog, the beautifully coordinated behavior of those long legs is not only directed by reflexes within the spinal cord but it is also under the overriding control of the brain. The simple picture of a reflex as an arc composed of a cutaneous sense organ, a sensory nerve passing into the spinal cord via a dorsal afferent root, making synaptic connections with a single interneuron within the cord which in turn stimulates a motor nerve leaving by a ventral efferent root and ultimately causing a muscle to contract, is a greatly over simplified picture; in fact, a fairy tale for children.

Understanding the reflex basis of action within the nervous system is well begun by reading Sir Charles Sherrington's classic *Integrative Action of the Nervous System*. Although originally published in the early part of this century, it has often been reprinted because this work brought knowledge of reflex action to a culmination which has not yet been appreciably altered, even though much more has been learned.

There are a number of rules, simple to state, which govern reflex action. To produce a reflex response, a stimulus must first attain a certain intensity; unless this **threshold** is achieved, nothing happens visibly. The threshold can be reached by a variety of means. The stimulus, whether a beam of light or a pinch on the toe, can be made more intense. A series of subthreshold stimuli may be applied until the animal responds. This phenomenon is called **temporal summation. Spatial summation** is also possible when an animal is given several stimuli, each subthreshold by itself, in different parts of the body. A most important point to remember is that the relationship between the intensity of the stimulus and the size of the response is a very indirect one.

It is important to remember that summation and threshold apply not only to the sensory nerve endings which receive the stimulus but also to the inter and motor neurons on which the synaptic end bulbs of other nerve cells impinge. For a nerve cell to send an impulse along its axon, the membrane of that cell has to be sufficiently electrically depolarized. Equally important, the synaptic end bulbs of many neurons raise the electrical polarization of the receiving cell membrane making it more difficult for that cell to fire off a nervous impulse. Such **inhibitory nerve fibers** are essential for without them not only would smoothly coordinated motion be impossible but a single impulse might spread throughout the organism producing general tetanus.

The Brain

The human brain has been called many things: "that great ravelled knot"; the "computer in the skull." And justifiably, because of all the structures known to man, the brain is the most complex, by several orders of magnitude. In fact, no computer remotely approaches the brain in complexity, much less in versatility. The number of neurons in the human brain has been calculated at about 12 billion. Vast numbers of them each receive branches from the axons of 10,000 other nerve cells. The number of possible combinations and permutations staggers the mind. Perhaps anyone familiar with Gödel's proof (that in mathematics there must always be an unproved axiom) may think it unfair to ask the brain to understand itself.

BRAIN SIZE AND INTELLIGENCE

A great deal has been made of the relation of brain size to intelligence. In general, the larger the brain, the greater the potential intelligence. A large absolute size appears necessary to provide for complex associative activities, including feedback mechanisms of self-adjustment. However, absolute size must be corrected for total body size. An elephant or a whale has a larger brain than a man in absolute terms, but not in proportion to body weight. The great size of an elephant's brain is presumably due to the large number of neurons essential to the sending and receiving of signals to the enormous body mass, and does not represent an increase in the higher associative centers. At the other end of the size scale, the proportionality rule again breaks down. Insectivores like moles, with extremely modest intellectual abilities, possess brains larger than man's in proportion to total body size.

Actual brain size varies widely in the higher primates. In chimpanzees and gorillas the brain ranges from 325 to 650 cm³ in volume. In fossil man-apes it ranged from 450 to 700 cm³; in fossil Java man 750 to 900 cm³; in Peking man, 800 to 1,200 cm³ and in Neanderthal man, 1,100 to 1,550 cm³. Although his brow was low and his features decidedly simian, brain size in Neanderthal man falls easily within the normal range of variation that is found in contemporary human populations.

Many attempts have been made to discover correlations between brain size or convolutions, on the one hand, and degree of intelligence and achievement in man, on the other. Except for obvious deformities such as those in microcephalic idiots, these efforts have not succeeded. Most men of outstanding intellectual ability have had brains in the upper half of the frequency distribution curve for size, but there have been a surprising number with brains well below average. It is not difficult to believe that along with size there may well be other factors—a slightly different twist of metabolism, for instance—that determine potential intelligence. And actual achievement is very clearly the result of many factors in addition to intelligence.

GENERAL BRAIN STRUCTURE

The basic structure of the brain is the same in all vertebrates whether fish, frog, bird, or man, because it always develops as five swellings at the anterior end of the embryonic neural tube. The names of these five major parts are the same in all vertebrates, and so are the basic structures developed from each. The difference is in the degree to which different parts are developed in different vertebrates. The inner central fluid-filled cavity connects all parts of the brain and continues down the spinal cord. This is why a bubble of air carefully injected into the central canal of the spinal cord of a patient who is sitting up will move up and into the cavities, **ventricles** as they are called, of the brain. Followed by X rays, such air bubbles can be used in the diagnosis and location of growths and other obstructions in the central nervous system.

The Cerebrum

The most anterior part of the brain develops two **cerebral hemispheres,** right and left. The roof or pallium of these hemispheres constitutes the cerebrum and is especially large in the mammals. In birds, the roof of the cerebral hemisphere is almost paper-thin. How then can birds learn so much? We now know that the characteristic song of most birds is somehow encoded in their brains. So also are the instructions about where and what kind of a nest to build, together with the appropriate construction methods. Perhaps even more remarkable, a small white-crowned sparrow carries within its skull navigational programming which will enable it to reach its own nesting grounds in southern

Alaska even after being transported in a closed cage from its winter home in California to Maryland. How a bird's brain can be programmed for such complex behavior is an unanswered question.

Discovering the answers to questions like these, not just for birds but for all animals, is motivating much of the research concerning brain structure. The difference between the brain of a fish and that of a fisherman, it should be recalled, does not lie primarily on the cellular level. The neurons of a jellyfish, a shark, and a dog are much alike. Far less does the difference lie on the molecular or atomic level. It is in its organization, its cytoarchitecture, that the characteristic features lie which confer on it such remarkable powers.

In primitive mammals like the insectivores (shrews, for example) and rodents, the surface or **cortex** of the brain is smooth, but in many others including the primates, the cortex is highly convoluted. In man a deep, more or less horizontal fissure, the **fissure of Sylvius,** separates the **temporal** lobe below from the **frontal** and **parietal** lobes above. The frontal lobe is separated from the parietal lobe by the vertical **fissure of Rolando** which runs upward from the fissure of Sylvius to the top of the cerebrum. The region at the posterior and lower end of each hemisphere is known as the **occipital lobe** (Fig. 20-2).

If the cerebral hemispheres are cut in half from left to right, it can be seen that the cortex, or superficial zone, is gray, whereas the inner portions are white with rounded gray masses here and there. The grayness of the gray matter is due to the presence of several billion cell bodies of neurons. The white of the white matter is due to the myelin sheaths of the axons of neurons. It is a mistake to regard intellectual ability as resident solely in the gray matter. The intercommunication systems represented by the white matter are equally essential. The rounded gray masses within the lower part of the brain are relay and shunting centers of neurons and are called **basal ganglia** or **basal nuclei.** (This, of course, is to use the word "nucleus" in its primary meaning of a "kernel" or central region.) The four largest of the basal ganglia make up the **corpus striatum,** which develops from the floor of the most anterior part of the brain, the telencephalon, and which constitutes the main bulk of the forebrain in birds.

The cerebral cortex of a higher mammal has six layers of cells each itself many cell layers thick. Most of the incoming fibers from other parts of the nervous system end in the fourth layer after branching extensively and making a wide variety of synaptic connec-

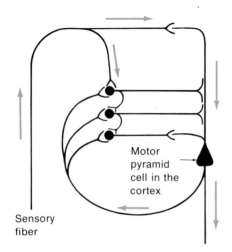

Motor pyramid cell in the cortex

Sensory fiber

Fig. 20-9. A simple feedback circuit between cells of the cerebral cortex. (From Moment, G. B., and H. M. Habermann. 1973. *Biology: A Full Spectrum.* The Williams & Wilkins Co., Baltimore.)

tions. It is now believed that a basic unit of cortical organization is a simple feedback loop (Fig. 20-9). The evidence is both visual from stained sections and physiological from electrical studies. The function of all these reverberating circuits is unknown.

The white matter immediately below the cortex contains innumerable fibers. Some of these axons pass from the cortex down into the basal ganglia or directly into the spinal cord. Others connect the two cerebral hemispheres. It is these right-left communicating fibers which make up the **corpus callosum**. Still other bands of fibers run at right angles to the right-left fibers and connect anterior and posterior cortical areas.

The corpus callosum long stood as one of the outstanding puzzles of brain organization. It is easily the most massive fiber tract of the brain. Yet when it was completely cut in human patients in the course of surgery for tumors, no observable defects of any kind, subjective or objective, resulted. Cats or monkeys which have had the corpus callosum cut are virtually indistinguishable from intact animals in normal daily activities. But if the optic chiasma is cut as well as the corpus callosum, so that the right eye is connected only to the right side of the brain and the left eye only to the left side, the answer to the old puzzle can be found. If a cat or monkey so prepared is taught to make some discrimination using only its right eye—for instance, press a round but not a square button to get food—it will be found when the right eye is covered and the left eye tested that the left

side of the brain knows nothing about choosing the round button. If only the optic chiasma is cut while the callosum is left intact, then tricks learned with one eye are easily performed using the other. So it was learned that the function of the corpus callosum is to keep each side of the brain informed about what the other experiences. If something is learned in one hemisphere, the corpus callosum enables a duplicate **engram** or memory trace to be established in the other hemisphere. (See Fig. 20-14.)

By means of electrical stimulation, surgical removal, and the recording of small electrical changes on the brain surface that accompany various activities, it has been possible to learn a fair amount about the general functions of different regions of the cerebrum (Fig. 20-10). Injuries in the occipital lobe region result in impaired vision. There are important speech areas along the temporal lobe and on the lower part of the frontal lobe. Motor and sensory centers for the skeletal muscles are lined up along the fissure of Rolando, the motor centers on the anterior ridge or gyrus, the sensory centers along the posterior gyrus. Curiously enough, the body is represented upside down. If a point close to the top of the brain just anterior to the fissure of Rolando is stimulated electrically, the foot will twitch. If points lower down are stimulated, muscles in the legs, trunk, arms, neck, eyelids, and other parts of the head will move, in that order. Most of the frontal lobes represent a "silent" area of indefinite function, concerned with the higher associations.

The Diencephalon

The structure of the second region of the brain, the **diencephalon,** is much the same in all vertebrates. Ventrally, there is the optic chiasma where the optic nerves enter the brain. Just posterior to the chiasma is a slight swelling, the **infundibulum,** to which is attached the pituitary gland. The cavity within the diencephalon is the third ventricle. It is connected anteriorly with the first and second ventricles that lie, respectively, in the two cerebral hemispheres, and posteriorly to the fourth ventricle, which is mostly in the medula oblongata. Dorsally there is a small **pineal body,** the function of which is only beginning to be known. The greatly thickened sides of the diencephalon are termed the thalami: the **epithalamus** above, the **thalamus** to the sides of the third ventricle, and the **hypothalamus** below. Most of the thalamus in mammals is in fact a neothalamus not present in reptiles. All the connections between thalamus and cortex—and they are legion—are reciprocal. We see here another feedback or reverberating system.

In the hypothalamus are centers for many of the emotional and subrational aspects of life. A center regulating sleep is here. Appropriate electrical stimulation of the hypothalamus in lightly anesthetized cats will provoke ferocious spasms of generalized rage, as though the cat were suddenly faced with a barking dog. Tumors of this region in man have produced long periods of inconsolable grief or continuous and equally inexplicable gaiety. A very narrowly delimited center in the hypothalamus of the rat controls appetite.

From the hypothalamus several fiber tracts run down into the posterior lobe of the pituitary gland. These neurons secrete a characteristic neurohormone which passes down the axons into the gland. It has been demonstrated that in this way the posterior pituitary may be brought into action by the nervous system.

The Mesencephalon

The **mesencephalon** of mammals includes four small lobes making up its roof, the optic tectum. This retains some visual functions as in the lower vertebrates but the main visual centers are now found in the occiptal lobes of the cerebrum.

The Cerebellum and Pons

The fourth division of the brain consists dorsally of the **cerebellum,** which contains centers controlling the subleties of muscular action. The **pons,** ventrally located, contains many fibers connecting the right and left sides of the brain. The cerebellum forms part of an important feedback system governing voluntary motion. Impulses from the cerebral cortex sweep down through the pons and up into the cortex of the cerebellum. Thence they pass back to a basal ganglion and then via the thalamus back to the cortex of the cerebrum. From the cerebral cortex the impulses may pass directly or indirectly into the spinal cord and thence to muscles, or again to the pons and cerebellum.

The Medulla Oblongata

The fifth and last division of the vertebrate brain forms the **medulla oblongata** in mammals, as in all other vertebrates. It consists mainly of fiber tracts passing to and from the rest of the brain and the spinal cord. Many of the fibers cross from one side of the brainstem to the other. Most of the spinal nerves (6 through 12) connect with the medulla. Re-flexes concerned with breathing and the heartbeat are centered here. Hence injury to the medulla can be quickly fatal.

Brain and Behavior

In mammals a remarkable series of experiments has revealed the presence of narrowly localized areas within the **hypothalamus** (the part of the brain associated with the pituitary gland) which govern appetite, thirst, sleep, sex drives, anger, and even generalized pleasure (Fig. 20-2).

In rats, and probably in all mammals including man, a pair of **nuclei** consisting of a very small number of cells in the hypothalamus controls satiety and another pair controls eating. The ventromedial nucleus contains cells that tell an animal that it has had enough to eat and should stop. If this **satiety center** or nucleus on both sides of the hypothalamus is carefully destroyed by an electric needle inserted into the brain, the rat becomes a ravenous eater and as a result becomes enormously obese. Such animals pass through three phases. As soon as they recover from the anesthetic—in fact, while they still may be groggy from it—they attack food and eat voraciously. After about a day, they settle down to a steady program of eating until they have more than doubled their weight. It is as though a 150-lb man came to weigh 300 or even 350 lb. The third phase is a static one in which the animal maintains its abnormal weight but cuts down on its eating. However, if it is starved for a day or two it can be induced to gorge itself again.

The question now arises as to how the satiety center "knows" when the animal has had enough to eat. It is common knowledge that the level of glucose in the blood rises after eating and falls during starvation. There is now good evidence that it is the action of glucose directly on the satiety centers in the ventromedial region of the hypothalamus that gives the stop signal. If animals are fed glucose to which a toxic atom of gold has been added by a sulfur link (i.e. gold thioglucose), the cells in the satiety center are selectively killed. Evidently these cells have a special affinity for glucose which is far greater than that of any other cells of the body. They take in the glucose and along with it the poisonous gold. As a result of their destruction, the satiety center is put out of order and the rat becomes obese.

Lateral and a bit ventral to each satiety

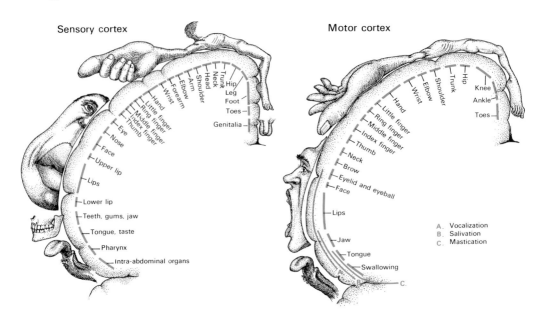

Sensory cortex

Motor cortex

A. Vocalization
B. Salivation
C. Mastication

Fig. 20-10. Sensory and motor homunculi showing the position of neural centers reaching different parts of the body and located along the ridges on either side of the central fissure. (Adapted from Penfield, W., and T. Rasmussen. 1950. *Cerebral Cortex of Man.* The Macmillan Co., New York; and various other sources.)

center is a small group of nerve cells. These make up an **eating center.** If both these ventrolateral nuclei are destroyed, a rat will stop eating more or less permanently.

Most remarkable of all is the discovery of a localized **pleasure center** in the hypothalamus. Like so many important scientific discoveries, from X rays to penicillin, this one was made accidentally by an astute observer. In rats it is possible to insert a fine metal electrode into a specific part of the brain and hold it there permanently by a plastic holder screwed into the skull. Such holders heal in place and apparently are unnoticed by the rat. The experimenter attempted to place the permanent electrode in the sleep center, but missed. The rat concerned was given a small electric shock via the electrodes whenever it approached a certain corner of its cage. Instead of going to sleep, the rat kept running over to the corner where it received the minute electric shock in its hypothalamus. The next step was to place the animal in a Skinner box (Fig. 21-3) where it could administer a shock to its own brain by pressing a lever. The rat quickly learned to press the lever to obtain a shock of 0.0004 amp or fewer lasting less than a second. Such rats pressed the lever anywhere from 500 to 5,000 times an hour! Control rats with the electrode in other parts of their brains pressed the bar only a dozen or so times an hour during random exploratory movements like those of any normal rat.

Microelectrodes implanted in the correct parts of the hypothalamus of goats, monkeys, and other mammals have confirmed the early discoveries on rats and mice. In monkeys, for example, it is possible to connect such microelectrodes to small transistor radios strapped to the animal. If one set of electrodes is positioned in the center for angry aggression and another in the center for friendly indifference, then the behavior of the monkey can be dramatically controlled from a distance by radio signals. Dr. Jose Delgado has done the same with bulls used for bullfights. By remote control, a charging bull can be made to "forget" his anger and turn aside peacefully. If and when chemicals are found which have the same effect, entire populations could be pacified by introducing such an agent into the public water supply—a possibility of enormous human import.

Equally spectacular results have been attained in birds. If microelectrodes are inserted into the correct region of the brain of a hen turkey, she will immediately perform typical male courting behavior, spreading and rotating her tail feathers, strutting, etc., when shown a model of a female turkey's head, *provided* a 0.1-milliamp current is applied in one of the electrodes.

Looking at the hypothalamus and the cerebral cortex in a larger perspective, students of animal behavior often divide behavior into two rather general types: appetitive and consummatory. **Appetitive behavior** is more or

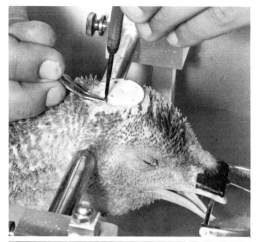

Fig. 20-11. Stages in the insertion of microelectrodes into precisely located points within the brain of a turkey. Note the stereotaxic instrument which holds the head in position very precisely. (Courtesy of R. S. Beese, Pennsylvania State University; from Moment, G. B., and H. M. Habermann. 1973. *Biology: A Full Spectrum*. The Williams & Wilkins Co., Baltimore.)

less random seeking behavior like the apparently aimless circling of the hungry hawk or the hurried walk of a mature caterpiller. The **consummatory act** is the dive of the hawk and the grasping of the unfortunate rabbit, or the spinning of a cocoon. The hypothalamus of the vertebrate motivates the appetitive phase.

Thus, it now seems certain that centers for appetitive behavior, drives, and general motivation lie in the hypothalamus, its associated region of the brain, and the temporal lobes. The cerebral cortex controls the consummatory acts, that is, the specific manner in which the motivations are expressed.

THE BIOCHEMISTRY OF LEARNING

For many decades men have sought the "memory trace" or "engram" in the brain, but completely without success. Now at last very promising clues are at hand as to the form in which learning is recorded and, in some animals, even where. In both flatworms and rats, several investigators seem to have transferred learning by transferring RNA from the brains of trained animals. Some of the experiments on flatworms could not be repeated by other workers in other laboratories, but some have been so confirmed. There are those who challenge the results on rats. However, there are enough confirmations to give many people the feeling — it is no more than that — that there is some real connection between RNA and learning. Furthermore, it has been shown that antimetabolites like azoguanine, which interferes with RNA synthesis, and puromycin, which is known to block protein synthesis, also block memory in goldfish and some other animals. Substances which enhance RNA synthesis are said to enhance memory also. Proteins are made according to the specifications of RNA, so it seems highly probable that all these experiments are dealing with the same mechanism.

The Sense Organs

Sense organs are transducers which translate various forms of energy — light, heat, sound, and certain molecular configurations — into the uniform language of the nervous system, that wave of negativity, the nervous impulse.

More than that, our sense organs, in conjunction with the nervous system of which they are a part, determine in the most fundamental way our conception of what our universe and we ourselves are like. If the genes necessary for the production of the pigments in the retina, essential to see color, are not present, the very concept of color remains completely outside of the experience of the beholder. The redness of red can by no means be described to the color-blind.

SMELL

The ability to discriminate odors is the dominant sensory ability of most vertebrates and many other animals. If the olfactory antennae of an ant are severed, it can no longer recognize the members of its own colony and will attack friend and foe alike. Olfaction is a dominant sense in all but a few mammals and is correlated with the fact that most are color-blind creatures of the evening and predawn. Pheromones, which are of great importance in insects and mammals, are mostly odors which attract males to females, are discussed elsewhere (see Chapters 1 and 8).

Many theories have been proposed; none is completely proven. There is much cogent evidence, however, for a theory which was suggested in primitive form by Lucretius in the 1st century B.C. The sensation recognized as a particular odor is due to the size and shape of volatile molecules which allow or disallow them to fit into like-shaped receptor sites on the nerve endings of the olfactory epithelium. Studies on men and frogs indicate that there are about seven primary odors: camphoraceous, musky, floral, peppermint, ethereal, pungent, and putrid. Seven primary odors would require seven differently shaped receptor sites, by no means an impossible demand. Elaborate investigations of the shapes of odorous molecules show that small roundish molecules, regardless of chemical constitution, give a camphor-like odor. All the molecules giving a floral odor are shaped like a key to a Yale lock, elongate with one large rounded end. Narrow elongate molecules give an ethereal odor, very small molecules with positive charges like formic acid are pungent, and so on for all the primary odors. By mixing various combinations of the seven presumably primary odors, it has been possible to simulate a large variety of other odors.

TASTE

The chemosensory ability called taste is far more restricted than is smell. In man there are only four recognized tastes: sweet, bitter, sour, and salty. Taste buds sensitive to each are restricted to special zones on the tongue—sweet and salty on the tip, sour along the sides, and bitter at the base. In fish what appear to be taste buds are distributed widely over the body. In butterflies the taste cells are located in the front feet; if the feet are placed in a sweet solution, the coiled tongue will be extended.

From the zoological point of view, one of the most interesting discoveries about taste is that animals far apart in the scale of life have closely similar tasting abilities, or perhaps "preferences." If a man and a blowfly are each asked to rank several sugars in order of sweetness, both will arrange them in the same order. Sucrose is sweeter than cellobiose which is sweeter than lactose which is sweeter than glucose. How do you ask a fly such a question? Dip its feet into a series of solutions of increasing concentrations of a sugar. When the fly reaches the first solution that tastes "sweet," it will lower its proboscis and suck up some of it (Fig. 20-12). By comparing the minimal concentrations of different sugars that will elicit a response, information can be obtained about their relative "sweetness" to the fly.

HEARING

Sound plays a very important role in the lives of many animals. The sonar of bats, whales, and dolphins is the envy of the navy. A blinded bat can locate the source of echoes of its own high-pitched squeaks so accurately that it can avoid piano wires stretched across a room. The function of the powdery scales on night-flying moths is to blur the echo of the bats' squeaks. The abilities of porpoises are equally amazing. A porpoise with opaque rubber cups over its eyes has no difficulty in locating and catching a fish and can even pick out the largest of a group.

The ears of mammals consist of three parts, the outer ear or **pinna** (absent in aquatic forms), the middle ear, and the inner ear. The **middle ear** develops from the first gill slit of the embryo and begins with the **eardrum** and extends down via the **Eustachian tube** into the pharynx. From the eardrum a chain of three tiny bones, the **malleus,** against the eardrum, **incus,** intermediate in position, and **stapes,** against the inner ear, conducts the vibrations picked up by the eardrum and carries them across the upper end of the eustachian tube to the inner ear.

The **inner ear** consists of the three **semicircular canals,** the organs of equilibrium, plus the cochlea, the organ of hearing. The **coch-**

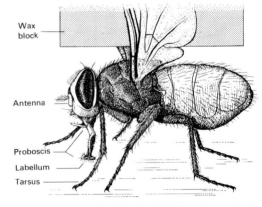

Wax block

Antenna

Proboscis

Labellum

Tarsus

Fig. 20-12. A blowfly with wings held firm in wax tasting sugar water with its feet. (After V. G. Dethier; from Moment, G. B. 1967. *General Zoology*, 2nd ed. Hougton Mifflin Co., Boston.)

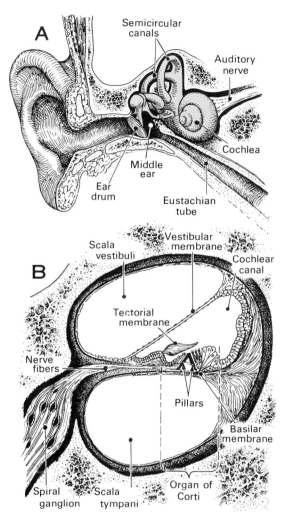

Fig. 20-13. Human ear. A, general topography. Note chain of three bones from ear drum across middle ear to the cochlea. B, cross section through one of the coils of the cochlea. Note so-called sensory hairs in the organ of Corti. (From Moment, G. B., and H. M. Habermann. 1973. *Biology: A Full Spectrum*. The Williams & Wilkins Co., Baltimore.)

lea resembles a snail shell made of a coiled tube which is itself divided along its length into three parts. A ventral **scala tympani** and a dorsal **scala vestibuli** enclose between them a **cochlear canal,** roughly triangular in cross section. Extending the length of the cochlear canal and resting on its basement membrane is the **organ of Corti.** It consists essentially of several parallel rows of **sensory hair cells** overhung by a **tectorial membrane.** Different sounds are thought to cause the basement membrane to vibrate to different extents in different regions along its course. This stimulates the sensory hairs of the organ

of Corti. The resulting nervous impulses pass along the eighth nerve into the brain where they are interpreted as sounds.

PROPRIOCEPTION

No sensory system is more important or so little known by most people as the proprioceptive system of sensory nerve endings in muscles, tendons, and joints. They tell us the position of our bodies, they give us our sense of physical unity as individuals. One of the most terrifying of psychological ailments is due to their malfunction which strips a person of the sense of being all together in one piece in one place. Without them it would be impossible to walk much less take part in athletic events or play musical instruments.

MINOR SENSES

There is a great variety of small sensory nerve endings in the skin and throughout the body. There are special pain endings in the form of irregularly branching naked nerve fibers. There are Meissner's corpuscles for touch, Krause's end bulbs for cold, Ruffini's corpuscles for heat. In fact, histologists have found more specialized nerve endings than psychologists can find sensations.

VISION

Merely to summarize what is known about the organs and organelles of vision would require several large volumes. Some light-sensitive structures are as simple as the eyespot of *Euglena*. Others are as complex as the eyes of the higher insects or the vertebrates. In the course of evolution, two major types of eyes have been developed. The differences and similarities of these two major types of photoreceptive organs conform to the requirements for vision imposed by the laws of physics. This fact makes it possible to offer some educated guesses as to what eyes might be like in animals evolving on another plant in another solar system.

The vertebrate retina consists of light-sensitive cells, the rods and cones. Color vision is a property of the cones. Areas of the retina where rods predominate, as they do around the periphery, are highly sensitive in dim light but lack color vision.

What happens when light falls upon the retina that sends a pattern of nervous impulses racing up the optic nerves to the brain? If a frog or other vertebrate is killed in the dark and its retina exposed and then observed, it will be seen that the retina is purplish but rapidly bleaches to yellowish-gray in

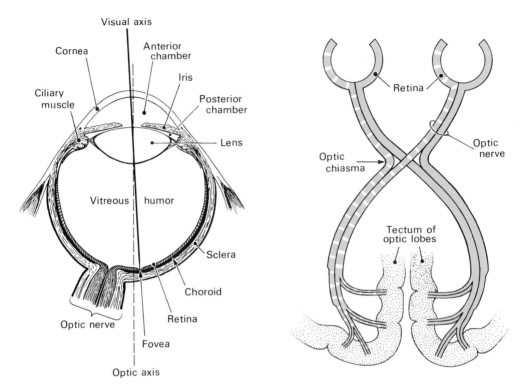

Fig. 20-14. Diagram of a section through the human eye and of the course of the optic nerves through the optic chiasma into the brain and to the tectum (literally "roof") of the optic lobes of the brain. (From Moment, G. B., and H. M. Habermann. 1973. *Biology: A Full Spectrum*. The Williams & Wilkins Co., Baltimore.)

the light. The original color is due to **rhodopsin** *(rhodon,* rose), also called visual purple, located in the **rods.** Rhodopsin itself is a compound molecule, a protein, **opsin**, plus a derivative of vitamin A, a carotenoid called **retinene**. It is for this reason that people whose diets are deficient in vitamin A have trouble seeing in dim light. That rhodopsin is in fact the pigment responsible for vision in dim light has been proven in a very precise and elegant way. It is known from a law of physics (Grotthuss-Draper law) that to be effective, to produce some change, radiation must be absorbed. Otherwise it merely passes through and nothing happens. It is also well known that different substances absorb light of different wavelengths. Rhodopsin best absorbs light of a wavelength of 500 nm. At either side of 500 nm, absorbance falls off rapidly. What is the curve of visual sensitivity in dim light of different wavelengths? Exactly the same as the absorption curve for rhodopsin.

USEFUL REFERENCES

Bullock, T. H., and G. A. Horridge. 1969. *Structure and Function in the Nervous System of Invertebrates*. W. H. Freeman and Co., San Francisco.

Cannon, D. F. 1949. *Explorer of the Human Brain: Santiago Ramon y Cajal*. H. Schuman, New York.

Delgado, J. M. R. 1971. *Physical Control of the Mind: Towards a Psychocivilized Society*. Harper Colophon Books, New York.

Dethier, V. G. 1962. *How to Know a Fly*. Holden-Day, San Francisco. (Humorous but basic.)

Gazzaniga, M. S. 1969. *The Bisected Brain*. Appleton-Century-Crofts, New York.

Gellhorn, ed. 1968. *Biological Foundations of Emotion*. Scott, Foresman & Co., Glen View and Palo Alto. (Paperback.)

MacNichol, E. F. 1964. Three-pigment color vision. *Sci. Am.*, 211:48–56, Dec.

Sherrington, C. S. 1947. *The Integrative Action of the Nervous System*. Cambridge University Press, Cambridge. (A great classic, often quoted for good reason.)

Unger, G. ed. 1969. *Molecular Mechanisms in Memory and Learning*. Plenum Press, New York.

Wald, G. 1968. Molecular basis of visual excitation. (Nobel Prize Lecture). *Science* 162:230.

Young, J. Z. 1964. *A Model of the Brain*. Clarendon Press, Oxford. (Thought provoking!)

Animal Behavior

The study of animal behavior is one of the major thrusts of contemporary biology. The present renaissance of research in this field grows out of the universal and enduring human interest in what animals do and how they come to do it. This intrinsic interest receives additional force from the theory of evolution. Animals are indeed our brothers in an undeniable biochemical and anatomical sense. Knowledge of animal behavior in general and especially of primate behavior will certainly throw light on the forces underlying human behavior. Gastric ulcers can be produced in monkeys subjected to continual severe worry. The biochemistry of memory is probably the same in a man as in a goldfish, but the overriding fact is that, while man is clearly both a vertebrate and a primate, like other species of primates, he is unique and fully as different from a baboon as a baboon is from either a lemur or a chimpanzee. Anthropomorphism, attributing human personal characteristics to animals, is an ever-present temptation to be avoided as is its opposite, an uncritical extrapolation from animals to man. Therefore, we must come to know ourselves and seek our own behavioral destiny.

The Modern Revolt

The modern explosion of research in animal behavior is the result of a threefold revolt.

One object of that revolt was an excessive concentration on the study of conditioned reflexes. Under the leadership of the pioneer Russian physiologist, Ivan Pavlov, investigators all over the world studied how a dog will secrete saliva or gastric juice at the sound of a bell or at the sight of a light if such a stimulus is presented several times immediately before the dog is given meat. It was readily admitted that such studies could reveal a great deal about learning and that a wide variety of reflexes in many different kinds of animals from pigs to goldfish can be conditioned. The studies in themselves are valid. The inadequacy lies in the fact that animals do far more than salivate or secrete gastric juice. To study only conditioned responses is to look at only a very small part of animal behavior.

The second object of this revolt was the psychiatric approach to the study of behavior initiated by Sigmund Freud and, in a somewhat changed form, by Carl Jung. From the scientific point of view there are two insurmountable difficulties here. The Freudian school does not provide a scientific methodology based on rigorous evidence. The use of controls is neglected and the statistical treatment of data ignored. Second, the method of collecting the basic data is not only subjective but entirely inapplicable to the study of animals. How can a bird tell you what it dreams about before it starts on the 1,000-mile migration to its distant breeding grounds? A spider, soon after hatching, will spin a complicated web in the precise way its parents spun their webs, even though both parents died and their webs were blown away by the winter winds months before the spider hatched. There may well be something here that in some way corresponds to Jung's "racial memory," although Jung died before the importance of DNA was recognized. In any case, neither bird nor spider can be placed on a psychiatrist's couch and asked to retell its infantile experiences.

The third facet of the revolt was against the popular American school of maze running

Fig. 21-1. Nesting terns, *Sterna hirundo,* engaged in a territorial dispute, an aspect of reproductive behavior in many fish, birds, and mammals. (Courtesy of Joan Stormonth Black, Gull Island Project, American Museum of Natural History, New York.)

and puzzle boxes. Here, as with conditioned reflex studies, the objection was not that it is wrong to investigate how animals learn, or fail to learn, to run a maze or escape from a Thorndike puzzle box. The maze today is still a very useful tool. The new widely used Skinner box, which has proved itself a powerful means to investigate such problems as learning, motivation, and concept formation, is a direct development of the old Thorndike puzzle box. The problem is that animals do much more than run mazes or press bars, to escape or receive a reward, when confined in a box.

The conditioned reflex and maze-running approaches came to be fused into extreme environmentalism, a general view of animal behavior still held by some people. Under the leadership of the late John B. Watson of The Johns Hopkins University, the idea grew that all behavior is due to learning, i.e. conditioning. The concept of the 17th century philosopher, John Locke, that the mind of every man begins as a clean slate, a *tabula rasa,* seemed to have been validated by modern experimental methods. Many writers, who were really familiar with only two animals, man and the white rat, extended this idea from humans, where it clearly has much truth, to include animals in general. Thus, it came to be widely held that all behavior is the result of a few simple drives and responses plus conditioning based on trial and error or imitation. Instinct especially was anathema, totally inadmissible; innate hereditary factors could safely be ignored because the only meaningful influences are environmental. That any such simplistic view is inadequate and wrong has become evident.

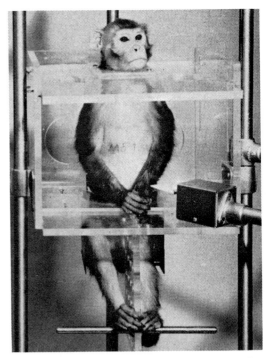

Fig. 21-2. Gastric ulcers in "executive" monkeys. If a monkey in this apparatus is given electric shocks at random over which he has no control, no ulcers develop. If he must interpret signals correctly in order to press the right lever to avoid shocks, then ulcers do appear. (From Brady, J. V. 1958. Ulcers in "executive" monkeys. *Sci. Am.* 199:95. Copyright © 1958. Scientific American, Inc. All rights reserved.)

Like most revolutions, the new era in the study of animal behavior had many forerunners. Charles Darwin devoted an entire chapter of his epoch-making *Origin of Species* to

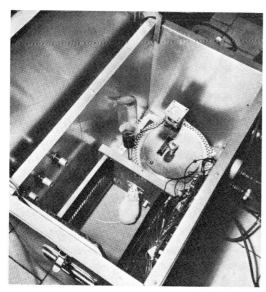

Fig. 21-3. A Skinner box in which an animal presses a bar to obtain a pellet of food and the number and timing of its actions is automatically recorded. (Courtesy of W. Fleischer, Harvard University News Office.)

the origin and evolution of instincts. Around the turn of the century, Jean Henri Fabre devoted much of his life to the detailed observation of insects and other arthropods, paying special attention to innate behavior patterns.

Winged ants, for example, emerging from their subterranean nurseries merely climb up a blade of grass, spread their wings, and fly up into the sky. There is no period of learning to fly that remotely resembles that of a child's learning to walk or to roller skate. In the United States, George and Elizabeth Peckham at Cornell made entomological history with their studies of steterotyped, species-characteristic, and complex behavior patterns in wasps. More recently, Oskar Heinroth in Germany published his now classic studies of species-specific courtship and reproductive behavior in different kinds of ducks.

The Modern Synthesis

FOUR BASICS

The present means of attempting to understand animal behavior is an ecumenical one of collecting, testing, and analyzing information obtained by many methods and by investigators of many schools. Four main components of the modern synthetic approach can be distinguished.

First, there is a real concern for observing

the whole behavior of animals in their natural surroundings and for attempting to understand the evolutionary and adaptive meaning of that behavior. A scientist who only observed how baboons respond in a Skinner test box might never suspect the complex social organization of a baboon troop. He could not be expected to discover that male baboons, so fierce and powerful that only lions do not fear them, nevertheless make devoted fathers. Furthermore, anyone who studies a wide variety of animals in nature can only be impressed with the complexity and uniformity of their innate behavioral patterns. This emphasis on the full spectrum of behavior as it occurs in nature was the contribution of pioneers like Konrad Lorenz in Germany, Niko Tinbergen in the Netherlands and later in England, and their followers who called their science of the whole behavior of animals **"ethology."**

Second, and equally important, modern investigators regard controlled laboratory investigations as an essential part of the study of the sum total of behavior. Anyone who observed a troop of baboons as it moved through the edges of the forest could not be expected to discover the extraordinary precision of the 32-day reproductive cycle of the

Fig. 21-4. Winged reproductive ants emerging from their underground home for their first and only flight. (Courtesy of W. H. Amos.)

females, nor the way hearing a fight between other baboons, even though unseen, will interrupt that cycle.

Third, it is basic to the modern view that all behavior is gene-dependent. This is because the DNA carries the blueprint according to which the animal develops. Within the DNA are the instructions by which each kind of spider constructs its own kind of web. Clearly there are many steps between the double helix and the web, but the pattern of the web is in the double helix nonetheless. Similarly, the DNA carries the information which results in the kind of brain that endows a dolphin or a person with such remarkable intellectual capacities.

No case more neatly shows the way genes can produce innate differences in behavior between closely related animals than a comparison between two subspecies of the deer mouse, *Peromyscus.* One lives in woods and climbs readily while the other lives on the plains and does not climb even when placed among bushes or trees. If, however, both are raised in cages with no opportunity to climb until fully grown, both climb well when given something to climb. If both are raised in cages with logs to climb, both species begin to climb as soon as their eyes open. The eyes of the plains species open before their leg neuromusculature can provide sufficient support. Consequently, they fall and apparently learn that climbing is a bad thing to do. The eyes of the woodland species do not open until after they can adequately support themselves. They climb successfully from the start and continue to do so.

The question of the role of heredity in behavior usually suggests clear-cut cases of sharply defined bits of behavior which appear to be independent of previous learning. For example, a newly hatched male mosquito does not have to learn what the "song" of a female mosquito is like. He does not begin by exploring many kinds of sounds until he discovers which is a female of the right species. Without previous sexual experience he will fly to a tuning fork emitting the same tone as that of a female mosquito.

Barking styles in dogs follow Mendel's laws. Some breeds of dogs, such as bloodhounds, beagles, and springer spaniels, are known as "open trailers." When following the fresh scent of a rabbit, they "give tongue," a characteristic loud baying and yelling sound. Other breeds, like the airedale, collie, and fox terrier, are "mute trailers," and will follow and even overtake their prey without making any noise. Eight different crosses have been made between these breeds, always with the same result, regardless of which breed was represented by the mother. The first generation always barked when following a "hot" trail, but the bark was always the yapping of the mute trailer rather than the rolling bay of the hound. The collie-hound was followed into the second generation. Random segregation of the factors for hair length and type of bark produced all four possible combinations: short hair and baying, short hair and yapping, long hair and yapping, and finally the double recessive, long hair and baying.

Fourth, the modern view is permeated with the realization that biological events, especially behavioral events, occur on many levels of organization, from the molecular and even atomic up to the levels of cells, organs, organisms, and populations. One goose cannot fly in a V formation; only a group of geese can do that. The instructions for geese to fly in that pattern are written in the DNA within the chromosomes of each goose. But between those purine and pyrimidine sequences and that honking, excited, flying V headed to the far north is a long sequence of steps by which each goose, with its characteristic nervous and muscular system, is constructed. The endocrine system must also be properly formed, but above all, the sense organs and the brain must be developed.

WHERE THE WHOLE IS MORE THAN THE SUM OF THE PARTS

All of these steps from the molecular to the anatomical level are important for a complete understanding, but it should never be forgotten that the essential difference between the behavior of animals high and low on the scale of life is not primarily a matter of biochemistry but rather of kind and degree of organization, specifically of cellular pattern. The brain of a man and the brain of a codfish are difficult indeed to distinguish biochemically. The neurons of both contain Nissl substance, which is the ribonucleic acid characteristic of nerve cells. In both cases the motor fibers are coated with lipid-rich myelin. A nerve impulse is a wave of electrical negativity of precisely the same kind in each case, although its speed is a bit slower in the cold-blooded fish. At the synapses where impulses pass from one nerve to another, the chemical compound acetylcholine functions alike in fish and man.

Thus, the difference in the behavior of the fish and fisherman is grounded in the enormous complexity of cellular organization in the man. This applies first and foremost to the nervous system. The cerebral cortex of a fish is a thin sheet; that of a human is a thick, convoluted structure with six layers, each many cells in thickness. While the nervous system is specialized as the determinant of behavior, many other systems play important roles—some more, some less obvious. A fish cannot bait a hook because it lacks hands; a man cannot obtain his oxygen from the water because he lacks gills. Without adrenal glands a man cannot become angry in any complete sense of the word. The most he can do is think about anger. All of these considerations have given new life to the old science of anatomy, especially the study of the gross structure and the cell architecture of the nervous system.

So it becomes evident once again that in passing from one level of organization to another, new properties emerge. When the cells are separated and their contents analyzed, valuable knowledge is obtained, but it throws only an indirect light on the problems of brain function. This is what biologists mean when they say that the whole is greater than the sum of the parts. To resort to a crude analogy, a chemical analysis of the metal sides and the leather head of a drum is inadequate by itself to explain the drum. The leather might as well be a shoe or an apron; the metal nails or hammers. The ultimate problems of the nervous system, which in some way make possible both artistic creation and conscious reason, are certain to be even more subtle and baffling than we can now suppose.

STIMULUS AND RESPONSE

In a nerve-muscle preparation, a stimulus is a relatively simple thing, perhaps a slight electrical charge applied to a nerve, and the response is the contraction of a muscle. The behavior of whole intact animals is built of just such events but in a far different and all-important context. Very commonly the character of the stimulus changes with the total situation. It is the entire pattern which is important, the **gestalt,** as psychologists say.

The classic case is the innate escape reaction of a goose to a hawk flying overhead. If a model shaped as shown in Figure 21-5 is pulled along a wire over the heads of ducks or geese, they pay little attention if it is pulled so that the long "neck" goes ahead. Perhaps it then resembles a goose. But if the same model is pulled in the opposite direction, ducks and geese scurry for cover. Apparently the long "neck" now appears to be a tail and the short "tail" as a head—and the whole object looks like a hawk.

Of all the potential stimuli impinging simultaneously on an animal (all the sights, sounds, smells, and "feels"), any individual only responds to one or two, presumably

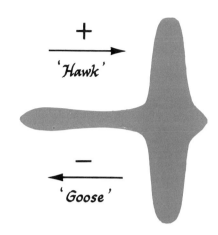

Fig. 21-5. Model which resembles a hawk when moving to the right and goose when moving toward the left. (From Moment, G. B. 1967. *General Zoology*, 2nd ed. Houghton Mifflin Co., Boston.)

those that have some special significance for the animal. When such a stimulus is standard and elicits a predictable response, it is commonly called a **sign stimulus.** For example, the red breast of an English robin is such a specific sign stimulus that another male robin will attack if it enters his territory. In fact, even a ball of red feathers will be attacked if presented to a male in the breeding condition within his own territory.

Note then three important characteristics of a stimulus situation which initiate behavior (1) The total context makes an essential difference. (2) Not all the physical forces impinging on an animal act as stimuli, but only a very limited fraction of those potential stimuli. (3) The hormonal or physiological condition of the reacting animal makes an important difference.

Sometimes it is difficult to untangle the role of various influences which result in a given response. For example, a billy goat raised in complete isolation from other males will, when he becomes sexually mature, prevent any other billy goat from mating with the she-goats in his flock. The sight of the other male mounting a female precipitates an immediate charge. However, a ram, which will mate with she-goats in the absence of ewes, can copulate with the she-goats without disturbance from the billy. The mating patterns of both goats and sheep are closely similar. However, the odor of the males is very different. Perhaps the proper odor is essential if the visual stimulus is to be effective. Since this whole behavior occurs only in sexually mature males, hormones are clearly involved. The point to remember is that a stimulus-response phenomenon is sometimes a very complex affair. Is the response just described instinctive? Is there a gestalt situation? Is there a sign stimulus? How should the role of odor, or some other difference between rams and billy goats, be regarded?

APPETITIVE VERSUS CONSUMMATORY RESPONSES

Many responses of animals fall into one of two general types: appetitive or consummatory. **Appetitive behavior** is more or less random, exploratory behavior. The stimulus is commonly internal, perhaps hunger, which means a lower concentration of blood glucose affecting a nerve center, and perhaps also certain stimuli from the stomach. **Consummatory behavior** consists of the specific acts that fulfill or consummate the drive or hunger behind the appetitive searching.

Examples abound in widely separated parts of the animal kingdom. A hungry hawk flies hither and yon scanning the ground in the appetitive phase of feeding behavior. The dive and subsequent killing and eating of the rabbit are the consummatory phase. A caterpillar ready to spin its cocoon experiences a compelling wanderlust. When it finds, more or less by chance, a suitable place, the consummatory act of spinning the cocoon is precipitated. In those animals where instinct predominates, consummatory acts are highly stereotyped while in those where intelligence is important, they are learned and highly variable. A hungry spider will always spin a web of a particular type, but who can predict the behavior of a hungry man?

The Way of Instinct

DEFINITIONS

The word "instinct" has acquired a very bad reputation among psychologists and other critical people, for several reasons. It has been used widely in a very loose sense as a synonym for any habitual or unconscious act, as when an American driving a car in England meets an emergency and suddenly steers over to the right-hand side of the road by "instinct." Worse yet, "instinct" has been used by journalists, literary writers, and others to lend a spurious air of scientific authenticity to ideas which may or may not contain some truth, but for which there is no solid scientific basis. The "work Instinct," the "death instinct," and the "leadership instinct" are all examples of this kind of pseudoscientific nonsense. And to some schools of psychology, instinct is a type of behavior which by definition is unanalyzable.

Charles Darwin, in his *Origin of Species,* began the chapter on instincts by saying: "I will not attempt any definition of instinct." he boldly claimed that everyone understood what was meant and then proceeded to give the example of the cuckoo. This bird lays its eggs in other birds' nests. When the young cuckoo hatches, it ejects the rightful eggs in a particular manner, which is really a gymnastic feat because the rightful eggs are almost as big as the young cuckoo. The North American cowbird also lays its eggs in other birds' nests and, as Lorenz has pointed out, mates when adult, not with the species by which it was raised, but with other cowbirds.

In biology, on the other hand, the word **instinct** carries a very useful meaning. Konrad Lorenz, one of the leaders in the modern study of instinctive modes of behavior, defines instincts as "unlearned, species-specific motor patterns." An American physiological psychologist, M. A. Wenger, has defined an instinct in essentially the same way as "a pattern of activity that is common to a given species and that occurs without opportunity for learning."

By instincts biologists mean types of behavior which fulfill the following four conditions: They are species-specific i.e. characteristic of a given species, and can be performed without an opportunity for learning although they may change somewhat with repetition. Instincts are more or less stereotyped motor patterns. The more complex ones usually consist of a series of rigid steps which must be followed in a certain unchangeable sequence. Instinctive acts are immediately adaptive, in contrast to the random trial-and-error type of behavior associated with learning. It should be added that although instincts are characteristic of given species, and some are unique to a single species, many are common to several species or even to a whole family, just as anatomical traits are.

The word instinct suggests a fairly complex behavior, like nest building, as compared to a **taxis,** which is merely the orientation of an organism in a field of force, for example, the caterpillar of the brown-tailed moth exhibits a taxis (a negative geotaxis) by climbing upward when hungry. In marked contrast it spins a particular type of cocoon in a rigid series of steps performed only once in its life with no possibility of learning.

IMPORTANCE OF INSTINCTS

There are several reasons why instincts are important. First, they form a very conspicuous part of animal life, vertebrate as well as invertebrate. The web weaving of spiders, the mating reactions of hens, the migrations of fish, the language of bees, the construction of mud nests by wasps and of hanging nests by Baltimore orioles, the courtship of salamanders—all these and a thousand and one other species-specific, unlearned, and stereotyped motor activities abound in nature.

Because instincts are species-constant they are an aid to zoologists in solving the twin problems of classification and evolution. A particular kind of spider can be identified as surely and far more easily by observing the structure of the web it spins than by looking at its anatomy. Spiders spin webs of four general types. Within each type each species has its own characteristic and uniquely different pattern (Fig. 21-6). Illustrators of detective stories are often rather careless about this but spiders never are. Bird songs are commonly used throughout the world to identify species.

ADVANTAGES OF INSTINCTS

What is the advantage of instincts over intelligence? Fifty years ago it was the belief of biologists that instinct was a kind of evolutionary precursor of intelligence. This idea has now been almost completely abandoned

in favor of the view that the animal kingdom has evolved in two major directions. In one, the arthropod line (insects and their relatives), instinctive actions predominate; in the other, especially the warm-blooded vertebrates, learning predominates, or at least plays an important role.

The efficiency of instinct is proved by the staggering biological success of insects. The secret is simple: instincts furnish an animal with prefabricated answers to its problems. This is especially important for an animal with a very short life span because no time need be lost in the inevitably wasteful trial and error of learning.

Birds also reveal the advantage of instincts. The type of nest any species builds is determined by instinctive acts. Weaver birds, for example, hand-raised without nests for four generations, will build typical weaverbird nests when given a chance. Try to imagine a pair of robins learning by pure trial and error without any previous experience how to incubate eggs. Should they construct a nest of fine twigs and plant fibers like a mockingbird, or a bulky nest of grass and rootlets like a hermit thrush? Or why not a hanging cup suspended from a forked branch like a vireo? The absurdity of these questions emphasizes the overwhelming importance of instinctive behavior in the lives of birds.

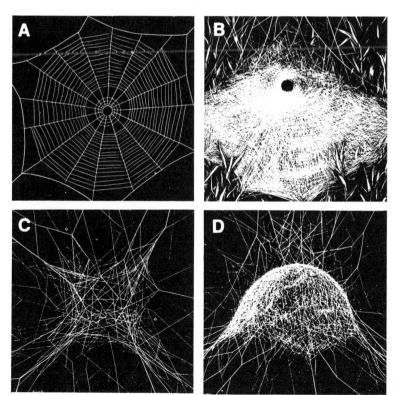

Fig. 21-6. Four important types of spiderwebs. A, orb; B, platform plus funnel; C, "crazy cat's cradle"; D, inverted dome. (Drawn by Elmer W. Smith. From Moment, G. B. 1967. *General Zoology,* 2nd ed. Houghton Mifflin Co., Boston.)

The only known alternative for instinct to enable a bird to build the correct kind of nest would be imprinting, a very early type of learning which will be discussed more fully later. This may play some role in the selection of the nesting site and of the material, although there is no evidence that it does. But as anyone who has tried to put a watch together will testify, knowing what materials to use is a very different thing from knowing how to put them together. In a robin's nest the beautiful bowl of dried mud that lines the rough outer layer is invisible to the nestlings in any case, since it is covered by the inner lining of fine grasses.

In brief, instincts are adaptations in the same way that structures are adaptations. Primarily they are reflections of the structure of the nervous system, although of course the proper organs—wings, silk glands, etc.— must also be present.

EASILY OBSERVED EXAMPLES OF INSTINCTS

Instinctive behavior patterns are easy to see but only if you are at the right place at the right time or can bring certain animals into the laboratory under the proper conditions and have the patience to sit and watch.

Among some of the familiar invertebrates, innate behavior patterns seem so simple that one hesitates to use the word instinct. Certainly the behavior of a jellyfish cannot be compared with that of an arthropod in complexity, yet it is equally species-specific and independent of any known conditioning. Lie on a wharf or on the deck of a boat and watch jellyfish. Unless injuried, most species will be doing one of two things. Those that live near shore, like *Aurelia* and *Chrysaora,* the common sea nettle of bays and estuaries, spend their time fishing by first swimming up to, or almost to, the surface of the water, then turning over so the sub-umbrellar surface is uppermost, and slowly drifting to the bottom. When they hit bottom, they turn over and muscular contractions begin which send the medusa upward again in a series of rhythmical contractions. This cycle is repeated hour after hour. Any small animals that become caught in the tentacles are dragged into the mouth and are digested. Jellyfish like the rapidly swimming *Liriope* could not behave in this way since they live where the sea is a mile or more deep. They swim, often in great swarms, but always stay near the surface.

Insects and spiders have long stood as the best examples of animals dominated by instincts. What about instincts in vertebrates, including humans and other mammals? Instincts which are complex, stereotyped, and prefabricated answers to life's problems are really the antithesis of intelligence or learning. Hence, it would not be expected that an animal well equipped with instincts like a spider would be very inventive. Nor would creatures with brains well able to learn be expected to come equipped with a set of highly developed inflexible instincts. In humans, and indeed most other mammals, there seem to be no complex instincts comparable to what can be found in arthropods. Nonetheless, the lives of vertebrates (including mammals) are permeated with innate responses. For example, the behavior of a pseudopregnant bitch (to be discussed later), the ways different breeds of dogs bark, and nest building by birds are familiar innate behavior patterns.

BEE TALK: BUT IS IT A LANGUAGE?

The most sensational case of instinctive behavior discovered in modern times is the language of honeybees. In fact, when Karl von Frisch first announced his discovery he was widely greeted by scepticism, but his observations have now been confirmed by all who have taken the trouble to look. A bee returning to the hive after a successful flight with her crop full of nectar does a little dance, a tail-wagging dance. This dance "tells" other bees both how far away the food is and its direction in relation to the position of the sun.

In performing the dance the bee makes a short run about twice its own length while vigorously wagging its abdomen. At the end of the run the bee turns in a half circle, right or left, to its starting point and then repeats the tail-wagging run. The number of turns per 15 seconds depends on the distance of the food source from the hive. The closer to home the food, the more dances per unit of time. Strong headwinds make the bees overestimate the distance. Perhaps the number of turns in the dance depends on fatigue, perhaps on some inner time sense, perhaps on other factors.

The surface of the honeycomb on which the bee dances is a vertical wall. If the tail-wagging run is directly upward, this indicates that the food is directly toward the sun. If the tail-wagging run is directly downward, it indicates that the food is directly away from the sun. If the wagging run is at an angle of 60° left of straight up, the food is to be found in a direction at 60° to the left of the sun.

The evolutionary origin and perfection of such a "language" must have taken millions of years of mutation and natural selection; but the fossil record of bees extends back

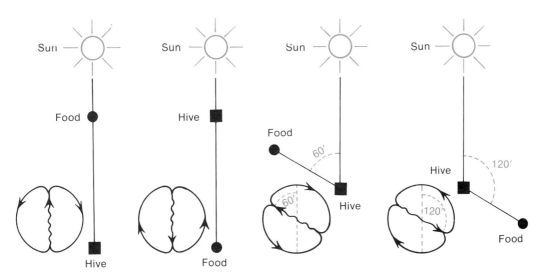

Fig. 21-7. Diagram to show the relation of the tail-wagging run of honeybees on the vertical surface of the honeycomb to the direction of the sun and the direction of the food source. Both the waggle dance and odor have been confirmed as ways in which bees communicate to each other the presence of food. (After von Frisch; from Moment, G. B. 1967. *General Zoology*, 2nd ed. Houghton Mifflin Co., Boston.)

millions of years, so there has been time enough. Moreover, it has recently been discovered that other insects, such as blowflies, which have no known language or social organization, nevertheless execute a little irregular sort of dance after eating. From such precursors the language of the bees probably evolved.

In more recent years Adrian Wenner has shown that honeybees also communicate the presence of flowers by carrying the odor with them into the hive. Thus, as in many biological processes from the development of the embryo to the digestion of food, there is some redundancy. It may not be very different from signaling anger both by how you look and what you say.

Neither the waggle dance nor the directional response to it by the bees seem to be learned. Can the dance then be considered a language?

The Way of Learning

Learning, which is virtually the same thing as intelligence, may be defined as a more or less permanent change in behavior due to past experience recorded in the nervous system. To specify that it is recorded in the nervous system excludes a change in behavior due to a mutilating accident—for example, the loss of a wing. To specify that it is more or less permanent excludes changes which oc-

cur after an animal has experienced a full meal or any similar temporary physiological change. A hungry animal and a satiated one usually act very differently, but the hunger will return and with it the appropriate behavior.

It is important to distinguish maturation from learning but it is by no means always easy to do so. **Maturation** can be defined as changed behavior caused by developmental changes within the nervous system. Before the neurons of a reflex arch have grown into place, clearly the reflex action cannot occur. The classic case is seen in the development of swimming in salamanders. Leonard Carmichael and George Coghill kept developing salamander eggs under complete anesthesia until the control salamanders had passed through all the beginning stages of swimming, from mere twitches to bending like the letter C, to ineffectual wiggles, and finally to swimming actively. The anesthetized salamanders were then placed in pure water and proceeded to swim off as skillfully as the control salmanders without passing through all the preliminary stages which resemble efforts to learn.

SIX MAJOR TYPES OF LEARNING

Habituation

Perhaps the simplest type of learning is called **habituation.** If an animal is repeatedly given a stimulus which is not followed by any reinforcement, i.e. any reward or punishment, the animal ceases to respond after a while. If a shadow is passed over an active barnacle, it will stop kicking its six pairs of double legs and will pull them into its shell. Each time the shadow passes the barnacle responds more slowly and opens out and begins kicking sooner until, after five more passes, it no longer responds. This type of learning is obviously related to habituation to drugs of various sorts. Habituation in this sense is defined as a change which makes necessary an increase in the stimulus, i.e. the dose, to obtain the same result.

Classical Conditioning

A basic type of learning studied with physiological precision is called **classical conditioning**. Its investigation was developed by Ivan Pavlov in Russia and by others in Europe and America. In this type of study the dog, pig, goat, or other animal remains standing in a large box and responds to a stimulus by salivating, raising a foot, or in some other manner. The primary or **unconditioned stimulus** is commonly food or a mild electric shock. If, just before meat is given, the dog sees a red light or hears a bell, after a few trials, he will salivate on seeing a red light or hearing a bell even though no meat is given. The light or the bell is then known as the **conditioned stimulus**. Note that the response is a relatively simple reflex type of action.

Instrumental Learning

A third type of learning is **instrumental learning**. The animal is called upon to perform some act, perhaps a rather complicated act, instead of merely exhibiting a relatively simple reflex. The animal must do something, i.e. operate. Hence this is sometimes called **operant conditioning**. Second, the animal is motivated, usually by hunger. The methods are simple. The animal is kept without food or without water to provide the motivation and is then placed in a maze, a Thorndike puzzle box, or a Skinner serve-yourself box. In the old puzzle boxes the animal had to discover a way to escape. The rat, cat, monkey, or other animal had to push, pull, turn, lift, and otherwise manipulate the various parts of the box, perhaps in a particular order, until he happened on the correct act which opened the door to freedom and food. In the Skinner box the animal learns that by pressing a lever, a simple type of slot machine will serve him a pellet of food or drink of water. The advantage of this device is that the number of times the lever is pressed can be automatically recorded and thus the rate of learning, the intensity of response, and other factors can be

Fig. 21-8. Canary which has learned under operant conditioning that the different object in a series, the one tablet among the screws, or the one screw among the tablets, is the object which covers the reward. (Courtesy of Nicholas Pastore.)

measured objectively. Note that both the Thorndike puzzle box and the Skinner serve-yourself box depend on an initial period of exploratory trial-and-error behavior.

Insight Learning

A fourth type of learning is called **perceptual learning** or, sometimes, **insight learning**. Use of these terms varies and is a reflection of our basic lack of understanding of the process of learning. Insight learning is said to occur when an animal does not go through a period of trial and error, but surveys the problem and then acts correctly on the first try. Insight learning is not regarded as a fourth type of learning by some people, but rather as a variety of trial-and-error learning with the trials and errors made mentally on the basis of previous experience.

Exposure Learning

Exposure learning has been known in a general way for centuries but is still little understood. There are no known rewards or punishments, i.e. no **reinforcements**. A rat will learn a simple maze simply by walking around in it as would a man. Consider the case of the English chaffinch so thoroughly studied by W. H. Thorpe in Cambridge. Only males sing and all have basically the same song but each male has his own particular variation. A male does not begin to sing until a year old (a long time in the life of a bird) yet when he does begin to sing, the song is like the one he heard the father singing when he was a nestling. There was certainly no reward, no "reinforcement" in any ordinary sense of the word. A similar case of the white-crowned sparrow has been investigated by Marler and Konishi. If baby sparrows are raised in soundproof rooms, they learn whatever "dialect" of white-crowned sparrow song is played to them from a tape recorder. If exposed to both the song of a white-crowned and of a song sparrow, they only learn the white-crown song. If played only the song of a song sparrow, they learn nothing and their vocalizations are mere avian gibberish. Evidently there is something very special about exposure learning in birds.

Imprinting

A sixth and very special type of learning has been discovered in recent years and has been named **imprinting**. Typical imprinting takes place only during a brief critical period occurring several hours after hatching or weeks after birth. It is very persistent and, as in exposure learning, there is no obvious reward.

In ducks, geese, and most breeds of chickens, the newly hatched bird becomes imprinted by the first large moving object that it sees within the first hours after hatching. Thereafter, it will follow that object to the exclusion of all others. Normally this first large moving object is the mother. However, if the eggs are hatched in an incubator and the first large moving object is a man, a male duck, or a mechanical floor polisher, the duckling is imprinted by this inappropriate object and will follow it with great and enduring persistence. A remarkable fact about imprinting, at least in duckings, is that if the duckling has to work—for example, is forced to scramble over little obstacles—to follow the box or whatever is being used for the imprinting, then the imprinting becomes more firmly established in less time than were the little duck merely walking over a smooth floor. This fact suggests a relationship between imprinting and operant conditions. Since there is no reward except the act itself, there seems also to be a relation between imprinting and exposure learning.

Fig. 21-9. Gosling following a box on which it has been imprinted. Note that it is ignoring the adult geese in the background. (Courtesy of A. Ogden Ramsey.)

Fig. 21-10. Emotionally disturbed young adult monkeys raised in isolation without either contemporaries or parents. (Photo by Sponholz for the University of Wisconsin Primate Laboratory.)

In puppies it has been found that there is a critical period centering around the 6th and 7th weeks of life when they become most readily and permanently tamed. If puppies have no contact at all with human beings until after they are 14 weeks old, it is virtually impossible ever to make them into friendly pets.

More important results emerged when Harry Harlow and Robert Zimmerman at the University of Wisconsin undertook to develop a completely healthy colony of monkeys free of the infections of various kinds which are passed on from parents to offspring. The obvious method was to rear monkeys from birth in isolation from their mothers (Fig. 21-10). Such monkeys were healthy but surprisingly abnormal in their behavior. They acted frightened and withdrawn. When they became sexually mature, which they did at the normal time, they showed no interest whatsoever in mating and only a few of the females could be bred even by placing them with patient and experienced males.

How widespread imprinting, in either a narrow or a broad sense, may be in the animal kingdom is not yet known. The indications are that the importance of early experience is widespread among vertebrates. In addition to occurring in birds and mammals it very probably takes place among certain fish. The ability of migrating salmon to return to the small tributary in which they were hatched, even though it is but one of many streams in a river system, may be due to imprinting with the characteristic smell of the water.

Hormonal Basis of Behavior

In addition to the neuromuscular system, behavior is profoundly influenced by the endocrine system, especially in vertebrates but also in worms, insects, and mollusks.

The migrations of birds, as well as their songs, are under hormonal control. Female canaries normally never sing. However, if a pellet of testosterone (male sex hormone) is inserted under the skin, a female will sing a typical canary song for several weeks until the pellet is completely absorbed. It is worth noting that the female has the neuromuscular and vocal apparatus for singing and knows what song to sing. What is lacking is the one key substance, testosterone.

The dominating influences of hormones are familar in mammals and are nowhere more dramatically revealed than in the pseudopregnancy which occurs in certain female dogs. Beagles seem especially subject to this affliction. Without mating, such bitches show many of the signs of pregnancy about 9 weeks after ovulation when puppies would be born. Such dogs not only have enlarged mammary glands and secrete milk but will steal a child's doll to defend and "mother." At other times such a dog is no more interested in dolls than in books. Pseudopregnancy can be brought to a quick end within 2 hours or less by an injection of male hormone.

Very convincing evidence shows that sex hormones induce sexual behavior by direct action on centers in the central nervous system. A female cat which has not only been spayed (ovariectomized), but has also had other reproductive organs (uterus and vagina) removed, can be caused to behave like a normal female in estrus (to welcome a male rather than to treat him with angry rejection) by the implantation of minute pellets of a female sex hormone, diethylstilbestrol, in the right part of the hypothalamus. If female sex hormone is radioactively labeled with carbon-14 and injected into the bloodstream, it will be found that the hormone is selectively localized by special cells in the hypothalamus.

Biological Clocks

Rhythms are common in both the plant and animal worlds. They may be annual like the migration of birds, lunar or monthly like menstruation, or daily i.e. **circadian** (*circa*, approximately, *dies*, day) like the folding down of bean leaves at night or the number of eosinophils among the white blood cells in the human circulation at different times of day. Observation in natural surroundings can tell much about such rhythms but not whether they are regulated by a biological "clock." To classify as "clock"—regulated, a rhythm must meet the following criteria. (1) It must persist under constant conditions of light or darkness, temperature, moisture, and other variables. (2) The period of the rhythm must be temperature—independent, neither speeded nor slowed by rises or falls in temperature. (3) The phase, i.e. the on or off, up or down of the rhythm, can be shifted in time by an appropriate stimulus. Noncircadian rhythms may be controlled by an entirely different mechanism. For example, the reproductive cycle of the female African Chacma baboon is governed by some kind of internal time-keeping process to a highly precise 32 days under many conditions. However, if a female hears and sees a fight between two baboons, her rhythm is completely turned off and some days are required before it recommences. The 4-day reproductive rhythm of ovulation in mice can be greatly altered by the odor of male mice. Probably such phenomena should be regarded as meeting the third criterion.

"Clock"-regulated circadian rhythms have now been found in organisms as diverse as potatoes (respiration) and fruit flies (activity), as well as in mice and men. It is these circa-

Fig. 21-11. Fiddler crabs, day (left) and night (right) adapted. (Courtesy of H. M. Webb and M. Berlinrood; from Moment, G. B., and H. M. Habermann. 1973. *Biology: A Full Spectrum.* The Williams & Wilkins Co., Baltimore.)

dian rhythms that are responsible for "jet lag" in people flying long distances either east or west. In their pioneering investigations of the fiddler crabs' change from a darker color at day to a lighter color at night, Frank Brown and Marguerite Webb found that these circadian rhythms are temperature-independent. Such a surprising discovery shocked some competent biologists into disbelief. All known biochemical processes (except photochemical ones) are temperature-dependent with a Q_{10} of about two. A chirping cricket makes a good example. Raise the temperature by 10°C and it will chirp twice as fast, lower it by 10°C and it will chirp half as fast. But temperature independence of circadian rhythms has now been confirmed in many organisms. The amplitude of the rhythms, the amount of swing up and down, can be changed by temperature, is less in the cold, but not the frequency.

The nature of the supposed clock itself is still a matter of speculation. J. W. Hastings has suggested that there may be a series of circular feedback reactions in which compound A is transformed into B and that into C and that into D, D into E, E into F, and F into A. There is also another compound, X, outside this circle of reactions which is converted into an inhibitor, I, of one of the steps in the circle. All these reactions are orthodox and are speeded up by increases in temperature, but when that happens the inhibitor is formed more rapidly and slows down the reactions in the circle which is the "clock." The result is that the reactions as a whole are temperature compensated.

The difficulties of penetrating to the source of a rhythm are well illustrated in the case of the daily rhythm of the eosinophil count in human blood. The number of circulating eosinophils is dependent on corticosterone from the adrenal cortex. When the concentration of this hormone rises, the number of eosinophils falls. So the eosinophil rhythm is the result of a rhythm of the adrenal cortex. But the release of corticosterone by the adrenal gland is controlled by ACTH, adrenocorticotropic hormone from the hypothalmus of the brain. Perhaps the trail ends there. Perhaps.

One persistent problem has so far defied final resolution. Is there really an internal endogenous timekeeper, or are all these rhythms the reflection of some exogenous influence that the organism can detect? It is well established that the earth's magnetism fluctuates on a daily cycle depending on whether the location on earth is facing toward or away from the sun. There is now evidence that animals, at least, can detect these weak magnetic changes which would not be screened out by the walls of constant temperature rooms. Possibly organisms tell time by sensing these rhythmic changes. After many negative results, it now looks as though migrating birds can detect the earth's magnetism. A firm answer to the question of whether the "clock" is endogenous within the organism or exogenous in the environment is yet to come.

Social Behavior

In the sense used here, social behavior is behavior in which two or more animals are responding to each other. Animals may aggregate because each responds in the same way to the same stimulus, but unless there is some interanimal reaction, we will not regard it as social. Daphnia will collect by the hundreds at the surface of water if the oxygen is depleted, but this is not a social response because there is no interaction between individuals. A single daphnia would move to the surface as would a hundred.

In marked contrast is the behavior of a newly hatched group of fish or squids. Both exhibit a very strong schooling instinct. If a stick is placed in the water and moved through the school, the little fish or squids can be made to separate into two groups as they swim past the stick. But the divided school snaps back into a single group as though drawn together by powerful magnetic forces just as soon as the leading individuals are a few inches beyond the stick. Many birds are gregarious and migrate and nest in flocks. The precise integration of all the members of a flock of pigeons as they maneuver in the air is a striking sight.

PECKING ORDERS

One of the most obvious social facts is the presence of dominance hierarchies or pecking orders among various kinds of birds and mammals. For example, among a group of hens, one will soon become dominant, able to peck any other hen in the henhouse and drive her from food. Hen number two feels free to peck any hen except hen number one. The third hen in the hierarchy can peck any hen except numbers one and two. And so the series continues. Similar series have been found in cows, mice, and other animals.

A number of facts should be noted about pecking orders. The first is that they can arise only among animals that can recognize each other as individuals and have an ability to learn. Nothing like a pecking order has been found among worms, ants, or even bees, where life is governed by innate reactions. Even the queen bee does not dominate the colony in the sense of giving directions.

The position of an animal in a pecking order depends on many variables, including his own past experience. Initially it will probably be the largest, most aggressive, and strongest individual which heads the pecking order. Animals low on the list can be raised to a higher status by several methods. Simple isolation from the group for a time may have this result. Perhaps isolation gives animals close to the bottom of the ladder a chance to get enough food and so build up their strength. Mice can be advanced in a pecking order by isolating them from the group and then arranging a series of encounters with other mice in which they will be victors. Apparently this increases their "self-confidence." Doses of sex hormone will cause chickens to rise in the pecking order. Starlings, however, rise in their pecking order after castration.

Knowledge of pecking orders has some practical value. If a cow who is number one in her herd is transferred to another herd where her position is unknown and where the other cows "presume" she is very low on the scale, she will become morose and fail to eat properly, and her milk production will fall. Chickens and other animals very low on their scales may not get enough to eat to develop properly.

On the other hand, for a species as a whole, a pecking order may possess definite advantages because it almost eliminates the wasted motion of continually fighting over food, drink, and space. Pecking orders cannot develop if the population is too large. Probably there are simply too many animals for any one individual to remember who stands where. The rigidity of the pecking order differs greatly among different species. There is much more give and take among canaries and pigeons than among hens.

TERRITORIALITY

Most vertebrates exhibit some degree of territoriality; that is, they will defend a given home area against intrusion by others of the same species. Usually this behavior is limited to the breeding season and to males. Territoriality is especially strong in birds where it was first discovered by Eliot Howard who noticed that male birds demark their territories by singing and by driving off other males. Apparently Howard was the first man in history to understand what bird song is all about. Territoriality has also been studied carefully in the little freshwater stickleback and can be readily observed in spring with sunfish or bass which protect "nests" in shallow water near shore.

Territoriality has not been extensively investigated in mammals, although some authors with literary rather than zoological experience regard it as the dominant force in human life. Perhaps the best studied case is the Eskimo dogs observed by Tinbergen. The dogs live in packs of 5–10 individuals and mark the boundaries of their domain with urine. Puppies have no sense of territory and are forever getting themselves in trouble by trespassing into the territory of other packs. When they become mature they very quickly learn the limits of their own and other territories. Some of the larger carnivores seem to mark off a home range by rubbing odors on trees.

In the primates, the order of mammals to which man belongs, there is much variation among genera and even among species. The much-studied tree-living howler monkey of Central America shows strong group territoriality. Each troop, of anywhere from a dozen to several dozen individuals, is acutely aware of the limits of its territory and defends it by sessions of howling, audible for long distances. Among the Old World monkeys, the rhesus monkeys; or macaques, are territorial. A group will move around within the same area day after day and fight off intruders with great ferocity, although threats usually suffice.

Baboons are rather different. Each troop has a vaguely defined home range, and there is much overlapping with the home ranges of other troops. Gorillas and chimpanzees, among anthropoid apes, have recently been extensively observed in the wild by Schaller and van Lawick-Goodall, respectively. In neither case is there any territoriality of the strict and intense sort seen in howlers, and nothing is really comparable to the situation in birds. Both gorilla and chimp troops have home ranges of considerable extent. Neither shows any defense of territory. Other groups may pass through and occasionally intermingle with impunity. On the basis of what is now known about primate behavior, there is no justification for the belief that territoriality is the dominant imperative of human life.

The primary biological function of territoriality, according to present knowledge, is to spread the members of a species more or less evenly throughout the available habitat. The result is that bluebirds or howler monkeys will achieve the optimum utilization of the habitat and its food supply. Conflict within the species will be minimized. In both teleost fish and birds, when a neighboring male enters another's territory and is challenged by the owner, the invader retreats. This is probably true of mammals as well. Dogs are certainly much less belligerent when away from home base than when in their own yards.

When two males meet at the boundary between their respective territories, there may be endless skirmishing but there may also be a very interesting phenomenon called **displacement activity**. A stickleback fish may neither attack nor retreat but perform some third and irrelevant activity such as beginning nest building. Apparently when two strong mutually exclusive responses are equally balanced, and the nervous system as a whole is raised to a high state of readiness (perhaps by adrenaline), some third response is set off or released from inhibition.

CONFLICT AND COOPERATION

Violent conflict and vigorous cooperation are commonplace in the animal kingdom. There is much talk about "aggression" among animals, but scientifically it is necessary to use a neutral term like conflict or **agonistic** behavior. "Aggressive" carries a strong moral overtone of wrongdoing, whereas **"agonistic"** can be freely used to describe a combat between two tomcats competing for the same female in a situation where aggressor and innocent victim seem quite irrelevant or an encounter where one animal only attempts escape.

One of the most important facts to remember is the danger, indeed the impossibility, of extrapolation from one species to another,

Fig. 21-12. Marching formation of a baboon troop. The central group consists of the clique of dominant males (color) plus mature females either with young or in heat. On the periphery are subordinate males, females, and juveniles. When the troop is threatened, the dominant males move to the periphery on the side of the danger. (From DeVore, I., after Eibl-Eibesfeld. 1970. *Ethology.* Holt, Rinehart and Winston, New York.)

even closely similar, species. Consider the contrasting behavior when a lion and a tiger get into a fight in a circus show involving several lions and tigers. All the lions cooperate and quickly gang up on the tiger. Meanwhile, the other tigers merely sit on their stools and look at the ceiling. As soon as the cooperating lions have killed the first tiger, they move to the second tiger, and then to the third. All this time the surviving tigers continue to remain aloof. Myrmicine ants (which are common under boards and on plants), if deprived of their antennae, attack friend and foe alike. Formicine ants, like the big black carpenter ant, will feed both friend and foe under the same conditions.

Nevertheless, there are a number of worthwhile points which can be made. The stimuli and situations which will elicit conflict are both varied and numerous. Sign stimuli are the most clear-cut. The male stickleback fish will attack almost any red object that appears within its territory (note the gestalt aspect), even though any resemblance to the red belly of another male stickleback is remote. For some carnivores, dogs and wolves for example, the sight of another animal retreating will precipitate an attack. In some circumstances pain or hunger stimulate conflict.

A highly informative experiment on conflict in the rhesus monkey (macaques) has recently been carried out in India by Charles Southwick and others. In a group of males, females, and juveniles in a large outdoor enclosure, the daily average of conflict encounters, fights, threats, hostile expressions, and submissions was established. The amount of food was then reduced from a superabundance to a somewhat restricted level. To everyone's surprise there was no general increase in conflict. The food supply was then reduced to a starvation level. The result was a reduction in conflicts. The macaques spent a large amount of time in slow explorations.

Keeping the food level constant, the effect of crowding was studied by reducing the area by half. This did not significantly increase the number of conflicts. What did produce a real pandemonium of aggressive conflict was the introduction of a strange individual, either male or female. Males took the lead in attacking strange males, and females in attacking strange females, but all members of the group joined in the attacks on the newcomers. It is too early to tell whether these results can be extrapolated to other primates.

There are many biological devices which reduce intraspecific conflict. Pecking orders and territoriality have already been mentioned. The ritualization of conflict is com-

monplace as in the yelling of howler monkeys, the chest beating of the gorilla, and many other stereotyped threat displays. There may be a correlation here with the degree of sociability of the species. Dogs, which are essentially pack animals, have a method of indicating to their opponents that they give in. A vanquished dog or wolf lies down on its back and assumes a more or less puppy-like posture. The cats of various kinds, which are solitary animals, are much more likely to fight until death because they lack a surrender sign.

Positive cooperation is familiar in the coordinated attacks of a pride of lions or a wolf pack, in the defensive circles of bison and musk oxen, or in the way male baboons work together to protect their troop from marauding leopards and other predators. Among invertebrates the cooperative work of the social insects is well known. The neural and physiological mechanisms underlying either conflict or aggression are little understood. As already noted, there are centers which control aggression in the hypothalamus of vertebrates. Pheromones, chiefly odors, play an important role in activating both conflict and cooperation. Remember the ants minus their antennae or the rams and billy goats.

Natural selection works to produce both conflict and cooperation. Where males have to fight for females, as among deer and lions, or have to protect the females and juveniles against predators as is true for baboons, natural selection selects the best fighters. Where family or group cooperation favors survival as in baboons, bison, and the social insects, natural selection favors the development of cooperative mechanisms. This is called **group** or **kin selection.** It is a biological basis of altruism.

REPRODUCTIVE AND PARENTAL BEHAVIOR

In both plants and animals, the primary function of sexual reproduction is the production of hereditary variation in the progeny. The basic biological objective of reproductive behavior in animals is to bring gametes together and, in many species, to nurture the young. With sessile animals which shed their eggs and sperms into the sea, synchronization of spawning is essential. Usually temperature or light is the coordinating stimulus. The famous palolo worms are timed by the moon to emerge from the coral reefs of the South Pacific and to spawn. Pheromones also coordinate spawning. The gametes of oysters, of *Nereis* the clam worm, and of tunicates carry with them some chemical which stimulates other individuals to shed gametes.

Among many teleost fishes, reptiles, birds, and mammals, the breeding period is controlled by photoperiod. Day length apparently plays an important role in controlling the time of sexual activity in those mammals which have only one breeding period per year, as with deer or sheep. In some animals, psychological factors also exert control.

Courtship is a prominent feature in the sexual behavior of animals as different as spiders, squids, parasitic wasps, birds, and mammals. Not infrequently, the courtship bears at least some resemblance to conflict. In birds of paradise, courtship involves elaborate displays of incredibly fantastic feathers. In some parrots, life-long pairing takes place between two young birds well before either has reached sexual maturity. The immature male merely gazes intently at an immature female and she gazes back. If this mutual exchange continues for more than a few minutes, the pair is mated for life.

In mammals there are the loners like tigers and rats, where males play no part in the care of the offspring and where there is no known permanent male-female bond. There is the dominant male plus harem as in seals, deer, and horses. There are family groups as in lions and wolves. Recent studies in both Siberia and northern Michigan show that the basic social unit of wolves is a family group of four to six individuals. Large packs are conglomerations of several such families. Likewise the amount of parental nurture, above and beyond the basic mammalian requirement of milk for the young, varies greatly, especially in the role of the male. Among elephants and whales one or more female adults, so-called aunts, attend the birth and help protect the newborn against tigers or sharks respectively. Male baboons are very protective of infants; male langurs, a common monkey of Indian woodlands, are very aloof.

Parental Emotions

It is impossible not to believe that parent apes, dogs, cows, dolphins, and other mammals share in some degree parental feelings of love which we experience. Birds likewise

Fig. 21-13. Lioness at the kill with two cubs. Male lions will play with their cubs but are not known to share food with them. (After Disney Productions; from Moment, G. B. 1967. *General Zoology;* 2nd ed. Houghton Mifflin Co., Boston.)

appear to share such emotions. However, a caution against anthropomorphism is warranted. The world certainly appears different to a bird. For example, a hen turkey will fight off any animal that approaches her chicks (more properly called poults). Her behavior resembles that of a human parent under similar circumstances, but a bit of study reveals profound differences. A turkey hen will attack a stuffed fox pulled past her nest. If a tape recorder of turkey poult calls is played within the stuffed fox her hostility completely disappears. Moreover, if a hen has never hatched chicks and is deafened, then when her eggs do hatch she will attack and kill her own chicks as though they were rats.

Animal Societies: Sociology

As Charles Darwin pointed out a century ago, E. O. Wilson in a highly controversial book very recently, and many thinkers in between, animal societies and the behavior patterns which make them possible form and evolve under the forces of natural selection and (we would add) mutation. Since any society whose members fail to reproduce ceases to exist, it is hardly surprising that the basis of a society is almost invariably a fertilized female or a family. Societies, i.e. complex social systems, have arisen independently many times among different groups of insects and several times among mammals.

Among insects, the termite pattern includes both sexes as workers, soldiers, and reproductives. The tie which binds termite society together is a potent pheromone secreted by the queen. It is licked off her body by the workers and distributed throughout the colony by mutual cross feeding. Not only does it ensure "loyalty," it is also an antifertility agent for both males and females which prevents them from becoming sexually mature. When termite colonies become extremely large, functional reproductive individuals appear on the distant fringes of the colony because they are so far from the source of the antifertility chemical that they escape from its inhibition.

In wasps, ants, and bees, the entire colony consists of the offspring of a single mating of one female. The workers which hatch from her fertilized eggs are underdeveloped females. Unfertilized eggs develop into haploid, i.e. monoploid, male drones. For some obscure reason, societal life works out well for insects, and there are many dozens of very different unrelated social species.

How does natural selection work in cases where the soldiers and workers are sterile and hence leave no offspring? After a swarm of winged functional male and female ants or wasps leaves the nest, each fertilized female finally settles down alone and begins a new colony. If she carries mutant genes which produce superior workers, her colony will be more successful than others and, although the superior workers transmit no genes because they are sterile, the winning genes will be transmitted through the fertile sons and daughters of the original queen.

In birds of many species, flocking occurs only outside of the breeding season. Nesting pairs of most songbirds are antisocial in proportion to the strength of the sexual bond between them. When this wanes, general social cohesiveness increases.

Many kinds of mammals live in social groups. Lions, porpoises, bison, sheep, wolves, deer — the list is a long one. Moreover, in many instances an individual separated from the group is helpless. A baboon separated from the troop is a dead baboon. Primates are especially notable for the variety of their social relationships. There may be only a loose hierarchy, a single dominant male or, as in the case of baboons, a ruling clique of several older males who cooperate with each other to maintain their authority. These dominant males protect the troops against lions.

Several writers have recently claimed that the biological basis of human societies is the endocrinological fact that the human female remains in a potentially reproductive condition all year. It is possible that this characteristic has played a role in maintaining continual social cohesiveness, especially in the beginnings of human societies; but there is no firm evidence for such a theory. Moreover, there is much evidence that a continuous reproductively active state in either sex is not a necessary precondition for permanent social organizations.

Compared with other mammals, different species of primates show little diversity in female reproductive cycles but very great diversity in social organization. One can hardly imagine more striking social contrasts than the retiring, small-familied gorillas, the gregarious and somewhat rambunctious chimpanzees, the solitary gibbon pairs, a baboon troop, or the loosely organized, female-led howler monkeys.

Among many mammals social bonds exist throughout the entire year, although the females experience estrus only briefly. Wolf packs consist of families and therefore have a sexual basis. However, the pack is a continuing unit for which the widely spaced estrous periods of the females are important but transient incidents. A Merino sheep separated from its flock any day of the year quickly becomes desperate to rejoin its fellows, yet ewes come into estrus only for a few days in early fall. Clearly in both birds and mammals, social cohesiveness transcends reproductive ties.

USEFUL REEFERENCES

Altmann, S. A., and J. Altmann. 1970. *Baboon Ecology.* University of Chicago Press, Chicago.

Beroza, M. 1970. *Chemicals Controlling Insect Behavior.* Academic Press, New York.

Dethier, V. G., and E. Stellar. 1970. *Animal Behavior: Its Evolutionary and Neurological Basis*, 2nd ed. Prentice-Hall Inc., Englewood Cliffs, N.J.

DeVore, I., ed. 1965. *Primate Behavior. Field Studies of Monkeys and Apes.* Holt, Rinehart and Winston, Inc., New York.

Frings, H., and M. Frings. 1964. *Animal Communication.* Blaisdell Publishing Co., New York.

Gaerluce, G. 1971. *Biological Rhythms in Human and Animal Physiology.* Dover Publications, Inc., New York.

Krishna, K., and F. M. Weesner. 1969/70. *Biology of Termites*, vols. 1 and 2. Academic Press, New York.

Lawick-Goodall, J. van. 1971. *In the Shadow of Man.* Houghton Mifflin Co., Boston.

Lorenz, K. Z. 1952. *King Solomon's Ring.* Methuen and Co., London. (Fascinating modern classic by one of the founders of ethology.)

McGaugh, J. L., N. M. Weinberger, and R. E. Whalen. 1967. *Psychobiology: The Biological Basis of Behavior.* W. H. Freeman and Co., San Francisco.

Southwick, C. H. ed. 1970. *Animal Aggression: Selected Readings.* Van Nostrand-Reinhold, New York.

Southwick, C. H. ed. 1963. *Primate Social Behavior.* Van Nostrand-Reinhold, New York.

Tinbergen, N. 1953. *Herring Gull's World.* William Collins Sons and Co., London.

Wilson, E. O. 1971. *Insect Societies.* Harvard University Press, Cambridge.

Wilson, E. O. 1975. *Sociobiology The New Synthesis.* Harvard University Press, Cambridge. (Monumental, controversial and fascinating.)

Wood, D. L., R. H. Silverstein, and M. Nakajima. 1970. *Control of Insect Behavior by Natural Products.* Academic Press, New York.

THE WEB
OF LIFE:
Interdependencies

[Photograph by S. B. Grunzweig.
Courtesy of Photo Researchers, Inc.]

The theory of the evolution of life from simple molecular beginnings to the phenomenon of man has made a greater impact on the intellectual life of mankind than has any other aspect of biological knowledge. It has profoundly influenced the thinking of many groups of people. The task of this chapter will be to present evolution and its implications from a biological viewpoint, leaving philosophy to the philosophers.

It is now over a century since 1859, when Charles Darwin published his epoch-making book, *The Origin of Species by Means of Natural Selection.* Since that event, new and certain knowledge has been won in many areas. Darwin's basic concept of evolution has been powerfully supported and, in fact, greatly broadened and deepened. At the same time much of the actual history of life on this planet has been spelled out in detail by new discoveries of fossils and new methods of dating them, especially by the use of such techniques as carbon dating. Even the old question of the origin of life from the non-living material of a planet seems close to solution. The methods of numerical taxonomy and biochemistry have also thrown new light on old problems of evolutionary theory.

The Classic Darwinian Theory

Charles Darwin was born in England on February 12, 1809, the very day and year of Abraham Lincoln's birth. When about 15, Charles was sent to Edinburgh with an older brother to become a medical student. However, he found the lectures boring and the surgery before anesthetics sickening. At the end of his second year, Darwin transferred to Christ's College, Cambridge, with the intention of studying for the ministry. There he became friends with several undergraduates who were avid beetle collectors and came to know Professor Henslow, a plant hybridizer and man of wide scientific interests, who held weekly open house for undergraduates. Through him Darwin was appointed as naturalist on the naval ship *Beagle* during its 5-year voyage around the world (1831–1836).

Darwin formulated his theory that evolution occurs by means of variation and natural selection at the age of 29, soon after his return from the voyage of the *Beagle*. Yet he did not publish the theory until 1859, after he had devoted 22 years to amassing factual evidence and studying all aspects of the problem. Darwin's great achievement, like that of Newton, lay in drawing together into a new synthesis several different lines of scientific advance. Darwin's presentation in his *Origin of Species* made it impossible for anyone who could read to miss the point. At the beginning of each chapter he tells you what he is going to say, and in the body of the chapter he says it clearly and with an overwhelming wealth of factual detail, and at the end of the chapter he tells you again exactly what he has said.

FOUR PRINCIPLES

Four primary groups of facts or principles form the basis of Darwin's theory of evolution.

The first is the geological principle of **gradualism**, or **uniformitarianism** as it is also called. This is the theory that the world, with all its mountains, plains, and valleys, was not suddenly created in 4004 B.C. or on any other particular date, but is the result of gradual changes over millions of years. Streams and rivers are constantly wearing down the mountains; upthrusts of land continue to produce mountains and earthquakes; sediment still settles on the ocean floor to be pressed into stone. Without this geological background 343

Fig. 22.1. Charles Darwin as a young man. (Courtesy of the Royal College of Surgeons, from Moorehead, A. 1969. *Darwin and the Beagle*. Harper & Row, Publishers, Inc., New York and Evanston.

of the long past, a theory of evolution is almost impossible.

The second principle on which Darwin based his theory was one made familiar by Cuvier and a long line of naturalists. It was the realization that many different kinds of plants and animals form a **graded series** from simple to complex, from generalized to specialized. From the time of Aristotle, these so-called ladders of being had been thought of as static, a part of the unchanging order of things. The important point, however, is not that the old naturalists like Linnaeus and Cuvier thought of plant and animal relationships as unchanging, but rather that they saw living things to be very clearly related to each other by their anatomy, and that these relationships permit us to arrange them in logical sequences. The arm of a man, the foreleg of a cat, the wing of a bat, or the flipper of a seal all show the same basic skeleton. Bone for bone their limbs are **homologous**. Among the insects, the crustaceans, or the coelenterates, everywhere similar basic patterns of structure had become evident. And as explorers and collectors brought in new kinds of animals and plants from the four corners of the world, it became increasingly clear that there are many graded series of types. The German poet and naturalist Goethe was an early proponent of the concept of homology. In his writings he noted and illustrated structural similarities in plants as well as animals. Goethe was the first to postulate that flowers are actually modified shoots and that the parts of a flower are morphologically equivalent (homologous) to leaves.

The third body of knowledge that Darwin utilized was from the practice of **plant and animal breeding**. By the early 19th century, plant and animal breeders had produced markedly improved races of wheat, oats, poultry, cattle, and the like by the method of selective breeding, that is, by **artifical selection**. Domestic plants or animals which showed the desired traits were selected by the breeder to be parents of the next generation. **Natural selection** is the obvious counterpart among wildlife of artificial selection among domesticated plants and animals.

The fourth and perhaps crucial principle that entered into the Darwinian synthesis was the theory of T. R. Malthus on **population growth**. Malthus published his famous *Essay on the Principle of Population* in 1798. The thesis of Malthus is really very simple. He pointed out the rather obvious fact that no population can increase to any significant extent without an increase in its means of subsistence, especially food. He then argued that history shows that whenever there has been an increase in food and living space, the human population has always also increased. His conclusion from all this was gloomy, for he held that only "moral restraint, vice, and misery" such as unwholesome occupations, or extreme poverty, diseases and epidemics, wars, plague, and famine could hold down the immense power of populations to grow if they are left unchecked.

In any finite environment, whether a test tube or a continent, there are limits to the size a population can attain. For example, one medium-sized female crab will carry over 4 million eggs. If 2 million of her eggs hatch into females, and if each were again to produce 4 million eggs, there would be 2 million times 4 million or 8×10^{12} offspring in the third generation. This prodigious number is the result of the reproductive potential of but three generations from one fertilized crab, yet three generations are as nothing in the life of a species that continues decade after decade, century after century. To illustrate this point Darwin characteristically used the slowest breeding animal known, the elephant, which does not become sexually mature until between the ages of 17 and 20, has a gestation period of 22 months, brings forth but one calf at a birth, and usually stops breeding when about 50 years old. Yet in time, if no forces checked their growth, Africa would be carpeted, coast to coast, with elephants.

The significant fact is that any living population (whether bacteria, elephants, or men) tends to increase geometrically, like compound interest, until checked. This insight of

Malthus gave Darwin his clue. In his own words:

I happened to read for amusement Malthus on population, and being well prepared to appreciate the struggle for existence which everywhere goes on from long continued observation of the habits of animals and plants, it at once struck me that under these circumstances favorable variations would tend to be preserved, and unfavorable ones destroyed. The result of this would be the formation of new species. Here than I had at last got a theory by which to work.

The theory of **evolution by natural selection** was proposed independently in 1855 in a short essay by Alfred Russel Wallace, a naturalist and explorer. Happily a joint publication led to mutual appreciation rather than bitterness. It is worth recording that Wallace also found his inspiration from reading Malthus.

In formulating Darwin's theory, two facts are basic. The first is the geometric or "compound interest" character of population growth just discussed. The second is the fact that living things vary in many ways and that many of the variations are inherited. These two factors lead to a struggle, which may be quite passive, for existence which results in the survival of the fittest plants and animals, and the ultimate extinction of the less fit. Thus, populations tend to change their character. No one in Darwin's time understood variation, although Darwin himself saw its crucial importance for his theory and discussed it. The important distinction between true hereditary variations due to changes in the chromosomes, and nonhereditary variations due to amount of food, exercise, sunlight, or other environmental factors, was not made clear until the work of a Danish geneticist, Johanssen, after the rediscovery of Mendel's laws. For Darwin's argument, this distinction was not crucial. It was enough to know that living things do, in fact, vary widely and that many if not all, of these variations are inherited. With no true understanding of the physical basis of heredity and mutation, it is little wonder that most of Darwin's immediate successors came to regard natural selection as the virtually self-sufficient cause of evolutionary change.

Lamarckianism

Before beginning a modern critique of Darwin's classic theory, there is a traditional view about evolution which has come down to us from before Darwin's day, and which is still important enough to require brief discussion. This view, which Darwin accepted and which is even today a hardy perennial among uninformed people in many parts of the world, is Lamarckianism, a belief in the inheritance of the effects of use and disuse. According to this view, the effects of good or poor nutrition, or any other environmentally induced condition in the adult, will somehow produce changes that are transmitted to future generations.

It is a fact that some scientific concepts persist long after they have been shown to be untenable. A notable example of such persistence occurred in Russia during the Stalinist era. Communist dogma at the time asserted that the environment influences heredity. No other views were permitted. Led by Lysenko, Russian geneticists also maintained

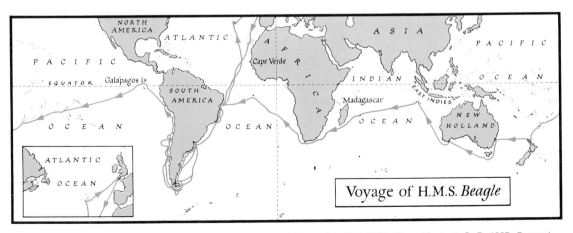

Fig. 22-2. The route of the *Beagle* during its voyage around the world, 1832–1826. (From Moment, G. B. 1967. *General Zoology*, 2nd ed. Houghton Mifflin Co., Boston.)

Fig. 22-3. Trofim Lysenko, Russian geneticist whose insistance that acquired characteristics can be inherited is a modern day example of the dangers of mixing science and politics. (From Huxley, J. 1949. *Heredity East and West.* Henry Schumann, New York.)

that hereditary changes could be brought about in plants by grafting or in animals by blood transfusion. Scientists in other countries have been unable to repeat Lysenko's experiments. In fact, Charles Darwin's cousin, Francis Galton, had earlier tried in vain to transmit hereditary traits by blood transfusion between different types of rabbits. It is now generally agreed even in Russia that Lysenkoism is scientifically false and was ultimately counterproductive in terms of agricultural practices. This period of 20th century scientific history should serve as a warning that the use of science to serve political ends can be self-defeating.

The Modern Synthesis

The modern, or neo-Darwinian, view of evolution sees four major causes which work together in producing evolutionary change. They are: (1) gene mutation and the subsequent and continual reshuffling of genetic factors through sexual reproduction; (2) natural selection; (3) the effects of chance, often called random genetic drift; and (4) isolation, especially temporal and geographical isolation, and its converse, gene flow between groups.

MUTATION AND SEXUAL REPRODUCTION

Mutation is so basic to the subject of evolution that certain relevant aspects merit restatement here. Mutations, whether naturally

occurring or produced by radiation or by chemical means, are random in two senses. There is no way of predicting specifically what mutation will occur next or of producing a specific gene change on demand. The ability to produce specific mutations artificially remains for the future, when much more is known about the chemistry of nucleic acids. Mutations are also random in that they bear no necessary relation to the needs of the organism. In fact, some 99 per cent of all mutations are harmful. This is not at all surprising, since existing animals and plants are the result of millions of years of adaptation. No wonder that the overwhelming probability is that no random change would be for the better. As with a ship or an airplane, making random changes in the blueprint is likely to be disastrous. It should also be remembered that whether or not a mutation is advantageous is relative to the environment.

Different genes possess different mutation rates in accordance with variations in their stability. Many cases of back mutations are known. Thus, a gene for normal pigmentation may mutate to a form resulting in albinism, and a gene for albinism may mutate into one for normal pigmentation. The net result of the difference between the mutation rate of a given gene in one direction and its mutation rate in the reverse direction is known as its **mutation pressure**.

NATURAL SELECTION

There are three different types of selection whether natural or artificial. **Stabilizing selection** tends to eliminate extremes in a population. A sea gull whose wings are too long is at a disadvantage and so also is a gull whose wings are too short. For a bird of a given size with a given mode of life in a given environment, there is an optimum wing length. Natural selection will shorten the lives of individuals which deviate very far in either direction from this optimum. Most selection seems to be of this stabilizing type, which is an anti-evolutionary type of selection pressure since it tends to prevent change.

The second type of selection is **directional**. It favors individuals which vary in one direction and works against those which vary in the opposite direction. This kind of selection can produce evolutionary change. Speed and maneuverability are advantages for a hawk; their opposites are serious handicaps. Increased ability to survive heat and drought is advantageous to desert animals and plants, their opposites disastrous.

The third type of selection is **disruptive** selection. This is probably the rarest type, but it can also be important in producing evolutionary change. In such selection, the extreme variants at each end of the distribution curve arc favored, while those in the intermediate range are at some disadvantage. Such a situation will tend to split a population into two parts, each of which may finally become a separate species. For example, in a population of tree-dwelling monkeys, the largest and heaviest individuals are able to defend themselves from attack by the others; the smallest and lightest, and therefore the most agile, can escape through the treetops, but the middleweights are neither fast enough to escape nor heavy enough to win against the somewhat larger members of the species.

SAMPLING ERRORS OR GENETIC DRIFT

In addition to mutation and selection, a very important factor in evolution is the effect of chance in very small populations. This effect is commonly called **genetic drift** or the **founder principle**. When this term is used it must be remembered that no definite "current" is implied, only the purely random sampling that occurs for many genes whenever the population is drastically reduced in size. For such reasons this aspect of evolutionary dynamics is sometimes called **sampling error**, or even the **bottleneck phenomenon**.

As an extreme case, suppose 10 finches happen to get caught in the hold of a grain ship in Argentina and do not escape until they arrive at the coast of China. There they may or may not form a new population, but whether they do or not, such a tiny flock can certainly not be expected to carry a completely representative sample of the genes common to the whole South American population from which they came. Suppose also that 20 per cent of the large South American population carries a dominant gene for an unusually wide white band on the wing feathers. If the flock of 10 were to be representative, two of the finches would have to carry this gene. If only one of the 10 happened to have it, then its frequency would have decreased by one-half. Actually, such a gene could easily be missing altogether in the little flock. Or a gene that was uncommon in the parent population might just have happened to be common among the 10, and so become common in the resulting new population. When all the thousands of genes present in a species are considered, it becomes evident that no very small group of individuals can be truly representative.

Besides the bottleneck phenomenon there are two additional points where pure chance can change the character of a small population. The first of these is in meiosis and fertil-

ization. Many sperms with their genes are lost; many eggs and 3 times as many polar bodies never form new individuals. In a large population these losses average out, but not so readily in a small population. For example, most of the genes for red hair might get lost in polar bodies. The second point where chance enters is in purely accidental death or survival among the members of a very small population. In large populations the effects of chance, good or bad, tend to cancel.

How small does a population have to be for genetic drift to occur? That depends on the force of natural selection and on mutation pressure, and so may be different for different genes. In putting an infinitely large population of equal numbers of black and white marbles through a bottleneck by drawing a random sample, the overwhelming probability is that a sample of a very few thousand would be representative. In the living world ideal mathematical situations are seldom, if ever, found. Hence, it is impossible to determine exactly how large any particular population of genes has to be in order to avoid the effects of random sampling, i.e. genetic drift.

How important has genetic drift been in the course of evolution? It is thought that in the remote human past, when the primates from which man arose, and the early men, were wandering about in small bands, genetic drift must have been important. Recent careful studies have compared the gene frequencies in a very small population which emigrated from a known part of Germany and settled in Pennsylvania about 200 years ago. Because of their religious beliefs, the Amish have not intermarried appreciably with other people (although over the years some have left the little community and joined the mainstream of American life). If genetic drift is a fact, some of the gene frequencies among this group should be different not only from those of the general population of Pennsylvania but also from those of the large parent population in Germany. This turned out to be true for the ABO and MN blood group genes, as well as some other minor hereditary characteristics including polydactyly, the possession of extra fingers or toes (Fig. 22-4).

It is also worth noting that natural selection, by drastically reducing a population, may put a species through a bottleneck where genetic drift becomes operative. For example, when an epidemic disease hits a population of wild ducks, there is a rigorous selection for any genes which make for resistance to this disease. All other genes, no matter how important in other situations (genes affecting speed in flight, correct migrating instinct, breeding ability, and success in es-

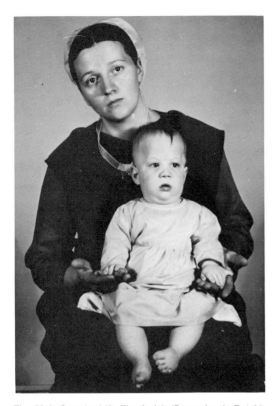

Fig. 22-4. Genetic drift. The Amish (Pennsylvania Dutch) have not intermarried with other people since they emigrated from Germany and settled in Pennsylvania about 200 yr ago. Because of genetic drift, their gene frequencies for blood groups and other characteristics such as polydactyly (shown here in the Ellis-van Creveld six-fingered dwarfism) differ from those of the general populations of Pennsylvania and Germany. (Courtesy of V. A. McKusick, The Johns Hopkins University.)

caping predators such as foxes, snapping turtles, and hunters) for the time become irrelevant. Only those genes that just happen to be in ducks which also carry the genes for disease resistance will escape through the population bottleneck. But the smaller the bottleneck the more certain that no representative sample of all these other genes will get through. Consequently, when the ducks increase again after the epidemic subsides, the new population will be genetically different.

ISOLATION

Isolation was regarded by Darwin and his immediate followers as so important that it was held to be a prerequisite without which no evolution was possible. With their pre-Mendelian view that heredity involved a blending of vaguely understood hereditary material ("blood") from the parents, it seemed inevitable that any deviation would be averaged out as the result of cross-breeding in subsequent generations. The discovery of the Mendelian factors of heredity, the genes, which enter and emerge from crosses unchanged, produced a modification of this view. If a gene or a certain combination of genes confers a selective advantage, it is not necessarily averaged out. On the contrary, it may be preserved and will tend to increase in frequency. In the case of a self-fertile population (such as we find in many plant species) or in a population that is very closely inbred, the proportion of homozygous individuals, whether homozygous for the dominant or the recessive gene, continually increases. This is because the homozygous individuals mated together can produce only more homozygotes, while a cross between two heterozygous individuals will give rise to only 50 per cent of heterozygous individuals like themselves; the other half of their offspring will be homozygous.

The evidence collected in recent years shows that isolation does play an extremely important role in the origin of new species. In the first place, different but related species are always separated in some way in nature so that they do not interbreed, even though they may be induced to do so under laboratory conditions. Second, wherever there is a barrier between two or more populations, these populations almost invariably show differences.

The Evidence for Evolution

For any science, the question of evidence must always be of central and overriding importance. The evidence that evolution has been a fact of history seems irrefutable to biologists, even though there are intelligent people who find it unconvincing. Each student will have to weigh all the evidence and make up his or her own mind.

GEOGRAPHICAL DISTRIBUTION

Of the evidence for evolution based on geographical distribution none is more striking than the relationship between the plants and animals on oceanic islands and those on adjacent continents. This kind of evidence forcibly impressed both of the originators of the theory of evolution by natural selection. Charles Darwin, in his voyage as a young naturalist on the *Beagle* visited the Cape Verde Islands off the West coast of equatorial Africa and the Galápagos Islands off the west coast of equatorial South America, and was profoundly struck by the different flora and fauna existing under similar conditions in geographically different areas. Alfred Russel Wallace studied the life of the long series of islands making up the East Indies.

Each oceanic island has its own native animals and plants. But, and this is the point, in every case they resemble the animals and plants on the nearest continent. The Galápagos Islands are inhabited by turtles, lizards, birds, and other animals that are **endemic** (native and unique), but are similar to lizards and birds on the adjacent shores of Central and South America. Likewise, the animals and plants of the Cape Verde Islands resemble neither those of South America nor of tropical Asia but those of the adjacent parts of equatorial Africa. These facts indicate that living things came from the nearest continent and then, under the influence of mutation, selection, genetic drift, and isolation, became different from the ancestral populations. There is some geological evidence that the Galpagos Islands were once connected to the South American mainland so that their populations and those on the continent were perhaps continuous in remote ages. Of course, it is also possible for islands to be populated by the descendants of a few land birds blown off course by unusual storm winds. Lizards, other animals, the seeds of plants can survive long sea voyages by clinging to floating logs, and the minute spores of lower plants can be carried great distances by air currents.

COMPARATIVE ANATOMY

The facts of comparative anatomy have always been regarded as basic supporting evidence for the fact of evolution, whatever the theory as to its mechanism. This interpretation is itself based on the twin concepts of homology and **analogy**. Homology is basic anatomical similarity, regardless of function, both in embryological development and in the adult. The arm of a man, the front leg of a horse, and the wing of a bird are homologous. The wing of an insect, although used for the same function as the wing of a bird, has a radically different basic structure; hence they are merely analogous to each other.

There are several kinds of homology. **Special homology** involves similar structures, like the forelimb, in two or more different animals while **serial homology** is homology between structures in an anterior-posterior series on the body of a single animal, such as the arm and leg of a man. Both of these homologies were made famous in T. H. Huxley's studies of the appendages of the lobster and the crayfish. The big claw of a lobster corresponds part for part with that of a crayfish or a crab (special homology) and there is also a basic anatomical correspondence between the big claw and all the other appendages along the lobster's body (serial homology). A basic similarity of structure and development without specific correspondence in the location or in the serial arrangement is known as **general homology**. An example is the anatomical correspondence between all of a shark's scales and its teeth.

The best examples of homologous structures in plants are flowers and vegetative shoots. There can often be seen a transition in a water lily flower, for example, between stamens and petals and between white petals and green leaflike structures. They, too, stand the test of developmental beginnings, both originating as primordia formed in the apical region of an existing shoot.

The facts of homology do not in themselves prove that evolution has occurred. However, the theory of descent with modification from

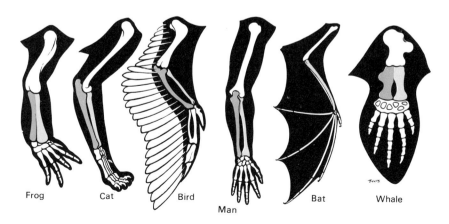

Frog   Cat   Bird   Man   Bat   Whale

Fig. 22-5. Skeletal homology in vertebrate limbs. Ulna in darker, radius in lighter color in each case. (From Moment, G. B. 1967. *General Zoology*, 2nd ed., Houghton Mifflin Co., Boston).

a common ancestor provides the only known scientific explanation.

COMPARATIVE EMBRYOLOGY

Embryology reveals many facts which can be explained easily only on the basis of evolution. How else can the presence of gill slits in the embryos of reptiles, birds, and mammals be explained except by postulating ancestors in which gill slits were functional? The development of the heart and the main arteries from it, and the development of the kidneys also show fishlike stages in the embryos of higher vertebrates.

The modern interpretation of such embryonic structures as gill slits in mammals is that they correspond to the gill slits of an embryo fish rather than of an adult fish. Thus, in a sense they are vestigial structures. There are several reasons which taken together seem adequate to explain their persistence. Changes in genes which affect the earliest stages of development will be very likely to produce much more drastic and even lethal results than mutations which affect the later stages when the major organs have been laid down. This in itself will tend to make the early stages of development very conservative. Equally or even more important, natural selection has only a very indirect action on the developmental stages within an egg or a uterus. Its action is largely restricted to such matters as rates of growth and means of obtaining food and eliminating wastes. It is highly significant that it is in this respect that vast changes have taken place; witness the evolution of the placenta and the other fetal membranes. A third factor is often cited as a cause of embryonic conservatism. At least some of the structures in the embryo of a man which resemble the embryo of a fish have a function in the machinery of embryonic development. For example, the primitive kidney or pronephros initiates the development of the pronephric duct, which becomes the vas deferens through which sperms pass to the exterior of the body. Likewise, the embryonic gill slits give rise to some of the endocrine glands (see Chapter 6).

VESTIGIAL ORGANS

Vestigial organs, such as traces of hind legs in certain kinds of snakes, small pelvic bones in whales and porpoises, ear and tail muscles in human beings, and vestiges of toes in horses, can be explained only on the basis of evolution unless one wishes to believe nature is capricious and deceitful. In fact, combined with the fossil record, vestigial organs like these furnish convincing proof of the fact of evolution.

In horses, which walk on one elongated and thickened toe on each foot (the hoof is the toenail), there are vestigial "hand" bones on either side of the main "finger" (the bones horsemen call splint bones). In the recent fossil ancestors of horses the splint bones are much larger and each is tipped with a small hoof, while in the remote ancestral fossils, there are four functional toes on each front foot and three on each hind foot. The geological record shows clearly that all the early primitive mammals were four-legged; therefore, whales and porpoises must be descended from ancestors equipped with hind legs. It is understandable then that whales and some snakes bear signs of hind legs and pelvic bones. The most cogent evidence for evolution is laid down in the fossil record and to this we now turn after a discussion of how life began.

Origin of Life

ALL LIFE FROM LIFE?

Until modern times it was commonly believed by ignorant and learned persons alike that life continuously and spontaneously arises from dead matter. Maggots were thought to generate in decaying meat, worms and even frogs from the mud and muck on the bottoms of ponds, clothing moths from a mixture of wool and grease. Only a little over a century ago zoologists believed that tapeworms arose spontaneously in the intestines of their hosts.

The first great blow to the theory of **spontaneous generation** came from the experiments of a 17th century Florentine physician, Francesco Redi (1626–1698), who showed conclusively that the age-old belief that maggots arise spontaneously in decaying meat was false. Instead, he proved, they develop from eggs laid by flies. His method is a classic example of scientific procedure, for it was clearly conceived to yield a definite answer and made good use of controls. Redi used a wide variety of meats, singly and in combinations, for (who knew?) perhaps some meat was unsuitable. Apparently he had access to a good butcher shop and a generous zoo because he records using beef, lamb, venison, chicken, goose, duck, dog, lion, swordfish, and even eel from the river Arno. He placed the meat in two series of jars, one covered with very fine cloth mesh, the other left uncovered. The meat in both covered and uncovered jars became stinking masses, reeking with putrefaction. Yet only in the uncovered jars where flies gained access and laid their eggs did maggots develop. Neither the kind of meat nor its degree of decay made any difference. Increasing knowledge about the life histories and embryological development of insects, frogs, and other macroscopic animals made it clear that animals always develop from previously existing animals, not from nonliving material.

As often happens in the history of science, an old question was reopened on a new level by discoveries in later centuries. In fact, the question of spontaneous generation has been reopened three times and is now once again the subject of intense scientific interest. In the 18th century it was the new knowledge of bacteria that raised the question again. Bacteria certainly seem "just to appear" almost everywhere. However, Spallanzani (1729–1799) finally succeeded in showing that if meat or broth is thoroughly boiled and the container sealed while still hot, bacteria never appear and the contents will remain unspoiled indefinitely. This was a discovery of major historical importance, because the canning of food, the control of contagious diseases, and the conduct of surgery are all heavily involved.

Again in the 19th century the question was raised, and given what seemed a final answer by Pasteur (1822–1895). The increasing knowledge of respiration made it seem possible, or even likely, that Spallanzani's heating had driven off the oxygen necessary for life. Moreover, the new chemical theories of fermentation seemed to support the idea of the spontaneous generation of life. But once again the results of Redi were confirmed for bacteria, this time in a long series of investigations by Pasteur culminating in 1864. In brief, Pasteur showed that if beef broth or other foodstuffs were thoroughly boiled in open vessels no bacteria would appear in them so long as the opening to the exterior was a long curved tubular neck bent horizontally so that no dust or dirt particles carrying bacteria could pass up it. From this time on, the phrase, *omne vivum ex vivo*, all life from life, has been a basic tenet of all branches of biology.

THE HETEROTROPH THEORY

In recent years the advancing knowledge of biochemistry and genetics, of historical geology, and of the possibilities of natural selection have combined to substantiate the heterotroph theory of life's origin from non-living matter. A **heterotroph** (*hetero*, other, + *trophos*, feeder) is an organism, such as an animal or a fungus, that is dependent on outside sources for complex nutrients, especially on a carbon source more complex than CO_2. An **autotroph** (*auto*, self, + *trophos*, feeder) is not dependent but can manufacture its own food from very simple inorganic materials, such as water and CO_2. A heterotroph theory of the origin of life has been discussed off and on for a good many years, and in fact was proposed in 1871 by Darwin himself. Much more recently the heterotroph theory has been set forth by J. B. S. Haldene in England and in great detail by A. I. Oparin in Russia. The evidence for this theory does not constitute proof, but it is nonetheless persuasive.

The heterotroph theory holds that, before the existence of any life on a planet, complex organic molecules would arise and, once formed, would remain in existence because there would be no bacteria, molds, or protozoa to devour them. The first living thing would be composed of these complex molecules built up into still more complex arrangements. A fundamental characteristic of the proteins and nucleic acids of protoplasm is their ability to duplicate themselves. Whether the first forms of life were "free genes" something like footloose **virus-like particles** or were colloid globules, the so-called **coacervates**, is not certainly known. Coacervation, that is, the formation of discrete colloidal globules separated from their environment by a distinct boundary membrane, is something which occurs readily in non-living material in the laboratory. Such a coacervate composed of self-duplicating nucleic acids and proteins would not be very different from an amoeba.

The Nonliving Basis

Evidence from both astronomy and geology indicates that in the remote past before the appearance of life, the earth's atmosphere lacked appreciable amounts of oxygen but was rich in just those carbon and nitrogen compounds, plus water vapor, that are required to form amino acids, proteins, and nucleic acids. Today the other planets, where it is certainly too cold or too hot for life, have atmospheres of this type. Spectroscopic analysis shows that the "airs" of Jupiter and Saturn are rich in methane (CH_4), ammonia (NH_3), and probably hydrogen (H_2). Space probes have recently confirmed that the atmosphere of Venus contains enormous amounts of carbon dioxide (CO_2); the atmosphere of Mars has both water (H_2O) and carbon dioxide. The earth's seas, after the first dozen million years (it is important to remember that evolution is a matter of some billions of years) contained salts of many kinds, including phosphates which seem essential for the energy transfers of living material and for building nucleic acids. Recently, two carbon compounds, formaldehyde and formic acid, have been detected in outer space.

The raw materials were certainly present. What were the chances that they would have combined into complex molecules like porphyrins, amino acids, and self-duplicating units like nuclei acids? There is, first, the enormous stretch of time during which the simpler compounds would have had opportunities to come together in all possible ways and combinations. Of course, if carbon, oxygen, nitrogen, and hydrogen could not form amino acids, no amount of time would avail; but these are the elements from which amino acids are made. These are undoubtedly possible reactions.

At least four conditions were present in addition to enormous stretches of time and the essential raw materials that would have greatly facilitated such synthetic reactions. One factor, perhaps the most important, was the intense ultraviolet radiation reaching the earth's surface before there was any appreciable free oxygen in the atmosphere. Free oxygen under the influence of solar radiation forms ozone which today constitutes an ultraviolet-absorbing blanket around the upper atmosphere of the earth. A second factor is lightning, which today is still producing nitrogen compounds in the atmosphere that are brought to the earth's surface in rain. Third, in the long course of geologic history, the earth's crust has undergone many upheavals. Mountains have appeared, continents have been submerged, and what is now dry land was once under the sea. In this process arms and bays of the sea were cut off from the ocean and may have dried up completely, leaving salt beds, or may later have been connected with the sea again. The Caspian Sea represents such a cutoff bay. This means that in the past there have been many types of places, from temporary tide pools to inland seas, where evaporation would slowly concentrate salts and organic compounds, thereby increasing the likelihood of new and complex chemical reactions.

A fourth condition has been noted recently by the biochemist and theoretician, Leslie Orgel, and that is freezing. When an aqueous solution of simple compounds is frozen, the water will separate out as ice and the solute molecules will be concentrated. This is what happens when the alcohol in hard cider is concentrated by the simple device of freezing it and throwing out the ice. Orgel finds that if the freezing occurs at the proper rate, the ice crystals will so concentrate and line up various kinds of simple molecules that they react with each other forming macromolecules which can be the ground material of life.

NATURAL SELECTION AT THE CHEMICAL LEVEL

As both Darwin and Oparin have emphasized, before any living things existed on our planet, complex compounds which today would be quickly metabolized by microorganisms of various kinds, would have remained. There would have been no decay in the usual sense because the organisms which cause decay did not exist. Under these conditions there would have been a natural selection among the various organic compounds mixing in the primeval seas and pools. Compounds that were less stable or more difficult to form would become less abundant, while

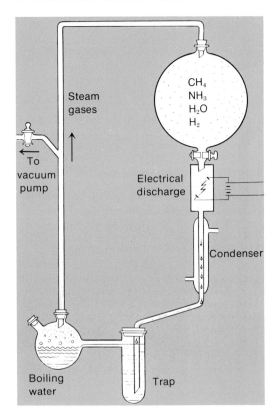

Fig. 22-6. The Urey-Miller apparatus. Water vapor, ammonia, methane, and hydrogen were circulated past an electrical discharge. When condensed liquid was analyzed it was found to contain a variety of complex chemicals including amino acids. (From Moment, G. B. 1967. *General Zoology*, 2nd ed. Houghton Mifflin Co., Boston.)

the more stable ones would accumulate. Compounds which had the ability to imprint their own organization on other compounds would be favored over those which lacked such an ability. Those well fitted to organize more molecules into configurations like themselves would be favored over those which did so only slowly. The slow ones would be "outbred," so to speak, and might even cease to exist.

EMERGENCE OF LIFE

By the time complex compounds appeared which had the property of organizing, or of catalyzing, the formation of other compounds like themselves, the shadowy borderland between the clearly nonliving and the obviously living had already been reached. In considering the probable events during this early period of life's origin, it is important to remember that many of the materials present in the primordial "soup," although relatively simple themselves, are nonetheless the substances out of which the highly complex macromolecules of life are built. Phosphoric acid, H_3PO_4, for instance, is an essential part of the structure of all nucleic acids and as part of ATP is also essential in most energy transfers within living systems. Glycine, composed of a carbon with an amino group and two hydrogens attached to it, plus one $-COOH$ group, is the simplest possible amino acid. Acetic acid, CH_3COOH, is almost the simplest possible organic acid. Yet glycine and acetic acid combine to form porphyrin, a key part of all cytochrome, chlorophyll, and hemoglobin molecules.

The existence of complex, more or less self-duplicating compounds and aggregates of compounds would not prevent natural selection from taking place. If anything, it would be intensified. At first the energy for the synthesis of these compounds would have been supplied primarily by ultraviolet light from the sun and by lightning. But once complex compounds had been formed in some abundance, those that could obtain the energy needed for synthesis by causing the breakdown of other compounds, rather than depending on solar radiation, would clearly have an adaptive advantage. There was probably no free oxygen present, certainly no more than trace amounts, so that these first energy-yielding reactions must have been anaerobic. At precisely what point these self-duplicating, energy-utilizing molecular entities can be said to have been alive is a semantic question.

The view that the first energy-yielding reactions to appear in the history of life were anaerobic is supported by the geological evidence that there was no free oxygen in the atmosphere of our planet in its youth. It is also supported by the fact that the initial stages in the utilization of foodstuffs to provide energy in animals, plants, and bacteria are always anaerobic. In other words, the Embden-Meyerhof glycolytic pathway, which does not depend on free oxygen, precedes the Krebs citric acid cycle and electron transport chain, which require free oxygen. In fact, there are good reasons for believing that the Krebs cycle evolved much later than the glycolytic pathway and that it was tacked on the end of it, so to speak.

Consequently, biologists no longer believe that the first living things must have been green plants, purple bacteria, or other autotrophs. Obviously, an autotroph must begin with a far more complex set of synthetic enzymes than a heterotroph, which depends on picking up and utilizing already existing compounds of some complexity. To begin with an autotroph is to begin at the wrong end.

Three Primary Life Styles Emerge

Once considerable numbers of primitive self-duplicating units had come into existence as indicated in Chapter 1, they would compete with each other for the dwindling supply of complex nutrients. In such a nutritional crisis, natural selection must have favored two very different lines of evolution. Any unit, organism if your prefer, that required fewer complex molecules in its "diet" would be favored over the unit which required more. For example, an organism which required compounds A, B, C, and D but did not need E because it could make its own E by transforming D would have had a selective advantage over the organism which required all five. This line of selection led to the photosynthetic plants, the autotrophs which can build themselves up from simple inorganic salts, CO_2, and water. Water and CO_2 were abundant from the start of life and the fermentations carried out by primitive organisms assured a continuing supply of CO_2. The porphyrins (specifically chlorophyll) provided the light-trapping colored compounds. Once the necessary enzymes were available for converting the trapped solar energy into chemically useful forms (high energy phosphate bonds and complex organic compounds), this line of living organisms and their descendants became nutritionally independent for all time.

Fortunately for us there was another evolutionary answer to the food crisis of those remote times. The second line of selection favored those units which either could move around and so get into a place where the necessary materials for their growth were present or else could engulf other organized units and secure the essential materials that way. This line of selection led to the animals. Once photosynthesis became widespread, it provided a supply of free oxygen. This made aerobic respiration possible for plants but it was of even greater importance for the evolution of oxygen-requiring animals which characteristically depend on vigorous activity, a highly energy-dependent trait. Before this could happen mitochondria had to evolve, either directly in the cells of plants and animals or as symbiotic aerobic bacteria.

Bacteria and other fungi have evolved in a direction which exploits a second nutritional opportunity open to heterotrophs. They obtain their complex carbon compounds from other organisms including dead plants and animals. Without the fungi, decay would not occur and ultimately the earth's supply of CO_2 would become locked in the corpses of plants and animals rather than being recycled. At what point in the history of life on this planet their activities became essential for the continuance of life is unknown.

LABORATORY CONFIRMATIONS

To some of these general arguments, Harold Urey and his student, Stanley Miller, have added laboratory proof. They set up an apparatus in which they circulated a sterile atmosphere of water vapor, ammonia, methane, and hydrogen past an electric discharge (to simulate lightning). The water vapor was then condensed, and in this liquid they found after some weeks a variety of complex substances including the amino acids glycine and alanine, two of the commonest in proteins. This experiment clearly ranks with the synthesis of urea by Wohler as a landmark in understanding the chemistry of life. The work of Urey and Miller has been confirmed and extended in other laboratories.

Strong support for the concept that successive mutations increase the synthetic abilities of primitive organisms has been obtained by students of mutations affecting the nutritional requirements of bacteria and fungi, especially the pink bread mold *Neurospora*. Essentially, mutations are occasional errors which occur in the duplication of nucleic acids, and there is no reason to suppose that such events were any rarer in the remote past then they are now.

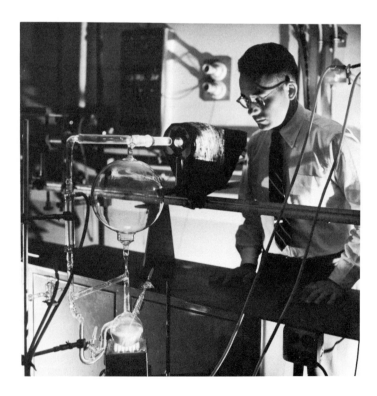

Fig. 22-7. Stanley Miller with the apparatus he and Harold Urey used to demonstrate the formation of simple amino acids from very simple compounds without the presence of living organisms. (Photograph by Fritz Goro. Courtesy of Time-Life Picture Agency.)

THE ROLE OF CHANCE

Does all this mean that the origin of life is a matter of chance? This often-asked question can easily lead into an exercise in semantics and scientific double-talk. In the sense that the particular time, place, and manner in which the first self-duplicating unit appeared would be no more and no less predictable than when the number five will turn up in a series of throws of dice, the origin of life was clearly a matter of chance. There is no reason why the first living unit should have been formed from compounds carried down by raindrops after synthesis by lightning rather than from molecules formed on the surface of the sea by ultraviolet radiation. At the same time, the potentialities for combining into amino acids and more complex self-duplicating units had to be present in the original methane, ammonia, and other molecules. If none of the dice used has a side with five dots on it, five will never come up regardless of how many times the dice are thrown. The appearance of life was a matter of chance, but the dice were loaded.

The History of Life on Earth

FOSSIL REMAINS

The vast drama of the evolution of life on this planet is at least partially recorded in the form of fossils in the rocks. Most fossils consist either of the body of an organism which has been gradually mineralized or of an imprint left by an organism. The sandstone of the Connecticut River valley and the Paluxy River in Texas are famous for fossilized footprints left by dinosaurs. Many fossils have not been completely mineralized and still retain traces of organic matter. All of the coal in existence is derived from forests which flourished tens of millions of years before the dinosaurs. All of the oil and all of the chalk cliffs are deposits formed from the oil droplets or skeletons of algae or protozoans. By contrast, some remains of extinct and prehistoric life are so recent that they have scarcely been changed enough to be called fossils. In many parts of the United States bones of extinct animals can be found. Preserved in ancient tarpits and in dry caves in the southwest are the remains of saber-toothed tigers, small and large species of elephants, and other animals no longer in existence. In the far north, in both Alaska and Siberia, the frozen remains of several woolly elephants known as mammoths have been found. Not only was the fur preserved but even the flesh was still recognizable as meat and was eaten by dogs.

Fossils are found only in sedimentary, that is, stratified, rocks such as slate and are formed in two principal ways. One is the slow accumulation of sand or other sediment on ocean or lake bottoms. The other is the accumulation of material on the land, perhaps in extensive bogs and marshes, which in later ages becomes submerged beneath the ocean by slow changes in the continents. Obviously older layers underlie more recent ones. Igneous rock, which is spewed out of volcanoes or formed the original crust of the earth, never contains fossils.

NEW METHODS OF CHRONOLOGY

The new ways of forcing the earth to reveal the age of her rocks and their fossils are continually becoming more precise. At the same time the older methods are still valuable. The rate at which Niagara Falls moves upstream, cutting its gorge as it goes, can be measured. By simple arithmetic it is then possible to estimate how long it has taken Niagara Falls to cut its gorge. Correcting for the probability that more water flowed over the falls in the past when the last glaciers were melting, the answer comes out to somewhat over 10,000 years, an estimate that agrees with estimates made by other methods.

Another method that gives reliable results in determining chronology of the recent past is to count varves in clay. **Varves** are layers of sediment formed where a river runs into a body of quiet water. This is especially true of rivers running down from snow-capped Alps or melting glaciers. Spring flood waters wash down coarse sediment. As the runoff slows down during the summer, the sediment becomes finer and finer. In the winter it is finest of all, to be followed the next spring by another coarse layer. Consequently, by counting the varves in a deposit of clay it is as easy to determine the number of years it was in forming as it is to tell the age of a tree by counting the annual rings revealed by a cross section of the trunk. By comparing the pattern of successive varves in one lake with those in another, it is possible to correlate their dates of origin.

Longer periods are dated by measuring the radioactivity of various embedded materials. Each radioactive element has a specific **half-life**, that is, the time it takes for the level of radioactivity in a sample to decay to one-half the original level. A crystal of uranium-238 will be half lead after 4.5 billion years. By determining the percentage of lead in such a piece of uranium, it is possible to tell its age.

TABLE 22-1
Time table of life on earth (After Moment, G. B. 1967. *General Zoology,* 2nd ed. Houghton Mifflin Co., Boston; and Fuller, H. J., and O. Tippo. 1954 *College Botany,* rev. ed. Holt, Rinehart and Winston, New York.)

Era and Duration (in Millions of Years)		Millions of years ago (from Start)	Period	Characteristic Life (Dominant Organisms)
Cenozoic	70	0.025	Quaternary Recent epoch	Herbs and man
		1	Pleistocene epoch	
			Tertiary	Angiosperms, mammals, and birds
		10	Pliocene epoch	
		25	Miocene epoch	
		40	Oligocene epoch	
		60	Eocene epoch	
		70	Paleocene epoch	
Mesozoic	160	130	Creataceous	Gymnosperms and reptiles
		165	Jurassic	
		200	Triassic	
Paleozoic	320	230	Permian	Lycopods, seed ferns, and amphibians
		250	Pennsylvanian	
		280	Mississippian	
		325	Devonian	Early land plants and fish
		360	Silurian	
		450	Ordovician	Algae and invertebrates
			Cambrian	
		550		
Pre-Cambrian	3,000	3,000		Unicellular organism Worms
Proterozoic				
Archeozoic				

Geological Events, Climate, etc.	Advances in Plant Life	Advances in Animal Life
	Increasing dominance of herbs	Man; rise of civilization
Periodic glaciation	Extinction of many trees; increase in number of herbs	Extinction of great mammals
Continued cooling of climate with temperate zones appearing; Cascades, Andes, Coast Ranges formed	Increasing restriction of plant distribution and of forests; rise of herbs	Appearance of man
Climate greatly changed—cool and semi-arid; marginal seas; Himalayas, Alps formed	Restriction of distribution of plants—retreat of polar floras; forest reduction	Culmination of mammals
Climate warm, humid, Pyrenees formed	World-wide distribution of tropical forest	Primitive mammals disappear; rise of higher mammals and birds first anthropoids
Climate cool, semi-arid; then warm, humid	Modernization of flowering plants; development of extensive forests—to polar regions. Sequoias prominent; tropical flora in Artic regions	Modern birds and marine mammals appear
Mountain glaciers		
Climate fluctuating, Rocky Mts. and Andes formed, Great Continental seas	Angiosperms dominant; gymnosperms dwindling, modern tropical plant families within Arctic Circle	Rise of primitive mammals
Climate fluctuating	Rapid development of angiosperms—many living genera present	Extinction of great reptiles
Climate very warm	Rise of angiosperms; conifers and cycads still dominant	
Climate warm, Sierras formed; great continental seas in Western N. America	First known angiosperms; Caytoniales; conifers and cycads dominant; cordaites disappear	Primitive birds and flying reptiles (pterodactyls); dinosaurs abundant; higher insects
Climate warm, semi-arid	Floras not luxuriant; higher gymnosperms increase (Cycadophytes, conifers, ginkgos); seed ferns disappear	First mammals; rise of giant reptiles (dinosaurs)
Climate dry with periodic glaciation; Appalachians, Urals formed; drainage of seas from continents	Dwindling of ancient groups, extinction of many; first cyads and conifers	Rise of land vertebrates
Period of crustal unrest, alteration of marine and terrestrial conditions	Dominant lepidodendrons, calamites, ferns, seed ferns, and other primitive gymnosperms (Cordaites); extensive coal formation in swamp forest	
Widespread shallow seas on N. America; Acadian Mts. formed	Dominant lycopods, horsetails, and seed ferns; early coal deposit	Rise of primitive reptiles and insects
Broad shallow seas on N. America during Silurian and Devonian	Early land plants (Psilophytales-Rhynia, etc); primitive lycopods, horsetails, ferns, and seed ferns; first forest	Rise of amphibians; fishes dominant
Taconic Mts. formed	First known land plants; algae dominant	Lungfishes and scorpions (air breathing animals)
Broad shallow seas on the N. American continent	Rise of land plants(?); marine algae dominant	Corals, star fishes, pelecypods, etc.; first vertebrates—armored fishes
Narrow seas within the borders of N. America; climate warm, uniform over earth	Algae—especially marine forms	Many groups of invertebrates—dominance of trilobites
Grand Canyon, Younger Laurentians formed		
Rocks, chiefly sedimentary	Bacteria and algae	Worms, crustaceans, brachiopods
Glaciation		
Older Laurentians formed	No fossils found; all organisms probably unicellular or very simple	
Rocks mostly igneous or metamorphosed; few sedimentary		

358   **Web of Life**

The greater the proportion of lead, the older the crystal. Although such crystals form only in molten igneous rock as it cools, it is possible to use them to date strata of sedimentary rock formed by the gradual deposit of silt or other material on the ocean floor. In volcanic convulsions molten igneous rock sometimes breaks through strata of sedimentary rock. Obviously all the strata through which such molten rock breaks must antedate the volcanic action. The molten rock often spreads out on top of the uppermost strata, forming an enormous pancake of igneous rock. Any sedimentary rock above this pancake must of course be younger. Obviously, this method is only useful in measuring long periods.

A recently introduced method is the radio-carbon technique. This provides an accurate clock for periods ranging from a few thousand to about 35,000 years ago, since the half-life of radioactive carbon-14 is $5,568 \pm 30$ years. This carbon is being formed in the earth's atmosphere by the action of cosmic rays on nitrogen and is also continually disintegrating, so that an enduring equilibrium exists age after age. Radioactive carbon built into plants by photosynthesis will not be replenished after they die. Consequently, the longer a piece of wood or part of any animal that ate plants exists, the less radioactive carbon it contains.

Human Origins

Our ancestry extends back to the origin of life, but our evolution as something very special among mammals began not earlier than sometime in the Paleocene, roughly 60 million years ago. At that time there was no distinction between the shrew-like insectivores and the monkey-like primates. But this primitive insectivore-primate stock possessed two characteristics which have left their marks on us to this day. These small mammals were generalized and tree-living.

When did the primates separate from the insectivores? The fossil record is missing, and even today it is impossible to make a clean-cut distinction between these two orders which span the enormous gulf between a man and a mole. Modern insectivores include not only moles but also shrews, which are common almost everywhere though seldom seen because of their small size and

Fig. 22.8. From the strange creatures in the foreground, Prochordates and Tunicates, began the long journey to mankind. (From Moment, G. B., and H. M. Habermann. 1973. *Biology: A Full Spectrum.* The Williams & Wilkins Co., Baltimore.)

nocturnal habits. The primates include man and the anthropoid apes, the Old World monkeys, the New World monkeys (which alone possess prehensile tails), and the lemurs, now largely restricted to Madagascar and adjacent lands.

The most ancient monkey-like fossils so far discovered were found in lower, i.e. earlier, Oligocene strata in Egypt. *Parapithecus*, as this creature was named, was perhaps close to the base population from which both Old and New World monkeys diverged; but no one can be certain. Our broad shoulders and well-developed collar bones are clearly adaptations for what anatomists call **brachiation**, i.e swinging from branch to branch. But exactly when our ancestors forsook the trees for the ground is unknown. However, it is practically certain that they were fairly large, since the habit of swinging from branch to branch in contrast to climbing and jumping like the little lemurs is limited to the larger primates.

It does not seem too surprising that our immediate subhuman predecessors left few fossils. Living either in forests or grasslands, opportunities for fossilization were scanty, except in occasional swamps, river quicksands, and tar pits. Evidently primitive man was too smart or too agile to get caught in these. However, anatomists have made some informed and plausible guesses about our early history.

Human evolution took place toward the end of the Cenozoic. This was a period when, over large areas of the earth, forests were giving way to grasslands as the climate became drier. With the succulent leaves and buds of forest plants and the diverse animal life of forests vanishing, many animals were faced with a harsh choice: to eat grass or to eat other animals that eat grass.

It is reasonable to suppose that these profound environmental changes affected the predecessors of modern man. The spread of the grasslands would necessitate either a changed way of life, with emphasis on hunting the animals (horses, deer, and mammoths) which ate the grass and shrubs, or a migration southward to keep within the dense forests. Perhaps both happened. Hunting of this kind would put an evolutionary premium on running and communication. Natural selection would then favor the further development of the upright posture and long straight legs, a good brain and ability to communicate, and use weapons and other tools. It would also favor the kind of disposition that makes for mutual cooperation. The breeding units were certainly small. Populations of many millions are found in insects, fish, and marine invertebrates, but surely did not exist in primates until human civilization became industrialized.

Under these changing conditions the primary evolutionary forces of mutation and natural selection produced as much change in the primates as they did in the horses. This must have been the period when genetic forces building on the juvenile form with its proportionately much larger brain and prolonged period of willingness to learn must have come into play. In very small populations such as existed in the past, random genetic drift would come into play and produce changes regardless of adaptive significance. Perhaps the nonadaptive differences, if there really are any, among different races of men are to be explained in this way.

The most ancient remains of man-apes, or ape-men, or primitive man have been found in the Olduvai Gorge east of Lake Victoria at the southern end of the Rift Valley in Tanzania. The anthropologist Louis S. B. Leakey and his co-workers have unearthed an abundant deposit of bones, primitive stone tools, and other evidence of a primitive culture. The skeletons in this East African site resemble others found in South Africa and in Australia but are different enough to be given a new name, *Zinjanthropus*. The character of the stone tools dates this settlement about 500,000 to 600,000 years ago. However, the potassium-argon dating method places the remains at the Olduvai site even earlier, about 1,750,000 years ago. The related *Australopithecus*, from both Taung in Bechuanaland in Africa and from Australia may be as ancient as *Zinjanthropus* or even more so. No one really knows.

Dating back into the Pleistocene some 500,000 years ago are the remains of another extremely primitive man or manlike creature called variously the Java man or *Pithecanthropus* or *Homo erectus*, and also the Peking man or *Sinanthropus pekinensis*. In the years since they were discovered, more remains of both have been found with intermediate characteristics so that it seems the Java and Peking groups represent only slightly different forms of the same population. These Java men averaged well under 2 meters (6 ft) in height with a low forehead, heavy bony ridge over the eyes, and heavy chinless jaw with large teeth.

Was *Homo (Pithecanthropus) erectus* truly human? If man is defined as a tool- and fire-using primate, the answer is yes. Along with the humanoid remains are many stones, apparently chopping stones, which seem shaped to fit into the right hand. In the

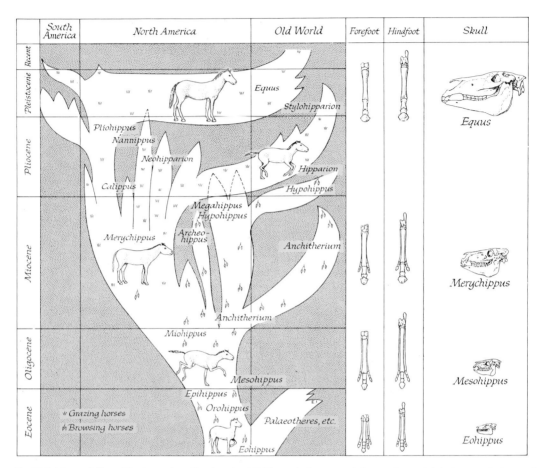

	South America	North America	Old World	Forefoot	Hindfoot	Skull

Fig. 22-9. The evolution of the horse from the dog-sized Eohippus to the present, the most completely documented case of the animal development in geological time. Note that there have been a number of branches to this evolutinary tree many of which have become extinct. (Adapted from Simpson, G. G. 1951. *Horses: The Story of the Horse Family in the Modern World and through Sixty Million Years of History.* Oxford University Press, New York.)

Chinese cave there are also circular charred areas, some deeply burned, which indicate the use of man's greatest tool, fire. There are many bones of extinct deer and other animals split open so as to expose the marrow, and also far too many human skulls in proportion to the scarcity of other human remains. These skulls have all been broken open. This rather disquieting fact is also evident in the caves of prehistoric man in Europe. The implication of cannibalism is strong.

NEANDERTHAL AND CRO-MAGNON PEOPLES

In marked contrast to all of these early men, who are known from a few scattered bones, are the Neanderthals and Cro-Magnons. The latter people have left numerous remains widely distributed over Europe and extending east into Israel and the Crimea.

The Neanderthal remains were first discovered in the Neander Valley near Düsseldorf, Germany, in 1856. At the time some experts thought that these bones merely represented a pathological maldevelopment of the skeleton of a modern man. But as similar bones were discovered deep in Pleistocene deposits in caves located in many different regions, it became clear that *Homo neanderthalensis* had indeed existed as a widespread population. The skull capacity and hence brain size (1300–1700 cm³) fall within the same general size range as that of modern man, but the proportions of the brain were slightly different. With the bones of Neanderthals have been found numerous hearths, chipped stone axes, flint spear tips, and other possible tools. There is evidence that Neanderthal man buried his dead with stores of food and implements. These men flourished for over 100,000 years during the third interglacial pe-

TABLE 22-2
Pleistocene glaciations and peoples (From Moment, G. B. 1967. *General Zoology* 2nd ed., Houghton Mifflin Co., Boston

Glaciation	Nonglacial Period and Duration	Human Population
	Postglacial 10,000 to 15,000 yr	Contemporary
Fourth, or Würm-Wisconsin		Cro-Magnon
	Third interglacial 120,000 yr	Neanderthal
Third, or Riss-Illinoian		
	Second interglacial 300,000(?) yr	
Second-, or Mindel-Karsan		
	First interglacial 100,000 yr	*Homo erectus* (Java and Peking populations
First, or Gurnz-Nebraskan		*Australopithecus: Zinjanthropus*

Fig. 22-10. Restoration of Cro-Magnon man, a prehistoric caveman who left magnificent wall drawings in caves in Europe and the Near East. (Courtesy of the American Museum of Natural History.)

riod, and into the fourth and most recent glaciation, the Würm-Wisconsin. This is a long time indeed, compared with the trivial 5,000–7,000 years of written history. How did they end? Perhaps they were exterminated by their successors, the Cro-Magnons, or perhaps they merged with them.

The Cro-Magnon people appeared in Europe and the Near East as the Neanderthals disappeared 25,000 to 50,000 years ago. Whether the Cro-Magnons had been pushed out of other regions by the increasing cold or actually evolved on location under the stress of the Ice Age, they were physically a superb race, with a brain capacity on the average equal to or greater than our own. They were tall and straight of limb and have left truly magnificent works of art on the walls of caves, so that we flatter ourselves in identifying them as *Homo sapiens.*

The culture of the Cro-Magnons was a Paleolithic, or Old Stone Age, culture in which they chipped but did not grind their stone axes, spear points, knives, lamps, and other implements. In many places they made such extensive use of reindeer bones and antlers that they are sometimes referred to as the reindeer men. From the bones around their fireplaces and deep in their caves, it is clear that they also ate horses, waterfowl, fish, shellfish, and probably, at least in some localities, other men. The most remarkable fact about their culture is the great number of magnificent wall paintings they have left illustrating reindeer, extinct mammoths, extinct bisons, and extinct types of horses. These people drew pictures of animals now limited to the far north, like the reindeer, on caves in Spain and Israel. They also left engravings and carvings in bone, soapstone, and ivory. They appear to have invented the bow and arrow, perhaps as the development of a stick with a leather thong attached to one end and used to throw a small javelin.

Exactly when man first arrived in North America is unknown, but it appears to have been during or even before the last glaciation. There is abundant evidence of a flourishing Stone Age culture in the southwestern United States on the shores of prehistoric lakes that have not held water since glacial times. On the coast of California there are prehistoric fireplaces containing charred bones of extinct elephants, camels, and horses. Only future investigation can tell who

Fig. 22-11. North American mammals characteristic of the Pleistocene epoch. All are drawn to scale with pointer dog (in box) to show relative size. 1, American mastodon; 2, saber-toothed tiger; 3, giant ground sloth; 4, stag moose; 5, giant beaver; 6, Colombian elephant; 7, Texas horse. These were some of the animals hunted by the first inhabitants of North America. (From Moment, G. B. 1967. *General Zoology*, 2nd ed. Houghton Mifflin Co., Boston.)

the firebuilders were and what was their relationship to Stone Age peoples in Europe and Asia.

Persistent Questions about Evolution

Recurring questions about evolution are of two sorts. The first are strictly scientific questions. What are the driving forces of evolution? What is the role of mutation, isolation, sexual reproduction, natural selection, random sampling? What has been the actual history of life on this planet? How adequate is the evidence for what biologists believe to be true? What are the thousand and one evolutionary stratagems which have produced and continue to maintain the wondrous forms of life — orchids, bumblebees, and heliozoans, the octopus and the columbine? These prob-

lems we have discussed. Two questions of very general importance remain and to these we now turn before considering the second type of questions which are philosophical.

THE PROBLEM OF EXTINCTION

It is an article of faith with most ecologists that all species and generous samples of all types of environments should be preserved for the enjoyment, study, and general benefit of present and future generations of mankind. Certainly our planet would be much the poorer without them and at the worst would become uninhabitable if pollutants kill most kinds of vegetation. Clearly the present ecological squeeze threatens many species and entire plant and animal communities with extinction. At the other extreme, there are species that North America at least could well do without. The European gypsy moth, *Porthetria dispar*, was introduced into Massachusetts a century ago by an amateur entomologist who "didn't know it was loaded." By the 1970's it is defoliating whole oak forests on the mountains of eastern Pennsylvania. The vicious Brazilian wild bee, which attacks and kills cattle and even people and is now rapidly spreading northward, is not an ecologically necessary member of the North American ecosystem. Consequently the search for the factors which produce extinction or even the drastic and permanent reduction in population size is now urgent both to preserve threatened species and to reduce species which threaten mankind.

Extinction is an ancient and common event in the history of life. Probably the most spectacular of all the many cases of extinction is that of the great reptiles of the Mesozoic. For over 150 million years the dinosaurs had flourished, experiencing an adaptive radiation which made them dominant on the land, in the swamps, in the sea, and in the sky with forms as diverse as the monstrous brontosaurs and the delicate little hop-skip-and-jump precursors of birds. What happened? One possibility is that they could not survive the widespread change in climate which took place at the close of the Mesozoic. Although the cooling and drying of the regions outside the tropics may account for the lack of the great reptiles in those areas, it is difficult to see why, on this basis, many of them could not still exist in the tropical regions of Central and South America, Africa, and South Asia or in the ocean.

A second possibility is competition with the primitive mammals. At the close of the Mesozoic the only mammals that had appeared were small and generalized, much like small mongrel dogs or opossums. They are supposed to have contributed to the downfall of the dinosaurs and other reptiles by eating reptile eggs, but such primitive mammals would have had to be ubiquitous and extraordinarily effective to eat all the eggs in so many environments.

A third possibility is that some epidemic viral, bacterial, or even protozoan disease wiped out the great reptile populations. This possibility is as hard to prove as to disprove. It is known that the tsetse fly, which transmits the trypanosome of African sleeping sickness, has made it impossible to keep domestic cattle in large areas of Africa, but (and this is equally important) the native relatives of cattle, like the various gazelles and antelopes, have enough immunity to prosper in these same regions and serve as "reservoirs" of the disease.

One possible explanation is a modern version of the old theory of racial senescence.

No modern zoologists believe that time, simply as duration, produces some inevitable racial aging. Too many different kinds of animals have existed unchanged, or with only the most trivial changes, for hundreds of millions of years (sharks, lungfish, some echinoderms, and others), but the natural course of evolutionary change resulting from mutation and natural selection does tend to make living things more and more narrowly specialized. And excessive specialization can act as a trap. When an organism becomes so highly specialized that it loses the versatility needed to meet changing conditions, in either the living or the non-living environment, then the highly specialized group succumbs to more generalized ones. For example, increase in size is as obvious in many of the mammalian lines of evolution as it was in the reptiles. What evolutionary forces lead to increased size? One obvious factor is competition, especially between males of the same species. Perhaps this leads to a disastrous spiral, in which only the largest males can win females but in which individuals become too large to function efficiently in food-getting and other ways, a situation which has happened to the lion. This disaster may have actually overtaken the big herbivorous dinosaurs such as the three-horned *Triceratops* and the 15-meter long carnivore, *Tyrannosaurus*, which preyed upon it.

One, all, or none of these four factors may have been important. All we can be certain of is that the evolutionary deployment of the reptiles was remarkably like that of the mammals at a later time. Both began with a small, generalized form, and both gave rise to hundreds of extremely diverse types—the mammals to tigers, rabbits, giraffes, bats, and baboons. The reptiles produced a corresponding diversity but, except for a few descendants, they have become extinct. Does a similar fate await mammals?

The study of present-day populations affords what seems to be more useful insights into extinction. Natural populations of animals and some plants continually undergo fluctuations depending on many factors—climatic and seasonal weather variations, predators and parasites, food supply, and the like. A prolonged cold rainy spell at the time when fawns are born can drastically reduce a year class of deer. When a population is large, the normal fluctuations in numbers do not threaten it with extinction. When a population becomes very small, there is always the danger that in the course of these fluctuations, one will dip to the zero point. Thus, it is not necessary for any single factor to destroy every breeding pair, only for it to lower the population to the danger level. The extinction of the enormous flocks of passenger pigeons which once darkened the skies of the U. S. Middle West is sometimes attributed solely to overhunting but it is to be noted that this occurred after the conversion of the extensive oak forests in which they nested into farm lands. This drastically reduced the population. Furthermore, the passenger pigeon was a social nester and, as is true of various sea birds which nest in dense colonies, very small colonies are commonly unsuccessful in rearing their young. The case of the national animal of New Zealand, the strange *Sphenodon* lizard, is instructive. All efforts to stop its decline in numbers failed until sheep were removed from the little islands on which it still existed. Without the sheep, the low shrubbery among which it lived returned and with the restoration of its appropriate environment, *Sphenodon* has increased in numbers. Without a proper environment no restriction on harvesting a plant or animal will prevent its permanent disappearance.

The Diversity of Life and the Problem of Relationships

The world of living things on this planet presents a bewildering jungle of thousands, tens of thousands, and in some groups, hundreds of thousands of different kinds; over 14,000 species of bony fish, nearly 2,000

species of tapeworms, a quarter of a million flowering plants, and well over half a million species of insects have been described. From whatever point of view biology is approached, whether that of agriculture, medicine, economics, ecology, or pure research, classification is necessary to bring manageable order out of this confusion.

Thus, **taxonomy**, as the science of classification is called, forms an indispensable framework of biological science.

THE BINOMIAL SYSTEM OF CLASSIFICATION

The system of classification in use today is the binomial system, established by Linnaeus, a contemporary of George Washington. According to this scheme each distinct kind of organism receives two names which together constitute the scientific name of the species. Thus all human beings belong to the species *Homo sapiens,* all wood frogs to the species *Rana sylvatica,* and all white oaks to the species *Quercus alba.* The name written first is the name of the **genus**, which is defined as a group of related **species**. Most frogs belong to the genus *Rana*, and oaks to the genus *Quercus.* Following the generic name is the specific name. The wood frog bears the specific name *sylvatica*, and the white oak bears the specific name *alba*. Note that the generic name is capitalized while the specific name is not. Both names are italicized. The word species is the same in both singular and plural. Specific names are commonly adjectives referring to some character-

Fig. 22-12. Flight of passenger pigeons shown in an 1875 print. Before their extinction, enormous flocks of these birds darkened the skies of North America.

istic of the species; thus *sylvatica* (of the woods) is the specific name for the wood frog, and *pipiens* (chirping) for the common laboratory frog. However, the specific name may be taken from a person, as in the case of *Rana catesbeiana,* the bullfrog (whether male or female) named after Mark Catesby, an early American naturalist. The system is an international one.

A TRUE BREAKTHROUGH?

Although scientists have been classifying living organisms for over 300 years and the theory of organic evolution has been well authenticated and generally accepted the world over for more than a century, no one knows whether an order of placental mammals corresponds to an order of insects or of flowering plants or perhaps to a class or even a family. The theory of evolution indicates that all mammals are more closely related to each other than any one is to a bird or a starfish, and that oaks are more closely related to grasses than to pine trees.

However, all these relationships are known only in vague general terms. Man is more closely related to the great apes than to the baboons or the monkeys, but how much more? What is the actual degree of evolutionary difference between any of the primates and a mushroom or a pine tree, or between the great apes themselves? A few years ago questions like these seemed too far-out to be taken seriously. Now quantitative answers are becoming available. Ultimately their answers will be in terms of DNA.

Meanwhile a true breakthrough has been achieved by the comparative study of amino acid sequences in one of the most ancient and universal proteins known, cytochrome *c*. It will be recalled that cytochrome *c* is a respiratory protein located in the mitochondria of all aerobic cells. There are about 110 amino acids in the cytochromes of all organisms tested from man, horse, whale, and chicken to turtle, fly, wheat, fungus, and yeast. The principle of comparison is basically simple. Each change in an amino acid at any point on the sequence represents a single mutation. Therefore, the greater the number of differences in the amino acids in the cytochromes of two organisms, the further apart they are. Or, to state this in a more logical way, the distance between any two or more organisms can be defined in terms of the number of locations in the cytochrome *c* sequences where the amino acids are different. It is of interest to observe that at 20 positions the same amino acid is present in all organisms from man to mold. By comparing the se-

TABLE 22-3
Classification of some familiar animals (From Moment, G. B., and H. M. Habermann. 1973. *Biology: A Full Spectrum* The Williams & Wilkins Co., Baltimore)

Taxonomic Categories	Amoeba	Man
Phylum	Protozoa	Chordata
Subphylum	Sarcomastigophora	Vertebrata
Class	Rhizopoda	Mammalia
Order	Amoebiformes	Primatiformes
Family	Amoebidae	Hominidae
Genus	*Amoeba*	*Homo*
Species	*proteus*	*sapiens*

TABLE 22-4
Classification of some familiar plants (From Moment, G. B., and H. M. Habermann. 1973. *Biology: A Full Spectrum* The Williams & Wilkins Co., Baltimore)

Taxonomic Categories	Edible Mushroom	White Oak
Division	Eumycophyta	Spermatophyta
Subdivision	Basidiomycophyta	Angiospermae
Class	Holobasidiomycetes	Dicotyledoneae
Order	Agaricales	Fagales
Family	Agaricaceae	Fagaceae
Genus	*Agaricus*	*Quercus*
Species	*campestris*	*alba*

quences of two organisms it is possible to arrive at a hypothetical sequence for a division point or "**node**" which represents the common ancestor from which the two organisms being compared have diverged. These nodes are computed on the assumption that the pattern requiring the minimum number of mutations is the correct one. Animal and plant relationships derived by this method are shown in Figure 22-13. Much remains to be done, but after three centuries of effort it is at long last possible to assign numbers to the realtionships of living things.

Additional questions of a basic nature come to mind. Why, of all the thousands of possible amino acids, are only two dozen found in living organisms on this planet? How did DNA come to be the basis of life? Would not any one of a dozen other macromolecules have done as well? What is the explanation for the relation between a given nucleotide sequence and a specific amino acid?

PHILOSOPHICAL QUESTIONS

The philosophical questions most frequently asked about evolution concern its meaning and the place of man in nature. These problems have been extensively explored by T. H. Huxley, Teilhard de Chardin,

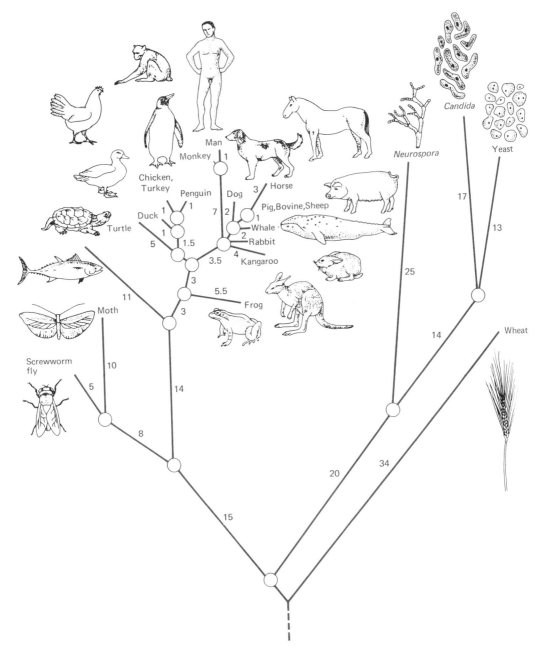

Fig. 22-13. Plant and animal phylogenetic tree based on amino acid analyses of cytochrome *c* proteins. The numbers indicate the minimum number of mutations required to produce the observed differences in the amino acid sequences of the cytochromes. A tree based on some other compound might be somewhat different. (From Dayhoff, M. O. *Computer analysis of protein evolution.* Copyright © *Scientific American*, July 1969 by Scientific American, Inc. All rights reserved.)

G. G. Simpson, and a host of other biologists. To the age-old question as to whether the appearance and evolution of life were due to chance, we have already given the answers which seem to us to be scientifically justifiable. The specific source of the organic molecules involved, for instance, whether they

were the products of reactions energized by ultraviolet radiation or by lightning, and the particular spot on the planet where these compounds became united in such a way that self-reproducing units were formed—all of this was a matter of chance. However, given the inherent properties of the elements of this universe, both chemical and physical, plus energy sources, plus reproduction with variation occurring throughout eons of time, then the emergence and evolution of life are so highly probable that statistically speaking they are as close as an asymptote to the inevitable. And underlying all of these questions is the ultimate mystery of the emergence of conscious intelligence out of the vastness of space, time, matter, and energy.

USEFUL REFERENCES

Briggs, D., and S. M. Walters. 1969. *Plant Variation and Evolution.* World University Library, McGraw-Hill Book Co., New York. (Paperback)

Darwin, C. 1859. *On the Origin of Species by Means of Natural Selection.* Doubleday & Co., Inc., New York.

Dobzhansky, T., M. K. Hecht, and W. C. Steere, eds. 1972. *Evolutionary Biology,* vol. 6. Appleton-Century-Crofts, New York.

Hanson, E. D. 1972. *Animal Diversity,* 3rd ed. Prentice-Hall, Inc., Englewood Cliffs, N.J. (Paperback)

Harris, R. J. 1969. *Plant Diversity.* Wm. C. Brown Co., Dubuque, Iowa. (Paperback)

Lack, D. 1961. *Darwin's Finches: An Essay on the General Biological Theory of Evolution.* Harper Torchbooks, New York. (Paperback)

Mayr, E. 1963. *Animal Species and Evolution.* Harvard University Press, Cambridge. (By the greatest living authority on birds.)

Moorehead, A. 1969. *Darwin and the Beagle.* Harper and Row, Publishers, Inc., New York. (Read it and you'll never forget it.)

Ponnamperuma, C., ed. 1972. *Exobiology.* Elsevier Publishing Co., New York. (Life in outerspace.)

Ponnamperuma, C. 1972. *The Origins of Life.* E. P. Dutton & Co., New York.

Romer, A. S. 1972. *The Procession of Life.* Doubleday-Anchor, New York.

Simpson, G. G. 1960. *The Meaning of Evolution.* Yale University Press, New Haven. (A modern classic.)

Solomon, M. E. 1969. *Population Dynamics.* St. Martin's Press, New York. (Paperback)

Stebbins, G. L. 1971. *Process of Organic Evolution.* 2nd ed. Prentice-Hall, Inc., Englewood Cliffs, N.J.

Volpe, E. P. 1967. *Understanding Evolution.* Wm. C. Brown Co., Dubuque, Iowa.

Wynne-Edwards, V. C. 1962. *Animal Dispersion in Relation to Social Behavior.* Oliver and Boyd, Edinburgh. (The beginning of sociobiology.)

23

Symbiosis and Parasitism

The human importance of parasitism has long been recognized. The Black Death which killed two out of every three students at Oxford during the 14th century was caused by the bubonic plague bacillus and is still a serious threat. Some historians believe that massive chronic infection of the population with the malarial parasite was an important factor in the decay of classical Greece and Rome. Even today, a wide variety of bacteria, viruses, and worms infect people and their domestic plants and animals. The annual damage inflicted on wheat, corn, and many other crops by rusts, smuts, and other fungi totals hundreds of millions of dollars.

In contrast, the importance of symbiosis, where two different kinds of organisms live in close association with mutual benefits, has been largely overlooked. Even to biologists the word often suggests a lichen, a double type of organism where a fungus and an alga live together, or some exotic crab which places sea anemones on its shell where they protect the crab and the crab, a messy eater, provides crumbs of food for the anemone. In fact most of our forest trees live in symbiosis with root fungi. There is good evidence that mitochondria and chloroplasts are the descendants of symbiotic bacteria and unicellar algae, respectively. Thus symbiosis may be basic to all the higher forms of life.

Definitions

The terminology applied to all the various types and degrees of association between individuals of two species varies somewhat from author to author. In many cases it is practically impossible to determine whether one species is benefited or harmed. We will follow the usage appearing in Pennak's *Collegiate Dictionary of Zoology*.

A **parasite** is an organism which lives in or on individuals of another species from which it derives nutriment—an animal or a plant which has a host that does not benefit from the association. Mistletoe growing on the branch of an oak tree and a copepod living inside the eye of a swordfish are both clearly parasites. Parasites may be external **ectoparasites** like lice or internal **endoparasites** like tapeworms. It should be remembered that it is often rather arbitrary whether a relationship is termed parasitic. Is a mosquito a parasite? Is an aphid which sucks the juices of a plant or a caterpillar that eats the leaves also a parasite?

Commensalism is usually defined as an association of individuals of two species where one derives some benefit and the other is neither helped nor harmed. Some biologists claim that such a relationship never exists because the host always is either injured or benefited, though perhaps very slightly. Firm proof is hard to obtain. Barnacles on whales are commonly cited commensals.

Mutualism is the term for a close relationship where both species are benefited. In addition to lichens and the crab-anemone partnership, one thinks of the hydra with green algae living within its endodermal cells and of the root nodule bacteria of legumes like clover.

Symbiosis, literally "life-together," is sometimes used to designate all of the classes of close association just discussed. Parasitism is antagonistic symbiosis.

Symbiosis is clearly the safest term to use

for all these relationships except in the most obvious cases of parasitism.

Positive Symbiosis: Mutualism

LICHENS

The most undeniable instances of symbiosis in the sense of mutualism are found in plants, notably the lichens. Many problems of very general theoretical importance lie hidden here. How is the growth of two such different organisms coordinated in both vegetative and sexual reproduction? What determines the form, so different in different species of lichens? Is the species concept truly applicable to a situation like this?

The component organisms of a lichen, an alga and a fungus, are members of diverse groups of the plant kingdom. The algal component may be either a prokaryotic blue-green or a eukaryotic green alga, while the fungus may be a Basidiomycete, an Ascomycete, or one of the fungi imperfecti, three very different groups of fungi. Both the algal and the fungal components of many lichens have been cultured separately in the laboratory. Thus, one cannot argue that their association in nature is a matter of absolute necessity. The photosynthetic products of the algal cells are utilized by the fungus while the fungus probably supplies water and inorganic nutrients to the alga.

Lichens are found on the surfaces of rocks, trees, soil, and wooden structures. They can be quite colorful, with gray-green, white, yellow, orange, yellow-green, brown, and black forms known. Lichen bodies exist in several shapes. **Crustose** lichens are hard, flat forms often growing in the surfaces of rocks. **Foliose** lichens are flattened, leaflike forms, and **fruiticose** lichens are usually erect, branched structures (Fig. 23-1). The best known fruiti-

cose lichen is "reindeer moss," *Cladonia rangifera,* a lichen that is the principal food of the reindeer, caribou, musk ox, and other animals inhabiting the tundra. Although sometimes used as fodder for animals, most of the lichens have a bitter taste and have very rarely been used as food for humans. There are reports of Iceland "moss" being eaten by Arctic explorers and the biblical "manna" may have been a lichen, *Lecanora esculenta,* that occurs in desert areas.

MITOCHONDRIA AND CHLOROPLASTS

The most surprising recent theory about symbiosis holds that mitochondria are really symbiotic bacteria. Certainly mitochondria and bacteria have much in common. They are about the same in size and shape, and both are self-duplicating by elongation and transverse fission. Mitochondria and chloroplasts contain their own DNA. Both contain phosphorylating assemblies of molecules in their complex membranes. If it turns out to be true that mitochondria are in fact symbiotic bacteria, the origin of this symbiosis was one of the major events in the history of life because mitochondria make aerobic respiration possible. They are found in the cells of all eukaryotic organisms that have been examined, higher animals and plants, fungi, algae, and protozoans. The significant exceptions are the very simple cells of prokaryotic bacteria and blue-green algae. The evidence that chloroplasts are also symbionts, essentially unicellular algae, rests on the presence of DNA in chloroplasts, which are self-duplicating, and on the close resemblance between these organelles and unicellular algae. This concept is made more plausible by the existence of symbiotic green algae in the green paramecium, *P. bursaria,* in the endoderm cells of the green hydra, and in various other organisms.

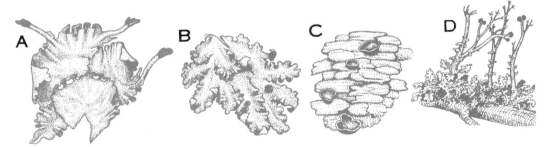

Fig. 23-1. Lichens. A, *Peltigera*, and B, *Parmelia*, both foliose lichens, × ½; C, *Xanthoria*, a crustose lichen, × 2; D, *Cladonia*, a fruiticose lichen, × ½. (From Moment, G. B., and H. M. Habermann. 1973. *Biology: A Full Spectrum*. The Williams & Wilkins Co., Baltimore.)

PROTOZOA, TERMITES, AND COWS

The most extensively investigated case of symbiosis among animals concerns the protozoans found in the guts of termites and wood roaches. Termites are completely dependent on these protozoans for the digestion of the wood which they eat, and these species of protozoans in turn are found only in termites and their relatives. Thus, the termite protozoan relationship is one of true commensalism. If termites are placed in an oxygen tent for 24 hours, subjected to high temperatures (36°C), or starved for 10 days, their intestinal microorganisms die while the termites themselves appear to be unharmed. After treatment, the termites eat wood as usual, but because without protozoa they are unable to digest cellulose, it passes unchanged through their digestive tracts.

Probably the most important case of symbiosis from the practical standpoint is that of the ciliates that swarm in the stomachs and intestines of cows and related animals. The number of these ciliates in a cow is truly prodigious, 100,000–400,000 per ml of intestinal contents. Are they harmful, useful, or even indispensable to the cow, in the way intestinal microorganisms are to the termite? A clear case of necessary symbiosis between ruminants and the microorganisms inhabiting their guts involves the production of vitamin B_{12}. This cobalt-containing vitamin (cyanocobalamin) is essential for animals. In cattle and sheep it is manufactured by their intestinal microflora. In order for the vitamin to be synthesized, cobalt has to be provided in the forage consumed by the host organism. Lush pastures can grow in soils deficient in cobalt because this is not an essential element for plants. However, inadequate amounts of cobalt in the diets of cattle causes a nutritional disease in their intestinal microorganisms. A consequence of this is a deficiency in vitamin B_{12} in the cattle. In parts of Australia and New Zealand where cobalt deficiency was discovered, cattle and sheep can be provided with continuing supplies of cobalt by lodging "bullets" of cobalt-containing materials in their throats. This is far less expensive than the alternative of supplying them with B_{12} injections.

PLANT MUTUALISM

An important but often overlooked mutualism exists between many of our familiar trees, pines, oaks, and willows, for example, and several kinds of fungi which grow either on or within their roots. The fungus may be an Ascomycete (the group to which mildews, black and green molds, and *Neurospora* belong) or a Basidiomycete (the group which includes mushrooms and puffballs). Such an intimate association between the root of a higher plant and a fungus is called a **mycorrhiza**. "Fungus roots," as mycorrhizae are sometimes called, tend to be short and swollen. Often only the lateral roots are invaded by the fungus and thus modified. Changes in roots include loss of regions where root hairs are located. The condition of such fungus-infected roots may be regarded as pathological, in which case the fungus would have to be classified as a parasite on the higher plant host. There is evidence, however, that both the fungus and the higher plant benefit from their association. The fungus gains organic nutrients in return for facilitating the uptake of water, minerals, and possibly organically bound nitrogen from the soil. When species in which mycorrhizae occur universally in nature are cultured in the laboratory, seedling growth in sterlized soil is notably slower than in soils inoculated with appropriate species of mycorrhizal fungi (Fig. 23-2).

Mycorrhizae are frequently found in and appear to be beneficial to plants growing in soils of forests or bogs containing much organic matter. The fungi, by metabolizing organic materials, release minerals that are then available to the higher plants. Thus, they make possible the eventual release of organically bound minerals that higher plants ordinarily cannot absorb from the soil.

HIGHER ANIMALS

Modern biologists share with the old 19th century naturalists an interest in relatively larger animals which live in some degree of symbiosis with each other. There is a fish which manages to find protection from larger fish by swimming among the tentacles of the Portuguese man-of-war, a fingernail-sized crab which lives within the mantle cavity of oysters, and insects which live in the darkness of ants' nests and beg for food from the ants with their antennae. Perhaps we have now crossed the line and are describing parasites, even though some writers call them commensals.

Negative Symbiosis: Parasitism

TRAITS CHARACTERISTIC OF PARASITES

Most parasites, whether plant or animal, share four important traits which adapt them for their way of life. First, there is a **loss of structures**, especially structures having to do

Fig. 23-2. Effects of mycorrhizal roots on growth of pine trees. A, 6-month-old seedling of *Pinus caribaea*, showing differences in growth between plant inoculated with symbiotic root fungi (left) and uninoculated plant (right), × ¼; B, 6-yr-old Honduras strain of *Pinus caribaea* trees growing in Puerto Rico. It is not uncommon for this tree to grow 6 ft in 1 yr. (Courtesy of E. Hacskaylo, Forest Physiology Laboratory, U.S. Department of Agriculture).

with obtaining food. For example, a dodder (Fig. 23-5) germinates from a seed and develops like a normal plant with root, stem, and leaves, but as soon as the young plant becomes attached to its host and sends its sap-absorbing holdfasts into the stems of its victims its roots die and the entire plant loses contact with the soil. Leaves are no longer formed except as mere vestiges. The comparatively simple structure of animal parasites is also notorious. Tapeworms and liver flukes are reduced to little more than a smooth muscular sac with a holdfast of some kind on the outside and a gonad on the inside, although rudimentary nervous and excretory systems and, in the case of the flukes, a gut, remain.

Second, many parasites possess **specialized structures** which aid in finding, entering, and remaining within their hosts. The adult has no use for eyes but larval parasites commonly posses them together with special inherited behavior patterns which increase the chances of their finding a suitable host. These behavior patterns may be as simple as crawling up a blade of grass where some host will graze. Other patterns, such as those found in parasitic wasps which seek out spe-

cial kinds of caterpillars on which to lay their eggs, involve complex sense organs as well as instincts. In such a parasite as an ichneumon wasp there is no loss whatever of the normal complement of organs—sensory, locomotor, digestive, and the like.

Third, most parasites are extremely **prolific**. The probability of any individual finding a suitable host is extremely low. Although greatly exaggerated reproductive powers are commonly thought of as characteristic of parasites, it should be remembered that many animals that are not parasites lay incredibly large numbers of eggs, e.g. oysters, most fish, and nematodes (whether parasitic or free living). Think of the enormous numbers of maple seeds shed year after year. Many parasites are hermaphrodites (e.g. the tapeworms and flukes), but again it is important to remember that many successful parasites, such as *Schistosoma* (a blood fluke) and most nematodes, have separate sexes while many snails and a great many plants are bisexual. The meaning of this difference therefore cannot lie in parasitism as such.

Fourth, although many parasites have only one host and are said to be **monogenetic**, many require two (or more) hosts and are termed **digenetic**. For example, the blood fluke *Schistosoma* lives its adult, i.e. sexually mature, life in humans. The larval forms live in snails. Many disease-causing parasites, including the rickettsias causing Rocky Mountain spotted fever and fungi such as the cedar-apple rust, have alternate hosts. Some parasites are very fastidious about the hosts they infest. The parasite of bird malaria cannot develop in humans, although it is transmitted and lives part of its life cycle in the common *Culex* mosquitos which must often inoculate people with the sporozoan of bird malaria. Other parasites are rather indiscriminate and can live successfully in many species. The trypanosome causing African sleeping sickness lives successfully in antelope, horses, and camels, as well as humans.

EVOLUTION OF PARASITES

The evolution of parasitism in most assuredly due to the same combination of mutation, selection, and isolation that underlies the evolution of all other organisms. One view holds that parasites began by being able to live on a single host, evolved to be able to live on several species and then finally became more limited in host specificity. Probably many patterns have occurred in the evolutionary history of parasitism. There are even parasites of parasites, so-called hyperparasites, among insects and the viruses which infect disease-causing bacteria. The well-adapted parasite does not kill its host. Ordinarily the parasite dies of starvation when the host dies. Consequently, it is a selective advantage for any parasite to injure its host as little as possible. Highly and quickly fatal parasites are believed to be recently evolved species.

HOST RESPONSES

Host responses to parasites show a wide range. In plants a pathological swelling called a **gall** results from the egg-laying sting of certain insects. Such golf ball-sized growths are common on the stems of goldenrod and the twigs of oaks and cedars (Fig. 23-3). Apparently the insects secrete some stimulus to produce these bizarre growth effects. The advantage to the insect which can develop inside such a gall and eat at the plant's expense is obvious. **Hypertrophy** (excessive growth of the host's tissues) also occurs in plant diseases caused by certain fungi such as the "club root" of cabbages and related plants infected with *Plasmodiophora brassicae*. In animals the host responses range from the production of antibodies to overt

Fig. 23-3. Insect galls on pecan twigs. These growths are caused by small insects (phylloxera) which are closely related to aphids. (Courtesy of the U.S. Department of Agriculture.)

behavioral traits which tend to lessen possible contacts with the parasite or its eggs.

PARASITES OF PLANTS

Plants, like animals, exhibit all degrees of interaction ranging from competition to states of dependency where a parasite is benefited while the host is harmed and states of interaction where the association of organisms is mutually advantageous. Such interactions occur among many diverse groups of plants ranging from bacteria and viruses to seed plants.

Because of the economic costs and human misery resulting from parasitic infections, subjects such as plant pathology, medical bacteriology, and mycology have been intensively studied. Organisms ranging from the viruses (which lack cellular structure) and the prokaryotic bacteria to flowering plants and a variety of animals can cause diseases of crops with very serious economic and social consequences. Some of the fungi parasitic on cereal grains produce chemical byproducts that are poisonous to humans or animals eating infected plants. **Ergotism,** a condition in which contraction of the smaller arteries and smooth muscle fibers can lead to gangrene, has been a serious disorder of epidemic proportions in certain European countries where rye bread is a staple of human diets.

There are fungus-caused plant diseases In which animals participate in the disease cycle as carriers of spores. The Dutch elm disease is spread by the bark beetle (Fig. 23-4). In other cases, man has undoubtedly been an unwitting accomplice in the introduction of parasite-caused plant diseases into new areas. Since 1900, a number of plant pathogens have been introduced into North America by man. The white pine blister rust was first discovered in 1906, chestnut blight in 1904, and Dutch elm disease in 1930. Marked deterioration of many environments and losses amounting to hundreds of millions of dollars have resulted. Strict quarantines have been imposed on the importation into the United States of plants, agricultural produce, and seeds in order to avoid future introductions that, like the chestnut blight and Dutch elm disease, might virtually exterminate or seriously threaten the survival of their host species. Often newly introduced parasites

that were not serious pests in the areas where they originated cause havoc when introduced into a new environment where they find more susceptible hosts, a lack of natural enemies, or conditions more favorable for infection or development.

There are some groups of seed plants that exhibit the full range of dependency on other species from a relatively harmless climbing habit to strict parasitism. In the morning glory family (Convolvulaceae) there are examples of both extremes. *Convolvulus*, the common bindweed, twines around any available support (including other plants) but is not dependent on the structure around which it climbs for anything but mechanical support. Its roots absorb water and minerals from the soil; its leaves are large and photosynthetic. In the genus *Cuscuta*, which also belongs to the Convolvulaceae, we find the dodders, obligate parasites that exhibit many of the kinds of structural modifications and adaptations characteristic of parasites in general (Fig. 23-5). The dodders are almost leafless and defi-

Fig. 23-4. Upper left, damage caused by the elm bark beetle is evident where a section of bark has been removed. Growth of the fungus carried by the beetle blocks conducting tissues of the tree. Lower left, elm bark beetle, × 18. (Courtesy of the U.S. Department of Agriculture.)

Fig. 23-5. Dodder, a parasitic seed plant, growing on lespedeza. Although their seeds germinate in the soil, dodders absorb nutrients from the plants they parasitize. (Courtesy of the U.S. Department of Agriculture.)

cient in chlorophyll (in color they range from pink to pale green or yellow), adult plants have no contact with the soil, and their reproduction can only be described as prodigious. They form many dense clusters of flowers. Seeds of the dodder germinate in the soil and their abundant food reserves are not wasted in the formation of large roots or leaves which later will not be needed. Instead, all available resources are utilized to form long stems that move whiplike in wide arcs until contact is made with another plant. If no contact is made, the seedling dies when its food reserves are exhausted. When a suitable host is found, the root disintegrates and the dodder continues its life as a twining parasite. At points of contact with the host, specialized suckers form.

Dodder seeds are disseminated with crop seeds, hay, movement of farm animals and implements, irrigation and surface drainage and in animal feces. Once farmland becomes infested with dodder it is necessary either to harvest crops before the dodder seeds have matured or to burn all infected plants. Control by means of herbicides or by rotation to other crops is also possible. Among the crop plants affected by these parasites are clovers, alfalfa, sugar beets, onions, and flax. In addition, many ornamental and wild species are affected.

The mistletoes provide further examples of true parasites among flowering plants. The mistletoes are, in addition strict **epiphytes.** Although similar, mistletoes belong to a number of genera, which means they have been evolving for a long time. In Europe a species of *Viscum* is found on hardwoods and conifers. In North America, all true mistletoes belong to the genus *Phoradendron.* They are restricted in their distribution to geographical regions where winters are mild. Their northernmost limits are Oregon, central Colorado, the Ohio River, and southern New Jersey. They are found southward to the West Indies and in central and northern South America. The over 70 different species of *Phoradendron* parasitize many hardwoods (including oak, elm, maple, sycamore, ash, poplar, walnut, and willow) and conifers (including juniper, cedar, and pine) (Fig. 23-6).

Mistletoes are **dioecious.** The female (pistillate) plants bear groups of white, yellow, or pink berries with sticky mucilaginous surfaces that easily stick to host plants or to birds that carry them to new hosts. Seeds germinate on the host where the developing shoot forms an attachment organ from which protuberances called **haustoria** penetrate the bark through lenticles or buds. Only young

Fig. 23-6. Female plant of western dwarf mistletoe growing on a pine. Note berries on mistletoe and spindle-shaped swelling on pine branch. (Courtesy of U.S. Department of Agriculture.)

shoots are invaded, but once they are penetrated, the haustoria spread through the bark region finally reaching the cambium and ultimately invading the xylem. Once established, these parasitic epiphytes can persist as long as the host tree remains alive. Most mistletoes have chlorophyllous leaves and stems and are dependent on their hosts only for a supply of water and minerals. Because they are photosynthetic, they are found primarily on the uppermost branches of forest trees. The standard technique for harvesting mistletoe for the Christmas market is to shoot it down with a shotgun.

ANIMAL PARASITES

Protozoa

The most primitive group of animals, the Protozoa, have developed some of the most devastating of parasites. *Endamoeba histolytica* the amoeba of **amoebic dysentery,** is only too well known to travelers. Among flagellates, trypanosomes are responsible for many important diseases. Among the best known is *Trypanosoma gambiense,* the causative agent of *African sleeping sickness.* (This is not to be confused with encephalitis lethargica, a form of sleeping sickness caused by a virus.) The trypanosome responsible for African sleeping sickness lives in the bloodstream, then in the lymph nodes, and finally in the central nervous system. In advanced stages the patient sleeps continually, becomes incredibly emaciated, is shaken by convulsions, and finally dies in a profound coma. Mortality is high, and large areas of Africa are rendered unfit for human habitation because of its toll. The parasite is transmitted from man to man and from man to the larger mammals, such as cattle, pigs, goats, and antelope, and back to man again by the bite of *Glossina,* the tsetse fly. Many species of trypanosomes parasitize fish, frogs, salamanders, birds, and, of course, mammals (Fig. 23-7). Bloodsucking flies are the usual **vectors,** although one species is known to be transmitted by a leech and another, *Trypanosoma equiperdum,* the cause of a disease in horses and mules, is transmitted from one horse to another during mating.

Most trypanosome diseases can be controlled by injecting compounds of the heavy metals mercury, antimony, or arsenic. Paul Ehrlich, who discovered that heavy metal compounds are effective against the organism of syphilis, also found as long ago as 1907 that heavy metals and some dyes are more poisonous to trypanosomes than to men. Yet much remains to be learned about how these agents work. Because heavy metal compounds will combined with the −SH (sulfhydryl) group of glutathione, a substance found in most cells, it is believed by some that this is the mode of action of the mercurials and other metal trypanocides. Others hold that they attack the −SH groups on enzymes.

Many of the most serious diseases in animals from earthworms to silkworms, rats, cows, and humans are due to protozoans called Sporozoa. They get their name from the fact that those which are not transmitted from one host to another by a bloodsucking insect usually pass from one victim to another in protective capsules called **spores** which contain either zygotes or juveniles. The most important such disease of domestic animals is **coccidiosis,** which afflicts poultry and cattle. The parasite lives within the cells of the host's intestine. **Malaria,** one of the longest known and most widespread of all human diseases is caused by a sporozoan. The discovery of its causative agent and how it is transmitted opened a new era in medicine although today more people still suffer from malaria than from any other single disease.

Malaria. The story of how the cause of malaria was discovered well illustrates the way science advances by the cumulative efforts of many individuals. A good point of departure is the arrival in London in the closing years of the last century of Ronald Ross, a young army surgeon from India. He went to the famous St. Bartholomew's Hospital determined to find out all he could about malaria. There he learned the varied facts that had been uncovered; later he drew them together with new ones of his own into a convincing explanation of the true cause of this great scourge. Following are some of the things he learned.

Some 14 years earlier, Alphonse Laveran, a French Army doctor stationed in North Africa, had found microscopic parasites in the blood of malarial patients. Laveran's discovery had been confirmed by Golgi in Italy who described how the parasites break out of the red blood cells into the blood plasma every time the patient has one of the periodic fever chills characteristic of malaria. Ross learned also of the speculations and observations of Patrick Manson and entered into correspondence with him. Manson practiced medicine along the China coast. He was credited with the discovery that elephantiasis, a disease characterized by great swelling of the legs and

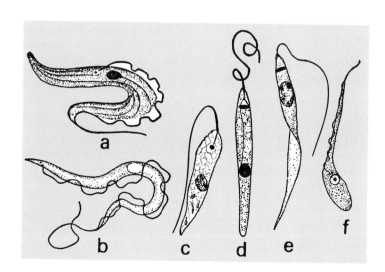

Fig. 23-7. Various trypanosomes taken from (a) a frog, (b) a newt, (c) a man with "oriental sore," (d) a house fly, (e) a milkweed plant, and (f) an insect. (Modified from Hegner, R. W. 1933. *Invertebrate Zoology.* The Macmillan Co., New York.)

Fig. 23-8. Ronald Ross, winner of the Nobel prize in medicine for his work on malaria. Ross demonstrated that this disease is transmitted by the bite of the *Anopheles* mosquito. (Courtesy of The Bettmann Archive.)

caused by a minute worm, was transmitted by the bite of a mosquito. He thought that malaria probably was similarly transmitted.

Several other lines of work contributed to Ross's synthesis. One was the development by Romanowksy of ways to stain blood cells and the malarial parasites in them. Another ingredient that contributed to the final result was the discovery by Danilewsky that many animals such as lizards and birds are subject to malaria.

Ross returned to India and fortified with all this knowledge, was able to discover how malaria is transmitted from one person to the next. He found that the infection is not carried by bad air but by the bite of a particular kind of mosquito, the *Anopheles,* or, as Ross said, the dapple-winged mosquito. He also discovered the life cycle of the malarial parasite within this species of mosquito. This discovery was the real turning point. Ross then succeeded in transmitting bird malaria from one bird to another by allowing mosquitos to bite first infected birds and subsequently bite healthy ones.

Various workers soon demonstrated the transmission of human malaria by mosquitos. An unpleasant controversy arose as to who really first discovered the cause of malaria. The report of an international commission resulted in a Nobel prize for Ross. But the controversy illustrates the fact that often when a scientific problem is ready for solution, i.e. when the theoretical and factual backgrounds exist, several people can reach the solution at about the same time. One important aspect of malaria remains to be considered, the social or **public health** aspect.

Ross himself made a major contribution with his malaria statistics. While the actual figures he used only hold for the special conditions under which he worked, they illustrate in simple fashion the nature of **epidemiology,** the science of occurrence and spread of diseases.

Ross found that only one out of four mosquitos succeeds in biting someone. The chance that a mosquito will live the 21 days required for the sexual phase of the parasite to take place so that sporozoites are in the mosquito's saliva is $1/3$, which is another way of saying that within 21 days two out of every three mosquitos are either killed by dragonflies, hit by raindrops, or meet their fate in some other manner. This means that the chance of a given mosquito both biting someone and living three weeks is $1/4 \times 1/3$, or $1/12$. This is an application of the **product law** discussed in Chapter 4. From such data the likelihood that a disease will increase or decrease can be predicted. For example, if there is one case of malaria on an island with 1,000 people and one mosquito, the chance that that one mosquito will bite the one malarial patient is $1/4 \times 1/1,000$, or 1 in 4,000. The further chance that this extremely unlucky mosquito will live 21 days and then bite someone else is $1/4,000 \times 1/3 \times 1/4$, or 1 in 48,000.

The first known therapeutic agent effective against malaria was quinine, an alkaloid derived from the bark of the cinchona tree. Its use was discovered by Peruvian Indians and was introduced into Europe by Jesuit missionaries. It is very effective against the asexual stages of the malarial parasite in the blood but ineffective against the stages in the fixed cells of liver and spleen. Consequently, patients treated with quinine are usually subject to relapses, sometimes every 9 months or so.

The first successful attempt to produce a substitute for quinine was achieved by a team composed of students of the great German pioneer in bacteriology, Robert Koch, and chemists of the I. G. Farben Trust. Other workers in England, the United Sates, and elsewhere have continued to synthesize possible antimalarial compounds. During World War II over 14,000 compounds were tested in the United States alone. There are now a dozen or more antimalarials available such as plasmochin, atabrine, chloroquine, daraprim, and others.

Parasitic Flatworms

The trematodes or flukes are parasitic flatworms known from ancient times. Some of the oldest Egyptian mummies contain embalmed trematodes and their eggs, yet our knowledge of their rather complex life histories has been achieved only within the past 75 years. They infect both man and his domestic animals as well as innumerable wild hosts. From the biological as well as medical and veterinary points of view trematodes may be divided into two major groups. One group consists of the **monogenetic flukes** which are external (or semi-external) parasites living on a single host and reproducing only sexually. The most readily found monogenetic trematodes are species which live in the mouths, nostrils, or urinary bladders of frogs. The gills of fish are also often infested.

The other group consists of the **digenetic flukes** which are internal parasites such as liver, lung, intestinal, and blood flukes. They have complex life histories involving both sexual and asexual reproduction and two or more hosts, commonly a vertebrate and an invertebrate. A digenetic trematode, the sheep liver fluke, *Fasciola hepatica*, was the first trematode known, and the first to have its life history worked out. It is prevalent in regions where sheep are raised, but fortunately is rarely found in humans. The Chinese liver fluke, *Clonorchis (Opisthorchis) sinensis*, regularly infests human beings and is widespread in Asia. Infection is contracted by eating uncooked fish. Many frogs that appear in vigorous health harbor lung flukes (*Haematoloechus medioplexus*), which are conspicu-

ous to the naked eye when the frog's lungs are cut open.

Most digenetic trematodes live in the intestines of their **definitive host,** i.e. the host in which the parasite becomes sexually mature. The most common is the giant intestinal fluke, *Fasciolopsis buski*, found throughout the Orient. The adults grow to be over 18 cm (7 in) long and produce an intestinal ulcer at each point of attachment. In addition to physical injury, they produce toxic substances which are absorbed in the bloodstream. Infection, especially in children, is incurred by eating uncooked aquatic plants.

The blood flukes, or schistosomes, have been the scourge of the Nile valley from time out of mind and remain so today in Egypt, over much of Africa, and in large parts of Asia and South America. Schistosomes live chiefly in the small blood vessels of the mesentery of the lower intestine; snails are the intermediate hosts.

Eggs of digenetic trematodes are laid almost continually and in enormous numbers during adult life in whatever definitive vertebrate host the worm becomes sexually mature. The size and shape of trematode eggs enable the species infesting a man or an animal to be diagnosed. The eggs are passed to the exterior with urine, feces, or sputum. If the egg falls into water, a microscopic ciliated larva called a **miracidium** hatches. To

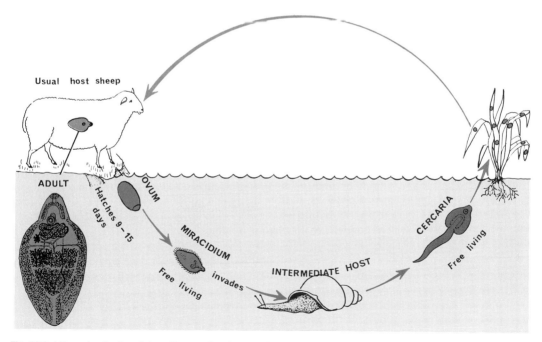

Fig. 23-9. Life cycle of a liver fluke with a snail as intermediate host. Adult fluke is drawn 1.4× actual size; other parts of figure are not to scale. (Modified from Jeffrey, H. D., and R. M. Leach. 1966. *Atlas of Medical Helminthology and Protozoology*. E. & S. Livingstone, Ltd., Edinburgh.)

survive, the miracidium must penetrate the tissues of a snail or other small invertebrate within 24 hours. Within the snail the miracidium loses its cilia and grows into a **sporocyst.** This is a small wormlike creature lacking mouth and gut but containing 8 or 10 groups of embryonic cells. Each group of cells forms a worm that may become another sporocyst or a larger worm having at least a rudimentary gut. The worms so formed are called **redias** for their 17th century discoverer, Francesco Redi (1626–1697), the Italian poet and naturalist known for his pioneer work on the origin of insects from eggs. A redia never leaves the body of the snail but produces, by a process of internal budding, tailed worms called **cercarias.**

A **cercaria** (*kerkos,* tail) superficially looks like a minute tadpole. The body is a small worm with a simple gut and no reproductive structures, but with several adaptations for its life. These include a pair of eyespots, the muscular tail, which is forked in many species, and either some cyst-producing glands or glands which pour out a powerful digestive enzyme that enables the cercaria to penetrate the host's skin. The cercarias break out of the redia in which they were formed, pass through the tissues of the snail, and finally swim away in the water.

The cercarias reach the definitive vertebrate hosts in a variety of ways. Those of the schistosomes reach the blood vessels of their victims by directly pentrating the skin of persons who work in irrigated rice fields or walk barefoot through puddles. If the proper host of a given kind of cercaria is not a man but a bird or other animal, the cercarias may still penetrate the human skin but are killed on the spot by a local inflammatory reaction, swimmer's itch in freshwater regions and clamdigger's itch around salt water.

In many species such as the sheep liver fluke, *Fasciola,* and the giant intestinal fluke, *Fasciolopsis,* the cercarias become encysted on some plant in or near the water and must wait until eaten by the sheep or man. The cercarias of the Chinese liver fluke encyst within the muscles of fish and there wait to be eaten. The cercarias of the frog lung fluke encyst in dragonfly larvae and other small aquatic animals until they are lucky enough to be eaten by a frog.

Once within the definitive host the cercaria develops directly into a sexually mature adult. Usually some migration through the body of the host is first necessary to reach the favored spot. The ingested cercarias of the sheep liver fluke burrow across the intestinal wall and into the coelom, finally reaching the surface of the liver. They enter the liver and lodge in the bile ducts. In the case of schistosomes, the cercaria are carried around by the bloodstream before they lodge in the capillaries of the inferior mesenteric and rectal veins.

The human cost of trematodes is great in terms of money and in terms of chronic ill health, sheer misery, or death. Control measures include eradication of snails as intermediate hosts, care not to eat uncooked food or drink unboiled water in regions where these plagues are endemic, and rigid enforcement of sanitary codes to prevent raw human excreta from contaminating bodies of water.

Tapeworms

Tapeworms (**cestodes**) are flatworms which lack both external ciliation and a digestive system and possess a **scolex,** or head, provided with suckers, hooks, or other adhesive organs. The fully mature adults vary according to the species from small worms 25 mm or less in length to giants over 10 meters long. The scolex is very small, the elongate ribbon-like body is usually composed of three or four to three or four thousand segments called **proglottids** (Fig. 23-10). With almost no exceptions, the sexually mature adult lives in the intestine of a vertebrate. No group of vertebrates from fish to man is immune from these debilitating parasites of which there are over 2,000 species.

The immature bladder worm or **cysticercus** stage is found in the muscles, livers, brains, and other organs of cattle, pigs, rabbits, men, and a wide variety of other animals. Until about a century ago it was believed that tapeworms arose spontaneously from eating too much of the wrong kind of food or from other vague causes. Mankind is indebted to a gynecologist named Kuchenmeister for demonstrating the actual origin of tapeworms. He showed that if you feed bladder worms from raw pork to dogs, tapeworms will appear in the dogs' intestines. He even demonstrated this in human beings by feeding bladder worms to a condemned criminal.

Nematodes

No group of animals that is so little known to the general public has anything like the economic, medical, or scientific importance of the nematodes, or roundworms. Nematodes are second only to insects in the damage they do to agriculture. They also cause numerous debilitating and fatal human diseases as well as diseases in wild and domestic animals. No vertebrate is known that cannot harbor parasitic nematodes, nor are crustaceans, mollusks, insects, or centipedes free

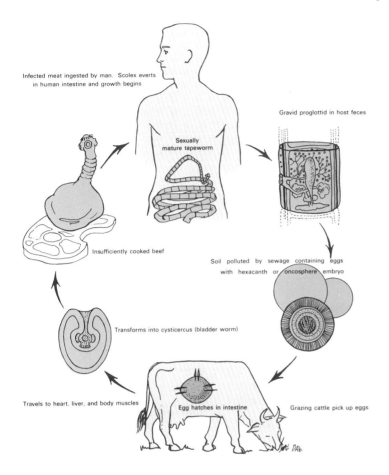

Fig. 23-10. Diagram of the life cycle of the beef tapeworm. (From Moment, G. B., and H. M. Habermann. 1973. *Biology: A Full Spectrum*. The Williams & Wilkins Co., Baltimore.)

Fig. 23-11. Potatoes grown in nematode-infested soil are too small to be worth harvesting. Nematodes that attack roots deprive plants of essential nutrients. (Courtesy of the U.S. Department of Agriculture.)

from them. The United States Department of Agriculture estimates the annual nematode damge to our agriculture at about $2 billion. The list of crops seriously injured by them includes beets, carrots, chrysanthemums, cotton, mushrooms, peanuts, potatoes (Fig. 23-11), rice, tobacco, tomatoes, wheat, and many others. Nematology deserves to rank with entomology in agricultural and medical importance.

Nematodes are among the most adaptable animals known, from an ecological standpoint. Not only are there some species that can live in almost any conceivable environment, but many single species live in widely different habitats, from ponds in northern Alaska to jungle lakes in Brazil, in both waters and soils of many types. Most soil and water-living species are semimicroscopic, many parasitic species grow to be 20 mm or more long, and some, like the common *Ascaris* which inhabits the intestines of humans and many of his domestic animals, grows to be 30 cm long. See Fig. 23-12 for examples of this highly varied group of organisms. In a few

cases it has been shown that a single species may have physiologically differentiated races which, although anatomically indistinguishable, nevertheless flourish under different conditions. Perhaps it is a matter of adaptive enzymes; nobody knows. Beyond the simple facts that their reproductive powers are pro-

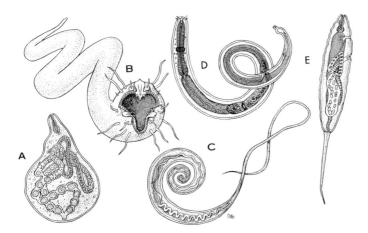

Fig. 23-12. Nematodes possess one of the most efficient body organizations known. Their skeletal, muscular, excretory, and, to a large extent, their nervous systems are all built into the body wall. A variety of nematodes are illustrated: A, gourd-shaped adult female of the rootknot nematode; B, a carnivorous species; C, the intestinal whipworm; D, a soil-dwelling species; E, a tailed nematode. (All are highly magnified.) (From Moment, G. B., and H. M. Haberman. 1973. *Biology: A Full Spectrum.* The Williams & Wilkins Co., Baltimore.)

digious, and that nematodes occur in enormous numbers, little is known concerning their population dynamics.

In diet many nematodes are carnivorous, eating protozoans, small annelids, rotifers, insect eggs, and even other nematodes. Some are herbivorous, living especially on roots, others live on decaying material, and many are parasitic. With many of the soil-living species the distinction between free living and parasitic becomes very fuzzy. Thus nematodes occupy many niches in the ecology of the soil.

Nematode Diseases of Plants. Among the best known and most destructive of the nematodes injurious to crops is the root knot nematode *Heterodera*. It produces swellings called galls on the roots of potatoes, turnips, tobacco, sugar beets, and some 75 other field and garden crops, vegetables, fruits, and weeds. The juveniles penetrate young roots, which respond by forming a swelling in which the worms mature (Fig. 23-13). The females swell up and when sexually mature become pear-shaped or egg-shaped, though only about 1 mm long. They never leave the gall, but merely spread their eggs in the soil when the root dies and disintegrates. The males resemble normal nematodes and wander about. In hot weather the life cycle requires less than a month. The eggs and juveniles within old galls can withstand drying but fortunately are killed by freezing.

Hookworm. The most widespread and important parasitic nematode is the hookworm, three species of which infest mankind. Adult hookworms live in the human intestine and are attached by their mouths to the villi. They ingest blood and secrete a poisonous anticoagulant that causes anemia and general weakness. The females are slightly over 1 cm long; the males are appreciably smaller. A

Fig. 23-13. Root knot on soybeans. These galls are caused by nematodes and are not to be confused with the nodules formed by nitrogen fixing bacteria living symbiotically in the roots of legumes. (Courtesy of the U.S. Department of Agriculture.)

mature female lays 5,000–10,000 eggs daily and usually lives 1 or 2 years.

The eggs already contain embryos when laid, and pass to the exterior in the host's feces. The eggs hatch in the soil, releasing larvae, which eat, grow, and molt twice in the soil before becoming infective. Optimum conditions are well-aerated topsoil, moist but not wet, absence of direct sunlight, and a temperature of about 33°C (90°F). The infective larvae appearing after the second molt are long, thin, so-called **filariform larvae.** They penetrate the skin of the feet, hands, or any other parts of a human body which come in contact with infected soil, and thus cause "ground itch." Within the human body the larvae are carried by the bloodstream to the lungs, where they may cause appreciable damage. From the lungs the growing worms make their way up the windpipe to the throat and then down into the intestine (Fig. 23-14).

Many people in the United States, mostly in rural areas in the South, are still afflicted with hookworm, *Necator americanus.* The number in Mexico, South America, and Africa also infected is said to run into the millions. Around the shores of the Mediterranean is found a slightly different species, *Ancylostoma duodenale.* In the Orient both species flourish. A United Nations report estimates a world total of over 600,000,000 cases. Control consists of sanitary disposal of human feces, wearing of shoes, and the general avoidance of contact with contaminated soil.

Trichinella. Trichinella spiralis is also a very serious threat to human health in the United States and in many other areas. The infection is contracted by eating insufficiently cooked meat of some omnivorous or carnivorous animal, usually pork. Severe cases, however, have followed eating bear steaks, and it would be a hazard of cannibalism. The larvae lie encysted in muscles or other tissue—brain, for example. When eaten and the cyst digested, the larvae burrow into the host's intestinal mucosa, where they grow and mature. Each female worm deposits about 10,000 larvae which migrate via the bloodstream and encyst in muscles and other

sites. It is this encystment that causes most of the damage, especially when the site is the heart or brain. Prevention is possible simply by the thorough cooking of all meat. This is an ecologically sound practice because it not only protects the human beings concerned but also prevents continuing the cycle by eliminating the possibility of infected garbage which might be eaten by dogs, pigs, rats, or other animals.

Other Animal Parasites

Among insects the ichneumon wasps parasitize other insects by laying their eggs on them, usually on the larvae. *Habrobracon* is one which has been much studied and is available for laboratory culture. Other species have been raised to control the numbers of various undesirable insects. The control of insect pests or disease-causing parasites by utilizing "parasites of parasites" is an ecologically sound strategy and provides an alternative to pesticides which may accumulate in the biosphere and become toxic to desireable organisms.

Even among vertebrates there are species which have been recognized as parasites. Two notable examples are the lampreys

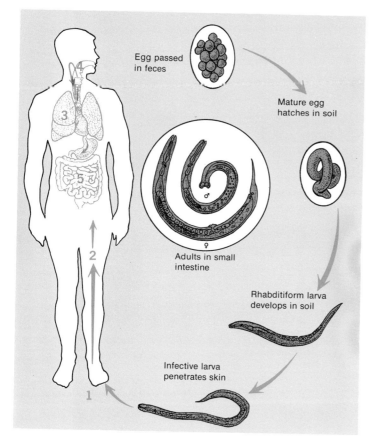

Egg passed in feces

Mature egg hatches in soil

Adults in small intestine

Rhabditiform larva develops in soil

Infective larva penetrates skin

Fig. 23-14. Life cycle of a hookworm. Infective larvae penetrate the skin (1). They are carried by the bloodstream (2) to the lungs (3). From the lungs, worms move up the windpipe to the throat (4) and then down into the intestine. (From Moment, G. B., and H. M. Habermann. 1973. *Biology: A Full Spectrum.* The Williams & Wilkins Co., Baltimore.)

which attack fish and suck their blood and the bloodsucking "vampire" bats.

USEFUL REFERENCES

Burton, M. 1970. *Animal Partnerships*. F. Warne and Co., New York.

Deverall, B. 1969/71. *Fungal Parasitism*. Institute of Biology Studies in Biology No. 17. Edward Arnold (Publishers), Ltd., London. Distributed in the U.S. by Crane, Russak and Co., Inc., New York. (Paperback.)

Leclercq, M. 1969. *Entymological Parasitiology*. Pergamon Press, New York.

Margulis, L. 1970. *Origin of Eukaryotic Cells*. Yale University Press, New Haven.

Noble, E. R., and G. A. Noble. 1964. *Parsitiology*. Lea and Febiger, Philadelphia.

Raff, R. A., and H. R. Mahler. 1972. The non-symbiotic orgin of mitochondria. *Science* 177:575–582.

Read, C. P. 1970. *Parasitism and Symbiology*. Ronald Press, New York.

Read, C. P. 1972. *Animal Parasitism*. Prentice-Hall, Inc., Engelwood Cliffs, N. J. (Paperback.)

Scott, G. D. 1969/71. *Plant Symbiosis*. Institute of Biology Studies in Biology No. 16. Edward Arnold (Publishers), Ltd. London. Distributed in the U.S. by Crane, Russak and Co., Inc., New York. (Paperback.)

Van Embden, H. F. 1975. *Ecological Aspects of Pest Control*. Institute of Biology Studies in Biology No. 50. Edward Arnold (Publishers), Ltd., London. Distributed in the U.S. by Crane, Russak and Co., Inc., New York. (Paperback.)

Wilson, R. A. 1967/71. *An Introduction to Parasitology*. Institute of Biology Studies in Biology No. 4. Edward Arnold (Publishers) Ltd., London. Distributed in the U.S. by Crane, Russak and Co., Inc., New York. (Paperback.)

24

Ecology

Our planet Earth is covered by a seamless garment, the **biosphere**, within which all plants and animals live. This biosphere is but a thin film of earth, water, and air scarcely two miles thick. Outside of this very special layer no living thing can exist unless encased in an airtight and temperature-controlled capsule. Within this film all living things are in a dynamic relationship with each other and with their nonliving environment.

Ecology is the science which deals with all of these complex interrelationships. Because the essential life-supporting qualities of the biosphere within which mankind must live are now threatened by the activities of man himself, no branch of science is more timely or important than ecology. The term **ecology** is derived from the Greek *oikos*, meaning house or household, and involves the study of the total economy of living organisms in the broadest sense of the word.

People are unavoidably a part of the world **ecosystem**. Thus, if we are to maintain our world environment as a desirable or even possible place for human life, it is an urgent necessity that we discover the laws of ecology and live within them, just as an airplane designer must work within the laws of physics. No animal, whether human or earthworm, can leave its environment untouched. Plants also affect their environment and do so in equally profound if sometimes less obvious ways. Thus, all the children of DNA, whether bacteria, plants, animals, or people, are bound together in a common destiny which is now threatened by an impending environmental crisis. Extinction has been a commonplace phenomenon in the long course of evolution and it may be that some of us will become extinct. If such should be the case, it certainly would not be the bacteria that would disappear first.

Ecology and Current World Problems

Before considering any of the details of the science of ecology, it is well worthwhile to look at some of the harsh facts that make this science of the interrelationships of plants and animals to their total environment one of the most important branches of biology in today's world. The problems of the environment which we face are largely the multiple results of industrialization and population growth. Consequently, most of these problems, rather than being strictly biological, are primarily political and economic problems although with important scientific dimensions. No great scientific breakthrough is needed to know that radiation is extremely harmful to all living things nor is any special scientific insight required to realize that mercury or other heavy metals are poisonous. It is only within recent decades, however, that the nightmarish consequences of massive dumping of industrial wastes into rivers and oceans have been recognized. Two dramatic and hideous "diseases" are widely publicized examples of the direct impact of rapid industrialization with its uncontrolled dumping of wastes in post-World War II Japan. Itai-Itai (ouch-ouch) disease results from poisoning by cadmium. The bones become brittle, a condition that is painful and eventually lethal. Minamata disease is caused by methyl mercury, which is formed by the action of microorganisms in mercury-polluted waters. Symptoms of Minamata disease include convulsions, deformed limbs, and mental derangement. An epidemic of this disease occurred in Minamata, Japan in the 1950's in families of fishermen who ate contaminated fish. Prenatal poisoning by

383

Fig. 24-1. A victim of mercury poisoning in Minamata, Japan. Tomoko Uemura, 16 yr old in this photograph, was exposed to methyl mercury from her mother's bloodstream. Because the fetus removed and concentrated mercury from contaminated fish, the child was born deformed while the mother showed no signs of mercury poisoning. (Courtesy of Magnum Photos, Inc., New York.)

methyl mercury can be especially insidious because this compound can be removed from the mother's bloodstream and become concentrated in the fetus. Consequently gross abnormalities in children can result even though the mothers who consumed the contaminated foods exhibited no apparent effects (Fig. 24-1).

The tragic epidemics of pollution-caused diseases that have occurred in Japan are in large part related to conditions which, while not unique to Japan, have been exaggerated in that country during the last 30 years: a high population density in a limited geographic area, a heavily industrialized society that is dependent on its industrial products for economic survival, and relatively unregulated dumping of industrial wastes. In some ways Japan has become an indicator society from which predictions of future hazards might be made for other parts of the world.

We will consider DDT later in this chapter, but it should be pointed out here that the World Health Organization has been the chief proponent of its use. Especially in the underdeveloped countries of the world, DDT has saved millions of people from death and suffering due to malaria and other insect-borne diseases. When you and your children are dying from disease and from starvation, you could not care less that the American eagle is laying soft-shelled eggs or that traces of DDT can be found in penguins at the South Pole. Although DDT has been banned in the United States, a stark question still confronts us all: what should be done until safer methods of insect control are available everywhere?

Birth control pills raise similar questions. Such hormonal preparations sometimes produce harmful and even fatal side effects. Is it enough to say that pregnancy also can produce harmful and even fatal side effects or that overpopulation results in disease and death?

The role of the ecologist is to provide the knowledge on which an intelligent and humane environmental management program can be based. In ecology, as in its sister science, medicine, knowledge can never be complete. Hippocrates, the father of medicine, emphasized long ago that "life is short and the occasion fleeting." In applied ecology, action must often be taken, as it is in medicine, on the basis of the best available information which will, in many cases, fall far short of a complete understanding of all the factors involved or a full foreknowledge of all the implications.

Environments

One of the first tasks of ecology is to bring some logical order out of the diverse kinds of environments in which organisms live and to discover how plants and animals are interrelated with their environments. The **range** of a species is the entire geographical area over which the animal or plant may be found. For example, the range of *Rana pipiens,* the meadow frog commonly used in laboratories, is from northern Canada to Panama, and from the Atlantic coast almost to the Pacific. The **habitat** of a species is the particular kind of environment in which it lives. The habitat of *R. pipiens* is damp meadows and along the edges of small streams or ponds. The ecolog-ical **niche** occupied by a species is usually defined as its ecological role in its environment, but it is also said that the niche occupied reflects the requirements for living of that organism. The ecological niche of *R. pipiens* is that of a small carnivore in moist locations. As commonly used in ecology, niche is an abstract, functional term.

Terrestrial environments change in a regular and similar way in passing from north to south and from mountaintops to valleys (Fig. 24-3). These changes are sometimes inconspicuous when one group of plants and animals gradually replaces another. Elsewhere the changes may occur in a series of fairly abrupt steps separating well-defined life zones. Each zone supports a characteristic living community called a **biome.** Thus there is a deciduous forest biome, a prairie biome, and so on. The line of contact between two biomes, large or small, is termed an **ecotone.** Theoretically, this is a zone of tension between two communities. Ecotones are often characterized by special animals and plants which thrive only in such transitional regions. The quail (bobwhite) is such a species which for successful breeding must nest close to the edge of a wooded area, preferably where forest, meadow, and shrub meet.

THE SEVEN MAJOR LAND BIOMES

Between the North Pole and the equator there are seven major biomes. The north and south polar regions, together with snow-capped peaks, constitute **Arctic zones,** which are similar in many ways though not identical. Except in a more or less sporadic form the Arctic zones are largely without life.

Tundra

South or below the Arctic zone is the **tundra.** This is a vast treeless region extending across northern Canada and northern Siberia. The **Alpine zone,** above the tree line on mountains, corresponds to the tundra. The characteristic plants are sphagnum and other mosses, reindeer moss and other lichens, grasses, and a few small shrubby bushes. The flowering herbaceous plants of the tundra are noted for their showy blossoms. Musk oxen, reindeer, snow hares, caribou, arctic foxes, and wolves are characteristic mammals. In the brief summer, insects (especially flies and mosquitoes) and nesting migratory birds abound. In New Hampshire, the tundra or Alpine zone begins on Mount Washington at less than 2,000 meters, while in the Rockies the tree line is up at the 3,500- to 4,000-meter

Fig. 24-2. Astronaut Edwin Aldrin walking near the lunar module during Apollo 11 extravehicular activity. Note the barrenness of a land without life. (Courtesy of the National Aeronautics and Space Administration.)

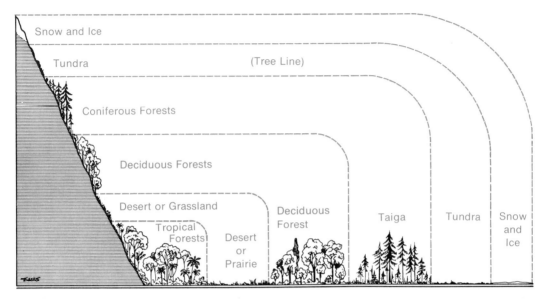

Fig. 24-3. Correspondence of vertical and horizontal biomes from mountaintop to sea level and from the pole toward the equator. (After Moment, G. B. 1967. *General Zoology,* 2nd ed. Houghton Mifflin Co., Boston.)

level. On Mount Popocatepetl in Mexico the tree line is at some 5,000 meters or more. Not all the factors that determine the position of the tree line are understood. Temperature is paramount, but its action is subject to considerable modification. Sometimes isolated areas of trees grow far north of the tree line. The line may be sharp, or there may be a transition zone of stunted trees.

Taiga

South of the tundra is a belt of conifers, spruce, firs, and pines called the **taiga.** Birch and aspens are scattered through the taiga and are usually the first trees to grow after a fire. The moose is the characteristic large mammal; rodents and mink are abundant small ones. Intermediate in size and also abundant are lynxes, black bears, wolves, and martens.

Deciduous Forest

South of the taiga is the familiar **deciduous forest** of hardwoods. In the northern part of the zone, beeches and maples tend to predominate; in the southern part, oaks and hickory. The Virginia deer, the black bear, and the opossum are characteristic, although the yellow-spotted black salamander, *Ambystoma maculatum,* of the woodlands is used as an "index animal" for this zone in the eastern United States. In some places, when soil and moisture conditions are right the deciduous zone is replaced by a southern coniferous region.

Grassland

In many parts of the world **grasslands,** variously termed prairies, steppes, veldts, or pampas, commonly occur south of the deciduous zone (north of this zone in the southern hemisphere). It is in this kind of environment that the evolution of horses, antelopes, and kangaroos took place. In some regions the prairie extends to the coniferous taiga, without any intervening deciduous zone. Rainfall is the major environmental factor governing the extent of grassland vegetation. Grasslands replace forests as annual rainfall decreases.

Tropical Rainforest

The **tropical rainforest** makes a sixth major life zone. Such regions are found in Central America, northern South America, equatorial Africa, and southern Asia and the East Indies. The pattern of trees is radically different from that in other types of forest. Few northern forests are composed of more than a dozen species, and great stands can be found composed of but one or two species. In tropical rainforests, however, several hundred kinds of trees are commonly found intermixed. Moreover, the vegetation is stratified. Canopies of great ironwood, banak, and other trees tower to a height of 40 meters (125 ft) or more. Below them, in partial shade, there is a second layer of trees, and beneath these, in turn, is a mass of small trees, bushes, and plants of many kinds. It is little wonder that in this environment many of the animals are tree dwellers. In British Guiana 31 out of 59 species of mammals are arboreal.

Desert

A seventh zone is **desert,** which may occur at almost any latitude. Thorny bushes with waxy leaves characterize deserts everywhere. In the Americas the typical plant is the cactus. Animal life is sparse and mostly nocturnal. A traveler going westward across the great plains of the United States can observe the transition from grassland to desert vegetation in regions of decreasing average rainfall.

Minor Biomes

Particular climatic and soil conditions have produced special biomes in various parts of the world. For example, the northwest coast of the United States has a nontropical rainforest. **Chapparal** biomes of dwarf evergreen oaks and other shrubby trees are found in temperate and subtropical regions with a rainy winter and a hot, dry summer. Some 6,000,000 acres of California are covered with chapparal. The shores of the Mediterranean

Fig. 24-5. Sagebrush and Joshua trees are typical plants of the Mojave desert, California. (From Moment, G. B., and H. M. Habermann. 1973. *Biology: A Full Spectrum.* The Williams & Wilkins Co., Baltimore.)

are also covered with chapparal in many places.

The character of a biome is determined by climate, which includes temperature, rainfall, amount of sunlight, and the character of the soil. The relative importance of climate and soil in controlling the living community varies from one region to another. No tropical rainforest can arise in the Great Plains area of the central United States, no matter what the soil conditions. In other regions, as in some parts of southeastern United States, soil seems to be the determining factor, within, of course, the limits set by climate.

MICROCLIMATES

The climate that matters for any animal or plant is obviously the climate in which it lives. This may be very different from the general climate of the geographical region constituting the range of the species. An insect living in the tree tops lives in a veritable desert compared with other species living on the forest floor where humidity is very high compared to that near the tops of the trees. Conditions under stones, on the north side of boulders, or a few inches under the sand of a beach are almost always quite different from those of the general surroundings (Fig. 24-6).

Fig. 24-4. Grazing land in Utah. Note coniferous taiga in background. (Courtesy of the U.S. Department of Agriculture.)

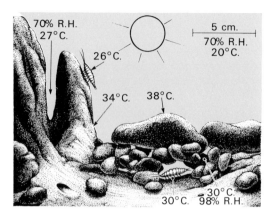

70% R.H.
27°C.

26°C.

5 cm.
70% R.H.
20°C.

34°C.   38°C.

30°C.
30°C. 98% R.H.

Fig. 24-6. Beach crustaceans encounter a wide range of microhabitats and microclimates. Note the differences in temperature and humidity above, at, and under the surface. (From Moment, G. B., and H. M. Habermann. 1973. *Biology: A Full Spectrum.* The Williams & Wilkins Co., Baltimore.)

This means that temperature and moisture readings recorded on equipment within a standard Weather Bureau louvered box set on a post one meter above ground level cannot be directly related to the animals, plants or microorganisms in the vicinity. Those species may and often do enjoy very different sets of conditions.

Atmosphere

All of the physical factors of the environment in which plants and animals live have been modified in greater or lesser degree by the presence of living organisms. We live at the bottom of an ocean of air surrounding our planet. Approximately 80 per cent of the air is nitrogen, which has persisted since the ages before life appeared on earth. The story of oxygen is very different.

It is now well established that atmospheric oxygen was not always present but instead is a product of photosynthesis accumulated since the first appearance of aquatic green plants. Few physical factors are of greater importance. Without oxygen active animal life as we know it would be impossible. Without ozone in the upper atmosphere the amount of ultraviolet radiation reaching the earth's surface would be so great that life on land would be impossible.

The amounts of moisture and CO_2 in the air are also influenced by plant and animal life. The composition of the ocean is continually being modified by the swarming myriads of organisms within it. Think of the millions of

tons of lime that have been removed from the sea by the protozoans whose shells formed chalk deposits. Even the amount of light and the temperature are modified, although to a slight degree, by the presence of plants and their effects on the amount of moisture, carbon dioxide, and oxygen in the atmosphere.

Soils

The soil is a very special case among the physical factors of the environment, partly because it has not merely been modified by living organisms but has actually been made by them. Without a covering of plants, the surface of the earth would be bare rock, washed by wind and rain. Because soil is of such crucial importance in agriculture and because it can be improved or easily damaged beyond repair, a special soil science, **pedology,** has developed. The Russians and Scandinavians have been leaders in this field and, consequently, a number of the terms are derived from their languages.

SOIL STRUCTURE

As can be seen in any freshly cut bank, soil consists of a series of layers, called **horizons.** The uppermost horizons are commonly known as **topsoil,** technically the **A horizons.** Topsoil contains a surface layer (A_{00}) of undecomposed debris consisting of leaves, twigs, animal remains, and the like called **litter.** Beneath this is a layer (A_0) of more or less decomposed organic matter, called **leaf mold.** True soil begins with a dark horizon (A_1) of **humus.** This layer has a high content of organic matter, thoroughly mixed with miscellaneous rock particles. Like the two layers above, it is rich in microscopic and semimicroscopic animals and plants. Below the humus is a lighter colored layer, the A_2 horizon, from which most of the soluble materials have been leached, i.e. washed out, by sinking rainwater. In prairie soils, the deep black **chernozems,** the leached A_2 layer is said to be missing or, if listed as present, is relatively unleached.

The **subsoil,** or **B horizon,** lacks appreciable amounts of organic matter and contains few soil organisms, whether animals, plant roots, fungi, or bacteria. It is said to be mineralized. The upper layer of the subsoil, termed the B_1 horizon, is dark with the minerals leached from the topsoil. The deeper B_2 horizon is a tightly packed mineral soil, which may show a bottom layer of lime or other mineral deposits just above the C_1 **horizon** of weathered bedrock. Major soil types are summarized in Fig. 24-7.

SOIL DEVELOPMENT

Changes in the character of the soil are an important factor that controls the kinds of vegetation that the soil can support. Some ecologists believe that deep burrowing earthworms are the chief agents in converting the mor soils of the taiga with their thin, sharply limited layer of humus into the mull soils of the deciduous forests with their thick and much less sharply delimited humus or topsoil. Darwin's extended account of the way earthworms cultivate the soil showed that all the humus of the soil of England passed many times through the bodies of earthworms. Recent measurements have substan-

tiated his findings. As already noted, earthworms burrow into both the topsoil and the subsoil, pass it through their bodies, and leave it in castings on the surface of the ground. In light dry soil, 2 to 3 tons of castings may be produced per acre per year; in moist pastures castings may weigh 10 tons; while in certain tropical localities amounts of up to 100 tons per acre per year have been measured. The soil also becomes finer and its colloidal properties modified in passing through earthworms. They continually deepen the topsoil, both by bringing subsoil to the surface and by dragging leaves into their burrows. Their burrows open the way for water and air to penetrate the soil. There can be no question that earthworms, once they have become abundant, play a very important role in soil dynamics.

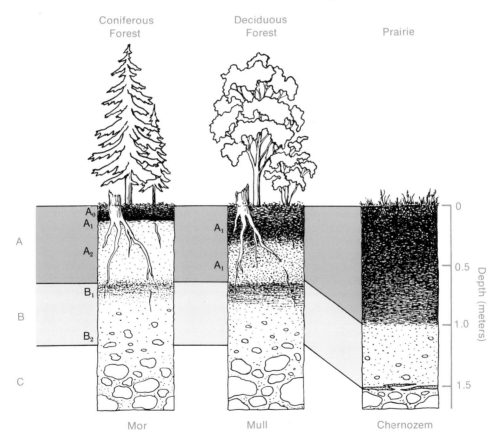

Fig. 24-7. Soil types. Soils are classified as podzols (mor and mull) or chernozems, depending on the amount of leaching that has occurred. Podzols are rich in aluminum and iron and deficient in calcium, and often red or brown in color. Such soils occur in the eastern United States. Chernozems are characteristic of the drier great plains and western regions and are rich in calcium. In soils of the coniferous forests (taiga) the humus or A_1 horizon is very shallow and sharply demarcated from the leached A_2 layer. Such a soil is called mor. Under temperate deciduous forests the A_1 horizon is much thicker and the A_2 less thoroughly leached and richer in organic matter. This type of soil is called mull. Prairie soil, or chernozem, is markedly different from both mor and mull. The A horizon of humus is extremely thick and fairly uniform. Because of limited rainfall, leaching is minimal. (Modified from Moment, G. B. 1967. *General Zoology*, 2nd ed. Houghton Mifflin Co., Boston.)

Earthworms are by no means the only animals of the soil. The fauna of the soil is at least as abundant and often more abundant than the fauna of freshwater lakes or most parts of the ocean. The A_0 horizons of litter and leaf mold abound with invertebrates. In addition to a vast horde of insects, adults and larvae, the soil is the normal environment for teeming populations of nematodes, many very destructive to crops, plus enormous numbers of minute spiders, mites, crustaceans, worms, rotifers, and even protozoans (especially amoebas and flagellates).

Also of great importance for soil building are the enormous numbers and kinds of microorganisms present. The fungi and bacteria play an essential role in the decay of dead plants and animals by releasing bound minerals and the end products of the breakdown of organic molecules, CO_2 and H_2O. Furthermore, a small number of bacterial species, by their ability to fix atmospheric nitrogen, provide a continuing supply of ammonia and nitrate, the only forms of nitrogen that can be absorbed through the roots of higher plants.

Ecological Succession

In most parts of the world today the community of plants and animals in a region remains constant decade after decade, presumably for many centuries. Such a stable fauna and flora constitute an ecological **climax** or a **climax community.** Each of the major biomes represents such a climax, which remains stable unless marked climatic changes take place. Whenever a climax community is upset or destroyed, as by cutting the trees and cultivating the soil, building a road, or flooding an area to make a freshwater pond, the climax may be restored through a regular series of stages known as a **sere.**

Each stage of a sere is characterized by its own typical plants and animals.

In the deciduous forest zone of the temperate regions of the Northern Hemisphere, annual herbaceous plants appear first after a severe fire as they did in the rubble areas of bombed cities after World War II. These are followed in a year or two by various kinds of grasses, then shrubs, and later by pines, birches, or perhaps scrub oaks. These are at last superseded by the beech-oak-hickory climax. Various other seres have been studied, notably the one observed when vegetation begins to grow over a sand dune or to fill up a freshwater pond.

The **Great Succession** in the Northern Hemisphere is the reforestation of the land following the last glaciation of the Pleistocene era. This process is only now beginning to draw to a close. The retreating glaciers left tundra conditions. Even in the southern United States, remains of musk oxen, mastodons, and other animals, now restricted to the far north, have been found. In time the tundra became coniferous taiga and this in turn gave way to the deciduous forest which is today's climax community in most of eastern and central North America and of Europe.

The causes of ecological successions are complex and are different in different cases. The situation following the destruction of a forest by a fire is not the same as after its destruction by a glacier. Climatic factors are important in all cases, but after a fire the climate commonly remains the same.

There has recently been a drastic change of policy by the U.S. Forest Service concerning fires in public forests. After decades of urging the public to prevent forest fires, a policy described by one park resources management specialist as the "Smokey Bear syndrome," and heroic efforts to put out all fires on public lands regardless of cause, there has been a shift in emphasis following the realization that forest fires can have benefi-

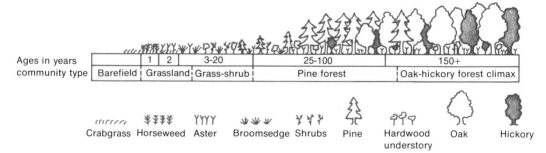

Ages in years	1	2	3-20	25-100	150+
community type	Barefield	Grassland	Grass-shrub	Pine forest	Oak-hickory forest climax

Crabgrass Horseweed Aster Broomsedge Shrubs Pine Hardwood understory Oak Hickory

Fig. 24-8. Stages of plant succession in a sere leading from a bare field abandoned after "overfarming" to a deciduous forest climax. The entire process requires at least 150 yr. (After Odum, E. P., and H. T. Odum. 1959. *Fundamentals of Ecology,* 2nd ed. W. B. Saunders Co., Philadelphia.)

cial effects. Past efforts to suppress fires have resulted in changed vegetation patterns and have even led to the instability of ecosystems. Some plants and animals need fires to reproduce or are benefited by nutrients in vegetation developing after a fire. Certain pine cones do not open to release seeds without fire. Careful observation of fires in the Grand Teton National Park during the summer of 1974 provided evidence that animals are attracted toward burnt off areas. Bears, coyotes, birds, moose, and even elk were observed in and near fires. Insects seem to be attracted toward the heat to lay eggs and they in turn attract birds. It is now apparent that preserving natural areas does not mean that all fires must be extinguished.

Lakes

The scientific study of lakes is known as **limnology.** Lakes are classified into three types on the basis of the amount of nutrition they provide for fish and other inhabitants. **Eutrophic** lakes are relatively shallow (10 meters or less) and contain an abundant supply of nutrients for algae and pond weeds, and therefore for crustaceans, and therefore for small fish, and therefore for larger fish. Lake Mendota in Wisconsin and many of the lakes in Scandinavia are well-studied examples of eutrophic lakes. The deeper parts of eutrophic lakes are poor in oxygen. **Oligotrophic** lakes are deep, relatively poor in nutrients and animals, and rich in oxygen. The Great Lakes of Middle North America, except Lake Erie, which is shallow, and the Finger Lakes of New York are typical oligotrophic bodies of water. **Dystrophic** lakes, usually parts of bogs, contain nutrients, but various organic acids and other substances inhibit the growth of plants and animals.

EUTROPHICATION

In the course of thousands of years oligotrophic lakes gradually become eutrophic. The rapid industrialization and urbanization of the United States in this century has led to massive and uncontrolled release of sewage, industrial wastes, and excess fertilizers into waterways and lakes. A consequence of this buildup of nutrients has been a pathological acceleration of eutrophication. Parts of Lake Erie have been severely damaged in this way. Population explosions of algae, so-called **blooms,** occur and can cause disastrous oxygen depletion in a lake. The algal cells in the

uppermost layers of the water screen out the sunlight so that those deeper down no longer produce a great excess of oxygen over and above their own respiration. Furthermore, such dense populations of algae contain progressively more and more dead cells which are decomposed by oxygen-using bacteria and fungi. Such oxygen depletion often results in massive fish kills which in turn decay, releasing even more potential nutrients for further overgrowth of plants. The resulting imbalance has literally "killed" many lakes and streams in the United States in recent decades.

FARMING A LAKE

Up to a point eutrophication can be highly beneficial. In recent decades the United States government has developed a farm pond program through the Soil Conservation Service. Over 350,000 ponds have been made by damming small streams. They are routinely stocked with bluegill sunfish and with bass.

Unfertilized ponds yield from 40 to 150 pounds of fish per acre, while fertilized ponds yield from 200 to 400 pounds. Both represent eutrophic conditions. The fish feed on each other and on plankton. In these ponds the fertilizer feeds the algae which constitute the phytoplankton. These tiny aquatic plants serve as food for *Daphnia, Cyclops,* and other crustaceans which are the most important members of the zooplankton. The crustaceans in turn are eaten by the smaller fish which are then eaten by the larger fish.

The amount of fertilizer added must be enough to produce a luxuriant growth, a so-called "bloom," of algae early in the summer. However, there must not be such a thick soup of algae that when they die off in the winter the processes of decay will use up all the oxygen and kill the fish. The number of fish initially introduced into the pond is also a very important factor. When bluegills are added at the rate of 1,500 per acre, they attain a weight of about 4 oz in 12 months, but if 180,000 per acre are added, they average only 0.02 oz. At the rate of 1,500 per acre *without* added fertilizer, the average weight is 1.1 oz.

ANNUAL TURNOVER

Most lakes in the temperate zone undergo a very important complete turnover of the water every spring and fall. Because this thoroughly mixes the dissolved oxygen and mineral nutrients in the lake, it stimulates plant and animal growth. This peculiar behavior of the water can be understood by remembering that water is most dense, i.e. heaviest, at 4°C (about 39°F). As water becomes still colder, it expands and becomes light. In the fall, as the surface water cools, it becomes cold, and in turn sinks. This process continues until the entire lake is at 4°C.

In the winter, as the water cools to below 4°C, it no longer sinks but remains on the surface and ultimately forms a layer of ice. Except for a very thin layer immediately under the ice, lake water never falls below 4°C. This is a doubly fortunate circumstance for living things. It means that in the water they are protected from freezing temperatures. It also means that the lakes are available as fluid environments, for if water continued to contract during cooling, ice would begin to form on the bottom of lakes and gradually fill them with ice. Under these circumstances ice would never leave the bottoms of lakes except in the southern edge of the temperate regions.

In spring, after the ice melts, the thin layer of water at the surface (which is below 4°C) warms, becomes heavier, and sinks until all the lake again becomes 4°C. As the surface water, heated by the sun, becomes warmer than 4°C, it does not sink but forms a warm surface layer. This is pushed by the prevailing wind against one shore of the lake. Water at that shore sinks (rather than piling up) while, at the same time at the opposite shore, cooler water rises to replace that blown away. As a result, a countercurrent is set up in slightly deeper water. Since water at the top is warmer, i.e. lighter, than deeper water in the lake, it forms a superficial layer called the **epilimnion.** Immediately below the epilimnion is a thin layer of water where the temperature falls very suddenly, the **thermocline.** The depth at which the thermocline occurs depends on season, winds, and other factors. In some lakes the thermocline begins within 2 meters of the surface and can be very noticeable to swimmers. In others, such as Lake Cayuga, one of the deep, narrow Finger Lakes in New York, it may be 10 meters below the surface. Below the thermocline is the **hypolimnion.** This is a zone of relatively uniform and cold temperature extending to the bottom. Just as the current along the undersurface of the epilimnion produces a current in the thermocline, so the current in the thermocline produces a slight current in the hypolimnion. The result is a very gradual and slight warming of the hypolimnion.

In deep oligotrophic lakes, the hypolimnion has a supply of oxygen and is hence inhabited by cold water species such as *Mysis,* a kind of shrimp, and by lake trout. In eu-

trophic lakes there is virtually no oxygen in the hypolimnion, partly because the algae in the epilimnion prevent the sunlight necessary for oxygen-producing photosynthesis to occur and partly because the continual "rain" of decaying dead organisms from above uses up the available oxygen. The fauna on the bottom of such lakes consists of insect larvae, some of which are red with hemoglobin, and other organisms adapted for low oxygen concentrations. Fish are scarce.

Streams, Rivers, and Estuaries

FLOWING WATER

Streams and rivers present special problems and opportunities for plants and animals. In the rapids, animals are adapted for holding firm to rocks and often possess flattened, streamlined bodies. Filamentous algae often abound attached to rocks on the bottom by means of holdfasts. Unless the waters are polluted with sewage, oxygen is abundant. Stream salamanders commonly lack lungs and breathe merely through skin capillaries.

ESTUARIES

Estuaries are tidal zones where fresh water enters the sea. In a sense, such regions are ecotones where tension exists between the freshwater and marine biomes. For a variety of reasons such environments are extremely favorable for life. Crabs, oysters, clams, fish, jellyfish, ctenophores, and many planktonic species abound. Plants are plentiful, a mixture of phytoplankton, larger algae, and aquatic higher plants, that provide an abundant input of organic materials for the food chain. Estuaries such as the Chesapeake Bay are rich sources of food for human consumption.

Oceans

Marine ecology has already yielded information of major importance for the fisheries industry. Enthusiasts believe that in the future we may be able to farm the oceans as we now farm the land. Because nearly 75 per cent of the earth's surface is ocean, this is indeed an exciting idea. Careful studies have been made of the yield of diatoms and other minute green plants which constitute the

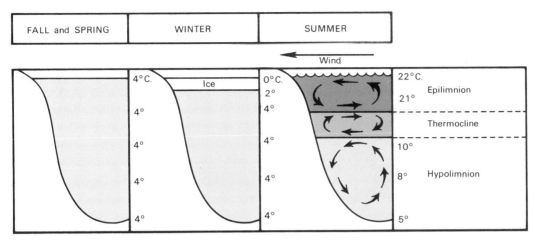

Fig. 24-9. Seasonal turnover of water in lakes. The density of water is greatest at 4°C and decreases as it cools below 4°C. Thus, ice forms at the top of a body of water. Wind and gravity are the major factors in producing the reverse temperature gradients characteristic of summer. (From Moment, G. B., and H. M. Habermann. 1973. *Biology: A Full Spectrum.* The Williams & Wilkins Co., Baltimore.)

basic food in the sea. The annual crop in terms of kilos of dry weight per hectare varies, according to the part of the ocean, from an amount equal to that of a forest to that of grass on a semi-arid plain. The large-scale scientific study of the oceans began with the world survey cruise of the *H.M.S. Challenger* from 1872 to 1876 with an international staff of scientists. The 50 enormous volumes which resulted from that voyage form the basis of our knowledge of marine ecology.

OCEAN CURRENTS

Key factors in the life of the sea are the wind-driven **ocean currents.** North of the equator the trade winds blow continually from the northeast toward the southwest. South of the equator the trade winds blow continually from the southeast toward the northwest. As a result great equatorial currents flow westward on either side of the equator in both the Atlantic and the Pacific oceans. When these westward currents strike the continents they are deflected north and south. This is the source of the Gulf Stream, flowing north along the east coast of North America, and of the Kuroshio (Japan) current, flowing northward from the Philippines along the coasts of China and Japan. The westerly winds in the temperate zones of both the northern and southern hemispheres help propel the Gulf Stream and the Kuroshio current, respectively, to the west coasts of Europe and of North America. The final result is a circular motion. In the Southern Hemisphere the Peru or Humboldt current, flowing northward along the west coast of South America, and the Benguela current, flowing northward

along the west coast of Africa, are due to similar forces. It is the cold Peru current that enables penguins to live close to the equator off the coast of South America.

There is another very different kind of current known as the **Arctic creep.** In the seas around both North and South poles the frigid waters sink and flow slowly along the bottom toward the equator. Here the water gradually rises to replace that which is lost by evaporation under the tropical sun.

OCEAN REGIONS

The oceans are divided into several general regions depending on depth. Most continents are surrounded by a **continental shelf,** a more or less flat plain under about 200 meters (roughly 600 ft) of water but in some places much less. Off the east coast of Florida directly opposite the Bahamas the continental shelf is only a few miles wide and relatively nonexistent when compared to the west coast of Florida and New England where the shelf extends approximately 300 km. The shelf itself and the waters over the shelf constitute the **neritic zone.** The part of the neritic zone near shore with water to about 50 meters deep, usually strong wave action, and enough light for plant growth is termed **littoral.**

Beyond the neritic zone is the **oceanic zone.** This is the part of the ocean which the navies of the world call "blue water." The bottom slopes rapidly down from the continental shelf to depths of 3,000 meters and more. The regions of these great depths are the **abyssal regions** where light never penetrates. The temperature is virtually constant at 3°C (about 37.4°F). Part of the ocean bottom in this abyssal region is a flat plain but there are great mountain ranges, mile-deep trenches exceeding the Grand Canyon of the Colorado River in dimensions, and extensive regions of hills and what resemble river valleys.

The animals and plants in each region of the oceans are characteristically different. The animals and plants that live near shore are called **littoral.** Those which live at sea, especially far out in the open ocean, are called **pelagic.** Those myriads of floating or feebly swimming organisms that live near the surface of the ocean, either in the neritic or oceanic zone, are called **plankton**—**zooplankton** if animals, **phytoplankton** if plants. Strong swimming animals are called **nekton,** but those which live on the bottom, especially at great depths constitute the **benthos.** Plants (mostly algae) are restricted to the upper layers through which sunlight can penetrate.

The most important members of the plankton are the diatoms, which constitute the basic plant food of the sea (Fig. 24-10), and the copepods, which feed on the diatoms. Every animal phylum is represented in marine plankton in the form of floating eggs, larvae, adults, or as all three. Very often members of the zooplankton are transparent and have special adaptations for floating.

Animals living at great depths in eternal cold and darkness are bizarre by any standards.

An interesting and economically very important difference between the marine populations near the poles and those near the equator is that in the polar seas there are enormous numbers of individuals but relatively few species. In tropical waters the situation is reversed; there are relatively few individuals but a great variety of different species. The explanation of this contrast is obscure, but it will be recalled that a similar contrast, insofar as number of species is concerned, exists between northern and tropical forests.

Although oceans cover nearly 75 per cent of the earth's surface, almost all the fishing done by man takes place over the continental shelves. Thus, only a tiny fraction of the vast ocean is used for the production of human food. The only significant exception is the whaling industry but many of the whales are captured near shore. It is worth noting that the baleen group of whales live exclusively on plankton, especially swarms of small oceanic crustaceans, while the toothed whales feed on giant squid and other members of the nekton.

INSTRUMENTATION OF OCEANOGRAPHY

The classic tools of the oceanographer are a set of nets and trawls (which can be towed through the water at fixed depths or along the bottom), a number of Nansen bottles (Fig. 24-11), and a good pair of sea legs.

In preparation for the coming modern age of ocean exploration and use, a wide variety of new instruments is being developed. Buoys adjusted to float at particular levels and equipped with powerful radio signaling devices make it possible to detect currents and record their temperatures. Transistor radios attached to whales and fish enable scientists to track these animals over long periods of time. With sonar and radar devices, bottom depths and contours can be mapped and schools of fish, swarms of crustaceans, and other animals can be located. Stable floating work "platforms" have been constructed which resemble elongated tubular boats, one end of which can be filled with water to up-end it to a vertical position. The upper end with the workrooms is then supported by a deep probe extending many feet down into the water so as to be free of wave action. Most important of all are the newly equipped ocean-going research ships and the large laboratories for analyzing their findings which have recently been built in the United States and abroad.

Matter and Energy Pathways

The major matter and energy pathways for animals begin with photosynthesis in the green plants, with the fixation of nitrogen by microorganisms, and with the absorption of various minerals from the soil and the sea by plants. In photosynthesis the energy of the sun is trapped and used to form carbohydrates. Carbon dioxide from the atmosphere and water are the raw materials for photosynthesis and free molecular oxygen is released as a waste product (see Chapter 10). The familiar over-all equation:

$$6CO_2 + 6H_2O \xrightarrow[\text{chlorophyll}]{\text{sunlight}} C_6H_{12}O_6 + 6O_2$$

gives the facts of primary importance to the ecologist, who is concerned with the sources of raw materials and energy, their amounts, and what happens to them after they have been processed by plants rather than with the biochemical machinery which enables chlorophyll to convert light into chemical energy.

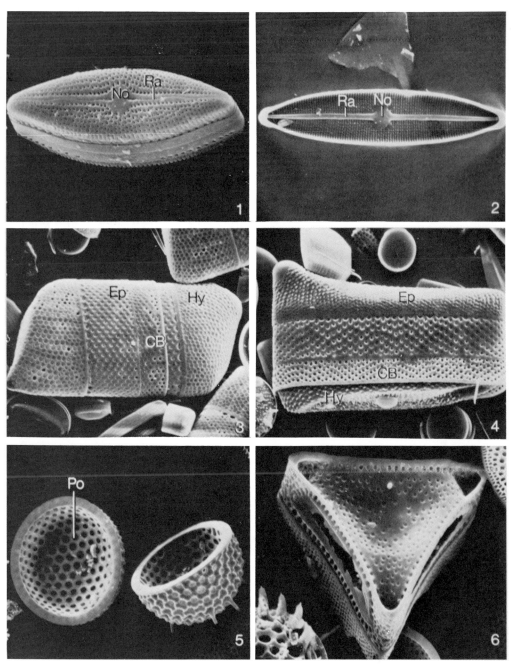

Fig. 24-10. Diatoms are plentiful in both oceans and fresh water. Their striking silica-containing cell walls consist of two overlapping pieces much like the halves of a box. The outer half (epitheca, Ep), inner half (hypotheca, Hy) and the connecting band (CB) joining the two halves can be seen in parts 3 and 4. Cell walls are ornamented with pores (Po), ridges, lines (including the raphe, Ra, running along the axis of bilaterally symmetrical forms), and nodules (No). Shown in parts 1, × 1585, 2, × 1100, 3, × 250 and 4, × 230 are bilaterally symmetrical species; in parts 5, × 535, and 6, × 725 are radially symmetrical species which appear round or triangular in surface view. (From Kessel, R. G., and C. Y. Shih. 1974. *Scanning Electron Microscopy in Biology*. Springer-Verlag, New York.)

He or she is interested in the effects that the withdrawal of all this CO_2 may have on the entire living community and what effects the release of oxygen may have and needs to know how much energy and matter a given community requires, how productive a given geographical region is, and how productive it might become. Finally, an ecologist wants to

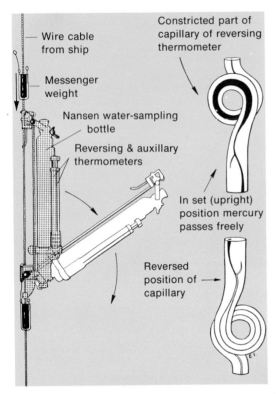

Wire cable from ship

Messenger weight

Nansen water-sampling bottle

Reversing & auxiliary thermometers

Constricted part of capillary of reversing thermometer

In set (upright) position mercury passes freely

Reversed position of capillary

Fig. 24-11. Nansen water sampling bottle. When the bottle is lowered, it is open at both ends and water passes through. Alongside the bottle is a reversing mercury thermometer. When bottle and thermometer have reached the desired depth, a metal weight is sent down the wire. When this "messenger" hits the catch holding the upper end of the bottle to the wire, the catch opens, the bottle falls over, and valves at each end of the bottle close, trapping a water sample. This action also causes the mercury column in the thermometer to break thus recording the temperature of the water at the depth of the sample. (After Vine, A., in *International Science and Technology.* December 1965.)

know where all the matter and energy come from, where they go, and how they get there.

A basic difference is at once apparent between the pathways for matter and for energy. The elements of which protoplasm is built, both the "big five" (carbon, oxygen, hydrogen, nitrogen, and phosphorus) and the various mineral elements (like iron, sulfur, sodium, calcium, potassium, iodine, and others required in small or even trace amounts), are used by different living things over and over again. Their pathways are cyclical.

In marked contrast, energy moves on a one-way street. It comes from the sun, is trapped by chlorophyll, and is utilized by living things in the three well-recognized ways, i.e. to construct complex molecules, to do the work of moving substances across membranes, and to bring about muscular and other types of protoplasmic motion. Ulti-

mately the complex large molecules are broken down, by one agent or another, and the energy is dissipated as heat. In terms of thermodynamics, living things therefore require a continual source of energy if they are to maintain their improbable structures and the activities that are part of them. Life on this planet is thus dependent on an outside source, the sun, for its energy, while it can reuse its supply of matter over and over again.

THE CARBON CYCLE

The carbon cycle begins with the formation of carbohydrates in photosynthesis. Carbohydrate may be built into the body of a plant directly or may be utilized to form lipids and proteins and other more complex substances, such as nucleic acids or hormones. If the plant is eaten by an animal, its carbon becomes incorporated (in the most literal possible sense of that word) into the substance of the animal. Carbon is returned to the atmosphere as CO_2 in the respiration of plants and animals. Any carbon caught in the corpse of a plant or animal is utilized by the microorganisms of decay and they in their respiration return it to the air as CO_2. The cycles of oxygen and of hydrogen are obviously closely linked to the carbon cycle.

There has been fear that we, including our industries, are using up our supply of oxygen so fast that we will all suffocate. However, recent careful and extensive independent studies at Columbia University and at the National Bureau of Standards show that during the period from 1967 to 1970, there has been no detectable decline in the amount of atmospheric O_2. Since 1910 there has been no change in concentration of O_2 from 20.996 per cent by volume of dry air, at least it is so very small that it cannot be detected by available methods. Yet the amount of industrialization and the number of automobiles (which account for about 50 per cent of the oxygen consumed in urban areas) have undergone an enormous increase. Actually, simple arithmetic will show that if we were to burn all known fossil fuel reserves (coal, oil, natural gas), we would use up less then 3 per cent of the atmospheric O_2 supply. The CO_2 released would stimulate rates of photosynthesis and thereby increase oxygen production. Under unusual circumstances pollution may cut down oxygen supplies to disastrously low levels. In fresh-water rivers and lakes the use of O_2 by bacteria metabolizing wastes may reduce the O_2 concentration far below the survival limits for many fish and other aerobic organisms.

Fig. 24-12. A group of female aquanauts 15 meters beneath the Caribbean Sea near St. John in the Virgin Islands. The TEKTITE II program was designed by the National Aeronautics and Space Administration to test the behavior of small groups living and working in a stressful environment resembling future space missions. Aquanauts carried on studies of marine life while submerged. (Courtesy of the National Aeronautics and Space Administration.)

THE NITROGEN CYCLE

The primary source of nitrogen is the atmosphere. Although the air is roughly 80 per cent nitrogen, most organisms cannot utilize the gaseous form of this essential element because it is extremely inert. Lightning fixes nitrogen, that is, forces it into chemical combination with other elements. This fixed nitrogen is washed out of the air by rain and reaches the earth as ammonia, NH_3, or nitrate, NO_2^-. However, most nitrogen fixation is carried on by special nitrogen-fixing bacteria in the sea or the soil. The nodules on the roots of peas, beans, and other legumes are specialized structures inhabited by nitrogen-fixing bacteria and are the most effective agents known for this purpose. Between 1 and 6 lb of atmospheric nitrogen are fixed per acre per year, depending on the type of soil and the number of legumes. Certain species of soil bacteria and blue-green algae also are nitrogen fixers.

The nitrates, from bacterial or other sources, are absorbed by plants and converted into amino acids and proteins. Plant proteins are the ultimate source of all animal proteins. Nitrogen is returned to the soil or sea water by the bacteria of decay, which convert animal and plant proteins and urea to ammonia, or by animals which excrete ammonia instead of urea or uric acid. The ammonia is utilized by nitrite bacteria both in the soil and in the sea, and they convert it to nitrites (NO_2^-). These, in turn, are converted into nitrates by nitrate bacteria, after which nitrogen is again available for plants. These are the main features of the nitrogen cycle. Certain denitrifying bacteria utilize fixed nitrogen as a source of energy and release free nitrogen as waste. Animals become part of the nitrogen cycle when they consume the nitrogen-containing compounds of plants or other animals.

THE PHOSPHORUS CYCLE

Phosphorus is a rare element compared with nitrogen, but because it is involved in the energy transfer mechanisms within cells it is equally important. The phosphates dissolved in soil and in the sea are derived from erosion of phosphate-containing rocks. Phosphates are absorbed by plants and built into plant compounds that are in turn consumed by animals. The phosphorus is returned to the soil and the ocean by decay bacteria. It seems unlikely that the phosphate content of agricultural soil is being maintained. Soils in the Mississippi Valley that have been under cultivation for 50 years show a 36 per cent reduction of their phosphates, compared with virgin soil of the same type. Soil fertility can be maintained by the addition of phosphate fertilizers. Up to a point such application can promote crop yields. Too much zeal in the application of fertilizers can have undesirable effects as we have already indicated in our discussion of eutrophication of lakes. Generally, crops respond to added phosphate and other fertilizers in a predictable way that was summarized over a century ago by the German chemist Liebig in his **law of the minimum.** According to this widely applicable rule of biology, it is the essential element present in the least amount relative to an organism's needs that limits its growth. Thus, if an adequate amount of phosphorus (or any other element essential for plants) is present, adding more will not increase yields.

FOOD AND ENERGY CHAINS

When **food chains** or **webs** are actually studied, they usually turn out to be pyramids leading from the basic plant source to herbivores and then up through a series of carnivores of increasing size but decreasing numbers. Such pyramids are often referred to as **Eltonian pyramids,** after the first scientist to point out this relationship. The various levels within the Eltonian pyramid are sometimes referred to as **trophic levels.** The total mass of protoplasm, the so-called **biomass** tends to become less and less in each ascending level of the pyramid. Because animals live within the laws of conservation of matter and energy, regardless of how efficient any animal is, catching food and all the other activities of living require expenditures of energy. Actually the amount of energy represented at any one trophic level of such a pyramid is only about $1/10$ of that represented at the adjacent lower level. Not only is there a pyramid of biomass and of energy but there is also a pyramid of individuals. All three factors narrow sharply as one trophic level is placed on top of another. A disconcerting fact about environmental pollutants, whether insecticides or heavy metals, is that from their point of entry into the biosphere, they tend to become more concentrated at each successive trophic level.

A good example of the **pyramid of numbers** of individuals is seen in an acre of Kentucky bluegrass. Here we find about 6,000,000 plants, 700,000 herbivorous invertebrates, 350,000 carnivorous invertebrates (such as spiders, ants, and predatory beetles), and three carnivorous vertebrates (such as birds and moles). A square mile of Arizona rangeland which was studied supported 1 coyote, 2 hawks and 2 owls, 45 jackrabbits, 8,000 wood

and kangaroo rats, and 18,000 mice. This case illustrates another aspect of nutritional relationship between animals. When two species occupy the same nutritional level in a food-population pyramid, the species with the smaller body size will generally have the larger number of individuals, but will not necessarily have the larger biomass.

A certain Wisconsin lake can serve as a more or less typical example of a **pyramid of biomass.** There were, one summer, about 1,000 kg of plants (the primary producers) per hectare (1 hectare = 2.47 acres), 114 kg of herbivorous animals on the second trophic level, and 38 kg of carnivorous animals on the third major trophic level.

A bog lake in Minnesota can illustrate the **pyramid of energy** represented in the three major trophic levels. The level of **primary producers,** actually the photosynthetic primary receivers of energy, yielded a total of 70 cal from the plants of a certain unit of volume of bog; the second major trophic level, that of herbivores, yielded 7.0 cal for the same volume; and the third major level, that of the carnivores, yielded only 1.3 cal per unit of volume.

Although these pyramids of numbers, of biomass and of energy have been shown to exist, there are circumstances where apparent exceptions have been found. A good example is the open ocean. Here the primary producers, primary and secondary consumers and ultimate predators can be determined (see Fig. 24-13). But the food web can branch and become complex. Baleen whales, the largest mammals ever to appear on earth, feed on a diet that is very close to the primary producers. Killer whales, on the other hand, can be regarded as being at the top of the Eltonian pyramid because they eat mammals, which in turn eat fish that eat planktonic animals which consume the primary producers, the planktonic plants. These plants at the base of the pyramid would be expected to have tremendous biomass compared to other components of the food web. In fact they do not. This apparent paradox does not negate the concept of trophic levels. It merely indicates that we must be aware of another aspect of the food chain, that is the rate of turnover, the productivity, i.e. the rate of energy flow through the organisms constituting that level. A relatively small biomass of planktonic plants in the ocean implies a tremendous productivity, a fact that is readily confirmed when rates of replication and solar energy capture in photosynthesis are actually measured.

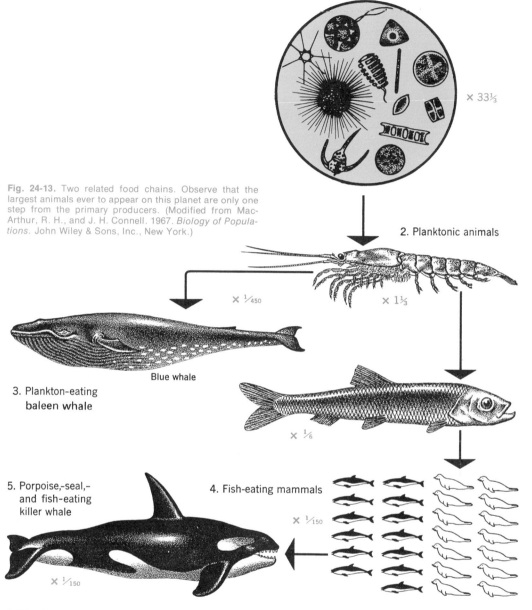

1. Microscopic planktonic plants

× 33⅓

Fig. 24-13. Two related food chains. Observe that the largest animals ever to appear on this planet are only one step from the primary producers. (Modified from MacArthur, R. H., and J. H. Connell. 1967. *Biology of Populations.* John Wiley & Sons, Inc., New York.)

2. Planktonic animals

× 1⅓

× ¹⁄₄₅₀

Blue whale

3. Plankton-eating **baleen whale**

× ⅙

5. Porpoise,-seal,- and fish-eating killer whale

4. Fish-eating mammals

× ¹⁄₁₅₀

× ¹⁄₁₅₀

Energy Flow and Productivity

MEASUREMENT OF ENERGY CONTENT

The amount of energy stored in a sample of coal or plants or animals can be measured in a **bomb calorimeter.** The plant or animal is homogenized and then completely dried; the resulting powder is weighed and placed within a metal container that can be immersed in water. The "bomb" is filled with oxygen, then immersed in a carefully measured amount of water at an accurately known temperature. The oxygen is ignited by an electric spark. The heat liberated by the resulting explosive total combustion of the dried material is absorbed by the water surrounding the metal container and as a result the temperature of the water is raised. Because it takes 1 calorie to raise the temperature of water 1°C, the number of calories present in a given weight of dried animal or plant can be calculated, once the weight of the water is known and its change in temperature is determined.

PRODUCTIVITY

An ecological question of paramount importance for mankind is how to determine the potential productivity of any particular type of environment (terrestrial, freshwater, or marine) and then to discover how the maximum productive potential can be attained. It is obvious from a consideration of the facts about ecological pyramids of numbers, biomass, and energy that the closer any species lives to the primary producers at the bottom of the pyramid, the greater its supply of matter and energy will be in any environment.

The actual productivity of an area of land or water in terms of the biomass of plants or animals it can or does produce turns out to be difficult and rather tricky to measure, except under rather special conditions. Four methods are in common use.

Methods for Determining Productivity

The **harvest method** is the most obvious and works well enough in certain simple situations. In a field where only one crop of the plant you happen to be interested in can be grown per year, 100 per cent of the crop can be harvested and its amount measured. The situation in the lake or forest or swamp, not to mention the ocean, is far different. To obtain a 100 per cent harvest is often extremely difficult. Fish in a lake may be killed with a chemical such as rotenone and those which float to the surface counted and weighed. Skin divers can attempt to make a census of the far larger numbers, perhaps the vast majority, which sink to the bottom. The activities involved in making such a census may profoundly disturb the water of the lake, the bodies of the dead fish will add nutrients, the oxygen and carbon dioxide content will be changed, so that a second census a year later may be taken under such different circumstances that it can hardly confirm the first.

An even greater difficulty lies in the fact that the mass of animals or plants existing at any one time is only a very rough indication of actual productivity and even less an indication of potential productivity. Two lakes or two parts of the ocean may contain the same biomass, but in one the standing population may have been reached only after a relatively long period of time and, if much were removed, would require a long time to regain its original size. In the other lake the turnover of population may be very rapid, or could and would become so if any significant number of animals or plants were removed. Hence, if the same number were harvested from each lake, it would require a far longer period of time for the population in the less productive lake to regain its original size than in the more productive one. How is one to know how often to harvest to get a true measure of productivity? How often to harvest raises the question, to be discussed in the following section on population dynamics, of the drastically different results of "harvesting" animals such as fish at different points on their population curves. In taking a census of plants in a given land area, one would expect the number of trees to remain constant but expect new tree seedlings and annual plants to germinate during each growing season.

It has been thought by some ecologists that differences in rate of turnover rather than differences in actual productivity can explain the ecological paradox that the biomass present in polar seas is so much greater than in the sunny tropical oceans where you might expect the greater productivity.

A second method of measuring productivity is to determine the rate of **oxygen production,** which is a measure of the amount of photosynthesis. This is more difficult than it sounds. Plants themselves as well as microorganisms and the animals in a lake use oxygen. Some ingenious workers have measured the rate at which oxygen is depleted from the hypolimnion during the summer. It will be recalled that during the summer there is no appreciable exchange between the water above the thermocline of a lake (where oxygen from the air and that produced by green plants in the water is available) and the deep water below the thermocline where there are no green plants and no access to atmospheric oxygen. The more productive the lake, the more organic debris will fall down into the hypolimnion, be metabolized by organisms there, and lead to exhaustion of the oxygen supply. This undoubtedly is true but only in certain special lakes.

The **disappearance of minerals** from a body of water has also been used as a measure of the rate of productivity. It has to be remembered that, in any closed body of water, the fauna and flora will reuse the minerals already present and that new supplies are being washed in by rivers and streams at various rates. The chief place where this method has been useful, and it is an important place, is in certain parts of the ocean. There are places where nitrogen and phosphorus accumulate in the water during the winter and are utilized by phytoplankton in the spring. Their rate of disappearance from the water can be determined.

Radioactive tracers, specifically phosphorus-32 which has a half-life of about 2 weeks and hence is fairly safe to use in the sense that it introduces no persistent radioactive contamination, have been introduced into ponds and their rates of incorporation into phytoplankton determined. This method seems to hold real promise for lake studies and possibly marine studies as well.

In summary, productivity depends on many very different kinds of factors. The amount of sunlight is, of course, basic. So also are temperature, moisture, sources of mineral nutrients, reproductive rates of the various kinds of organisms present and the position of the organisms on the population curve when harvesting is done. Easily the most important of all is the **genetic constitution** of the members of the population. This has been dramatically demonstrated in the case of such crops as hybrid corn and new varieties of wheat, but it is most certainly true of animals as well. Those who would cultivate the oceans must look at the genetic make-up of the plants and animals as well as at currents and temperatures.

The Dynamics of Populations

The study of populations has been a major concern of biologists every since Charles Darwin enunciated his theory of evolution based on natural selection. The tremendous reproductive potential of plants and animals is one of the most powerful of evolutionary forces. The study of populations is central to all the problems of ecology, and no part of ecology has more important practical aspects than a knowledge of the causes and characteristics of the rise and fall of populations. From a practical point of view, the problem is how to minimize undesirable populations and optimize desirable ones.

The major factors controlling population size can be more or less arbitrarily divided into three categories—**habitat, predation,** and **competition** within species. Some ecologists emphasize the role of population density on population growth and speak of density-dependent and density-independent factors. Others deny the existence of truly density-independent factors.

EFFECT OF HABITAT

The **habitat** includes both the non-living features, such as climate, periodicity and in-

tensity of sunlight, soil, and topographical features, as well as the resident plants and animals. Tree squirrels do not live on treeless plains. Often a whole ecological community of animals depends on a plant community. When the plant biome disappears, the animal community must also disappear. This happened in shallow waters along the Atlantic coast of the United States when eelgrass was killed by disease. The populations of scallops, certain polychaetes, jellyfish, and many other animals shrank almost to zero when the eelgrass meadows where these animals lived or bred disappeared.

EFFECT OF PREDATION

An important factor influencing population size is predation by carnivores and by parasitic plants or animals. It is difficult to determine accurately how important predation by carnivores or even by disease organisms really is. It is entirely possible that there are cases in nature, perhaps many cases, where the number of the individuals eaten is so small in proportion to the total population of the prey that predation makes no measurable difference. This seems highly probable when the predator population is limited by some factor other than the supply of prey. For example, ospreys live by catching fish out of the ocean. The supply of possible prey is therefore enormous, if not absolutely unlimited, but ospreys are held in check by the limited number of suitable nesting sites which must be isolated from the birds' own enemies and fairly close to the water. In some populations, such as those of fish in lakes, it has been shown that removing small fish will permit those left to grow faster to a larger size than would have been possible under the initial crowded conditions. Recent studies of several lakes in Oklahoma show that it is practically impossible to deplete a healthy population of fish by predation with a hook and line. Of course if the lake is fished with a seine, the entire population could be caught.

There are other well-established cases where predation has drastically reduced and virtually exterminated the prey. This can be rather easily demonstrated in the laboratory. If a ciliate called *Didinium,* which attacks and eats parameciums, is placed in a culture of its prey, the population of didiniums will increase and, after a time, the population of parameciums will decrease until final extinction. At this point, of course, the predators are faced with an irremediable famine and extinction also. If the fingerbowl world of parameciums and didiniums is complicated by the introduction of obstacles which will partially hide prey from predator, then irregular fluctuations of population abundance and scarcity will be observed. The more complex the environment, the longer the cycles continue. Clearly, the more complex situation more closely resembles the condition of nature.

The likelihood, however, is that in the vast majority of prey-predator relationships, a relatively stable steady state has been attained in the course of evolution. Were this not so, either the prey, the predator, or both would have become extinct. This is why the introduction of new predators produces such spectacular results. It is also why the elimination of the restraints imposed by predators may produce disastrous results for the prey.

Public attention is often focused on populations which fluctuate at more or less regular intervals. The classic case is that of the snowshoe hare and the Canadian lynx (Fig. 24-14). Records kept from about 1800 by the Hudson's Bay Company reveal oscillations with about a 10-year periodicity. In general, when hares are abundant so are the lynxes and, as would be expected, the population peaks for the lynxes tend to come slightly later than those for the hares. A number of such prey-predator cycles are known among lions and zebras, snowy owls and lemmings, and other animals.

It was long supposed that there was a causal relationship between the numbers of the prey and their food. Volterra and Lotka described these **prey-predator** cycles in mathematical terms; so they are commonly referred to as **Volterra-Lotka** cycles. The oscillations certainly exist, but there is now good reason to question whether in all cases the number of the predators merely rides up and down with the fluctuations in the numbers of the prey. For example, on Anticosti, a large island about 100 miles long situated in the Gulf of St. Lawrence, the population of snowshoe hares rises and falls as it does on the mainland of Canada, despite the fact that there are no lynxes at all on Anticosti. The fluctuations in the hares are perhaps due to several cooperating factors such as competition for food plants, increased incidence of disease with crowding, or the effects of fighting among themselves.

COMPETITION BETWEEN SPECIES

Many different species of animals and plants live in the same habitat—a spruce forest, a grassy meadow, or a pond. Do any two species occupy the same ecological niche within their habitat? Darwin suggested long ago that this was highly unlikely because either one or the other species would, in all probability, be at least slightly better adapted and hence, over a long period of time, able to outbreed the less well adapted and finally eliminate it. In more recent years J. Grinnell, on the basis of a long study of California birds, came to the conclusion that no two species did in fact occupy the same ecological niche. Within the following decade the same idea was supported by V. Volterra in Italy and G. F. Gause in Russia and came to be called the **principle of competitive exclusion.**

It is commonplace to find two or many species living together and apparently competing, i.e. occupying the same niche. Careful

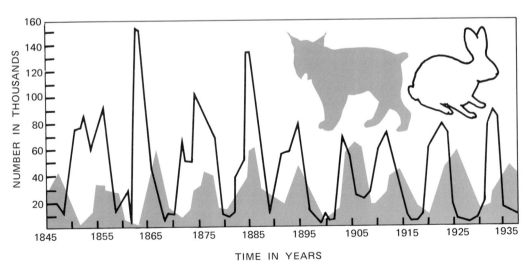

Fig. 24-14. Volterra-Lotka (prey-predator) cycles. The oscillating changes in population densities of lynx and snowshoe hares are classic examples of these cycles. (After MacLulich, D. A. 1937. *Fluctuations in the Numbers of Varying Hare (Lepus americanus).* University of Toronto Studies in Biology Series No. 43, University of Toronto Press, Toronto.)

scrutiny in every case so far has shown that the different species actually occupy different although sometimes slightly overlapping niches. A well-known case investigated by R. H. MacArthur is that of the four species of warblers which live during the nesting season in the spruce forests of the north. He found that each species had a different zone of the tree where it specialized in hunting for food, although there was a certain amount of overlap. One species hunted for insects almost exclusively among the fresh growth at the tips of branches near the top of the tree. Another species specialized in the foliage over a year old beneath the new spring growth, another specialized in hunting on the leafless branches close to the trunk and in the upper part of the tree, and another chose a similar zone but close to the ground. Not only is the location of food hunting different, but there certainly would also be a difference, though not an absolute one, between the insects and spiders available in these various regions of the trees.

One of the most spectacular examples of species occupying separate niches within the same area of the earth's surface is found in the intertidal zones at the edges of the oceans. Here there is a distinct zonation of the algae with the greens, browns, and reds located in that order from average high to low tide levels. Such zonation is a consequence of the adaptations of each species to specific conditions of light, exposure to drying, temperature, and osmotic fluctuations.

COMPETITION WITHIN SPECIES

In addition to the general habitat and predation, competition within a species can control population size. Students of evolution commonly believe that this type of competition is the most severe of all. Sparrows do not compete with ducks, much less with turtles or sunfish. The more different any two organisms, the less they compete. Nor do all seed-eating birds, for example, compete with all other seed-eaters. As anyone who has fed wild birds or has raised finches or canaries knows, birds have very strong food preferences, often based on bill size. Only two individuals of the same species will choose exactly the same seeds.

As a result of recent studies, intraspecific competition in house mice is better understood than in any other species. The method is to place two or three breeding pairs in rooms with nesting boxes and food. As long as there is plenty of food and enough nesting sites, the population increases with virtually no emigration from the room. When the population reaches a level where all the food is eaten up every day, then enough mice emigrate to keep the population stationary at this high level. The birth rate remains as before. It is a curious fact that the mice which emigrate represent a typical cross section of the colony, not just the young, the old, or the most vigorous, but some of both sexes and all ages except nurslings.

If there is no escape from the room when the population catches up with the food supply, the picture is very different. First, the nurslings die. Then the birth of new litters stops. The older mice gradually die off, presumably of old age. As this happens, the total number of mice in the colony falls, but the average weight of the individuals gradually increases so that the total weight of the colony remains constant. The amount of food given per day is adequate for a certain biomass of mice, irrespective of whether it is divided among a larger number of thin mice or a smaller number of fatter ones.

What mechanism brings reproduction to a standstill in such a mouse colony? Apparently a very important factor is the emotional effect of the continual fighting caused by the food crisis. The adrenal glands become noticeably enlarged, and the reproductive endocrines are in some way thrown off balance. Thus there is in these mice a kind of **negative feedback** mechanism which sets limits to the population when it begins to press against the food supply. Signs of adrenal damage have also been found in crowded deer populations as well as in woodchucks and other species. Negative feedback is the kind usually found among living organisms. Many actions produce some counteraction which checks the results of the first and restores the organism to equilibrium.

Positive, or "runaway," **feedbacks** have also been reported in animals. It is as though a thermostat, once the temperature had risen to a given point, instead of turning the heat down, turned on more heat, and then when the temperature climbed still higher, turned on the heat still more. Positive feedbacks soon lead to disaster. For example, in several sea birds such as gulls and guillemots, the smaller the nesting colony, the fewer eggs hatch and the fewer young are raised. This leads to a vicious downward spiral. It is as though every time the temperature fell, the thermostat turned the heat down even more.

POPULATION GROWTH CURVES

The free growth of a population in an open environment can be regarded as a positive feedback situation. The more individuals, the faster the population grows, and the faster the population grows, the more individuals, and so on. As Malthus pointed out in 1798, populations tend to grow exponentially, like money at compound interest. As a population grows, it tends to increase faster and faster, and then in a finite environment (and all environments, whether test tubes or continents, are finite) growth begins to slow down more and more. The result is an **S-shaped** or **logistic curve** (Fig. 24-15).

This curve of population growth has the widest possible application, for the very simple but basic reason that all animals reproduce in a geometrical rather than an arithmetical series. In other words, the new individuals become part of the reproductive capital and add their own share of new individuals. The curves illustrated here are taken from studies of yeast cells and fruit flies, but they hold similarly for mice and deer and people. More precisely, the increase in population is proportional to N, the total number of individuals in the population. The actual rate of increase depends on the reproductive ability, r, of the individuals. This factor is very large for oysters, moderate for mice, and smaller for elephants. The rate of increase is not only determined by $r \times N$, but is correlated with how close N is to the upper possible limit in any finite environment—test tube, milk bottle, or geographical area. This factor is $(K - N)/K$, where K is the maximum possible population. In the beginning, when N is very small, this factor is approximately equal to one. As N increases, this factor becomes smaller and smaller until $N = K$, then $(K - N)/K = 0$ and increase in population size ceases. In terms of calculus, the growth rate at any point can be expressed as follows:

$$dN/dt = rN(K - N)/K$$

Practical Applications of Growth Curves

One important practical application of the facts expressed by this formula concerns the best time to harvest from a population of fish or deer or other organisms. Removing a certain number of individuals when a population is close to the lower end of its curve will produce a marked depressing effect. Removing the same number at the upper end of the curve will either have no appreciable effect or will allow those left to attain a somewhat greater size. Because of this principle emigration from overpopulated countries furnishes no permanent cure for the overcrowding.

Not only can much larger harvests be reaped when the harvesting is done close to the top of the population curve, but a harvest of any given size can be taken with far less effort. This has been demonstrated in the case of the codfish. If one population is about three times as large as another, it will require only about one-third the fishing effort (time towing nets, etc.) to harvest the same amount of fish from the large as from the small population (Fig. 24-16).

From the human point of view the objective is often harvests in reverse, the diminution or elimination of an undesirable population such as insect pests in a garden or greenhouse. Here the strategy must be to press hard with the control measures when the population curve is at its lowest, if permanent results are to be achieved.

Three types of methods of pest control deserve special mention. Perhaps the most ingenious is the use of **sterilized males**. Unfortunately, although this method has been a spectacular success, it is expensive and is effective only against species in which the females mate just once. The U.S. Department of Agriculture used this method, for example, against the highly destructive screwworm fly which produces maggots that attack the flesh of cattle. In one campaign they raised and sterilized by X rays 100 million male flies per week. These were set loose from planes flying

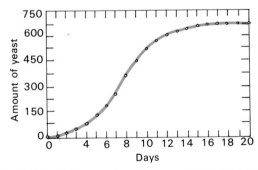
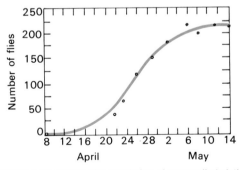

Fig. 24-15. Logistic curve of population growth in a finite environment. Left, growth of a population of yeast cells (relative number of cells measured as optical density of suspension). Right, growth of a population of fruit flies. (From Moment, G. B., and H. M. Habermann. 1973. *Biology: A Full Spectrum.* The Williams & Wilkins Co., Baltimore.)

Harvest 90% per year and
natural mortality 5% per year

Harvest 30% per year and
natural mortality 5% per year

Standing Population

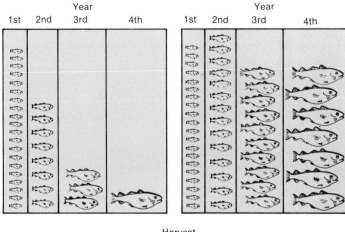

Harvest

Relative wt. (units)

| 1/4 | 1 | 3 | 6 |

Relative wt. (units)

| 1/4 | 1 | 3 | 6 |

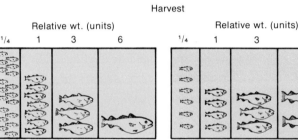

Weight of catch 26¹/₂ units
(approx. 90% of each group)

Weight of catch 26¹/₂ units
(approx. 30% of each group)

Fig. 24-16. The principle of maximal yield and minimal effort. Upper symbols represent the populations of 1-, 2-, 3-, and 4-yr-old fish in stock populations. Lower symbols represent the fish caught. The same number of pounds of fish are harvested in both cases, but 3 times as many hours are required to catch this weight of fish from the smaller population and more of the fish are small in size. (After T. Park in Allee, W. C., A. E. Emerson, O. Park, T. Park, and K. P. Schmidt. 1949. *Principles of Animal Ecology.* W. B. Saunders Co., Philadelphia.)

over the infested country. Within 2 years the screwworm fly was completely eliminated over wide areas. Over $200 million worth of damage to cattle is reliably estimated to have been averted within a 3-year period. The recent resurgence of the screwworm appears to be due to a change in cattle breeding by which the highly vulnerable calves are born throughout the summer instead of only in the colder months.

A second method is a very neat combination of chemical and biological techniques. **Sex attractants** have been extracted from female insects, analyzed chemically, and their molecular structures determined. Such compounds have been manufactured commercially and used to lure male insects into traps or to poisoned bait, a far safer method than spraying poisons on whole plants or whole forests. The sex attractant of the female gypsy moth, a species highly destructive to both coniferous and deciduous forests and to fruit trees, has been identified and is sold under the name of Gyplure. It will be of great

interest to see how successful its use will be. The gypsy moth was introduced into Massachusetts in about 1868 by an entomologist said to be an amateur. In any case, he "didn't know it was loaded." Millions of dollars of destruction has been the result and more millions have been spent spraying orchards with lead arsenate, introducing a small beetle which eats the gypsy moth larvae, and spraying forests by plane with DDT. Perhaps sex attractants will be the answer.

A third method of pest control is the introduction of **bacterial** or **viral diseases** in order to devastate an unwanted population. This carries obvious dangers. One must be certain that the disease will not attack other, desirable species. This method has been very successful against the great rabbit plagues of Australia where all other known methods had failed. The pathogen used was the myxomatosis virus which is transmitted from one rabbit to another by mosquitos. Two bacterial diseases have been very successful in controlling and all but eradicating the Japanese beetle in the United States after the various insects which parasitize this beetle in Japan had been introduced to no avail.

The older chemical pesticides such as DDT and various neurological and enzymatic poisons can be very effective in controlling insects, rodents, and sometimes even nematodes. But they have very serious and undesirable side effects. Most of these poisons are very long-lasting, if not indestructible. They get into food chains and are transmitted from plants to herbivores and then up the chain of carnivores. The bald eagle has picked up enough of these poisons so that its eggs often fail to hatch even though the adults are not killed. Moreover, these poisons, when used over a forest or a swamp, indiscriminately kill many if not all, wildlife species. DDT in the ocean could eventually depress the oxygen-forming capacity of oceanic plants with disastrous results to the atmosphere. Very promising results have been achieved in developing compounds that are toxic to insects but will disintegrate spontaneously on contact with water and air.

MATHEMATICAL MODELS

The development of computers has provided a means for predicting the future behavior of complex systems more accurately than in the past. The business of "futureology" has at its basis the extrapolation of past and current trends into the future. For example, demographers, who study trends of population size and distribution, can predict more or less accurately the future population of a given city or geographical area if they know the past and present size of the population, the rates of birth and death, and the rates of immigration and emmigration. Assuming no drastic changes in these factors and the absence of natural disasters it is possible to predict what size a given population will be 5 or 10 years hence.

As any demographer will agree, however, the business of extrapolation is a tricky one because human behavior is exceedingly unpredictable. To be convinced, one has only to examine trends of birth rates in the United States over the past 3 decades. Yet it is becoming increasingly important to be able to predict the future, or more particularly the consequences of specific human activities. In an overcrowded world of limited resources it becomes necessary to ask and find answers to questions such as: How much grain can the prairies of the western United States produce each year and how much can be exported? How much electrical power will be needed in the area served by a utility company 5 or 10 years in the future? What are the consequences in terms of air and water pollution of doubling the capacity of a steel mill? How will the application of automobile pollution controls affect air quality in the Los Angeles basin? All of these questions involve complex circumstances influenced by economic, political, sociological, and perhaps even philosophical as well as scientific factors; and a change in any single factor can affect all the others. To make predictions concerning such complex circumstances requires the use of computers.

Some startling predictions about the prospects of human survival have been made and published by a group of scientists and economists known as the Club of Rome. They have charted future trends of industrial activity, availability of resources, levels of pollution, and numbers of people. According to their predictions (based on computer generated mathematical models), global collapse seemed inevitable within the next 100–200 years. Recently some errors in the data that were fed into the computer programs were recognized and as a result of these modifications of the initial assumptions, doomsday has been postponed by some years. There has been considerable controversy concerning the prognostications of the Club of Rome and the ways in which they should be used to avoid sure disaster. Should industrial growth be restricted to avert ever increasing levels of pollution? Should there be worldwide enforcement of pollution controls and, if so, at whose expense? Certainly the option of returning the world to a simpler agrarian way of life is no longer open to us. The future involves hard choices and it is essential that decisions be based on the facts and reasonable extrapolations from them rather than ignorance and wishful thinking. Mathematical models and computer simulation make it possible to make predictions from recognized trends or to select from a large number of options the "best" possible course of action. How to decide what is "best" of course depends on the values of those making the decisions. "Best" might be defined as most beneficial to the greatest numbers, the least harmful to the greatest numbers, or the most effective relative to cost. It is to aspects of our planetary crisis, which mathematical modeling may help ressolve, that we now turn.

Our Planetary Crisis—Is There a Way Out?

Except for nuclear explosions which could make the human race itself extinct, no modern problem even approaches the population explosion in magnitude. What contribution can biologists make toward a solution of this complex crisis of pollution, overcrowding, malnutrition, and death? Probably the most useful and effective thing that biologists can do is to try to see all the relevant facts as clearly as possible and to present them fully and honestly regardless of whether they agree with established ideas or disagree with the desires of any group. The ultimate worth of science lies right here. It is an attempt to see the world as it is.

Why the population explosion? The answer is easy. Advances in the biomedical sciences have resulted in a very large measure of death control by the conquest of many diseases. What blocks a solution? The answers to that question extend far beyond the dimensions of biological science. However, several fixed points and guidelines are visible to help navigate these seas so full of human tragedy.

For over a century biologists have known that in a finite environment, the number of plants and animals is limited by the inescapable laws of mathematics. Every year every female oyster in the Chesapeake Bay lays enough eggs to supply an entire year's harvest if all grow to maturity. If all the eggs of all the oysters were to grow to maturity and reproduce, it would take less than a decade for the bay to be transformed into a solid mass of oysters. Impossible you say. Yes, because the environment with its food for oysters and its ability to carry away metabolic wastes is limited. Charles Darwin made the same point with African elephants, the slowest breeding animals.

Our spaceship Earth is limited in area and in resources, but those limits do allow a certain amount of choice. What is regarded as under- or overpopulation depends in part on philosophical outlook and life style. The frontiersman Daniel Boone found the scattered cabins of eastern Kentucky overcrowded and moved further west. The writer Chekhov, dissatisfied with what he called "the idiocy of village life" moved into the big city. Likewise the quality of life is a very slippery and subjective matter once the subsistence level is safely passed. Those who rejoice in a simple life in tune with nature where they can raise their own food and hand craft their own goods without the tinsel of the neighborhood cinema, find much of the United States just as underpopulated as the slums of Los Angeles and Rio de Janeiro are overpopulated. Nor should anyone imagine that crowding inevitably causes criminal violence by some scientific law. Too many people forget the great cruelty and violence of the sparsely

populated Old West. On the other hand, visitors recently returned from Singapore report that it is apparently true that the crime level is very low despite great crowding. Suggested explanations include the large percentage of Buddhists, the low economic expectations of most people combined with slightly more prosperous conditions, and even police effectiveness. The point is that anyone must be very cautious in making dogmatic assertions about the effects of population density on other parameters of human life. Remember the negligible results of crowding found by Southwick among monkeys in India (Chapter 21). France is about 40 times as densely populated as the United States. Is life there only $1/40$ as good?

But the inexorable limits remain. The world is faced with limited options. Organic gardening is fine as far as it goes but there is no reason to believe that the world can be fed that way, as helpful as it would be in saving resources and diminishing pollution. The agriculture of Bangladesh is largely organic and the result is a nightmare of famine. What are our options?

A study by Dennis Meadows and others at the Massachusetts Institute of Technology and by the Population Research Bureau in Washington, D.C. indicates three major possibilities for the future. One is the "**Population Crash Curve**" based on "letting nature take its course." Nature's course can be seen in what happens to the lemmings and the snowshoe hares where the population increases to a peak about every ten years and then crashes to a low level (carrying with it their chief predators the lynx and the snowy owl). After a low the cycle starts over (see Fig. 24-14). No green revolution offers a cure for this because the population catches up with its food supply. At present the human population doubles 3 times every century, so if, as predicted, there are 6 billion people in the year 2000, and present rates of population growth continue, there would be 48 billion in 2100! (see Fig. 24-17).

A second possibility is a "**Gradual Transition to Zero Population Growth.**" This seems unlikely partly because there is such powerful momentum in population growth in the undeveloped countries where about half the population is under 15 years of age.

The most probable course seems to be a "**Modified Irish Curve,**" one which undergoes a tragic but not completely catastrophic crash followed by a series of undulations and economic and social readjustments until finally leveling off. It will be recalled that in the 1840's there was a severe famine in Ireland caused by a failure in the potato crop. More

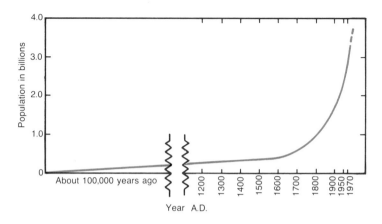

Fig. 24-17. Curve of human population growth. (After Turk, A., J. Turk, and J. T. Wittes. 1972. *Ecology, Pollution, Environment.* W. B. Saunders Co., Philadelphia.)

than 1 million people died, some 2 million emigrated, and the remainder responded by changing their agriculture from a single crop system to a diversified one and by holding population growth down by late marriage, nonmarriage, or small families.

There are several reasons for believing that the leveling off will be somewhat bumpy. Rapid slowing of population growth brings complications. Both West Germany and France import workers from Turkey, Poland, and elsewhere. The proportion of the population who are aged and need support from social security programs or other sources increases at the expense of the younger sector, imposing a heavier and heavier burden on them. For both these and perhaps other reasons a recent premier of Japan has called for a slight rise in the birth rate. This may not be as bad as it seems because the birth rate of Japan is not only the lowest in Asia but is almost as low as that of the United States, which is barely at the replacement level.

National, racial, and religious ambitions often work against population limitation. The government of Ceylon, for example, has withdrawn its support from a successful birth control program because of the fear that the native Ceylonese would be overwhelmed by the rapidly increasing Tamils who were imported as cheap labor and were not interested in birth control. Dictators and power-hungry organizations of all types characteristically attempt to promote population growth among those they control.

There is room for guarded but real optimism. Not only have the birth rates of the United States and Japan fallen to close to the replacement level but the governments of both China and India are supporting birth control programs. Latin America now has the highest birth rate in the world, but reports from workers in family planning clinics there indicate that Latin American women are desperately anxious to be able to limit the size of their families.

Just as important, it is not realistic to expect the developed countries to continue to feed nations like Bangladesh or others indefinitely if they permit their populations to grow without limit. The presently well-fed nations could not continue to feed others which continued to grow indefinitely even if they tried. More important, it is no genuine or thoughtful compassion to continue decade after decade giving food to any nation that does not curb its own population because the end result will be catastrophe for everyone.

An inevitable conclusion is that because modern medicine has given a large measure of death control to the whole world, birth control has become a necessity. Once the population explosion is brought under control, the pressures for ever-expanding industrialization will be somewhat lessened and there will be more energy and resources to attack the problems of pollution which presently hang over us all like the smog over our great cities.

In addition to problems linked to the absolute numbers of people, some present-day crisis conditions are related to patterns of distribution of people. Throughout the world, this century has been a time of urbanization with an accelerating migration of people from farm to city. Over 70 per cent of the population of the United States now live in less than 2 per cent of its area. This has accentuated urban problems that have existed for centuries: atmospheric pollution, waste disposal, congested transportation systems, shortages of housing, crime, and other forms of social disorganization. The decay of the central city and the flight of the middle class to the suburbs is only a minor perturbation in a massive movement of humans into urban areas. It may

Fig. 24-18. Smog over Los Angeles. A warning that will be heeded or a vision of the future? (Courtesy of the Los Angeles County Air Pollution Control District.)

be that cities have become too large to be governable according to existing political systems. Whether these problems can best be solved by newly developed regional systems of government, by limiting the size of cities and building new ones, or other means, remains to be seen.

The problems created by the accelerating growth of human populations and the increasing disruption of our planetary ecosystem are based in biological facts but extend far beyond them into the areas of economics, political theories, and social philosophies. Who is to say what the optimal population is and when zero growth is desirable? Is it the same in all parts of the world and in all cultures? What life styles should be encouraged or even enforced? Obviously biologists alone cannot provide the answers but it is their urgent task to proclaim loud and clear what the dangers and the opportunities are. Biologists should discover and make available to all people the knowledge necessary to make possible humane and wise decisions. Only then can we continue to be a successful form of life.

USEFUL REFERENCES

Bajema, C. J. 1971. Genetic implications of population control. *Bioscience* 21:71–75. (Discussion of an una-voidable consequence of population control.)

Berg, A. 1975. To save the world from lifeboats. *Nat. Hist.* 84:4–6. (Author was on the staff of the World Bank and famine relief coordinator in India.)

Canby, T. Y. 1975. Can the world feed its people? *National Geographic* 148.2–31. (An informative and handsomely illustrated article on the world food crisis.)

Ehrlich, P. R., and A. H. Ehrlich. 1972. *Population, Resources* and *Environment,* 2nd ed. W. H. Freeman, San Francisco.

Ehrlich, P. R., J. P. Holdren, and R. W. Holm, eds. 1971. *Man and the Ecosphere.* W. H. Freeman, San Francisco. (Readings from the *Scientific American.*)

Handler, P. 1975. On the state of man. *Bioscience* 25:425–432. (A reasoned and reasonably optimistic assessment by a President of the National Academy of Sciences.)

Hardin, G. 1972. *Exploring New Ethics for Survival.* The Viking Press, New York.

Hardin, G. 1974. Living on a lifeboat. *Bioscience* 24:561–568. (Some unpleasant facts about the consequences of trying to feed the world.)

McLaren, I. A., ed. 1971. *Natural Regulation of Animal Populations.* Aldine Publishing Co., Chicago. (Thought-provoking.)

National Research Council Committee on World Food, Health and Population. 1975. *Population and Food: Crucial Issues.* National Academy of Sciences, Washington, D.C.

Odum, E. P. 1971. *Fundamentals of Ecology.* 3rd ed. W. B. Saunders Co., Philadelphia.

Population Reference Bureau. 1971. Man's population predicament. *Pop. Bull.* 27:2–6.

Southwick, C. H. 1976. *Ecology and the Quality of Our Environment,* 2nd ed. D. Van Nostrand, New York.

Ward, B., and R. Dubos. 1972. *Only One Earth, the Care and Maintenance of a Small Planet.* W. W. Norton and Co., Inc. New York.

Index

Page numbers in **boldface** in the index indicate the location of definitions or major discussions of terms and concepts. An *f* after a page number indicates a figure or a table.

Abeles, F., 149, 290
ABO blood groups, inheritance, 75
Abscisic acid (dormin), 279, 291
    antagonism with cytokinin, 291
    antagonism with gibberellin, 291
    induction of dormancy, 291
    inhibition of germination, 291
    structure, 291*f*
Abscission of leaves, 149
    in explants of cotton, 149*f*
Abscission zone, 290*f*
Absorbance (optical density), 184
Absorption, intestinal, 229*f*
Absorption spectrum, **137**
    carotenoids, 284*f*
    chlorophyll, 185*f*
    cytochrome *c*, 212*f*
Abyssal regions, of oceans, **393**
Acetabularia, 139, 140*f*
    grafting experiments, 140*f*
Acetic acid, 41, 48, 353
Acetylcholine, **307, 310,** 311*f*
Acetylcholinesterase, 301
Acetyl-CoA (coenzyme A), 230, 269
Acid-loving plants, 135
Acidity, 26
Acidophils
    blood, 242*f*, 243
    pituitary, 275
Acquired characteristics, 71
Acrosome
    reaction, 105
    sperm, 101, 102*f*
ACTH (adrenocorticotropic hormone), 273, 274
Actin, 299
Actinomycin D, 144, 270, 291
Action spectrum, **137**
    chlorophyll synthesis, 137*f*
    light-dependent seed germination, 137*f*
    photosynthesis, 137*f*, 184*f*, 185*f*
    photosystems I and II, 190*f*
    phototropism, 137*f*
Active transport, **26***f*
    intestinal, 229
    renal, 261
Actomyosin, 299
Adams' apple, 295, 296*f*
Adaptations, reproductive, 151
Addison, Thomas, 273
Adenine, 81, 82*f*, 259*f*
Adenosine 5' monophosphate, 48*f*
Adenosine triphosphate (ATP), **19**
Adenyl cyclase, 269, 270
ADH (antidiuretic hormone), 261
Adrenal glands, 272–273, 274*f*
    cortex, 272
    medulla, 272

Adrenaline (epinephrine), 5, 36, 43, 307
Adrenergic nerves, **307**
Adrenocorticotropic hormone (ACTH), 273
Adventitious roots, 166
    auxin promotion of, 282
Aerobe, **18**
Afferent sensory neuron, 303, 305*f*
African sleeping sickness, 372, 374
    transmission of, 374
African violets, propagation of, 167
Agave (century plant), 174
Agglutination of blood cells, 75
Aging
    in plants, 147–149
    life span, 123
Aging-ethylene hypothesis, 290
Aggression, 335–336
Agonistic behavior, **335**
Agassiz, L., 154
Ainu, 78
Air pollutants, 135
    effects on plants, 135*f*
Akaryons, 17
Alanine, transfer RNA for, 86*f*
Albinism, 95*f*
Aldehyde group, 38
Aleurone layer, **286,** 287*f*
Alfalfa, pollination by honey bee, 176*f*
Algal "blooms," 224, 391
Alkalinity, 26, 27
Alkapton (homogentisic acid), 88, 95*f*
Alkaptonuria, 88, 95*f*
All-or-none effects, 307
Allantois, 120*f*
Allard, H. A., 138, 152, 171*f*, 172
Alleles, **58**
Allergy, 263
Alpha cells, 275
Alpine zone, **385**
Alternation of generations, 125
Altricial chicks, 154
Amino acids, **42**
    "essential," 232
    fate, 230*f*
    sequences, 365, 366
    structure of, 43*f*
Amino group, 38
Amino sugar, 39
Aminobenzoic acid, 234
Amniocentesis, **97,** 97*f*
Amish, 347, 348*f*
Amnion, 113, 115*f*, 120*f*
Amniote, 120*f*
AMO-1618, 287

Amoebic dysentery, 374
Amplification, neural, 308
α-Amylase, 286
Anabolic hormones, 274, **278**
Anabolism, **47**
Anerobe, **18**
Analogy, **349**
Anaphase, mitotic, **29,** 30*f*
Anatomical temperature controls, **263**
Ancon sheep, 70
Anderson, J., 193
Androgens, **157, 273,** 274
Androsterone, **274**
Anemias, 117
Ångstrom unit (Å), **16, 136**
Animal pole, egg, 102
Animal societies, 338
Anisotropic band, 298*f*, **299,** 300*f*
Annual plants, **174**
Annual rings, 129, 129*f*, 355
Anopheles mosquito, 376
Antagonistic muscles, **297**
Antenna pigments, in photosynthesis, 189, 189*f*
Anterior horn, neural, 305
Anterior lobe cells, pituitary, 275
Anterior lobe, pituitary, 274
Anther, 174, 175*f*
Antheridia, **125**
    fern, 170*f*
Antheridium-inducing factor in ferns, 170
Anthers, haploid tobacco from, 145, 145*f*
Anthocyanin, 27
    synthesis, 138
Anthracene acetic acid, 284*f*
Anthropomorphism, 321, 338
Antibiotics, 92, **235**
Antibody, 75, **264**
Anticholinesterases, 310
Anticoagulant, 295
Antidiuretic hormone, 261, 274
Antigen, **264**
Antigibberellins, 287
Antimalarial compounds, 376
Antimetabolites, 235
    and memory, 317
Antisenescence effects of cytokinins, 289
Ants
    formicine behavior, 336
    myrmicine behavior, 336
    winged, instinct, 323
Aortic arches, 114*f*, 239
Aphids, 106
    feeding on phloem cell, 254*f*

Apical dominance, **282**
Apical meristems, 128, 131
   and aging in plants, 147
   root, 131f
   shoot, 131f
Apoenzyme, **204**, **234**
Appetitive behavior, 316–317, **326**
Aquanauts, 397f
Archegonia, **125**
Archenteron, **111**
Archimedes, 227
Aristotle, 55, 154
Arnold, W., 188
Arteries, vertebrate, 239, 240f, 241f
Arctic creep, 393
Arctic zones, **385**
Ascaris, 379
Ascidian, 110
Ascorbic acid, 233
Asexual reproduction in plants, 164–
   168
Asexual spores, 165
   conidia of *Neurospora*, 88, 89f
Askenasy, E., 253
Aspergillus, conidia, 166f
Association neuron, **303**
Aster, mitotic, 30f
Asymmetry, molecular, 40, 40f
Atabrine, 376
Atmosphere before life, 351
Atmospheric gases, 134, 388
ATP (adenosine triphosphate), **19,
   21**
   muscle, 299, 300f
   synthesis
      oxidative phosphorylation, 212–
         214
      photosynthetic   phosphoryla-
         tion, 196
Atrium, heart, 140, 141f
Atropine, 311f
Australopithecies, 359
Autografts, **264**
Autonomic nervous system 305,
   306f, 307
Autosomal linkage, 66
Autotroph theory, 351, 353
Auxin, 279, 280–285, 281f
   and adventitious roots, 282
   and apical dominance, 282
   and tropisms, 280–282
   bioassay, 281
   chemical specificity of, 284
   concentration effects, 282f
   structure, 281f
   tracer studies, 283
Auxin to cytokinin ratio, 288
   effects on plant tissue cultures,
      289f
   in control of apical dominance,
      289
Avery, O., 92
Axil of leaf, 128
Axon, **302**, 303f
Azotobacter, and nitrogen fixation,
   224

Baboon
   parental care, 154
   reproductive cycle control, 158
   troop, 335f
Backbone, 295f
Bacteria, generation time of, **93**
Bacteriophages, 93
Bailey, L., 56
Bakanae (foolish seedling) disease

   of rice, 285
Balbiani, E., 68
Bands, muscle, 300
Barley seeds
   effect of gibberellin, 287f
   structure, 287f
Barnes, R., 234
Barr bodies, 76, 77f
Barren women, 159
Basal ganglia, 313
Basal metabolic rate (BMR), **215**
Basal nuclei, 313
Basophils, 275
Bayliss, W. 268
Bacteriophage, 33, 34f
B-cells, 265, 266f
Beadle, G., 88
Beagle, H. M. S., 343
   voyage of, 345f
Beal, W., 72
Bean, protein, 232
Bears, 5
Bee(s) (*see* Honeybees)
Bee orchid (*Ophrys apifera*), 176f
Beefalo, 3f
Beer-Lambert law, **137**
Beerman, W., 117
Begonias, propagation of, 167
Behavior, 321–329
   embryonic, 122
   gene-dependent, 324
   hormones, 332
   temperature control, **263**
Bell, C., 305
Bell-Magendie law, 305
Benguela current, 393
Benthos, **394**
Benzene, 38, 39f
6-Benzylaminopurine, 288f, 289
Bergson, H., 108
Beriberi, 233f
Bernard, C., 50
Bernstein, J., 308
Beta cells, 275
Biddulph, O., 255
Biddulph, S., 255
Biennial plants, 134, **174**
Bile
   duct, 228
   pigments, 260
Biharziasis, 4, 371, 337–378
Biloxi soybean, 171, 172
Binary fission, **165**
Bindweed (*Convolvulus*), 373
Binomial system, 364, 365
Binomial theorem, 60
Bioassay for auxin, 281
Biochemistry of smell, 318
Biomass, **398**
Biomes, **385**
   major land biomes, 385–387
   minor land biomes, 387
   vertical and horizontal corre-
      spondence, 386f
Biosphere, **383**
Biotin, 234
Bird, 153f, 322f
   migration, 333
   song, 152, 327
   testosterone, 332
Birth
   control, 408
   rates, human, 408
   stages, 160, 161f
Bison, 3
Blaauw, A., 281
Bladder, urinary, 157f

Black death, 4, 368
Blackfellows, 78
Blackman, F., 185, 186
Blackman reaction, 185
Blade, of leaf, 128
Blastocoel, 111, 116f
Blastocyst, 111
Blastoderm, 111f
Blastomere, **104**
Blastopore, 111f, 112f
Blastula, 101f, 112f
"Blind staggers," 220
Blood
   cells, 242f
   fetal, 120
   groups, human, 74–76
      A, 159
      ABO, 75, 75f
      MN, 75
      racial differences, 76
      Rh, 75
   osmotic pressure, 259
   pH, 259
   platelets, 242f, 243
   transfusions, 76
Blooms, of algae, 224, 391
Blue Andalusian fowl, 57, 57f, 58f
Blue-green algae, 16, 19f
Blushing, neural control, 305
Boardman, N., 193
Body temperature controls, **263**
Bomb calorimeter, 231, **399**
Bonds, 38, 44, 82
   sulfhydryl, 85, 269
Bone formation, 297
Bonner, J., 143
Boone, D., 407
Borlaug, N., 235
Boron, in plant nutrition, 219, 223f
Borthwick, H., 138, 171f
Bottleneck phenomenon, 347
Boveri, T., 110
Bowman, W., 260
Bowman's capsule, 260, 251f
Boyle, R., 47, 216
Boysen-Jensen, P., 281
Brachiation, 359
Brachycardia, 242
Brain
   development. 121
   functional centers
      aggression, 316
      appetite, 315
      breathing, 315
      courtship, 316, 317f
      motor, 314, 316f
      pleasure, 316
      satiety, 315
      sensory, 314, 316f
      sleep, 314
      speech, 314
   human, 304f
   organization, 313
   size, 79
      and intelligence, 312, 313
Brazilian wild bees, 363
Breathing, 214–216
   center, 216
   mechanics of, 215
   regulation of, 216
Briggs, R. W., 110
Bromodeoxyuridine (BrdU), 84
Bronchus, 215
Brown, F., 333
Brown, R., 13
Bryophyllum, photoperiodic control
   of reproduction in, 167, 172

Bubonic plague, 368
Buchner, E., 50, 204
Budding, **165**
in yeast, 165f
Buds, 128, 128f
auxin concentration effects, 282f
Buffalo, 3, 4f
Bulbils, 167
Bulbs, **166**
Bundle sheath, **199**
Bursa of Fabricius, 265
"Bush disease" in sheep, 37
Butler, W., 138
Butyric acid, 41f

C_3 and C_4 plants, 199
leaf structure, 200f
photosynthetic pathway, 201f
photosynthetic rates, 202f
Cabbage, treated with gibberellin, 286f
Calciferol, 234
Calcitonin, 271
Calcium
bone, 297
in plant nutrition, 219, 223f
metabolism, 271
Callus, **140**, 288
Calorie, **47**, 230–231
Calvin cycle, 197–200, 199f
Calorimeter, 214
Calvin, M., 197, 197f
Cambrian period, 356–357
Canadian lynx, population cycles, 402f
Canary grass, 281
Canary, learning, 330f
Cancers, possible role of viruses, 94
Cannibalism, 360
Capillaries, 239
Capsid, **33**, 34f
Capsomere, **33**, 34f
Carbohydrates, 39–41
in diet, 230
metabolic pathways of, 49
Carbon
atom, structure of, 38f
cycle, **396**
of photosynthesis, 197–200, 200f
dating, 343
in plant nutrition, 218
Carbon-14 tracer in photosynthesis, 197
Carbon dioxide
and photosynthesis, 187–188, 187f
concentration in atmosphere, 187
stimulation of breathing, 216
Carbon monoxide inhibition of cytochrome, 210
Carbon tetrachloride, 38, 38f
Carbonic anhydrase, 243
Carboniferous era, 187
Carboxyl group, 38
Cardiac muscle, 298, 299
Cardiology, 241
Carmichael, L., 122, 329
Carnivore, **227**
β-Carotene, absorption spectrum, 284f
Carotenoids, 185, 195
absorption spectra, 284f
as accessory pigments, 195
possible involvement in phototropism, 283
protective function of, 195
Carotid arteries, 239, 240f, 241f

Carotid bodies, 216
Carpals, 295f, 296
Carrier, genetic, **58**, 65
Carrot cell cultures, 141f, 142f
plants from, 141f, 142f
Cartier, J., 233
Castle, W., 71
Castration, 268
Cat, skull, 296f
Catabolism, **47**
Catecholamines, 272
Catesby, M., 365
Cathey, H., 169f
Caul, 120
Cave men, 359–361
CCC, 287
Cedar-apple rust, 372
Cell
division, 28–32
G_1, S_1, G_2 stages, **30**, 32f
fixation, 14
membrane cyclic AMP, 269
plate, 29, 31f
research, methods, 14–17
size, 27f, 28f
structure, **18**
eukaryotic, **19**, 20f
prokaryotic, **18**f
theory, 13
Cellulose, 40
Cenozoic Era, 356–357
Central nervous system (CNS), 304
Centrifugation, 16, 17f
Centriole, **22**, 29, 30f, 101, 102f
Centrolecithal eggs, 104f
Centromere (kinetochore), **29**, 30f, 61
Century plant (Agave), 174
Cercaria, 377f, **378**
Cerebellum, 315
Cerebrum, 304f, 313
cortex, 313
hemispheres, 004f, 313
Cervix uteri, 157f
Chaffinch, English, 331
Chalk cliffs, 355
Chalazas, 119
Chance, 59
in evolution, 347
in origin of life, 355
Chang, T., 265
Chapparal biome, 387
Chardin, T. de, 365
Chekhov, A., 407
Chemical control of sexuality in plants, 170
Chemical fertilizers, 135
Chemical regulators of plant growth, 170
Chernozem soil, **388**, 389f
Chestnut blight, 373
Chimpanzee, 335
brain size, 312
Chitin, **40**, **294**
Chlorine, in plant nutrition, 219, 223f
2-Chloroethanol, 170
Chloroform, 38, 38f
Chlorophyll, 243
Chlorophyll a, absorption spectrum, 185f
Chlorophyll b, absorption spectrum, 185f
Chloroplast, 20f, 22f, 23f
as symbionts, 369
structure, 182, 183f
lamellae, 182

Chloroquine, 376
Chlorosis, **221**
Cholesterol, 41
Choline, 42
sulfate permease, 93
Cholinergic nerves, **307**
Chondroitin, 40
Chondrodystrophic dwarfs, 70
Chorion, fetal, 120f
Chromatids, 62
sister, 15f, **29**, 30f
Chromatography, paper, **197**, 198f
Chromophobes, 276
Chromosome, **29**
aberrations, 69
and heredity, 60
giant banded, 68, 69f, 117f
human karyotype, 85f
mapping of, 67–68, 68f
salivary, 117f
semiconservative replication, 85f
Chronology, methods 355
Cilia, 22, 23
Cinchona tree, 376
Circadian rhythms, 278, 332
Circulation
annelid-arthropod, 238, 239f
blood, 237
fetal, 120f
hepatic portal, 241f
open, 139
pulmonary, 241f
Cis-aconitic acid, 209, 211f, 212
Citric acid (Krebs cycle), 43, 207–209, 211f
Clamdigger's itch, 378
Classification
animal, 365f
general, **364**
new methods, 365, 366f
plants, 365f
Cleavage
frog egg, 109f
types, **107**, 108
Clever, W., 118
Climacteric, **290**
Climax community, **390**
Clinostat, 139
Clitoris, 118
Clocks, biological, 332
Clonal life span, 123
Clonal selection theory, 265
Cloned cells, **66**
Clostridium, and nitrogen fixation, 224
Clotting, blood, 243, **244**
Club of Rome, 406
Club root disease of cabbages, 372
CNS (central nervous system), 304
Coacervate, **351**
Coal, 355
Cobalt, in plant nutrition, 220, 223f
Coccidiosis, 375
Cochlea, 318, 319f
Cocklebur, 172, 173
Coconut milk, 140
cell division factor in, 287
Code, genetic, 83, 87, 87f
replication of, 83
Codon, **81**, 83, 87f
Coelom, 111, 114f
Coenzyme, **204**, **234**
Coenzyme A, 48, 207
Coenzyme Q (CoQ), 211, 212f
Coghill, G. E., 122, 329
Cohesion-tension theory, 252–253, 252f

Colchicine, 144
Cold sensors, 319
Colds, common, 34
Coleoptile, 127*f*
Coleorhiza, 127*f*
Colon, 228
Color blindness, 65
Color vision, 319, 320
Columbus, M., 238
Commensalism, **368**
Companion cells, of phloem, 253, 253*f*
Compensation point, photosynthetic, **186**
Competition, effect on populations, 402–403
  between species, 402
  competitive exclusion, **402**
  within species, 403
Competitive exclusion, principle of, **402**
Competitive inhibition, **206**
Condensation, **245**
Conditioning
  classical, 330
  operant, 330
Cones, 126
  visual, 319
Conflict, conditions, 336
Conidia,
  of *Aspergillus*, 166*f*
  of *Neurospora*, 88, 89*f*
Conjugation, 92, **93**
  in *E. coli*, 93*f*
  tube, 93
Conklin, E. G., 110
Consciousness, 305
Conservation of
  energy, law of, **47, 214**
  matter, law of, **47, 214**
Constitutive enzymes, 90
Consummatory behavior, 317, **326**
Continental shelf, **393**
Contour plowing, 135
Contraception, hormonal
  female, IUD, 159
  male, 160
Contraction, muscle, theories, 299, 300*f*
Convoluted tubules, renal, 261*f*
Cook, Capt. J., 233
Cooperative behavior, 335, 336
Copepods, 394
Copper, in plant nutrition, 219, 223*f*
Corepressor, **91**
Cori, C., 50*f*, 51
Cori, G., 50*f*, 51
Cork cambium, **128**, 130
Corms, **166**
Corn (maize)
  in diet, 232
  leaf blight, 4
  plant, elements in, 219*f*
Corpus, **131**
Corpus callosum, 304*f*
  function, 314
Corpus luteum, 103*f*, 107*f*, 157
Corpus striatum, 313
Correns, C., 56
Cortex, plant, 128, 129*f*
Cortex, adrenal, 272
  cerebral, 304*f*, 313
  renal, 260*f*
Corticosterone, 333
Cortisone, 272, 273

Cotyledon, 127*f*
Courtship, 153, 337
Covalent bonds, **38**
Cow (ruminant)-protozoan symbiosis, 370
Cowbird behavior, 326
Cowper's glands, 155
Crabgrass, 199
Crafts, A., 251
Cranial nerves, 305
Craviots, 231
Creatinine, 259
Cretin, **270**
Cretinism, goitrous, 95*f*
Crick, F., 83, 84*f*
Cristae, 19, 21*f*
Cro-Magnon people, 360, 361*f*
Crop
  rotation, 135, 225
  yields, effects of fertilizers, 222
Cross-breeding, 73
Crossing over, 62, 66–67
  in *Drosophila*, 68*f*
Crotonic acid, 41*f*
Cryptorchid testes, 156
Crypts of Lieberkühn, 229
Cuckoo instinct, 326
Curry, G., 283
Cuticle, 128
Cuvier, G., 344
Cyanide inhibition of respiration, 210
Cyanocobalamin, 234
Cycles, reproductive, animal, 119
Cyclic AMP (adenosine monophosphate), 269
Cycloheximide, 291
Cysticercus (bladder worm) stage of tapeworm life cycle, **378**, 379*f*
Cytochrome, 243
  discovery of, 210
  in photosynthesis, 185, 194, 194*f*
  oxidase, 210, 211, 212*f*
  taxonomy, 365, 366*f*
Cytochrome *a* (Cyt a), 211, 212*f*
Cytochrome *b* (Cyt b), 211, 212*f*
Cytochrome *c* (Cyt c), 211, 212*f*
  absorption spectrum, 212*f*
Cytokinesis, 13, **29**, 31*f*, 288
Cytokinins, 140, 279, 287–289
  and cell division, 287
  antisenescence effects, 149, 289
  bioassay for, 289
  structures of, 288*f*
  synergism with auxin, 288, 289*f*
Cytoplasm, plant cell, **13,** 14, 18, 20*f*
Cytosine, 81, 82*f*

2,4-D (2,4-dichlorophenoxyacetic acid), 284, 284*f*
  toxicity to monocots and dicots, 284
DaGamma, Vasco, 233
Damsel fly, 294*f*
Danielli, J. F., 24*f*
Danilewsky, 376
Daraprim, 376
Dark (Blackman) reactions of photosynthesis, 186–187
Darwin, C., 56, 72, 175, 279, 281, 326, 338, 344*f*, 351, 352, 389, 401, 407
  life of, 343
  evolutionary theory, 343–345
Darwinian theory, 343

Dating, radiocarbon method, 358
Davenport, C., 78
daVinci, Leonardo, 237*f*
Davson, H., 24
Daylength, seasonal changes in, 172
Day-neutral plants, **172**
DCMU (dichlorophenyldimethyl urea), 196
DDT, 384, 406
Deamination, 230
Decarboxylation, of pyruvic acid, 209*f*
Deciduous forest, **386**
  index animal of, 386
Deer mouse behavior, 324
Definitive host, of parasite, **377**
Defoliation, 290
Dehydration synthesis, 40
Dehydrogenase, 210–211
Deletion, chromosomal, 69
Demography, 406
Dendrite, **302,** 303*f*
Denitrifying bacteria, 225, 397
Deoxyribonucleic acid (DNA), 81, 82*f*
  replication of, 83, 83*f*
Deoxyribose, 39*f*, 81
Derepressed genes, 142
Descartes, René, 311
Desert, 387
  Mojave, 387*f*
Determinate cleavage, 108
Detoxification, by liver, 230
Development in plants, 124–150
  gene control of, 142–143
  hormonal regulation of, 143
  phytochrome regulation of, 143–144
  role of light, 136–139
deVries, H., 56
Dextrorotary optical isomers, **40**
Diabetes mellitus symptoms, 272
Diabetics, 259
Diaper test, for phenylketonuria, 95
Diaphragm, 215, **227**
Diastole, **241**
Diatoms, 394, 395*f*
Dichlorophenyldimethyl urea (DCMU), 196
Dictyosome (*see* Golgi apparatus)
Diencephalon, parts, 314
Diener, Theodor, 35
Diet
  human, 230–236
  mental health, 233
  vegetarian, 232
Differential gene activation, 144
Differentiation
  embryonic, 108–110
  in plants, 132
    effects of light, 136–139
    role of nucleus, 139–143
  reversible, 116
  role of cytoplasm, 108
Diffusion, **246**
Digenetic flukes, 377
  definitive hosts of, 377
Digenetic parasite, 372
Digestion, 41, 228, 229
  enzymes, 228*f*
  products of, 229*f*
Digitalis, 39, 269
Diglyceride, 42
Dihybrid cross, **59,** 60, 61*f*
Dihydroxyacetone phosphate, 198,

199*f*
6-(γ,γ-Dimethylallylamino)-
purine, 288*f*
Diminishing returns, law of, **224**
Dinosaurs, 357*f*, 363
Dioecious plants, **175**, 374
Diphosphoglyceric acid, 207, 208*f*
Diploid cells, **27**, 28*f*
Disaccharide, **39**
Diseases, virus-caused, 33, 264
Displacement activity, **335**
Disulfide bonds, 44, 269
Diurnal thermoperiodicity, 169
effects on flowering, 169
effects on plant growth, 169
Divergence, animal-plant, 9
Dixon, H., 252
DNA (deoxyribonucleic acid), 17, 81,
82*f*
and behavior, 324
polymerase, 83
replication of, 83, 83*f*
Dodder (*Cuscuta*), 371, 373, 373*f*,
374
Dodge, B., 88
Dog
packs, 334
vs. dogfish, 298*f*
Dogfish (shark), 297, 298*f*
Dolk, H., 281, 283
Dominant genes, **58**
Dormancy
in buds, 169
temperature effects, 169
in seeds, 285
effect of gibberellin, 285
Dormin (*see* Abscisic acid), 291
Dorsal, **239**
Dorsal aorta, 239, 240*f*
Dorsal lip, blastopore, 111*f*, 116*f*
Dorsal sensory ganglion, 305*f*
Dosage compensation, in heredity,
77
Double fertilization in seed plants,
175
Double helix, of DNA, 81
Down's syndrome (mongolism), 55,
96
Driesch, H., 55, 108
Drosophila
eye pigments, 88
gynandromorph, 119*f*
Dubos, R., 231
Duchenne muscular dystrophy, 300,
301*f*
Duckweed (*Lemna*), asexual repro-
duction, 167
Duodenum, 228, 268
Duplication, chromosomal, 69
Dutch elm disease, 373
damage to conducting tissues,
373*f*
elm bark beetle, 373*f*
Dwarf
achondroplastic, 297*f*
cretin, 271
pituitary, 276, 277*f*
Dwarfism
Ellis-van Creveld, 348*f*
in plants, reversal by gibberellin,
285
Dystrophic lakes, **391**
Dzierzon, J., 56

Ear, human, **319***f*

Eardrum, 318, 319*f*
Earth
atmosphere, elements in, 219*f*
crust, elements in, 219*f*
magnetic field fluctuations, 333
waters, elements in, 219
Earthworms
and soil development, 389
circulation, 239*f*
neurosecretory cells, 277*f*
East, E., 72
Ecdysis, 294*f*
Ecdysone, 118, 268, 269
Ecological succession, 390–391
field to forest, 390*f*
Ecology, **383**, 383–409
Ecosystem, **383**
Ecotone, **385**
Ecstatic display, 153*f*
Ectoderm, 110, 111*f*
Ectoparasite, **368**
Efferent motor neuron, **303**, 305*f*
Eggs, 101
plant, 125
types, 103, 104*f*
Ehrlich, P., 235
Electromagnetic spectrum, 136*f*
Electron transport, 49, 210–212,
212*f*, 213*f*
Elements in plant tissues
essential, 218–220, 223*f*
non-essential, 220
Elephant, 355, 362*f*
breeding, 344
Elephantiasis, 375
Eltonian pyramid, **398**
Embden-Meyerhof pathway (glycoly-
sis), 48, 206–207, 208*f*, 299,
353
Embryo
frog, 112*f*, 113*f*
human, 115*f*
plant, 120
sac, 175
vertebrate, diagram, 114*f*
Embryology, comparative, 350
Embryonic conservatism, 350
Emerson enhancement effect, **189**
Emerson, R., 188, 188*f*, 189
Emigration, effect on populations,
403
Encephalitis, 33
End product inhibition, 116
Endemic species, **349**
Endocrine gland, **267–268**
human, 268*f*
Endocrinology, methods, 268
Endoderm, 110, 111*f*
Endodermis, **130**, 251
Endomitosis, **145**
Endoparasite, **368**
Endoplasmic reticulum, 20*f*, 21
Endoskeletons, 293, 295–297
Endosperm, 126, 127*f*, **175**, 286,
287*f*
Energy
chains, 398–399
of combustion, of glucose, 213
pathways of, 394, 398–400
Engelmann, T., 183, 183*f*, 184
experiment of, 184*f*
Engram (memory trace), **314**, 317
Entelechy, 55, 108
Environment, 385–387

effect on heredity, 71
Enzymes, **42**, 203–206
constitutive, 90
digestive, 228*f*
function, 204–205
inducible, 90
induction, 90
synthesis, control of, 92*f*
terminology, 204
theories of action, 205–206, 207*f*
tools for studying, 205
Enzyme-substrate interaction, 207*f*
Eosinophil (acidophil), 242*f*, 243
circadian rhythm, 333
Ephrussi, B., 88
Epicotyl, 127*f*
Epidemiology, **376**
of malaria, 376
Epidermal layer, of stems, 128, 129*f*
Epididymis, 155, 157*f*
Epilimnion, **392**
Epinasty, **290**
of tomato petioles, 290*f*
Epinephrine, **272**, 274*f*
receptor sites, 269
Epiphytes, **374**
Epistasis, **89**
in dog coat color, 90
EPSP, 310
Equilibrium, 318, 319*f*
Erepsin, 228
Ergosterol, 232
Ergotism, **372**
Erlich, P., 375
Erythroblastosis fetalis, 76
Erythrocytes, 242*f*
Erythropoietin, 262
Escherichia coli, 33, 90
conjugation, 93*f*
Esophagus, **227**, 228*f*
"Essential" amino acids, 232
Estradiol, 269*f*, 273
Estrogen, 156, **157**, 273
effects on bone, 297
effects on uterus, 270*f*
Estrone, 269*f*, **273**
effects, 159
Estrous cyles, 158*f*
Estrus, **157**
Estuaries, 392
Ethical problems, smallpox, 4, 263
Ethology, **323**
Ethylene, 149, 279, 290–291
and defoliation, 290, 291
and leaf abscission, 290, 290*f*
characteristics of, 38
epinasty, caused by, 290, 290*f*
in illuminating gas, 290
induced climateric, 290
structure, 39*f*, 290
Etiolation, 138
Eugenics, 98–99
negative, 98
positive, 98
Eukaryotic cell, **16**, **18**
Eunuchs, 268
Eustachian tubes, 114, 318, 319*f*
Eutophic lakes, 391
Eutrophication, **391**
Evocator, embryonic, 114
Evolution, 343–367
evidence, 348–362
kidney, 262*f*
organelles, 19, 369
theories, 345–347

Ewart, A., 192
Excitatory postsynaptic potential (EPSP), 310
Excretion, **259**
  anatomy of, 260f, 261f, 262f
  waste nitrogen, 259
Exercise, violent, 299
Exocrine gland, **267**
Exophthalmia, 271
Exposure learning, 330
Exoskeletons, 293, 294f
Experimentation, animal, 4, 363
Explants, 149
  petiole abscission of, 149f
Extensor muscles, **297**
Extinction
  animals, 362, 363
  environmental factor, 364
  theories, 362, 363
Eye spot; *Euglina*, 319
Eyes
  human, 32f
  transplantation, 122

Fabre, J. H., 323
Fabricius, H., 100, 237
FAD (flavin adenine dinucleotide), 211, 212
Fallopian (uterine) tube, 104, 107f, 118, 157f
Fatigue, muscular, 299
Fats, 41
Fatty acids
  saturated, **41**
  unsaturated, **41**
Feedback
  cerebellar, 314
  hormone control, 267, 275f, 276, 277, 403
  neural, 314f
  O_2 and red cells, 262
  repression, **91**
Femur, 295f, 296
Fermentation, 48
Fern gametophyes, 170f
  antheridium-inducing factor, 170
Ferredoxin, 185, 194, 194f, 195
Fertilization, 60–61, **104**, 105f, 156
  artificial, 105
  in plants, 125
  physiology, 106
Fertilizers
  effects on crop yields, 222–224
  terminology, 222
Fertilizin, 106
Feulgen, G., 46
Fibril, muscle, 300f
Fibrin, 243
Fibrinogen, 243
Fibula, 295f, 296
Fiddler crabs, photoperiod, 333f
Filament, of stamen, 174, 175f
Filariform larvae, of hookworm, **380**
Final common pathway, neural, 309
Fireflies, 152
First filial generation (F_1), 57
Fischer, E., 44, 206
Fiske, V., 278
Fission, 18
Fissure of Rolando, 304f, 313
Fissure of Sylvius, (lateral f.), 304f, 313
Fixative, 14
Flaccid plant cells, 246, 248f
Flagellum, 18f, 22, 23f
Flatworms, 153
  parasitic, 376–378

Flexor muscles, **297**
Florigen, 174, 279
Flowering
  chemical control, 168f, 173
  control by daylength (photoperiod), 168f, 171–174
  effect of temperature, 134
Flowers, 126
  adaptations for pollination, 176–177
  bisexual, 175
  color and soil pH, 135
  development, 174–175
  double fertilization in, 175
  parts, 174, 175f
  pistillate (female), 175
  staminate (male), 175
  structure, 174, 175f
Flukes (trematodes), 376–378
Fluorescence, 193
Fly taste test, 318f
Foard, D., 132
Folic acid (folacin), 234
Follicle
  cells, 101
  graafian, 103f
  Follicular fluid,156
Food, 230–236
  and aggression, 336
  chains (webs,) 398–399, 399f
  plants, geographic origins of, 178
Forest fires, 390
Formic acid, 41
Fossil fuels, 187
Fossils, 355
Founder principle, 347
Franklin, B., 24
Frederick the Great, 96
Frequency, of light (ν), **137**
Freud, S., 321
Frobisher, M., 92
Frog, embryo, 111f, 112f, 113f
Fructose-1,6-diphosphate, 206, 208f
Fructose-6-phosphate, 206, 208f
Fruit, 127, **175**
FSH (follicle-stimulating hormone), 158, 159
Fuel of life, calories, 230
Fumaric acid, 206, 208, 209, 211f
Functional groups (radicals), **38**
Fungi, 10
Funk, C., 233
Futureology, 406

G_1 and G_2 stages, cell division, **30, 32f**
β-Galactosidase, 91
Gall, **372**
  on pecan twigs caused by insects, 372f
Galston, A., 249
Galton, F., 56, 346
Gametangia, **125**
Gametogenesis, 31, 101, 102f, 103f
Gametophyte plants, **125**
  ferns, 170f
  Marchantia (liverwort), 167f
Gamma plantlet, 133, 133f
Gamma radiation, effects of, 133
Gamma globulins, 263, 264f
Garner, W. W., 138, 152, 172
Garrod, A., 88
Gas chromatography, 290
Gases, atmospheric, 134
Gastric juice, 228
Gastrocnemius muscle, 307

Gastrulation, **110**, 111f
Gause, G., 402
Geese, imprinting, 331f
Geiger counter, 15
Gemmae, of Liverworts, 167f
Genes, **58, 81**
  action
    sequential, 116
    visible, 117f
  activation, 144
  and enzymes, 88–89
  dominant, **58**
  exchange, between races, 79
  frequencies, small populations, 347
  interaction, 89–90
  pool, human, 74
  recessive, **58**
  regulation, 90–92
  therapy, 94–95
Gene-dependent behavior, 324
Generation time, of bacteria, **93**
Genetic code, 83–87, 87f
  replication of, 83
Genetic diseases, 95–98
Genetic drift, 347
Genetic engineering, 92–94, 235
  of plants, 145–147
Genetic improvement, 71, 92–94
Genetic variability, 63
  in plants, 168
Genital folds, 118
Genital tubercle, 118
Genome, **59**
Genotype, **59**
Genus, **364,** 365f
Geographical distribution, 348–350
Geological times, 356–357f
George III, 96
Geotropism, 139, **280**
  auxin and, 139, 281
  inhibition by morphactin, 292f
Germ layers, **110**, 111f
Germ plasm theory, **108**
German measles, 114
Germinal epithelium, 103f, **156**
Germinal vesicle, **103**
Germination, of seeds, 127f
  radish, 130f
Gestalt, **325f**
Gestation periods, 160f
Gibberella fujikuroi, 285
Gibberellins, 170, 279
  and derepression of genes, 286
  and elongation in cabbage, 286f
  and seed germination, 285, 287f
  and sex of flowers, 287
  reversal of dwarfism, 285
  structure, 285f
Gills, 215
  slits, 113, 114f, 115f
Girdling, **250,** 253
Glacial ages, 361f
Glial cells, **304**
Glick, B., 265
Globulin, of pea cotyledons, 143
Glomerulus, 260, 261f
Glucagon, 269, 272
Glucosamine, 30, 40, 295
Glucose, control, 272
Glyceraldehyde phosphate, 207, 208f
Glycerides, mono, di, tri, 229f
Glycerine (glycerol), 38, 39f, 41f
Glycerol (glycerine), 38, 39f, 41f, 228

Glycine, 42, 43*f*, 353
Glycogen, 40, 50, 206
 liver, 230
 muscle, 299
Glycolic acid pathway, 199
Glycolysis (Embden-Meyerhof Pathway) 48, 206–207, 208*f*, 353
 muscle, 299
Glycoside, **39**
 linkage, **39**
 in sucrose, 39*f*
Gnat giant chromosome, 117*f*, 118
Gnotobiotic animals, **235**
Gödel's proof, 312
Goebel, K., 148
Goethe, J. von, 295, 344
Goiter, 271*f*
Golgi, C., 375
 complex (apparatus), 20*f*, **21***f*
Gonads, 273–274
Good, R., 265
Gorilla, 335
Graafian follicle, 107*f*, 156, 158*f*
Grafting, 167–168
 techniques, 168*f*
Grafts, animal organs, 264
Grana, **22**, **182**
 lamella, **22**
Grassland, 386
Gravity
 and plant growth, 139
 and size, 7
Gray, A., 72
Gray matter, **304***f*, 313
Great Succession in the northern hemisphere, 390
Green hydra, symbiotic algae in, 369
Green paramecium (*P. bursaria*), symbiotic algae in, 369
Green revolution, 3, **235**, 236, 407
Griffith, F., 92
Grinnell, J., 402
Grobstein, C., 115
Grotthus-Draper law, 137, 320
Ground itch, 380
Ground tissue, 128
Growth hormone, 276
Growth
 animal, 100–123
 plant, 124–150
Guanine, 81, 82*f*
Guard cells, **248**
Gulf stream, 393
Gums, 40
Gurdon, J. B., 110
Guthrie test, for phenylketonuria, 95
Guttation, **251**, 251–252
 of strawberry leaf, 251*f*
Gynandromorph, **118**, 119*f*
Gyplure, 405
Gypsy moth, 362
 control of, 405
Gyrus, brain, **314**

Habituation, 329
Haber, A., 132
Haberlandt, G., 192, 286, 287
Habitat, **401**
 of a species, **385**
Habrobracon, 381
Haldane, J. B. S., 351
Hales, S., 247, 251, 311
Half-life, **15**
 radioactive **355**, 358
Hallucinogens, 5*f*, 311*f*
Halophytes, **220**

Haploid (monoploid) set of chromosomes, **27**, 28*f*, 59
 plants, 114–115, 145*f*
Harden, A., 50, 207
Hardin, G., 6
Harlow, H., 331
Harrison, R. G., 113, 122
Hartig, T., 253
Harvest method, for determining productivity, **400**
Harvey, E. B., 106, **108**
Harvey, W., 100, 138*f*, 237*f*, 247
Hastings, J. W., 333
Haustoria, of mistletoes, 374
Health, 3, 230, 263, 267, 374
 mental, 4, 233, 234
Hearing, 318, 319*f*
Heart, 5
 embryonic, 114*f*
 mammalian, 140, **141***f*
 muscle, 298, 299*f*
 rate, 242
  neural control, 305
  vagus inhibition, 310
  voluntary control, 306
"Heat" (estrus), **157**
Heat sensors, 319
Heinroth, O., 323
Helianthus annuus (sunflower)
 aging in, 148
 heteroblastic development, 148
Helmholtz, H., 307
Hemocyanin, 243
Hemoglobin, 243
 fetal, 117
Hemophilia, 65, 96
Hendricks, S., 138, 172
Henle's loop, 260, 261*f*
Hensen's band, muscle, **299**, 300*f*
Hensen's node, 111
Henslow, J., 343
Heparin, 40, 295
Hepatic portal vein, 229, 230*f*, 241*f*
Hepatitis, 90
Herbicides, 236, 284
Herbivore, **227**
Hereditary diseases, prevention of, 94, 97
Heredity, and environment, 71
Hertwig, O., 105
Heteroblastic development, **148**
 in ivy, 147*f*
 in sunflowers, 148
Heterografts, **264**
Heterokaryocytes, **145**
Heteropolysaccharides, 40
Heterosis, 73, 74
Heterotroph theory, 351, 353
Heterozygous genotype, **58**
Hevesey, G., 255
Hexose, **39**
Hierarchies, behavioral, 334
"High efficiency" C_4 plants, 198–202
 leaf structure, 200*f*
 photosynthetic pathway, 201*f*
 photosynthetic rates, 202*f*
Hill, R., 192
 reactions, **192**
  oxidants for, 192
Hilum, 127*f*
Hippocrates, 3, 55, 384
Histocompatibility genes, **265**
Histones, 118
History of life, 355, 356*f*, 358*f*, 360*f*, 361*f*, 362*f*
Hodgkin, A. L., 308

Holism, 324
Holm, R., 149
Holoblastic cleavage, 107
Homeostasis, 258
Homogentisic acid (alkapton), 88, 95*f*
 oxidase, 88
Hemoiotherms, 9, **215**, **262**
 respiration and body weight, 215*f*
Homologous chromosomes, 62
Homology
 evolution of, 344
 general, 349
 serial, **349**
 skull, 295, 296*f*
 special, **349***f*
 vertebrate limbs, 349
Homozygous genotype, **58**
Homunculus, neural, 316*f*
Honeybee(s), 106, 363
 on alfalfa blossom, 176*f*
 talk (dance), 328, 329*f*
Hooke, R., 13
Hookworm, 380–381
 anticoagulant secreted by, 380
 filariform larvae of, 380
 ground itch caused by, 380
 life cycle of, 381*f*
 species of, 381
Hopkins, F., 50
Hopkins, G., 233
Hormonal control, sex traits, 156
Hormone(s), **143**
 animal, 267–292
 behavior, **332**
 interrelations, 275*f*
 table of, 276*f*
Horse, evolution, 350, 360*f*
Horseshoe "crab," 4, 5*f*, 243
Homografts, **264**
Howard, E., 334
Huang, R. C., 143
Human
 birth, stages, 100, 101*f*
 brain, 304*f*, 312–317
 diversity, 151
 embryo, 115*f*
 evolution, 358*f*, 359, 360
 gene pool, 74
 genetic diseases, 95–98
 karyotype, 62*f*
 musculature, 298*f*
 society, biological bases, 338
Humerus, 295*f*, 296
Humus, **388**
Huntington's disease (chorea), 98
Huxley, A., 308
Huxley, H. E., 299
Huxley, T. H., 295, 349, 365
Hyaluronic acid, 40
Hybrid vigor, **72**, 74
Hybridization, 71–72
 in corn, 72, 73*f*
 of somatic cells, 66, 94, 147, 235
Hydathodes, 251
Hyde, J., 191
Hydra, 297
Hydrogen
 atom, structure of, 38*f*
 bonds, 44
  in DNA, 82
 in plant nutrition, 218
Hydrolysis, **41**, **245**
Hydroponics, **221**
Hydroxytryptamine (serotonin), 310, 311*f*

Hyperparasites, 372
Hyperthyroidism, 271
Hypertonicity, 26
Hypertrophy, **372**
Hypocotyl, 127f, 130f
Hypolimnion, **392**
Hypophysis (see pituitary)
Hypothalamus, 304f, 314
  behavior control, 315, 316
  control of pituitary, 274
  sex effects, 119
  sex hormones, 154f
  temperature control, 263
Hypothyroidism, **270**
Hypotonicity, 26

Ichneumon wasp, 371, 381
Ileum of intestine, **228**
Ilium bone, 295f
Immune system cells, 266f
Immunity, 263–266
  kinds of, 263, 264
  suppression, 265
Immuno competent cells, origin, 265, 266f
Immunoglobulin, 264f
Immunological "memory," 264
Implantation, 107f
Imprinting, 153, 331f, 332
Inbreeding, 72, 74
Incipient plasmolysis, 247, 248f
Incus, 296f, 318, 319f
Independent assortment, law of, 60
Indeterminate cleavage, 107
Index animal, 386
Indians, Algonkian diet, 232
Indole-3-acetic acid (IAA) (see Auxin), 281f
Indolealanine (tryptophan), 43f, 44
Inducer substance, 91
Inducible enzymes, 90
Induction, embryonic, 114–116f
Influenza, 33
Ingen-Housz, J., 47, 182
Inhibition
  end product, 116
  feedback, 116
  neural, 312
Inhibitors, of enzymes, 205
Inhibitory postsynaptic potential (IPSP), 310
Inman, O., 192
Inner ear, 318, 319f
Inner membrane spheres, **214**, 213f
Inositol, 234
Insect societies, 338
Insectivore, **358**
Insight learning, **330**
Instinct, 323, **326**–329
  examples, 323f, 328
Instrumental learning, **330**
Insulin, 44, 230, 269, 272
Intelligence, 10
  brain size, 312, 313
Interferon, 264
Intermediate lobe, pituitary, **275**
Intermedian, 275
Interneuron, **303**, 305f
Internode, of stem, 128, 128f
Interphase, mitotic, 28, 30f, 32f
Interstitial cells, testicular, 155f
Intestinal flora, 234
Intraspecific competition, 403
Intrauterine devices (IUD's), 159
Inversion, chromosomal, 69, 69f

Involuntary muscles, 298, 299f
Iodine, thyroid, 271
Irish curve, of population growth, 407
Iron, in plant nutrition, 219, 223f
Irrigation, of soils, 135
Islets of Langerhans, 271
  alpha cells, 272
  beta cells, 272
Isocitric acid, 209, 211f
Isolation, evolution, 348
Isolecithal eggs, 103, 104f
Isomers, **40**
Isotonicity, 26
Isotropic band, 298f, **299**, 300f
Itai-Itai (ouch-ouch) disease, 383
  poisoning by cadmium and, 383
IUD's (intrauterine devices), 159

Jacob, F., 90, 91, 116
Japanese beetle, control of, 405–406
Java man, 359
Jenner, E., 263
Jellyfish, 328
"Jet lag," 333
Johannsen W. L., 168, 345
Joly, J., 252
Jones, D., 72
Jonkel, C., 5
Joshua trees, 387f
Jung, C., 321
Jupiter, 351
Juvenile morphology of plants, 147, 147f

Kamen, M., 191
Karyokinesis, 13
Karyotype, **29**, **61**
  human, 62f
Keilin, D., 208, 210
Keto group, 38, 269
α-Ketoglutaric acid, 209, 211f, 212
Khorana, H., 38
Kidney
  energy requirement, 261
  evolution, 262f
  function, theories, 260–261
  human, 260f
Kinetin (6-furfuryl amino purine) 288, 288f, 289
Kinetochore (centromere), **29**, 30f, 61
Klinefelter's syndrome, 77
Knight, T., 56, 139
Knop, W., 221
  solution, 221
Koch, R., 280, 376
Konishi, M., 331
Krause's end bulbs, 319
Krebs, H., 208, 210f, 230
Krebs (citric acid) **cycle**, 48, 207–209, 211f, 353
  in muscle, 299
Krenke, N., 148
Kruschev, N., 72
Kuchenmeister, 378
Kurosawa, E., 285
Kuroshio (Japan) current, 393
Kwashiorkor, 231f

Label, radioactive, 15f
Labia major, 118
Labor, birth stages, 160, 161f
Lactate dehydrogenase, 299

Lactation, control, 161, **162**f
Lactic acid, 48, 208
  in muscle, 299
Lactogenic hormone (see Prolactin)
Lakes, 391
  annual turnover, 391–392, 393f
  dystrophic, **391**
  eutrophic, **391**
  farming of, 391
  oligotrophic, **391**
Lamarkianism, 345
Lampreys, 381
Landsteiner, K., 74, 75f
Langurs (woodland monkey), 337
Laplace, S., 214
Larynx, 228f
Laveran, A., 275
Lavoisier, A., 47, 214, 216
Layering, **166**
Leaf, 127, 128
  abscission, 149, 290, 290f
    in explants of cotton, 149f
  as unit of physiological time, 148
  internal structure, 132f
  mold, **388**
  primordia, 131
  veins of, 132, 132f
Leakey, L. S. B., 359
Learning, **329**
  biological prerequisites, 154
Lecithin 42f
Lederberg, J., 93
Leeuwenhoek, A. van, 13, 239
Leghemoglobin, **225**
Lemna (duckweed), asexual reproduction, 167
Lenticels, 246
Lerner, A. B., 278
Lesch-Nyhan syndrome, 96, 97f
Lethal mutations, 69
Letham, D., 288
Leukocytes, 242f
Leukoplast, 22
Levels of organization, 11, 325
  atomic, 37
  molecular, 37
  subatomic, 36
Levorotary optical isomers, **40**
Lewis, W., 114
Leydig's cells, testicular, 155f
Lichens, 369, 369f
  crustose, 369, 369f
  foliose, 369, 369f
  fruiticose, 369, 369f
Liebig, J., 204, 224
Life
  cycles, of plants, 124–126
    generalized, 125f
    flowering plants, 126f
  expectancy, 160
  on other planets, 9, 10
  origin of, 9, 10, 350–355
  styles, three primary, 353–354
Ligament, **297**
Light
  effects on plant growth, 136–139
  energy per quantum, 137
  intensity, and photosynthetic rates, 185, 186f
    compensation point, **186**
    saturating, 186
  nature of, 136
  reactions of photosynthesis, 188–197
  wavelengths of, 136

Limb girdles, 295*f*, **296**
Limnology, **391**
Limulus, 4, 5*f*, 243
Linkage
  analysis, 66
  autosomal, 66
  groups, 66
Linnaeus, C., 344, 364
Linoleic acid, 232
Lion family, 337
Lipids, 41–42
  composition of, 41
  diet, 232
  function of, 41
  membrane **24**, 25*f*
Lipmann, F., 207, 208
Litmus, 26
Litter, soil, **388**
Littoral zone, of oceans, **393**
  organisms of, 394
Liver, 141*f*
  functions, 229–230*f*
Liverwort gametophytes, 167*f*
Lobes of brain, 304*f*, 313
Locke, J., 322
"Loco" weed, 220
  Astragalus, 220
  Oxytropus lambertii, 220*f*
Locomotion, 294
Locus
  genetic, 58
  on chromosome, 81
Loeb, J., 106
Logistic curve of population growth,
  404, 404*f*
Long-day plants, **172**
  response to photoperiod, 172*f*
Long-night plants, 173
Lophophora cactus, 5*f*
Lorenz, K., 323, 326
Lovebird courtship, 153
"Low efficiency" C₃ plants, 198–202
  leaf structure, 200*f*
  photosynthetic pathway, 201*f*
  photosynthetic rates, 202*f*
LSD (lysergic acid diethylamide),
  310, 311*f*
Lucretius, 318
Ludwig, C., 260
Lungs, 215
  circulation of, 241*f*
  excretion in, 259
Lutein, absorption spectrum, 284*f*
Lymph nodes, 242
Lymphatics, 229, 242
Lymphocytes, 21, 242*f*, 243, 265,
  278
Lyon, M., 76, 77*f*
  hypothesis, 76
Lysenko, T. 346*f*
Lysergic acid diethylamide (LSD),
  310, 311*f*
Lysine, maize, 232
Lysis, of bacterial cells, 94
Lysosome, **22**

Macaque, 335
MacArthur, R., 403
MacMunn, C., 88
Magendie, F., 305
Magnesium, in plant nutrition, 219,
  223*f*
Magnifications, 15, 16
Major elements needed by plants,
  219, 223*f*
Malaria, 368, 375–376

Maleic hydrazide, 170
Malic acid, 208, 209, 211*f*, 212
Malleus, 296*f*, 318, 319*f*
Malonic acid, 206
Malpighi, M., 13, 160, 253
Malthus, T. R., 344, 345, 404
Mammalian reproduction, hor-
  mones, 157–160
Mammary glands, **162***f*
Mammoths, 355, 362*f*
Manganese, in plant nutrition, 219,
  223*f*
Manson, P., 375
Mapping chromosomes, 67–68, 68*f*
Marasmus, 231
Marchantia (liverwort) gameto-
  phytes, 167
Marigold (*Calthra palustris*), 176*f*
  ultraviolet reflection patterns,
  176*f*
Mary Queen of Scots, 96
Maryland mammoth tobacco, 171,
  172
Mass flow (Münch) hypothesis,
  255*f*, 256–257
Mathematical models, 406
Mather, C., 72
Mating, 153
Matter, pathways of, 394–398
Matthaei, G., 185
Maturation, behavioral, 330
McCarty, M., 92
McCleod, C., 92
Meadows, D., 407
Measles, 34
Medulla oblongata, 304*f*, 315
Megaspore, **175**
Megasporocyte, **175**
Meiosis, **28**, **30**, 31, 60–64, 63*f*, 64*f*
  65*f*, 101*f*, 102*f*, 103*f*, 155*f*
  in plants, 125
Meiotic division
  first, 103*f*
  second, 103*f*
Meissner's corpuscles, 319
Melanin pigments, 43
Melatonin, **278**
Membrane(s)
  cellular functions, 24, 25, 26
  egg, 119
  fertilization, 104
  fetal, 120*f*
  of plant cells, 246*f*
  permeability, control by phytochrome, 144
  plasma, 24, 25, 26
  potential, 308
  protein mosaic, 24, 25*f*
  selective permeability of, 245
  structure, 24, 25*f*
  "unit membrane," 24, 25*f*
  vitelline, **119**
Memory, 317
  biochemistry of, 317
  "trace" (engram), 314
Mendel, G., 56, 56*f*, 74
  laws, 6, 56–60
  independent assortment, 60
  segregation, 56–58
Menstruation, 158, 159
Mental health

general, 4, 234
  vitamins, 233
Meristems, 127, 131
  apical, **128**, 131*f*
Meroblastic cleavage, 107
Mertz, E. T., 233
Mescaline (peyote), 5*f*, 36, 311*f*
Mesencephalon, 315
Mesoderm, 110, 111*f*
  parietal-visceral, 110
Mesosome, 17
Mesonephric ducts, 118
Mesozoic, 356–357
Messenger RNA (mRNA), 84, 118
Metabolism (*see* Respiration), **47**,
  51, 230–231
Metamorphosis, Frog, 117
Metaphase, mitotic, **29**, 30*f*
Metchnikoff, E., 123, 243
Methane (CH₄, marsh gas), 38, 38*f*,
  351
Methionine in maize, 232
Methyl group, 38
Meyerhof, O., 50
Microclimates, 387, 388*f*
Microhabitats, 388*f*
Micronutrients, **37**
Micropyle, 127*f*
Microscope
  electron, 15, 16*f*
  light, 16
  Nomarski optics, 14, 63*f*
  phase contrast, 14
Microspore, **174**
Microsporocyte, **174**
Microtome, 14
Microtubules, 22, 23*f*
  in mitosis, 29, 30*f*
Middle ear, 318, 319*f*
Middle lamella, 20*f*, **30**, 31*f*
Midgets, 276, 277*f*
Midpiece, sperm, 101, 102*f*
Miescher, F., 36, 45, 46*f*
Migration, bird, 152
Milk secretion, 157, 158, **162***f*
Miller, C., 287
Miller, S., 352, 354*f*
Mimosa pudica (sensitive plant), 144
  control of sleep movements by
  light, 144*f*, 249
Minamata disease, 383, 384*f*
  methyl mercury and, 383
  symptoms of, 383
Mineral(s)
  deficiency in plants
  and crop yields, 222–224
  symptoms of, 221, 222*f*, 223*f*
  disappearance and productivity,
  400
  movement in plants, 256*f*
  nutrition, plant, and reproduction
  171
Minimum, law of, **224**
Minkowski, O., 272
Minor elements needed by plants,
  219, 223*f*
Miracidium, **377**, 377*f*
Mistletoes, 374
Mitochondria, 17*f*, **19**, 21*f*, 23*f*, 49
  as symbionts, 369
  evolution, 354
  muscle, 299
  sperm, 102*f*
  thyroxin effect, 271
Mitosis, 13, **28**, 29, 30*f*, 31*f*, 32*f*
MN blood groups, 75

Mohr, H., 144
Moles, brain size, 312
Molisch, H., 192
Molting, 294*f*
Molybdenum
  in nitrate reductase, 225
  in plant nutrition, 219, 223*f*
Mongolism (Down's syndrome), 55, 96
Monkey
  howler, 334
  rhesus, 335
  ulcers, 332*f*
Monocarpic plants, **174**
Monocytes, 242*f*, 243
Monod, J., 90, 91, 91*f*, 116
Monoecious plants, **175**
Monogenetic
  flukes, 377
  parasite, 371
Monohybrid cross, **59**
Monoploid (haploid), **27**, 28*f*, 59
  plants, 144–145, 145*f*
Monosaccharide, **39**
Montagu, Lady Mary, 263
Mor soil, 389*f*
Morgan, T., 6, 66, 67*f*, 71, 81, 88
Morphactins, 291–292
  effects on geotropism, 292*f*
  effects on phototropism, 291*f*
  structure, 292*f*
Morula, 107*f*
Mosaic eggs, 103
Motivations for research, 3
Motor neuron, 303*f*
mRNA (messenger RNA), 84
Mucosa, **228**
Mull soil, 389*f*
Muller, H., 70
Müller, J., 307
Mullerian ducts, 118
Münch (mass flow) hypothesis, 255*f*, 256–257
Muscle
  action, kinds, 297, 299
  arrangement, 297
  contraction
    biochemistry, 299, 300
    mechanism, 300*f*, 301*f*
  dystrophy, 300, 301*f*
  fibril, 300*f*
  metabolism of, 51
Muscular system, 297–301
Mutagenic agents, 70
Mutation, 56, 69–70
  in *Drosophila*, 70*f*
  induced, 70
  lethal, 69
  pressure, 346
  randomness of, 70
  rate of, 70
Mutualism, **368**, 369
  fungus roots, 370
  lichens, 369
Mycorrhizae, 370
Myelin sheath, **303**
Myoneural junction, **309**
Myosin, 299
Myxomatosis, cause of rabbit plague, 405

NAD (nicotinamide adenine dinucleotide) 207, 212, 234
NADP (nicotinamide adenine dinucleotide phosphate), 194, 195

Nanometer (nm), **136**
Nansen bottles, 394, 396*f*
Naphthaleneacetic acid, 170
Neanderthal people, 360, 361
  brain size, 312
Negative feedback (see also Feedback)
  in reproduction, 403
Nekton, **394**
Nematode(s) (roundworms), 378–381, 380*f*
  damage to potatoes, 379*f*
  diseases of plants, 380
  ovary of, 156
  physiological races of, 379
  testis of, 155
Neo-Darwinism, 346
Nephron, 260, 261*f*
Neritic zone, of oceans, **393**
Nerve poisons,
  anticholinesterases, 309
  drugs, 309–311*f*
Nervous system, 302–320
  development, 121–123
Nervous impulse, 307, 308*f*
Neural circuits, behavior, 121, 122
Neural
  crest, 113
  folds, 113*f*
  plate, 113*f*
  temperature control, **263**
  tube, 114*f*
Neurons, types, 302, 303*f*, 304
Neurosecretion, 152
Neurosecretory cells, 275*f*, 277*f*
Neurospora (pink bread mold), 88, 354
  biochemical mutants, 90*f*
  life cycle, 89*f*
Neurulation, 113
Niacin, mental health, 233
Niche, of species, **385**
Nicotiana tabacum (tobacco), photoperiodic control of flowering, 171
Nicotinamide adenine dinucleotide (NADP), 194, 195
Nirenberg, M., 87*f*
Nissen, P., 93
Nissl body, **302**, 303*f*
Nitrate reductase, 143, 225
  control by ammonium ions, 143
Nitrogen
  balance, 232
  cycle, 225, 226*f*, **397**
  fixation, 224–225, 397
  in plant nutrition, 218, 223*f*
    effects of excess, 224
Nitrogen-fixing bacteria, 94
Nitsch, C., 144
Nitsch, J., 144
Node of Ranvier, 303*f*
Node, of stem, 128, 128*f*
Nomenclature, **364**
Noncompetitive inhibitors, **206**
Norepinephrine, **272**, 274*f*
Nostoc, and nitrogen fixation, 224
"No-till" farming, 236
Notochord, 111*f*, 114*f*, 294
Nuclear transplantation, 110
Nucleic acids, 45–46
Nucleocapsid, **33**
Nucleolus, 18
Nucleotide, 46, **82**
Nucleus, 14

plant cell, 20*f*
Nutrient solution for plants, 221*f*
Nutrition, 227–236
  human requirements, 230–236
Nyctinasty (see Sleep movements)

Oat (*Avena sativa*), assay for auxin, 281
Occipital lobe, 304
Oceanic zone, **393**
Oceans, 392–394
  currents, 393
  instruments for study of, 394
  regions, 393
Octopus, 293*f*
  poison, 310
Odors, biochemistry, 318
Oils, 41
  "mineral," 355
Olduvai Gorge, 359
Olfaction, 317
Oligotrophic lakes, **391**
Omnivore, 227
Oogenesis, 101, 103*f*
Ootid, 103*f*
Opaque-2 maize, 232
Oparin, A. I., 351, 352
Operator, **91**
  gene, 91
Operon, **91**
  theory, 91–92, 92*f*
Opie, E., 272
Opsin, 320
Optic chiasma, 304*f*, 314, 320*f*
Optic vesicle, 114*f*
Optical density (absorbance), 184
Optical isomers
  dextrorotary, **40**
  levorotary, **40**
Order, taxonomic, 365
Organ of Corti, 319*f*
Organelles, **18**
Organic compounds, **37**
Organic gardening, 221, 407
Organizer, embryonic 111, 114, 116*f*
Orgel, L., 352
Origin of life, 9, 10, 350–355
Ornithine cycle, 230
Osmometer, 246*f*
Osmosis, **246**
Osmotic concentration, cellular, 259
Osmotic pressure, **246**
  blood, 259
Ospreys, 401
Osteoporosis, 297
Outcrossing, 73
Ovarian cycles, photoperiod, 152
Ovary
  human, 101, 103, 107, 157*f*
  of flower, 174, 175*f*
  types, 156
Oviduct (Fallopian), 107*f*, 157*f*
Ovulation, 103*f*, 104, 107*f*, 158*f*
  mammalian 107*f*, 157
Ovule, **174**, 175*f*
Owen, R., 227
Oxalacetic acid, 207, 208, 209, 211*f*
Oxalosuccinic acid, 209, 211*f*
Oxford University, 4
Oxidation, **37**
Oxidative phosphorylation, 49, 212–214
Oxidizing agent, **37**
Oxygen
  atom, structure of, 38*f*

"debt," 299
deprivation, symptoms of, 216
evolution, photosynthetic
 origin of, 191
 role of manganese, 191
 production and biological productivity, 400
 supply homeostasis, 262
 toxicity, 216
Oxygen-carbon dioxide transport, 243
Oxytocin, 156, 157, 274
Ozone
 layer (in atmosphere), 182, 388
 protective, 352

Paal, A., 281
Pacemaker, heart, 240
Paleg, L., 286
Paleolithic culture, 361
Paleozoic, 356–357
Palisade layer, of leaf, 132f
Palolo worm spawning, 336
Pallium, 313
Pampas, 386
Pancreas, 269, 271
Pancreatic juice, 229
Pantothenic acid, 234
Paper chromatography, **197,** 198f
Paramecium, 19
Parasexual hybrids, 147, 235
Parasites, **368**
 animal, 374–382
 digenetic, **372**
 evolution of, 372
 host responses, 372
 monogenetic, **371**
 plant, 372–374
 traits of, 370–372
Parasitism (negative symbiosis) 370–382
Parasympathetic nervous system, 006, 306f, 307
Parathyroids, 271, 272f, 273f, 297
Parental behavior, 154, 336
Parental care, 154, 336
Parietal lobe, 304
Parthenogenesis
 artificial, 106
 in plants, **167**
 natural, 106
Parturition, stages, 160, 161f
Passage cells, 251
Passenger pigeon, 364f
Pasteur, L., 33, 34, 40, 204, 263, 351
Pavlov, I., 311, 321
Pea (*Pisum sativum*), 174
 root nodules, 225f
 self-pollination, 175
Peckham, E., 323
Peckham, G., 323
Pecking orders, 334
Pectoral girdle, 295f
Peking man, 359
 brain size, 312
Pelagic organisms, **394**
Pellagra, 234, 235f
Pelvic girdle, 295f
Penetrance, of genes, 74
Penguins, 153f, 154
Penis, 118, 156, 157f
Pentose, **39**
Pepsin, 228, 229
Pepsinogen, 229
Peptide bond, 44, 44f

Perennial plants, **174**
Pericycle, **130**
Perikaryon, **302**
Peripheral nervous system, 304–307
Permeability, nervous impulse, 307
Permease, 26f
Peroxidase, horseradish, 229
Peroxisomes, 199
Peru (Humbolt) current, 393
Pest control methods, 404
 bacterial diseases, 405
 newer, 236
 sex attractants, 405
 sterilized males, 404
 viral diseases, 405
Petal, **174,** 175f
Petiole, of leaf, 128
Peyote cactus, 5f, 36, 311
Pflüger, E. F.W., 110
PGA (3-phosphoglyceric acid), 198, 199f
pH, 26, 27, 259
 of soils, 135
  effect on flower color, 135
Pharyngeal clefts (gill slits), 113, 115f
Pharyngula, 113
Pharynx, **227,** 228f
Phenotype, **59**
Phenylacetic acid, 284f
Phenylalanine, 42, 43f, 88, 95f
 metabolic pathways from, 95f
Phenylketonuria (PKU), 95
 tests for, 95
Pheromones, **7,** 8f, **152,** 336
Phloem, 128
 contents, 255
 function, 250
 structure, 253f
 transport, theories of, 256
3-Phosphoglyceraldehyde (PGAL), 198, 199f, 207, 208f
2-Phosphoglyceric acid, 207, 209f
3-Phosphoglyceric acid (PGA), 198, 199f, 207, 209f
Phospholipid in membranes, 24, 25f
Phosphon, 287
Phosphorescence, 193
Phosphorus
 cycle, **397**
 distribution in plants, 254f
 in plant nutrition, 218, 223f
Phosphorylation
 oxidative 49, 212–214
 photosynthetic, **195,** 196
Photochemistry of photosynthesis, 193–197
Photolysis, **191**
Photomorphogenesis, **136**
Photon, 137
Photoperiod
 animal, 152
 control of animal reproduction by, 172
 control of plant reproduction by, 169
 seasonal changes in, 172
Photoperiodic control of flowering, 171–174
 classification of plants according to, 172
 mechanism of, 173
Photoperiodism, in plants, **138**
 control of flowering, 138
Photophosphorylation, **195**

cyclic, 195
non-cyclic, 196
Photoreceptor, **137**
 of photosynthesis, 183
Photorespiration, 199
Photosynthesis, 181–202, 394
 action spectrum, 184f, 185f
 environmental factors and rates of, 185–188
 estimated world total, 182
 oxygen evolution, 191
 photoreceptor, 183
 photosystems I and II, **193**
 primary reactions of, 188–197
 stoichiometry, 197
Photosynthetic oxygen evolution
 origin of, 191
 role of manganese, 191
Photosynthetic phosphorylation, **195**
 cyclic, 195
 non-cyclic 196
Photosynthetic unit, **188-189,** 189f
Photosystems I and II, **193**
 action spectra, 190f
Phototropism, 138, **280**
 action spectrum of, 137f, 283f
 auxin and, 138, 281
 inhibition by morphactin, 291f
 possible photoreceptors, 284
Phrenic nerves, 216
Phyllotaxy, of leaves, 128f
Phylum, 365
Phytochrome, 138
 control of gene activation, 143
 control of membrane permeability, 144
 in photoperiodic control of flowering, 173
 in plant photomorphogenesis, 173
 photoconversion of, 173
Phytoplankton, **394**
Phytylmenaquinone, 234
Pilus (pili), 18
Pilobilus, spore dispersal toward light, 177
Pine leaf section, 250f
Pineal gland (body), 277, 278, 304f
Pistil, **174,** 175f
Pistillate (female) flowers, **175**
Pitcairn islanders, 79
Pith, 128, 129f
Pits, of tracheid cells, 251
Pituitary (hypophysis), 274–277, 304f
 anterior lobe, 275f
 hormones, 101
 lactation, 162
 portal vein, 275f, 277
 posterior lobe, 274, 275f
  blood supply, 275f, 277
Pituitary-ovary-uterus axis, 158f
PKU (phenylketonuria), 95
Placenta, 120f, 159
Plague, bubonic, 4
Planetary crisis, 406–409
Planets, chemical compounds on, 351
Plankton, **394**
 phytoplankton, **394**
 zooplankton, **394**
Plant
 cell
  eukaryotic, 20f
  membranes, 246f

Plant—*cont.*
  dispersal
    agents of, 177
    human role, 177–178
  geography, 178
  growth regulators, 170
  hormones, 279–292
    early studies of, 281f
    experimental approaches, 280
    interactions of, 289
    sequential action of, 289
  life cycles, 124–126
    flowering plants, 126f
    generalized, 125f
Plasma
  blood, 242
  cells, 243, 278
  membrane (plasmalemma), 20f, 245, 246f
Plasmids, 18
Plasmochin, 376
Plasmolysis, 26, 246, 248f
Plastids, 22
Plastocyanin, 185, 194, 194f, 195
Plastoquinone, 185, 194, 194f, 195
Plato, 98
Platypus, 103
Pleistocene, 356–7
  animals, 362f
Plumule, 127f
Podzol (mor or mull), 389f
Poikilotherms, **258, 262**
Poisons, as enzyme inhibitors, 205
Polar body, 103f, 104f
Pollen, **174**
  dispersal, 175–176, 176f
  embryo development from, 146f
  haploid plants from, 144–145, 145f
Pollinating agents, 176–176
  bees, 176f
Pollutants, concentration at successive trophic levels, 398
Pollution
  air, 406, 409f
  and disease, 384
  water, 406
Polycarpic plants, **174**
Polydactyly, 347, 348f
Polymers, 40
Polymorphonuclear leukocytes, 242f, 243
Polynucleotide, 46
Polypeptides, **44**
  and genes, 81
Polyribosomes (polysomes), **21**, 84, 85f
Polysaccharide, **39**
Polysomes (polyribosomes), 84, 85f
Polyunsaturated lipids, 232
Pons varolii, 304f
Population
and food supply, 235, 236
  crash curve, 407
  dynamics of, 401–406
    competition effects, 402–403
    growth curves, 403–406, 404f
    predation effects, 401
  explosion, human, 407
  growth, 160, 344
    under stress, 273
  growth curves, 403, 404f
    human, 408f
    practical applications of, 404, 405f

level of, 11
  optimal, 409
  size, role in extinction, 363, 364
Porphyria, 96
Porphyrin, 353
Porphyropsin, 117
Porpoise, 258f
Portuguese man-of-war, 370
Postganglionic neurons, 306
Positive feedback, in reproduction, 403
Potassium
  in plant nutrition, 219, 223f
  metabolism, 271
  role in stomatal opening, 249
Potato ,
  new variety, 235
  origin of, 177
  tubers, 166
  variability of, 164
Prairies, 386
Pre-Cambrian Era, 356–357
Precocial chicks, **154**
Predation, effect on populations, 401–402
  Volterra-Lotka cycles, 402, 402f
Preganglionic neurons, 306
Prey-predator cycles, 402, 402f
Priestley, J., 47, 182
Primary endosperm nucleus, **175**
Primary growth, in plants, **128**
Primary phloem, 129f
Primary producers, **398**
Primary reactions, of photosynthesis, 188–197, **193**, 194f
  and quantasomes 188, 195–196
Primary structure, of protein, **85**
Primary xylem, 129f
Primate, 358, 359
  patterns of living, 338
  vision, 153
Primitive gut (archenteron), 113
Primitive streak, 111
Primordia, leaf, 131
Probability, laws of, 59
Process, neural, **302**, 303f
Prochordates, 358f
Product law
  and epidemiology of malaria, 376
  and probability, 59, 63
Productivity, determination of, 399–401
  harvest method, 400
  minerals, disappearance, 400
  oxygen production, 400
  radioactive tracers, 401
Progeny test, **72**
Progesterone, 157, 159, 162, **274,** 269f
Proglottids, of tapeworm, **378**
Prokaryotic cells, **16, 17,** 19f
  division, 19f
Prolactin, 157, **162**f
Promoter gene, 91
Prophase, mitotic, 29, 30f, 32f
Proplastids, 22
Proprioception, **319**
Prostaglandins, **278**
Prostate gland, 155, 157f
Prosthetic group, **204, 234**
Protein(s), 42–45
  bean, corn, 232
  globular, 24, 25f
  hormones, 268
  realities, myths, 231

sources, 232
  succotash, 232
  structure of, **44, 85**
  synthesis of 84, 86f
    egg, 106
Protoplasm, 13
Protoplast, 14
  of plant cells, 145
  fusion of, 147f, 235
Protozoa, parasitic, 374–376
Pruning, effects on plant growth, 282
Pseudocopulation, 175
Pseudopregnancy, 332
Psychokinesis, 302
Psycho-pharmacology, 310
Ptyalin, 228
Puffing of chromosomes, 269
Pulvinus, **249**
Puppies, socialization, 331
Purkinje, J., 13
  cell, 303f, 304
Puromycin, 144
Pylorus, 228
Pyramid
  of biomass, **398**
  of energy, **398**
  of numbers, **398**
Pyramidal cell, 303f, 304, 314f
Pyridoxin, 234
Pyruvic acid, 48, 48f, 207, 209f
  in muscle, 299

Quantasomes, 184–185, **185**
  composition of, 185
Quantum
  of electromagnetic radiation, 137
  energy of, 137
  requirement, of photosynthesis, **190**
  yield, of photosynthesis, 190, 190f
Queen Victoria, 96
Quinine, 376

Rabies, 33
Races, human, 78–79
Racial crossing, 78–79
Radar, in oceanography, 394
Radicals (functional groups), **38**
Radicle, 127f
Radioactive tracers, and productivity measurement, 401
Radioautograms, distribution of ^{32}P, 254f
Radioautography, **197**
Radioisotope, methodology, 255
Radius, 295f, 296
Rainforest
  non-tropical, 387
  tropical, 386–387
Raleigh, Sir Walter 177
Ramon y Cajal, S., 113, 122
Ramos-Galvan, 231
Randall, M., 191
Random, mutations, 346
Range, of a species, **385**
Rat uterus, hormone effects, 270f
Ray parenchyma cells, of phloem, 253, 253f
RDA (recommended daily allowance), of nutrients, **232**
Reaction centers, of photosynthesis, 193
Réaumer, R., 204

Receptor sites
  drug, 310
  hormonal, 269
  muscle, 300
  smell, 318
Recessive genes, **58**
Recommended daily allowance
  (RDA), of nutrients, **232**
Recycling in nature, 225–226, 354
Red blood cell (see Erythrocyte),
  96 f
  count, altitude, 262
  formation, 294
Red drop, in photosynthesis, **189**
Redi, F., 350, 378
Redia, **378**
Redox potential, **211**
Reducing agent, **37**
Reduction, **37**
Reflex
  action, rules of, 312
  neural, 311–312
  pupillary light, 311
Regulative eggs, 103
Regulator gene, **91**
Reindeer
  moss (Cladonia rangifera), 369
  people, 361 f
Reinforcement, behavioral, **330,** 331
Relationships, plant and animal,
  265, 266 f
Renal (see Kidney) corpuscles, 260
Rennin, 229
Repressed genes, 142
Repressor, genetic, **91**
Reproduction
  animal, **151**
  in plants, 164–178
    asexual, 164–168
    sexual, 168–177
Reproductive cycles, 119
  baboon, 332
  control, in animals, 158
  mouse, 332
  neural control, 332
Reproductive potential, 344
Reproductive system
  anatomy, 155–157
  female, 156, 157 f
  hormones, 158 f
  male, 155, 157 f
Reserpine, 311 f
Resolving power, 16
Respiration, 47–51
  aerobic, 21, 48–50
  anaerobic, 18, 48
  classic laws of, 214–215
  efficiency of, 49
  eggs, 106
  in animals, 214–216
  terminal, 49
Respiratory assembly, in mitochondria, 213 f
Respiratory enzyme, of Warburg,
  210
Respiratory quotient, **214**
  of carbohydrate, 214
  of fat, 214
  of protein, 214
Retina, 319, 320 f
  dedifferentiation, 116
Retinene, 320
Retinol, 234
Revelle, R., 235
Reverse transcriptase, 94

Rh blood groups, 75
Rh immunoglobulin (Rho-GAM), 76
Rhesus
  factor, 75
  monkey (Macaca), 75, 335
Rhizobium, and nitrogen fixation,
  224
Rhizoids, fern, 170 f
Rhizomes, **164,** 166
Rhodopsin, 117, 320
Rhodospirillum, and nitrogen fixa-
  tion, 224
Rho-GAM (Rh immunoglobulin), 76
Riboflavin, 233
Ribonucleic acid (RNA), 81
  messenger (mRNA), 84
  ribosomal RNA, 84
  transfer (sRNA or tRNA), 84, 86 f
Ribose, 39 f
Ribosomes, 17, 21, 84, 85 f
Ribulose diphosphate, 198, 199 f
Rickets, 297
Ringer, S., 242
Ripeness-to-flower, **169**
RNA (ribonucleic acid), 81
  messenger (mRNA), 84
  polymerase, 91
  ribosomal RNA, 84
  transfer (sRNA or tRNA), 84, 86 f
Robertson, J. D., 24
Rocky Mountain spotted fever, 372
Rods, visual, 319
Romanowsky, D., 376
Root(s), 127
  auxin concentration effects, 282 f
  barley, 130 f
  cap, 130 f, 131
  hairs, 130, 130 f, **250,** 251 f
  knot nematode (Heterodera), 380
    on soybeans, galls caused by,
    380 f
  nodules, 225, 225 f
  pressure, 251–252
  radish, 130 f
  structure, 130
Rooting, of cuttings, 167
  effect of plant hormones, 167
Rose, M., 116
Rosette growth patterns in plants,
  173
  effect of giberellins, 285, 286 f
Ross, R., 375, 376 f
Rowan, W., 152
Ruben, S., 191
Rubenstein, B., 149
RuDP (ribulose diphosphate), 198,
  199 f
Ruffini's corpuscles, 319
Ruminants, symbiotic protozoans
  in, 370
Runners, of plants, 166
Russell, B., 7

Saber-toothed tigers, 355, 362 f
Sachs, J., 221, 279, 280 f
Salivary gland chromosome, 69 f
Salmon, migration by odor, 332
Salmonella, transduction in, 93
Sampling errors, 347
Sanger, F., 44, 45 f
Santa Gertrudis cattle, 73
Sap, ascent in plants, 252
Sarcomere, 298f, **299,** 300f
Satter, R., 249
Saturated fatty acids, 41

Sawyer, R., 235
Scala tympani, vestibuli, 319 f
Schaller, G. B., 335
Schistosomiasis, 4, 377–378
Schizophrenia, 310
Schleiden, M. J., 13, 28
Schooling, fish, 334
Schrödinger, E., 8, 10
Schultze, M., 13
Schwann, T., 13, 28
  cell, 113, 303 f
Science, nature of, 5, 6
Scion, **167**
Schistosoma, 371, 372
Schistosomes (blood flukes), 377,
  378
Scolex, of tapeworm, **378**
Screwworm fly, control of, 404
Scrotum, 118, 156, 157 f
Scurvy, 233
Sea nettle, 328
Sea urchin embryo, 111f
Secondary growth, in plants, **128**
Secondary roots, origin of, 130
Secondary sexual characteristics,
  119
  male, 156
Secondary structure, of protein, **85**
Secretin, 229, 268
Secretion, **259**
Seed(s), 126, 127, **175**
  coat, 126, 127 f
  dormancy, 126
  germination, 127 f
  plants
    life cycles, 126
    structure, 127–132
  structure, 127 f
Segregation, law of, 56–57
Selection, 72
  artificial, 344
  directional, 346
  disruptive, 346
  group, 000
  natural, 344, 346
  stabilizing, 346
Selective herbicides, 284
Selenium, in plants, 220
SEM (scanning electron micro-
  scope), 16
Semicircular canals, 318, 319 f
Seminal fluid, 155, 156
Seminal vesicle, 155, 157 f
Seminiferous tubules, 101, 155 f
Semipermeable membrane, and osmosis, 246
Semipermeability, 26
Senescence, 123
  in plants, 147–149, 174
  racial, 363
Sense organs, 317
Sensory, afferent, neuron, 303 f,
  305 f
Sepal, **174,** 175 f
Sere, **390**
  field to forest, 390 f
Serotonin (5-hydroxytryptamine),
  310, 311 f
Sertoli cells, 155 f
Servetus, M., 238
Sex
  anomalies, in humans, 77–78
  attractants, 7, 8 f, 152
    as a means of pest control, 405
  biological function, **151**

Sex—*cont.*
 chromosomes, 65
  and human abnormalities, 76–78
 determination, 64, 104
  artificial, 64
  natural, 64
 differentiation, **118**
 identification, 152
 neural mechanisms, 154 *f*
 of flowers
  effect of auxin, 287
  effect of gibberellin, 287
Sex-linked genes, 65
 in *Drosophila,* 65 *f*
Sexual reproduction
 biological function, 10
 in plants, 168–177
  factors affecting, 168
Shark muscle pattern, 297, 298 *f*
Shaver, J., 106
Sheep liver fluke (*Fasciola hepatica*), 377, 378
 life cycle, 377 *f*
Sherrington, C., 312
Shoot, 127 *f*, 128 *f*
Short-day plants **172**
 response to interrupted dark period, 173 *f*
 response to photo period, 172 *f*
  Short-night plants, 173
Shrew, 9
Shull, G., 72
Sickle cell anemia, 96
Sickled red blood cells, 96 *f*
Siegelman, H., 138
Sieve
 plate, 253, 253 *f*
 tubes (elements), 253, 253 *f*, 254 *f*, **255**
Sign stimulus, **325**, 336
Silicon, in plant nutrition, 220, 223 *f*
Simpson, G. G., 366
Sinanthropus, 359
Size(s), 16
 and metabolism, 9
 of organisms, 7, 8, 9
Skeletal muscle, 298, 299 *f*
Skeleton, 293–297
 appendicular, 295 *f*
 axial, 295 *f*
 functions, 294
 human, 295 *f*, 296 *f*
Skinner box, 316, 322, 323 *f*
Skoog, F., 282, 287
Skulls, 295, 296 *f*
Sleep movements (nyctinasty)
 in *Albizzia julibrissin* (mimosa tree), 249
 in *Mimosa pudica* (sensitive plant), 249
 role of potassium ions,249
Sliding filament theory, 300 *f*, 301 *f*
Slime molds, 132
Slobodkin, L. B., 7
Smallpox, 33
 vaccination, 263
Smell, biochemistry of, 318
Smooth muscle, 298, 299 *f*
Snowshoe hare, population cycles, 402 *f*
Social behavior, 333–339
Social groups, biological bases, 338, 339
Socrates, 79
Sodium

in plant nutrition, 220, 223 *f*
 pump, 309
Soil(s), 388–390
 animals of, 390
 consistancy of, 136
 development, 389–390
 effect on plant growth, 135
 horizons, **388**
 organic matter in, 136
 pH of, 135
 structure, 388–389
 types, 389 *f*
Somatic cell hybridization, 66, 94, **145**, 145–147
 by protoplast fusion, 147 *f*
Somatic nervous system, **305**
Somatotropin, 301
Sonar
 animal, 318
 in oceanography, 394
Southwick, C., 336, 407
Soybean (*Glycine max*), 171
Space-Age perspectives, 7
Spallanzani, L., 351
Spawning synchrony, 151
Spaying, **270**
Specialization, excessive, 363
Species, **364**, 365 *f*
Specificity, of enzymes, 204
Spectrophotometer, 184
Speed of nervous impulse, 307
Spemann, H., 114
Sperm
 "bank," 105
 nuclei, in pollen tube, **175**
 plant, 125
 structure, 101
Spermatid, 102 *f*, 102 *f*, 155 *f*
Spermatocyte, 102 *f*, 103 *f*, 155 *f*
Spermatogenesis, 101, 102 *f*
Spermatogonium, 155*f*
Spermatozoa, 155*f*
Sperry, R. W., 122
Sphenodon, 364
Sphinctor muscles, 297
Spider
 reproduction, 153
 webs, 327 *f*
Spinal animal, man, 311
Spinal nerves, 305 *f*, 306
Spindle fibers, 29, 30 *f*
Spiral cleavage, 103, 108
Spongy layer, of leaf, 132 *f*
Spontaneous generation, 350–353
Sporangia, 125
Spore(s), 125
 formation, 165
 mother cells, 125
Sporocyst, **378**
Sporophyte, **125**
Sporozoa, diseases caused by, 375
Squalus, muscles, 297, 298 *f*
Squid, 4, 308
S stage, cell division **30**, 32 *f*
St. Thomas Aquinas, 98
Stadler, L., 70
Stamen, **174**, 175 *f*
Staminate (male) flowers, **175**
 promoted by gibberellin, 287
Stanley, W. M., 33
Stapes, 296 *f*, 318, 319 *f*
Starch, 40, 206
Starling, E. H., 268
 "law of the heart," 242
Stearic acid, 41

Stem(s), 127, 128 *f*
 auxin concentration effects, 282 *f*
 cells, immune, 266 *f*
 functions of, 128
Steppes, 386
Stereoisomers, 40 *f*
Sterilized males, as a means of pest control, 404
Steroid(s), 41, 268–269 *f*
 hormone, 118
Sterols, 232
Steward, F., 140, 287
Stickleback, 152, 334
Stigma, **174**, 175 *f*
Stimulus-response, **325**
Stock, **167**
Stolons, **164**, 166
Stomach, **228**
Stomatal apparatus, 248–249
 diurnal periodicity of, 248
 of *Vicia faba* (broad bean), 248 *f*
 response to environmental factors, 249 *f*
 sunken, in pine, 250 *f*
Stomates, 132, 132 *f*, 246, 248 *f*
 response to enviornmental factors, 249 *f*
Stone, L., 122
Streptomycin, 295
Stress, effect on hormones, 273
Striated muscle, 298 *f*–301 *f*
Stroma, **22**, **182**
 lamella, **22**
Style, **174**, 175 *f*
Subsoil (B horizon), **388**, 389 *f*
Substrate, of enzyme, **204**
Succinic acid, 208, 209, 211 *f*, 212
 dehydrogenase, 206, 206 *f* 211, 212, 212 *f*
 inhibition by malonate, 206
Succotash, protein in, 232
Sucrose, 39 *f*
Sugars, 39
Sulfa drugs, 206
Sulfanilamide, 234
Sulfhydryl bonds, in protein, 85
Sulfur, in plant nutrition, 219, 223 *f*
Summation, neural
 spatial, 310
 temporal, 310
Sumner, J., 204, 204 *f*
Sunflower (*Helianthus annuus*)
 aging in, 148
 heteroblastic development, 148
 determinate development, 148
Surface-volume ratios, 9, 27 *f*
Sutton, W., 66
Sweetbreads, 271
Swift, J., 9
Swimmer's itch, 378
Symbiosis, **368**
 protozoan-termite, 370
 protozoan-ruminant, 370
Symbiotic bacteria, 19
Sympathetic nervous system, 305, 306 *f*, 307
Synapse, types, **309** *f*
Synapses
 chemical control 310, 311 *f*
 functions, 309–311
 in meiosis, 62
Synaptic knobs, end bulbs, 309 *f*, 310 *f*
Systole, **242**

Szent-Györgyi, A., 207, 208, 209 f, 299

Tabula rasa, 322
Taiga, 386, 387 f
Tapeworms (cestodes), 378
  beef, life cycle, 379 f
Target organ, **267**
Tarsals, 295 f, 296
Taste, 318
Tatum, E., 88
Taxis, **326**
Taxonomy, **364**
Tay-Sach's disease, 97
T-cells, 265, 266 f
Tectorial membrane, 319 f
Teeth, adaptations, 227
Telolecithal eggs, 103, 104 f
Telophase, mitotic, **29**, 31 f
Temin, H., 94
Temperate viruses, 94
Temperature
  effects on plants
    biochemical reactions, 134
    development of biennials, 134
    flowering, 134
    photosynthesis, 187 f
    reproduction, 169
    tomato growth, 134 f
  neural control, **262**, **263**
  reproduction, 152
Template, for DNA synthesis, 83
Temporal lobe, 304 f
Tendon, **297**
Teratogen, 285
Termite society, 338
Termite-protozoan symbiosis, 370
Terns, territorial dispute, 321f
Territorial display, 152, 153 f
Territoriality, 334, 335
Tertiary structure, of protein, **85**
Testes
  general, 101
  grasshopper, 102 f
  human, 155 f, 156 f
Testosterone, 155, 269, **274**
  action site, 154 f
Tetrad, in meiosis, 62, 102 f
Tetraploid, 28
Tetrose, **39**
Thalamus, 304 f
Thalidomide, 114
Thermocline, **392**
Thermoperiodicity, diurnal, 169
  effects on flowering, 169
  effects on plant growth, 169
Thermoperiodism, **133**
Thiamine (vitamin B₁), 233
Thimann, K., 282, 283 f, 284
Thiourea, 170
Third eye, 277
Thorndike puzzle box, 322
Thorpe, W. H., 331
Threshold, neural, 307
Thrombase, 244
Thrombin, 244
Thrombus, 243
Thymine, 81, 82 f
Thymus gland, in immunity, 265, 278
Thyroid
  gland, 272 f, 273 f
  hormone, 270, 271
Thyrotropin, 271
Thyroxin, 43, 271
TIBA (tri-iodobenzoic acid), 236

Tibia, 295 f, 296
Tinbergen, N., 323
Toadfish, kidney, 261, 262 f
Tobacco (Nicotiana tabacum)
  haploid, 145 f
  mosaic disease, 33 f
  photoperiodic control of flowering, 171
Tocopherol, 234
Tom Thumb midgets, 276, 277 f
Tonoplast (vacuolar membrane) 245, 246 f
Tonsils, 228 f
Topsoil (A horizon) **388**, 389 f
Totipotency (embryonic), **140**
  evidence from animals, 108, 110
  evidence from molecular biology, 142-143
  evidence from plant tissue culture, 140
Tracheas, 215
Tracheids, of xylem, 251
Trade winds, 393
Tranquilizers, 311 f
Transaminase, 232
Transamination, 209
Transcellular strands, of sieve cells, 255
Transcriptase, reverse, 94
Transcription, of genetic code, **84**
Transduction, 92, **93–94**
  role of viruses, 93
Transformation, **92–93**
  role of DNA, 92
Translation, of genetic code, **84**
Translocation
  chromosomal, 69
  in plants, 253–257
Transmitters, neural, 311
Transpiration, **247**
  adaptations that influence rates, 249–250
Transpirational pull, 251
Tree line, 386
Trematodes (flukes), 376–378
  digenetic, 377
  control measures, 378
  monogenetic, 377
Tricarboxylic acid (Krebs) cycle, 207–209, 211 f
Trichinella spiralis, 381
Triglycerides, 42, 229 f, 232
Trihybrid cross, 60
Tri-iodobenzoic acid, 236
Trinitroglycerine, 41 f
Triose, **39**
Triplets, in genetic code, 83
Triploid cells, **27**, 28f
Trisomy, **96**
Tritium, 15, 46
Trophic levels, **398**
Tropical rainforest, 386–387
  stratified vegetation in, 386
Tropisms, **280**
  and auxin, 281
  geotropism, **280**
  phototropism, **280**
  receptors, 283
Trypanosomes, 375 f
  diseases caused by, 375
  heavy metal trypanosides, 375
Trypsin, 228
Tryptophan (indolealanine), 43 f, 44, 278
Tsetse fly (Glossina), 375

Tube nucleus, **175**
Tubers, **166**
Tundra, **385**
Tung, T. C., 110
Tunica, **131**
Tunicates, 358 f
Turgor
  of plant cells, **246**, 248 f
  pressure, gradients of, 257
Turkey, brain and behavior, 316, 317 f
Turner's syndrome, 77
Turnover rate, in ecosystems, 400
Twins
  dizygotic (fraternal), 121
  fetal membranes, 121
  monozygotic (identical), 121
Twitty, V. C., 115
Tyrannosaurus, 363
Tyrosine, 43 f, 271, 274 f
Tyrosinosis, 95 f

Ubiquinone (coenzyme Q), 211, 212 f
Ulna, 295 f, 296
Ultraviolet radiation, 352
Ulva taeniata, action spectrum of photosynthesis in, 185 f
Umbilical cord, 120 f
Uniformitarianism, geological, **343**
"Unit membrane," 24, 25 f
Unsaturated fatty acids, 41
Uracil, 81, 82 f
Uranium-238 dating, **355**
Urea, synthesis, 230, 259 f
Urease, 204
Ureter, 260 f
Urethra, 157 f
Urey, H., 352 f, 354
Urey-Miller apparatus, 352f, 354f
Uric acid, 259 f
Urinary bladder, 260
Urochrome, 260
Use and disuse, genetic effects, 345
Uterus (womb), 118, 156, 157 f

Vaccine, 34
Vacuolar membrane (tonoplast), 245, 246 f
Vacuole
  food, **22**
  plant cell, 20 f
Vagina, 118, 156, 157 f
Vagus nerve, 216, 239, 310
Value judgments, 6
Vampire bats, 382
van Helmont, J., 220
van Lawick-Goodall, J., 335
Van Niel, C., 18, 191, 191 f
  theory of photosynthesis, 191
Variability in plants, 168
Variation
  biological, 151
  genetic, 63
Varner, J., 286
Varves, **355**
Vas deferens, 155, 157 f
Vascular bundles
  in dicot stems, 129, 129 f
  in monocot stems, 128, 129 f
Vascular cambium, **128**
Vascular ray, 129 f
Vascular tissues, of plants, 128, 129 f
  arrangement of, 247 f

Vascular tissues, of plants—*cont.*
  phloem, 246, 253 *f*
  xylem, 246, 247 *f*
Vasopressin, **275**
Vectors, of parasitic diseases, 375
Vegetal pole, egg, 102
Vegetive propagation, of plants, 166–168
Veins
  of leaves, 132, 132 *f*, 250
  vertebrate, 240 *f*, 241 *f*
Veldts, 386
Venom, insect, 310
Ventral, **239**
Ventricles
  brain, 304 *f*, 313
  heart, 244 *f*
Venus, atmosphere, 351
Vernalization, **134**, **169**
Vertebrae, cervical, 295
Vesalius, A., 237, 293, 298 *f*
Vessels, of xylem, 251
Vestigial structures, 350
Vicia faba (broad bean) stomates, 248 *f*
Villi, intestinal, 229 *f*
Vilmorin, L., 73
Viral diseases, 264
Virchow, R., 28
Virion, **33**
Viroids, 34 *f*, **35**
Virulence, of viruses, 94
Viruses, 32–35
  DNA, 33
  possible role in cancer, 94
  role in transduction, 93
  structure, 34 *f*
  temperate, 94
  virulent, 94
Visceral muscles, 298, 299 *f*
Vision, 319
Visual cones, 277
Visual purple, 117, 320
Vitalism, 108
Vitamin(s), 233–235
  analogues, 206
  B group, 234 *f*, 235
  discovery, 233

functions, 234 *f*
mental health, 233
role in explorations, 233
sources, 234 *f*
Vitamin A, 101, 234, 320
Vitamin B$_{12}$ (cyanocobalamin), 370
  cobalt and synthesis, 370
  production by symbiotic proto-
    zoans in ruminants, 370
Vitamin C, 233
Vitamin D, 297
Vitamin E, 101
Vitelline (yolk) membrane, 101
Viviparity, 120
Volterra, V., 402
Volterra-Lotka (prey-predator) cycles, 402, 402 *f*
Voluntary muscles, 298, 299 *f*
von Frisch, K., 328
von Mering, J. F., 272
von Mohl, H., 290
von Tschermak, E., 56

Wallace, A. R., 345, 349
Warblers, food supplies of, 403
Warburg, O., 189, 189 *f*, 208, 210
  apparatus, 205, 205 *f*
Water
  as photosynthetic substrate, 194
  movement in plants, 245, 250–253, 256 *f*
  role in plants, 245
Watson, J. B., 322
Watson, J. D., 83, 84*f*
Wavelength of light, units of, **136**
  Ångstrom (Å), **136**
  nanometer (nm), **136**
Waxes, 40, 41
Webb, M., 333
Weed killers, 284
Weismann, A., 30, 61, 108
Wenger, M. A., 326
Wenner, A., 329
Went, F., 279, 280 *f*, 281, 283
  experiments of, 282 *f*
Wetherell, D., 142
Whales
  Baleen, 398, 399 *f*

killer, 398, 399 *f*
Wheat, new variety, 235
White blood cells, 21, 242 *f*
White matter, **304** *f*
White pine blister rust, 373
Wiesner one-hormone theory, **118**
Wilting, **245**
Wilson, E. O., 338, 339
Wittwer, S., 286 *f*
Wohler, F., 37, 354
Wolf packs, 338
Wolffian ducts, 118
Womb (uterus), 156, 157 *f*
Woodhall, J., 233
Woods Hole, Cape Cod, 106
Woody stem, cross section, 129 *f*
Wound hormones, 279

X chromosome, 64
Xenopus, African toad, 110
X-linked coat color in mice, 77
X-linked genes, 65
  in *Drosophila*, 65 *f*
Xylem, 128
  function of, 250

Y chromosome, 64
Yabuta, T., 285
Yam estrogen from, 274
Yamada, K. M., 116
Yellow fever, 33
Yolk plug, frog, 112
Yolk sac, 120 *f*
Yomo, H., 286

Z lines (disks), 298*f*, **299,** 300*f*
Zeatin, 288, 288 *f*
Zero population growth, 407
Zimmerman, R., 331
Zinc, in plant nutrition, 219, 223 *f*
Zinder, N., 93
Zinjanthropus, 359
Zona pellucida, 105*f*
Zooplankton, **394**
Zwitterion, 42, 44 *f*
Zygote, 104, 175
Zymase, 204

The frontiers are not east or west, north or south, but wherever a ma██████████ a fact.
Henry David Thoreau

about the use of color in this book

Color has been used to emphasize and
clarify particular aspects or parts of complex figures
or reaction schemes. A further important use of
color is to code for haploidy vs. diploidy, notably
in the chapters about plants. The symbol $\frac{n}{2n}$
indicates that color has been used to designate
haploid (monoploid) structures while diploid
structures are shown in black.